Theory and Applications of Microbiology

Theory and Applications of Microbiology

Editor: Jake Burns

MURPHY & MOORE
www.murphy-moorepublishing.com

Murphy & Moore Publishing,
1 Rockefeller Plaza,
New York City, NY 10020, USA

Visit us on the World Wide Web at:
www.murphy-moorepublishing.com

ISBN: 978-1-63987-534-4 (Hardback)

Cataloging-in-Publication Data

Theory and applications of microbiology / edited by Jake Burns.
p. cm.
Includes bibliographical references and index.
ISBN 978-1-63987-534-4
1. Microbiology. 2. Microorganisms. 3. Biology. I. Burns, Jake.
QR41.2 .T44 2022
579--dc23

Table of Contents

Preface

The study of unicellular, multicellular and acellular microscopic organisms is known as microbiology. The various sub-fields studied under the umbrella of microbiology are bacteriology, virology, mycology, protistology and parasitology. Bacteriology is the branch of microbiology which deals with the study of the morphology, physiology, genetics and biochemistry of bacteria whereas the study of various aspects related to viruses is termed as virology. Microbiology can be classified into in diverse sectors such as food microbiology, medical microbiology, industrial microbiology and soil microbiology. It studies the pathogenic microorganisms which cause diseases in humans, as well as the useful microorganisms which are responsible for beneficial processes such as fermentation and antibiotic production. Some of them are also used as vectors for transferring foreign DNA into plants and animals. This book outlines the theories and applications of microbiology in detail. It provides significant information of this discipline to help develop a good understanding of microbiology and related fields. This book will serve as a valuable source of reference for graduate and post graduate students.

This book is a result of research of several months to collate the most relevant data in the field.

When I was approached with the idea of this book and the proposal to edit it, I was overwhelmed. It gave me an opportunity to reach out to all those who share a common interest with me in this field. I had 3 main parameters for editing this text:

1. Accuracy - The data and information provided in this book should be up-to-date and valuable to the readers.
2. Structure - The data must be presented in a structured format for easy understanding and better grasping of the readers.
3. Universal Approach - This book not only targets students but also experts and innovators in the field, thus my aim was to present topics which are of use to all.

Thus, it took me a couple of months to finish the editing of this book.

I would like to make a special mention of my publisher who considered me worthy of this opportunity and also supported me throughout the editing process. I would also like to thank the editing team at the back-end who extended their help whenever required.

Editor

1

Antistaphylococcal and Antibiotic Resistance Modulatory Activities of Thirteen Cameroonian Edible Plants against Resistant Phenotypes

Brice E. N. Wamba, Armelle T. Mbaveng, Paul Nayim, Joachim K. Dzotam, Ornella J. T. Ngalani, and Victor Kuete

Department of Biochemistry, Faculty of Science, University of Dschang, Dschang, Cameroon

Correspondence should be addressed to Victor Kuete; kuetevictor@yahoo.fr

Academic Editor: Clemencia Chaves-López

Background. In this study, 18 methanol extracts from Cameroonian edible plants were tested for their antibacterial activities against 26 strains of *S. aureus*; the role of efflux pumps in the resistance of tested bacteria and the antibiotic resistance-modulating activities against selected multidrug-resistant (MDR) phenotypes were also investigated. *Methods.* Broth microdilution assay was used to evaluate the antibacterial activity, the role of efflux pumps, and the antibiotic resistance-modulating effects of plant extracts. *Results.* Extracts from *Dacryodes edulis* seeds (DES) and *Dacryodes edulis* bark (DEB) were active against all 26 tested bacterial strains, within the minimal inhibitory concentration (MIC) range of 256–1024 μg/mL. MIC values varied from 64 to 1024 μg/mL against 96.2% of the 26 tested bacteria for *Phaseolus vulgaris* leaves (PVL), 92.3% for *Azadirachta indica* bark (AIB), *Dacryodes edulis* leaves (DEL), and *Ricinodendron heudelotii* leaves (RHL). The lowest MIC value of 64 μg/mL was obtained with the extract from *Cucurbita maxima* beans (CMB) against MRSA4 strain and from *Uapaca guineensis* bark (UGB) against MRSA9 strain. Bacterial efflux pump inhibitor (EPI), carbonyl cyanide *m*-chlorophenyl hydrazone (CCCP), improved the activity of DES and UGB as well as that of extracts from *Hibiscus esculentus* leaves (HEL) and *Uapaca guineensis* leaves (UGL) against resistant *S. aureus* strains. Antibiotic-modulating effects against more than 70% of the *S. aureus* strains tested were obtained when RHL (at MIC/2) was combined with CIP, ERY, and KAN (88.89%), CHL (88.89%), TET (77.78%), and STR (88.89%). *Conclusion.* The present study demonstrated that the 13 tested plants had antistaphylococcal effects and that DES, HEL, UGL, and UGB could be used in combination with EPI to combat resistance to *Staphylococcus aureus*. Also, it demonstrated that some studied extracts and mostly RHL could be used as antibiotic resistance modulators to fight against resistant strains of *S. aureus*.

1. Background

Bacterial infections caused by *Staphylococcus aureus* are globally responsible for 7–10% of deaths annually [1]. It is the most virulent species of the genus *Staphylococcus* and has emerged as one of the most important human pathogens in the last decades, being one of the main causes of hospital and community infections [2]. This bacterium causes a wide range of clinical infections, ranging from common infections such as skin and soft tissue infections to septicemia, pneumonia, and toxinosis [3, 4]. The fight against *S. aureus* is hindered by the development of resistance of various strains to antibiotics [5–9]. The multidrug resistance (MDR) observed in Gram-positive bacteria is mostly attributed to overexpression of efflux pumps and antibiotics-degrading enzymes. This MDR of *S. aureus* propels the search of new antibacterials with more efficiency and low toxicity. Plant kingdom contains a variety of pharmacologically active secondary metabolites, and some of them have been reported for their antibacterial activities [10, 11]. Their use to combat *S. aureus* antibiotic resistance is an attractive strategy. In regard to the loss of efficacy of several antibiotics and the scarcity of new antibacterial agents, it is also important to search for substances capable of restoring the activity of antibiotics. Antibacterial screenings of African plants have yielded promising results in the past [12–15]. The present study was set up to evaluate the antistaphylococcal potential of 13 Cameroonian food plants: *Azadirachta indica* A. Juss (Meliaceae), *Citrus grandis* (L.) Osbeck (Rutaceae), *Cucurbita maxima* Duch. (Cucurbitaceae), *Dacryodes edulis* (G. Don) H. J. Lam (Burseraceae), *Hibiscus esculentus* L. (Malvaceae), *Ipomoea batatas* (L.) Lam (Convolvulaceae), *Irvingia gabonensis* (Aubry. Lec. ex O. Rorke) Baill. (Irvingiaceae), *Phaseolus vulgaris* L. (Fabaceae), *Ricinodendron heudelotii* (Baill.) Pierre ex Heckel (Euphorbiaceae), *Saccharum officinarum* L. (Poaceae), *Spondias mombin* L. (Anacardiaceae), *Theobroma cacao* L. (Sterculiaceae), and *Uapaca guineensis* Muell. Arg. (Euphorbiaceae). The study was extended on the role of efflux pumps in resistance to some plant extracts as well as the ability of extracts to potentiate the activity of selected antibiotics.

2. Materials and Methods

2.1. Plant Materials and Extraction. The thirteen edible plants were collected from different localities of Cameroon, namely, Obala (Centre Region), Muyuka (South-West Region), and Dschang (West Region) from March to April 2016. The collected plant samples included the bark of *Azadirachta indica* A. Juss (Meliaceae), leaves of *Citrus grandis* (L.) Osbeck (Rutaceae), beans of *Cucurbita maxima* Duch. (Cucurbitaceae), leaves, bark, and seeds of *Dacryodes edulis* (G. Don) H. J. Lam (Burseraceae), leaves of *Hibiscus esculentus* L. (Malvaceae), beans of *Ipomoea batatas* (L.) Lam (Convolvulaceae), leaves of *Irvingia gabonensis* (Aubry. Lec. ex O. Rorke) Baill. (Irvingiaceae), leaves of *Phaseolus vulgaris* L. (Fabaceae), bark and leaves of *Ricinodendron heudelotii* (Baill.) Pierre ex Heckel (Euphorbiaceae), leaves of *Saccharum officinarum* L. (Poaceae), leaves of *Spondias mombin* L. (Anacardiaceae), leaves and beans of *Theobroma cacao* L. (Sterculiaceae), and leaves and bark of *Uapaca guineensis* Muell. Arg. (Euphorbiaceae). Plants were identified at the National Herbarium in Yaoundé (Cameroon) where the voucher specimens were conserved under the registration numbers (Table 1). The dried and powdered material (100 g) of each plant was macerated in 300 mL of methanol at room temperature for 48 h and then filtered using Whatman filter paper No. 1. The filtrate obtained was concentrated using a rotary evaporator under reduced pressure to obtain the crude methanol extract, which was kept at 4°C until further use.

2.2. Chemicals. Eight reference antibiotics (RA) purchased from Sigma-Aldrich (St Quentin Fallavier, France) were tested: ampicillin (AMP), cefepime (CEF), chloramphenicol (CHL), ciprofloxacin (CIP), erythromycin (ERY), kanamycin (KAN), streptomycin (STR), and tetracycline (TET); *p*-iodonitrotetrazolium chloride (INT) (Sigma-Aldrich) was used as bacterial growth revelator; carbonyl cyanide *m*-chlorophenyl hydrazone (CCCP) and chlorpromazine (CPZ) (Sigma-Aldrich) were used as efflux pump inhibitors (EPIs); and dimethylsulfoxide (DMSO) was used to dissolve the plant extracts.

2.3. Bacteria, Culture Media, and Growth Conditions. The strains of *Staphylococcus aureus* used included a reference strain obtained from American Type Culture Collection (ATCC; ATCC 25923), 8 methicillin-resistant *S. aureus* (MRSA) strains (MSSA1, MRSA3, MRSA4, MRSA6, MRSA8, MRSA9, MRSA11, and MRSA12) (obtained from the culture collection of the Laboratory of Microbiology, Graduate School of Pharmaceutical Sciences, The University of Tokyo, Japan, and provided by Dr Jean P. Dzoyem, University of Dschang) [8, 9], and 17 resistant clinical laboratory strains of *S. aureus* (SA01, SA07, SA18, SA23, SA36, SA39, SA56, SA64, SA68, SA88, SA114, SA116, SA124, SA126, SA127, SA135, and SA139) available in our laboratory collection and previously isolated from patients in Ad-Lucem Hospital in Banka-Bafang (West Region of Cameroon) [54]. Their bacterial features are summarized in Table 2. They were cultured at 37°C overnight on Mueller Hinton agar 24 h prior to any assay. The Mueller Hinton broth (MHB) was used as liquid culture medium for susceptibility tests.

2.4. Preliminary Phytochemical Screenings. Potential classes of potential antibacterial phytochemicals such as alkaloids (Dragendorff's and Mayer's tests: 5 mg plant extract in 10 mL methanol; a portion of 2 mL extract + 1% HCl + steam, 1 mL filtrate + 6 drops of Mayor's reagents/Wagner's reagent/-Dragendorff's reagent; cream precipitate/brownish-red precipitate/orange precipitate indicated the presence of respective alkaloids), saponins (frothing test: 0.5 mL filtrate + 5 mL distilled water; frothing persistence indicated the presence of saponins), steroids and triterpenoids (Liebermann–Burchard test: 5 mg plant extract in 10 mL chloroform, filtered; a 2 mL filtrate + 2 mL acetic anhydride + conc. H_2SO_4; blue-green ring or pink-purple indicated the presence of steroids or triterpenoids), phenolics: anthraquinones (5 mg plant extract in 10 mL methanol; a portion of 2 mL + 2 mL ether-chloroform 1 : 1 v/v + 4 mL NaOH 10% (w/v); red color indicated the presence of anthraquinones), flavonoids (5 mg plant extract in 10 mL methanol; a portion of 2 mL + conc. HCl + magnesium; ribbon pink-tomato red color indicated the presence of flavonoids), polyphenols (ferric chloride test: 5 mg plant material in 10 mL methanol; a portion of 2 mL + 2 mL $FeCl_3$; violet-blue or greenish color indicated the presence of phenols), and tannins (5 mg plant extract in 10 mL distilled water; a portion of 2 mL + 2 mL $FeCl_3$; blue-black precipitate indicated the presence of tannins) (Table 3) were investigated according to the described phytochemical methods [11, 55].

Table 1: Information on the studied plants.

Species (family); voucher number	Traditional uses	Bioactive or potentially bioactive components	Known antimicrobial activities of plants
Azadirachta indica A. Juss (Meliaceae); 4447/SRFK	Antimalarial, anticancer, anti-inflammatory, antidiabetic, antihyperglycaemic, antiulcerous [16]	Alkaloids, glycosides, flavonoids, and saponins [17]	Antibacterial activity of ethanol leaf extract: *Ec*, *Kp*, *Pm*, *Sa*, *Pa*, *Ef* [18]; ethanol and methanol extract of leaves: *Ec*, *Pa*, *St*, *Sa*, *Bp* [17]
Citrus grandis (L.) Osbeck (Rutaceae); 25,860/SRFC	Cancer [19, 20]	Alkaloids, cardioglycosides, saponins, tannins, terpenoids, flavonoids, and steroids [21]	Antibacterial activity of petroleum ether, ethyl acetate, chloroform, ethanol, and leaf water: *Ec*, *Sa*, *Pa*, *Pm* [21]
Cucurbita maxima Duch. (Cucurbitaceae); 42,449/HNC	Diabetes, cancer, antihypertensive, anti-inflammatory, immunomodulating, and bacterial infections [22, 23]	Tannins, saponins, alkaloids [24], cucurbitaxanthin, gibberellin, and α-tocopherol [25]	Antibacterial activity of aqueous seed extract: *Ec*, *Sa*, *Kp*, *Ef*, *Pa* [24], ethanol and aqueous extract of flowers: *St*, *Ec*, *Ef*, *Bc* [26]
Dacryodes edulis (G. Don) H. J. Lam (Burseraceae); 1874/SRFK	Gastrointestinal disorders, toothache, earache [27], dysentery, anemia, leprosy, [28], skin diseases, and sickle cell disease [29]	Ethyl gallate and quercitrin [27]	Antibacterial activity of hydromethanolic, butanol, aqueous, ethanolic extract, ethyl acetate of bark: *Ec*, *Pa*, *Bc*, *Sa* [27]
Hibiscus esculentus L. (Malvaceae); 42,823/HNC	Inflammation, pain [30], cancer, hypoglycemic [31]	Not reported	Not reported
Ipomoea batatas (L.) Lam (Convolvulaceae); 55,594/HNC	Antidiabetic, anti-inflammatory, reducing risk of cardiovascular disease, anticancer, reducing stomach stress, nausea, diarrhea [32]	Caffeoylquinic acid [33], vitamin E, beta-carotene, lutein, saponins [32], flavonoids, and chitinases [34]	Not reported
Irvingia gabonensis (Aubry. Lec. ex O. Rorke) Baill. (Irvingiaceae); 52,936/HNC	Diarrhea, hernia, yellow fever, dysentery, antipoison [35, 36], gonorrhea, gastrointestinal and hepatic disorders, wound infections, diabetes, analgesic [37]	Saponins, tannins, phenols, and phlobatannins [35], alkaloids, cardiac glycosides, anthraquinones, tannins, flavonoids [36], 3-friedelanone; betulinic acid; oleanolic acid; 3,3,4-tri-*O*-methylellagic acid; 3,4-di-*O*-methylellagic acid; hardwickiic acid [38]	Antibacterial activity of aqueous and ethanol extract of leaves and bark: *Sa*, *Ec* [36]; S: *Bst*, *Ca*, *Cf*, *Ea*, *Ecl*, *Mm*, *Ng*, *Pa*, *Pm*, *Pv*, *Sa*, *Sd*; *Bc*, *Bm*, *Bs*, *Ck*, *Ec*, *Kp*, *Sfl*, *St*, *Sf* [38]
Phaseolus vulgaris L. (Fabaceae); 42,587/HNC	Antioxidant [39], cancer, estrogenic, antidepressant [40], bacterial infections, tuberculosis [41]	Alkaloids, steroids, and flavonoids [42]	Antibacterial activity of aqueous, alcohol, chloroform, ether extract of seeds: *Sa*, *Pa*, *Sf*, *Bs*, *Kp*, *Ec* [41]
Ricinodendron heudelotii (Baill.) Pierre ex Heckel (Euphorbiaceae); 50,852/HNC	Cough, antidote, intestinal diseases [43], yellow fever, malaria, stomach pain, headache, dysentery [44]	Tannins, polyphenols, alkaloids, glycosides, flavonoids, steroids, and saponins [43], aleuritolic acid, labda-8(17), 13-diem-3β, 15-diol, *E*-ferulic acid octacosylate [44]	Antibacterial activity of methanol extract of the leaves: *Sa*, *Sf*, *Pv*, *Cf*, *Mm*, *Ko*, *Kp*, *Ec*, *Pa*, *St* [43], *Pf*, *Bs*, *Sa*, *Ec*, *Ca*, *Af* [44], *Ec*, *Ea*, *Pa*, *Ps*, *Kp*, *Ecl* [18]
Saccharum officinarum L. (Poaceae); 42,958/HNC	Jaundice and liver problems, hemorrhoids, dysentery, menorrhagia [45]	Flavonoids, saponins, tannins, and alkaloids [46]	Antibacterial activity of methanol extract of the stems: *Ec*, *Kp*, *Sa*, *Pa* [46]
Spondias mombin L. (Anacardiaceae); 21,249/SRFK	Diuretic, febrifuge, diarrheal diseases, dysentery, hemorrhoids, gonorrhea, leucorrhea [47]	Saponins, tannins, flavonoids, alkaloids, and glycosides [48]	Antibacterial activity of ethanol, methanol, water, and acetone extracts of the leaves: *Kp*, *Sa*, *St*, *Ea*, *Sm* [48]
Theobroma cacao L. (Sterculiaceae); 60,111/HNC	Cardiovascular, gastrointestinal, and nervous diseases [49]	Alkaloids, anthraquinones, cardiac glycosides, and saponins [50]	Antibacterial activity: *Sa*, *Ec*, *Sd*, *Kp*, *Sm*, *Pa*, *Pm* [51]

Table 1: Continued.

Species (family); voucher number	Traditional uses	Bioactive or potentially bioactive components	Known antimicrobial activities of plants
Uapaca guineensis Muell. Arg. (Euphorbiaceae); 53,136/HNC	Leprosy, epilepsy, edema, rheumatism, aphrodisiac, fever, inflammation, absorption [52]	Steroids, alkaloid, terpenoids, and gallic acid [53]	Not reported

HNC: Herbier National du Cameroun; SRFC: Société des Réserves Forestières du Cameroun; SRFK: Société des Réserves forestières du Kamerun; Af: *Aspergillus flavus*; Bp: *Bacillus pumilus*; Bc: *Bacillus cereus*; Bm: *Bacillus megaterium*; Bs: *Bacillus subtilis*; Ca: *Candida albicans*; Cf; *Citrobacter freundii*; Ck: *Candida krusei*; Ea: *Enterobacter aerogenes*; Ec: *Escherichia coli*; Ecl: *Enterobacter cloacae*; Ef: *Enterococcus faecium*; Kp: *Klebsiella pneumoniae*; Ko: *Klebsiella oxytoca*; Mm: *Morganella morganii*; Ng: *Neisseria gonorrhoeae*; Pa: *Pseudomonas aeruginosa*; Pf: *Pseudomonas fluorescens*; Pm: *Proteus mirabilis*; Ps: *Providencia stuartii*; Pv: *Proteus vulgaris*; Sa: *Staphylococcus aureus*; Sd: *Shigella dysenteriae*; Sf; *Streptococcus faecalis*; Sf l: *Shigella flexneri*; Sm: *Serratia marcescens*; St: *Salmonella typhi*.

Table 2: Bacterial strains and features.

Bacteria	Features	References
ATCC 25923	Reference strain	—
S. aureus MSSA1	Clinical isolate: Met susceptible; Nis^r, Chl^r	[8, 9]
S. aureus MRSA3	Clinical isolate: $Ofxa^r$, Kan^r, Tet^r, Erm^r	[8]
S. aureus MRSA4	Clinical isolate: $Ofxa^r$, Kan^r, Cyp^r, Chl^r, Gen^r, Nis^r, Amp^r	[8, 9]
S. aureus MRSA6	Clinical isolate: $Ofxa^r$, Flx^r, Kan^r, Tet^r, Cyp^r, IM/Cs^r, Chl^r, Gen^r, Nis^r, Amp^r	[8, 9]
S. aureus MRSA8	Clinical isolate: $Ofxa^r$, Flx^r, Kan^r, Erm^r, Cyp^r, Im/Cs^r, Chl^r, Gen^r, Nis^r, Amp^r	[8, 9]
S. aureus MRSA9	Clinical isolate: $Ofxa^r$, Flx^r, Tet^r, Erm^r, Cyp^r, Im/Cs^r, Chl^r, Gen^r, Nis^r, Amp^r	[8, 9]
S. aureus MRSA11	Clinical isolate: $Ofxa^r$, Kan^r, Erm^r, Cyp^r, Im/Cs^r, Chl^r, Nis^r, Amp^r	[8, 9]
S. aureus MRSA12	Clinical isolate: $Ofxa^r$, Flx^r, Kan^r, Erm^r, Im/Cs^r, Chl^r, Gen^r, Nis^r, Amp^r	[8, 9]
SA01	Clinical isolate: Erm^r, Amp^r	[54]
SA07	Clinical isolate: Erm^r, Dox^r	[54]
SA18	Clinical isolate: Amp^r, Dox^r, Vm^r	[54]
SA23	Clinical isolate: Imi^r, Aug^r	[54]
SA36	Clinical isolate: Dox^r, Vm^r	[54]
SA39	Clinical isolate: Amp^r	[54]
SA56	Clinical isolate: Amp^r, Dox^r	[54]
SA64	Clinical isolate: Amp^r, Dox^r	[54]
SA68	Clinical isolate: Amp^r, Vm^r	[54]
SA88	Clinical isolate: Erm^r, Vm^r	[54]
SA114	Clinical isolate: Amp^r, Dox^r	[54]
SA116	Clinical isolate: Erm^r	[54]
SA124	Clinical isolate: Erm^r	[54]
SA126	Clinical isolate: Amp^r, Dox^r	[54]
SA127	Clinical isolate: Amp^r, Dox^r	[54]
SA135	Clinical isolate: Erm^r	[54]
SA139	Clinical isolate: Erm^r	[54]

Chl^r, Cyp^r, Erm^r, Flx^r, Im/Cs^r, Kan^r, Met^r, $Ofxa^r$, Tet^r, Vm^r, Amp^r, Dox^r, Aug^r, Gen^r, and Nis^r resistance to chloramphenicol, ciprofloxacin, erythromycin, flomoxef, imipenem/cilastatin sodium, kanamycin, methicillin, ofloxacin, tetracycline, vancomycin, ampicillin, doxycycline, augmentin, gentamicin, and nisin, respectively, SA: *Staphylococcus aureus.*

2.5. INT Colorimetric Assay for MIC and Minimal Bactericidal Concentration (MBC) Determinations. The MIC and minimal bactericidal concentration (MBC) determinations on various strains of *S. aureus* were performed using the rapid INT colorimetric assay [56] with some modifications as previously described [14, 38]. The samples were dissolved in DMSO/MHB. The final concentration of DMSO was lower than 2.5%. The twofold dilutions of the samples were made in a 96-well microplate, and the tested bacterial concentration was 1.5×10^6 colony-forming unit (CFU)/mL. The microplates were incubated at 37°C for 18 h. All assays were performed in triplicate and repeated thrice. Wells containing MHB, 100 μL of inoculum, and DMSO to a final concentration of 2.5% served as negative control. The MIC of each sample was detected after 18 h incubation at 37°C, following addition (40 μL) of 0.2 mg/mL of INT and incubation at 37°C for 30 minutes as the lowest sample concentration that prevented the color change of the medium and exhibited complete inhibition of microbial growth [56]. The MBC was determined by adding 50 μL aliquots of the preparations, which did not show any growth after incubation during MIC assays, to 150 μL of MHB. These preparations were further incubated at 37°C for 48 h. The MBC was regarded as the lowest concentration of samples, which did not produce a color change after addition of INT as mentioned above [57, 58].

2.6. Evaluation of the Role of Efflux Pumps in the Resistance of Selected Bacteria. To evaluate the involvement of efflux

Table 3: Extraction yields and phytochemical composition of the plant extracts.

Plant extract and part used		Yields (%)	Alkaloids	Polyphenols	Flavonoids	Anthraquinones	Tannins	Triterpenes	Steroids	Saponins
Azadirachta indica	Bark	10.3	+	+	−	−	+	−	+	+
Citrus grandis	Leaves	2.6	+	+	−	−	+	+	+	−
Cucurbita maxima	Beans	2.6	−	+	−	−	+	+	+	+
	Leaves	6.2	−	+	+	+	+	+	+	+
Dacryodes edulis	Bark	9.1	−	+	−	+	+	+	+	+
	Seeds	6.9	−	+	+	+	+	+	+	+
Hibiscus esculentus	Leaves	1.9	−	+	−	−	+	−	+	−
Ipomoea batatas	Beans	3.3	+	+	+	+	+	+	−	+
Irvingia gabonensis	Leaves	6.7	−	+	−	−	+	−	+	+
Phaseolus vulgaris	Leaves	1.2	−	+	−	−	+	−	+	+
Ricinodendron heudelotii	Bark	2.9	−	+	+	+	+	+	+	−
	Leaves	7.2	−	+	+	+	+	+	+	+
Spondias mombin	Leaves	21.4	−	+	−	−	+	+	+	−
Saccharum officinarum	Leaves	8.4	−	+	−	−	+	−	+	+
Theobroma cacao	Leaves	3.1	−	+	−	−	+	+	+	+
	Beans	6.2	+	+	+	+	+	+	+	+
Uapaca guineensis	Leaves	7.3	−	+	−	−	+	+	+	+
	Bark	6.1	+	+	−	−	+	+	+	+

−: absent; +: present; yield calculated as the ratio of the mass of the obtained methanol extract/mass of the plant powder.

pumps in the resistance of selected bacterial strains to some of the active plant extracts, *Dacryodes edulis* seeds (DES), *Hibiscus esculentus* leaves (HEL), *Uapaca guineensis* leaves (UGL), *Uapaca guineensis* bark (UGB), and CIP (reference drug) were tested in the absence or presence of EPI (CCCP (0.5 μg/mL) or CPZ (25 μg/mL)). MICs of samples alone or in combination with EPI were determined as above, and the increase in activity was determined as the ratio of MIC of sample alone versus sample in combination with EPI. All assays were performed in triplicate and repeated thrice.

2.7. Antibiotic Activity Modulation Assays. To evaluate the antibiotic resistance modulation activity of the most active extracts: *Azadirachta indica* bark (AIB), *Dacryodes edulis* seeds (DES), *Dacryodes edulis* bark (DEB), *Dacryodes edulis* leaves (DEL), *Phaseolus vulgaris* leaves (PVL), *Ricinodendron heudelotii* leaves (RHL), and *Uapaca guineensis* bark (UGB), a preliminary assay was performed in order to assess the MICs of antibiotics in the absence and presence of these extracts using the broth microdilution method as previously described [14, 38, 56]. *S. aureus* SA88 was used for the preliminary assay, and the samples were tested at various subinhibitory concentrations (MIC/2, MIC/4, MIC/8, and MIC/16). Results allowed to select DEB, DEL, DES, RHL, and UGB to be tested further against *S. aureus* ATCC 25923, and 8 resistant strains of *S. aureus* (MRSA3, MRSA4, MRSA9, MRSA11, MRSA12, SA18, SA36, and SA64) at MIC/2 and MIC/4. Briefly, after serial dilutions of antibiotics, extract was added to each well at its subinhibitory concentrations, the bacterial inoculation was done, and the MIC was determined. Rows receiving antibiotic dilutions without extracts were used for the determination of the MICs of the antibiotics. The modulation factor was defined as the ratio of the MIC of antibiotic alone versus that of antibiotic in the presence of extract. Modulation factor $\geq$2 was set as the cutoff for biologically significance of antibiotic resistance-modulating effects [59].

3. Results

3.1. Phytochemical Composition of Plant Extracts. The major classes of phytochemicals were screened in the 18 studied plant extracts (Table 3). It appears that all extracts contained polyphenols and tannins. Other classes of phytochemicals were selectively present. Only the extract of the beans of *Theobroma cacao* contained all the investigated classes of secondary metabolites.

3.2. Antibacterial Activity. The antibacterial activities of the 18 tested extracts and ciprofloxacin against 26 strains of *S. aureus* are summarized in Table 4. It appears that extracts from *Dacryodes edulis* seeds (DES) and *Dacryodes edulis* bark (DEB), within a MIC range of 256–1024 μg/mL, and ciprofloxacin (MIC below 4 μg/mL), were active against all 26 tested bacterial strains. Other extracts were selectively active, and MIC values varied from 64 to 1024 μg/mL against 25/26 (96.2%) tested bacteria for *Phaseolus vulgaris* leaves (PVL), 24/26 (92.3%) for *Azadirachta indica* bark (AIB), *Dacryodes edulis* leaves (DEL), and *Ricinodendron heudelotii* leaves (RHL), 23/26 (88.5%) for *Hibiscus esculentus* leaves (HEL), 19/26 (73.1%) for *Uapaca guineensis* leaves (UGL), 18/26 (69.2%) for *Ricinodendron heudelotii* bark (RHB) and *Uapaca guineensis* bark (UGB), 17/26 (61.5%) for *Saccharum officinarum* leaves (SOL), 16/26 (61.5%) for *Ipomoea batatas* leaves (IBL) and *Theobroma cacao* leaves (TCL), 15/26 (57.7%) for *Citrus grandis* leaves (CGL), 14/26 (53.8%) for *Theobroma cacao* beans (TCB), 12/26 (46.2%) for *Cucurbita maxima* beans (CMB), 10/26 (38.5%) for *Spondias mombin* leaves (SML), and 7/26 (26.9%) for *Irvingia gabonensis* beans

Table 4: MIC and MBC of the plant extracts and ciprofloxacin against *Staphylococcus aureus* strains.

Bacterial strains[a]	Samples[b], MIC and MBC in μg/mL (in parentheses)																		
	Plant extracts																		Antibiotic
	AIB	CGL	CMB	DES	DEB	DEL	HEL	IBL	IGB	PVL	RHB	RHL	SML	SOL	TCB	TCL	UGL	UGB	CIP
ATCC 25923	1024 (–)	–	–	1024 (–)	512 (–)	1024 (–)	–	–	–	512 (–)	1024 (–)	512 (–)	–	1024 (–)	–	–	–	–	<0.5 (16)
SA01	512 (–)	1024 (–)	–	512 (–)	512 (–)	512 (–)	512 (–)	–	–	512 (–)	–	512 (–)	–	1024 (–)	512 (–)	–	512 (–)	512 (–)	<0.5 (4)
SA07	1024 (–)	–	–	512 (–)	512 (–)	512 (–)	512 (–)	1024 (–)	–	1024 (–)	–	512 (–)	–	1024 (–)	–	–	512 (–)	–	<0.5 (1)
SA18	512 (1024)	1024 (–)	–	512 (–)	512 (–)	512 (–)	512 (–)	–	1024 (–)	512 (–)	1024 (–)	512 (–)	–	512 (–)	512 (–)	–	512 (–)	1024 (–)	<0.5 (8)
SA23	1024 (–)	512 (–)	–	512 (–)	512 (–)	1024 (–)	512 (–)	1024 (–)	1024 (–)	512 (–)	1024 (–)	512 (–)	1024 (–)	1024 (–)	512 (–)	512 (–)	512 (–)	512 (–)	<0.5 (<0.5)
SA36	512 (–)	–	1024 (–)	256 (1024)	1024 (–)	1024 (–)	512 (–)	1024 (–)	–	1024 (–)	512 (–)	512 (–)	1024 (–)	1024 (–)	1024 (–)	–	512 (–)	512 (–)	1 (8)
SA39	1024 (–)	1024 (–)	–	512 (–)	1024 (–)	1024 (–)	1024 (–)	–	–	1024 (–)	1024 (–)	–	1024 (–)	–	–	512 (–)	512 (–)	512 (–)	<0.5 (16)
SA56	1024 (–)	–	–	1024 (–)	1024 (–)	1024 (–)	1024 (–)	–	1024 (–)	512 (–)	–	512 (–)	–	1024 (–)	–	–	–	1024 (–)	<0.5 (4)
SA64	512 (–)	512 (–)	1024 (–)	512 (–)	1024 (–)	512 (–)	512 (–)	1024 (–)	–	512 (–)	256 (–)	1024 (–)	1024 (–)	–	–	1024 (–)	512 (–)	1024 (–)	4 (8)
SA68	1024 (–)	–	–	1024 (–)	1024 (–)	1024 (–)	1024 (–)	1024 (–)	–	1024 (–)	1024 (–)	512 (–)	–	–	1024 (–)	–	–	–	<0.5 (<0.5)
SA88	512 (–)	1024 (–)	1024 (–)	512 (–)	1024 (–)	1024 (–)	512 (–)	1024 (–)	–	512 (1024)	–	512 (–)	–	1024 (–)	1024 (–)	1024 (–)	1024 (–)	1024 (–)	<0.5 (2)
SA114	1024 (–)	–	–	1024 (–)	1024 (–)	1024 (–)	256 (–)	512 (–)	–	512 (–)	512 (–)	–	1024 (–)	1024 (–)	–	512 (–)	512 (–)	512 (–)	<0.5 (<0.5)
SA116	–	1024 (–)	–	512 (–)	1024 (–)	512 (–)	512 (–)	–	1024 (–)	1024 (–)	–	1024 (–)	1024 (–)	1024 (–)	–	–	512 (–)	–	<0.5 (<0.5)
SA124	–	1024 (–)	–	512 (–)	1024 (–)	–	–	1024 (–)	–	–	–	1024 (–)	–	–	–	512 (–)	–	–	<0.5 (<0.5)
SA126	1024 (–)	–	1024 (–)	512 (–)	512 (–)	1024 (–)	512 (–)	1024 (–)	–	1024 (–)	1024 (–)	1024 (–)	–	–	–	1024 (–)	1024 (–)	–	<0.5 (<0.5)
SA127	512 (–)	–	–	1024 (–)	1024 (–)	1024 (–)	512 (–)	1024 (–)	–	512 (1024)	–	1024 (–)	1024 (–)	–	–	–	–	–	<0.5 (<0.5)
SA135	1024 (–)	–	–	512 (1024)	1024 (–)	1024 (–)	1024 (–)	–	–	1024 (–)	–	512 (–)	–	1024 (–)	–	–	–	–	<0.5 (1)
SA139	512 (–)	–	–	512 (–)	512 (–)	–	1024 (–)	1024 (–)	–	512 (–)	1024 (–)	512 (–)	–	1024 (–)	–	1024 (–)	–	1024 (–)	<0.5 (<0.5)
MSSA1	512 (–)	–	256 (512)	512 (1024)	1024 (–)	256 (–)	512 (1024)	1024 (–)	–	512 (–)	256 (1024)	512 (1024)	–	1024 (–)	256 (–)	512 (–)	256 (1024)	128 (1024)	2 (16)
MRSA3	512 (1024)	1024 (–)	256 (–)	512 (1024)	1024 (–)	512 (–)	1024 (–)	–	1024 (–)	512 (–)	256 (–)	1024 (–)	512 (–)	–	256 (–)	256 (–)	256 (–)	128 (–)	2 (16)
MRSA4	512 (–)	512 (–)	**64** (512)	512 (512)	512 (1024)	256 (–)	1024 (–)	1024 (–)	–	128 (–)	128 (–)	1024 (–)	1024 (–)	1024 (–)	256 (–)	512 (–)	256 (–)	256 (–)	1 (16)
MRSA6	512 (–)	512 (–)	128	512 (–)	1024 (–)	1024 (–)	–	512 (–)	–	1024 (–)	256 (–)	256 (–)	–	–	128 (–)	256 (–)	256 (–)	128 (–)	2 (8)
MRSA8	512 (–)	1024 (–)	128 (1024)	256 (–)	256 (–)	512 (–)	512 (1024)	–	–	512 (–)	256 (–)	512 (–)	–	512 (–)	128 (–)	512 (–)	256 (–)	128 (–)	2 (8)
MRSA9	512 (–)	1024 (–)	128	512 (–)	512 (–)	1024 (–)	512 (–)	256 (–)	1024 (–)	512 (–)	256 (–)	512 (–)	256 (–)	512 (–)	128 (–)	512 (–)	128 (–)	**64** (–)	2 (16)
MRSA11	1024 (–)	1024 (–)	256	512 (–)	512 (–)	1024 (–)	512 (–)	1024 (–)	1024 (–)	512 (–)	256 (–)	512 (–)	–	–	128 (–)	512 (–)	256 (–)	128 (–)	2 (16)
MRSA12	512 (–)	1024 (–)	128	512 (–)	256 (–)	256 (–)	512 (–)	–	–	1024 (–)	256 (–)	512 (–)	–	1024 (–)	128 (–)	512 (–)	256 (–)	128 (–)	2 (4)

[a]Bacterial strain (SA: *Staphylococcus aureus*; MRSA: methicillin-resistant *Staphylococcus aureus*), [b]samples (AIB: *Azadirachta indica* bark, CGL: *Citrus grandis*, CMB: *Cucurbita maxima* beans, DES: *Dacryodes edulis* seeds, DEB: *Dacryodes edulis* bark, DEL: *Dacryodes edulis* leaves, HEL: *Hibiscus esculentus* leaves, IBL: *Ipomoea batatas* leaves, IGB: *Irvingia gabonensis* beans, PVL: *Phaseolus vulgaris* leaves, RHL: *Ricinodendron heudelotii* leaves, RHB: *Ricinodendron heudelotii* bark, SML: *Spondias mombin* leaves, SOL: *Saccharum officinarum* leaves, TCB: *Theobroma cacao* beans, TCL: *Theobroma cacao* leaves, UGL: *Uapaca guineensis* leaves, UGB: *Uapaca guineensis* bark, and CIP: ciprofloxacin); R: MBC/MIC; (–): >1024(MIC); nc, not calculated; MIC: minimal inhibitory concentration; MBC: minimal bactericidal concentration; CIP: ciprofloxacin; –: MIC and MBC at up to 1024 μg/mL; MIC in bold: significant activity [37, 60].

TABLE 5: MIC of extracts and ciprofloxacin in the absence (−) and presence (+) of carbonyl cyanide *m*-chlorophenyl hydrazone (CCCP) against selected strains of *Staphylococcus aureus*.

Bacterial strains[a]	Samples[b] and MIC in μg/mL and fold increase of activity (in parentheses)									
	DES		HEL		UGB		UGL		CIP	
	−	+	−	+	−	+	−	+	−	+
ATCC 25923	256	**32 (32)**	256	**<8 (>32)**	−	**<8 (>128)**	1024	**<8 (>128)**	2	1(2)
MRSA3	256	**<8 (>32)**	256	**<8 (>32)**	512	**<8 (>64)**	512	**<8 (>64)**	1	<0.5 (>2)
MRSA4	256	**16 (16)**	128	**16 (8)**	1024	**128 (8)**	1024	**64 (16)**	2	<0.5 (>4)
MRSA6	128	**<8 (>16)**	256	**<8 (>32)**	1024	**<8 (>128)**	512	**<8 (>64)**	1	<0.5 (>2)
MRSA8	256	**128 (2)**	128	128 (1)	−	512 (>2)	512	128 (>4)	<0.5	<0.5 (nd)
MRSA9	128	**<8 (>16)**	64	<8 (>8)	1024	**<8 (>128)**	256	**<8 (>32)**	<0.5	<0.5 (nd)
MRSA11	256	**<8 (>32)**	128	**<8 (>16)**	1024	**<8 (>128)**	256	**<8 (>32)**	1	1(1)
MRSA12	256	**<8 (>32)**	128	**<8 (>16)**	1024	**<8 (>128)**	256	**<8 (>32)**	<0.5	<0.5 (nd)
SA01	128	**<8 (>16)**	128	**<8 (>16)**	1024	**<8 (>128)**	256	**<8 (>32)**	<0.5	<0.5 (nd)
SA07	512	**<8 (>64)**	512	512 (1)	512	512 (1)	256	**<8 (>32)**	<0.5	<0.5 (nd)
SA18	1024	**128 (8)**	256	**128 (2)**	1024	**512 (2)**	512	**<8 (>64)**	1	<0.5 (>2)
SA88	512	512 (1)	−	−	1024	1024 (1)	1024	256 (>4)	1	<0.5 (>2)
SA114	1024	**<8 (>128)**	−	−	1024	**128 (8)**	512	**<8 (>64)**	<0.5	<0.5 (>2)
SA135	512	**64 (8)**	−	−	−	**128 (>8)**	512	**<8 (>64)**	1	<0.5 (>2)

[a]Bacterial strain (SA: *Staphylococcus aureus*; MRSA: methicillin-resistant *Staphylococcus aureus*), [b]samples (DES: *Dacryodes edulis* seeds, HEL: *Hibiscus esculentus* leaves, UGL: *Uapaca guineensis* leaves, UGB: *Uapaca guineensis* bark, and CIP: ciprofloxacin); MIC: minimal inhibitory concentration; CCCP was tested at 0.5 μg/mL; (-): >1024 μg/mL; values in bold represent increase of activity ≥2.

TABLE 6: MIC of extracts and ciprofloxacin in the absence (−) and presence (+) of chlorpromazine (CPZ) against selected strains of *Staphylococcus aureus*.

Bacterial strains[a]	Samples[b] and MIC in μg/mL and fold increase of activity (in parentheses)									
	DES		HEL		UGB		UGL		CIP	
	−	+	−	+	−	+	−	+	−	+
ATCC 25923	512	512 (1)	−	−	512	512 (1)	128	1024 (0.13)	2	2 (1)
MRSA3	256	256 (1)	1024	1024 (1)	512	512 (1)	−	−	4	4 (1)
MRSA4	512	512 (1)	−	−	256	256 (1)	1024	1024 (1)	2	2 (1)
MRSA6	256	256 (1)	512	128 (2)	512	512 (1)	256	−(0.25)	2	2 (1)
MRSA8	256	256 (1)	1024	1024 (1)	256	256 (1)	−	−	1	2 (0.5)
MRSA9	256	256 (1)	1024	1024 (1)	512	512 (1)	256	−(0.25)	1	2 (0.5)
MRSA11	512	512 (1)	1024	1024 (1)	512	512 (1)	128	−(<0.13)	4	4 (1)
MRSA12	512	512 (1)	−	−	512	512 (1)	−	−	2	2 (1)
SA01	512	512 (1)	1024	1024 (1)	512	512 (1)	512	512 (1)	1	1 (1)
SA07	512	512 (1)	1024	1024 (1)	512	512 (1)	1024	1024 (1)	4	4 (1)
SA18	512	512 (1)	−	−	1024	1024 (1)	−	−	1	1 (1)
SA88	512	512 (1)	−	−	1024	1024 (1)	−	−	4	4 (1)
SA114	1024	1024 (1)	−	−	1024	1024 (1)	−	−	1	1 (1)
SA135	512	512 (1)	−	−	1024	1024 (1)	−	−	1	1 (1)

[a]Bacterial strain (SA: *Staphylococcus aureus*; MRSA: methicillin-resistant *Staphylococcus aureus*); [b]samples (DES: *Dacryodes edulis* seeds, HEL: *Hibiscus esculentus* leaves, UGL: *Uapaca guineensis* leaves, UGB: *Uapaca guineensis* bark, CIP: ciprofloxacin); CPZ: chlorpromazine at 25 μg/mL; CIP: ciprofloxacin; MIC: minimal inhibitory concentration.

(IGB). The lowest MIC value of 64 μg/mL was obtained with CMB against MRSA4 strain and UGB against MRSA9 strain.

3.3. Role of Efflux Pumps in the Resistance of Strains of S. aureus. Four extracts (DES, HEL, UGL, and UGB) and CIP (reference drug) were tested in the absence or presence of CCCP (0.5 μg/mL) and CPZ (25 μg/mL) to evaluate the role of efflux pumps in the resistance of 14 tested *S. aureus* strains. The results are summarized in Tables 5 and 6. It appears that CCCP significantly improved the activity of the 4 tested extracts against the majority of *S. aureus* strains (Table 5). The increase of activity in the presence of CCCP ranged from 2-fold to >128-fold. The highest increase of activity (>128-fold) was obtained when DES, UGB, and UGL were tested in the presence of CCCP on at least one *S. aureus* strain. In contrast, in the presence of CPZ, no improvement in the activity of the four extracts was observed (Table 6). This is clear indication that CCCP was the appropriate EPI of the studied *S. aureus* strains.

3.4. Antibiotic Resistance Modulation Activity of Extracts. Seven plant extracts, AIB, DES, DEB, DEL, PVL, RHL, and UGB, at their various subinhibitory concentrations (MIC/2, MIC/4, MIC/8, and MIC/16) were first tested in combination with 8 antibiotics: CHL, TET, CIP, AMP, CEF, ERY, STR, and KAN against *S. aureus* SA88 strain. The results summarized in Table S1 (Supplementary Materials) show that better modulation of the activity of antibiotics was obtained with all extracts at MIC/2 and MIC/4. At their MIC/2, 2-fold or more increase of antibiotic activities was obtained with PVL, AIB, DEB, DES, DEL, UGB, and RHL and 2, 4, 4, 6, 6, 7, and 8 of the 8 tested antibiotics, respectively (Table S1; Supplementary Materials). Consequently, the most active extracts, DEB, DEL, DES, RHL, and UGB, were further tested in combination with the above 8 antibiotics against the reference strains (ATCC 25923) and 8 resistant strains of *S. aureus* at MIC/2 and MIC/4 (Tables S2–S6; Supplementary Materials). Results showed that 2-fold or more antibiotic-modulating effects against more than 70% of the *S. aureus* strains tested were obtained when DEB was combined with CHL at MIC/2 (77.78%; Table S2; Supplementary Materials), when DEL was combined with CHL and STR at MIC/2 (77.78%; Table S3; Supplementary Materials), when DES was combined with CIP (77.78% at MIC/2), CHL (100% and 88.89 at MIC/2 and MIC/4 resp.), TET (77.78% at MIC/2 and MIC/4), and STR (88.89% and 77.78% at MIC/2 and MIC/4 resp.) (Table S4; Supplementary Materials), when RHL was combined with CIP, ERY, and KAN (88.89% and 77.78% at MIC/2 and MIC/4 resp.), CHL (88.89% at MIC/2 and MIC/4), TET (77.78% at MIC/2 and MIC/4), and STR (88.89% at MIC/2) (Table S5; Supplementary Materials), and when UGB was combined with CHL, KAN, and STR (77.78% at MIC/2 and MIC/4) (Table S6; Supplementary Materials).

4. Discussion

4.1. Phytochemical Composition of Extracts. Polyphenols and tannins were detected in all extracts. The role of several molecules belonging to polyphenols as antibacterials has been demonstrated [10, 11, 38, 61]. Tannins also belong to a class of polyphenols, and its presence in all extracts could in part explain the fact that all the tested extracts were active in at least one strain of the tested bacteria [10]. However, it should be made clear that the presence of a class of secondary metabolite with reported antibacterial effect is not a guarantee of the good activity of a plant. The antibacterial effect depends on the structure and the amount of a particular phytochemical in the plant or possible interactions with other compounds. This could explain why the extract from the beans of *Theobroma cacao* that contained all the investigated classes of secondary metabolites (Table 3) was not the most active sample (Table 4).

4.2. Antibacterial Potential of Extracts. Resistance of bacteria to antibiotics propels the search of new agents to fight against MDR phenotypes. In the present study, clinical strains of *S. aureus* used were previously reported as resistant to at least one commonly used antibiotic [8, 9] (Table 2). Several locally isolated strains of *S. aureus* [54] were used herein, to better adapt the study to our environment. According to established criteria, MIC values in the range of 100–1000 μg/mL are indication that plant extracts bear antimicrobial activities [62]. Also, the antibacterial activity of botanicals is considered significant if MIC values are below 100 μg/mL, moderate if $100 \leq$ MICs ≤ 625 μg/mL, and weak if MICs > 625 μg/mL [37, 60]. On these bases, it can be deduced that all the tested plant extracts had antistaphylococcal activities, except *Irvingia gabonensis* beans (IGB), with MICs above 1000 μg/mL against all tested strains of *S. aureus* (Table 4). This activity was significant for CMB against MRSA4 strain and UGB against MRSA9 strain (MIC: 64 μg/mL). Most of the recorded MIC values ranged from 512 to 1024 μg/mL, indicating that extracts rather exhibited moderate to low antistaphylococcal effects. However, this activity could be considered important because the clinical strains of *S. aureus* used were resistant phenotypes while extracts were from edible plants. In effect, it was suggested that if botanicals are food plants, as they are allegedly nontoxic or less toxic than other medicinal plants, their antibacterial activity is significant in a range of $100 \leq$ MIC ≤ 512 μg/mL and moderately active in a range of $512 <$ MIC ≤ 2048 μg/mL [63].

4.3. Role of Efflux Pumps in Susceptibility of S. aureus Strains to the Extracts. Bacterial efflux systems are associated with major human health concerns as they are involved in the resistance of pathogenic bacteria such as *S. aureus* [64–66]. Previously, inhibition of efflux pumps by natural products has been found to improve the activity of antibiotics against *S. aureus*. For example, inhibition of the TetK efflux pump was reported with the essential oil of *Chenopodium ambrosioides* and its constituent α-terpinene against *S. aureus* IS-58 strain [66]. In the present study, two well-known EPIs (CCCP and CPZ) were used to assess the implication of efflux pumps in the resistance of the studied *S. aureus* strains to plant extracts. CCCP is an inhibitor of the proton-motive force of ATP-binding cassette (ABC) transporters of several Gram-negative and Gram-positive bacteria, including *S. aureus* [5–7]. CPZ is capable of reversing or reducing the antibiotic resistance of bacteria including *S. aureus* [67], due to its indirect effects on ATPase activity that is dependent upon Ca^{2+} [68]. In the present study, it was found that CCCP contrary to CPZ improved the activity of the four extracts (DES, HEL, UGL, and UGB) (Tables 5 and 6). This indicates that ABC transporters are involved in the resistance of the studied strains of *S. aureus* and that combination of extracts such as HEL, UGL, and UGB with an inhibitor of ABC transporters could improve the antistaphylococcal fight.

4.4. Antibiotic Modulation Effects of Extracts. The antibiotic resistance-modulating effects of several botanicals and phytochemicals against resistant bacteria have been documented [12–15, 59, 69]. It has been suggested that extracts capable of potentiating the activity of antibiotics on more than 70% of bacteria could be potential efflux pump inhibitors [70]. In this study, antibiotic modulation activity of extracts at their MIC/2 on more than 70% tested strains of *S. aureus* was obtained with the association of DEB and DEL and with 1/8 (12.5%) tested antibiotics (Tables S2 and S3; Supplementary Materials), UGB with 2/8 (25%) antibiotics (Table S6; Supplementary Materials),

DES with 4/8 (50%) antibiotics (Table S4; Supplementary Materials), and RHL with 6/8 (75%) antibiotics (Table S5; Supplementary Materials). Hence, the tested extracts and mostly RHL may act as efflux pump inhibitors [70]. The use of CCCP indicated that ABC transporters were the efflux pumps involved in the resistance of the tested bacteria, suggesting that the above extract could be the inhibitors of such pumps. The potential of the *R. heudelotii* leaf extract (RHL) to reverse antibiotic resistance in Gram-negative MDR bacteria was previously reported [18]. The present study therefore provides more information about the ability of this plant to modulate the activity of antibiotics against resistant strains of *S. aureus*.

5. Conclusion

In conclusion, the present work provides informative data about the antistaphylococcal potential of 13 Cameroonian food plants. It also indicates that some extracts such as DES, HEL, UGL, and UGB could be used in combination with EPI to combat resistance of *Staphylococcus aureus* to antibiotics. Finally, this study also demonstrates that some studied extracts and mostly RHL could be used as antibiotic resistance modulators, providing a new weapon against the resistance of *S. aureus* to antibiotics.

Abbreviations

AIB: *Azadirachta indica* bark
AMP: Ampicillin
ATCC: American Type Culture Collection
CCCP: Carbonyl cyanide *m*-chlorophenyl hydrazone
CEF: Cefepime
CFU: Colony-forming unit
CGL: *Citrus grandis*
CHL: Chloramphenicol
CIP: Ciprofloxacin
CMB: *Cucurbita maxima* beans
CPZ: Chlorpromazine
DEB: *Dacryodes edulis* bark
DEL: *Dacryodes edulis* leaves
DES: *Dacryodes edulis* seeds
DMSO: Dimethylsulfoxide
EPI: Efflux pump inhibitors
ERY: Erythromycin
HEL: *Hibiscus esculentus* leaves
IBL: *Ipomoea batatas* leaves
IGB: *Irvingia gabonensis* beans
INT: *p*-Iodonitrotetrazolium chloride
KAN: Kanamycin
MBC: Minimal bactericidal concentration
MDR: Multidrug-resistant
MHB: Mueller Hinton broth
MIC: Minimal inhibitory concentration
MRSA: Methicillin-resistant *Staphylococcus aureus*
PVL: *Phaseolus vulgaris* leaves
RA: Reference antibiotics
RHB: *Ricinodendron heudelotii* bark
RHL: *Ricinodendron heudelotii* leaves
SA: *Staphylococcus aureus*
SML: *Spondias mombin* leaves
SOL: *Saccharum officinarum* leaves
STR: Streptomycin
TCB: *Theobroma cacao* beans
TCL: *Theobroma cacao* leaves
TET: Tetracycline
UGB: *Uapaca guineensis* bark
UGL: *Uapaca guineensis* leaves.

Conflicts of Interest

The authors declare that they have no conflicts of interest.

Authors' Contributions

Brice E. N. Wamba, Paul Nayim, Joachim K. Dzotam, and Ornella J. T. Ngalani carried out the study. Armelle T. Mbaveng and Victor Kuete designed the experiments. Armelle T. Mbaveng and Victor Kuete wrote the manuscript. Armelle T. Mbaveng and Victor Kuete supervised the work and provided the bacterial strains. All authors read and approved the final manuscript.

Acknowledgments

The authors are thankful to the Cameroon National Herbarium for identification of plants and also thank Dr. Jean P. Dzoyem (University of Dschang, Cameroon) for providing the MRSA strains of *S. aureus*.

Supplementary Materials

Preliminary evaluation of antibiotic resistance modulatory activity of extracts against *S. aureus* SA88 (Table S1); antibiotic resistance modulatory activity of the bark methanol extract from *Dacryodes edulis* bark (DEB) (Table S2); the leaf methanol extract from *Dacryodes edulis* leaves (DEL) (Table S3); the seed methanol extract from *Dacryodes edulis* seeds (DES) (Table S4); the leaf methanol extract from *Ricinodendron heudelotii* leaves (RHL) (Table S5); and the bark methanol extract from *Uapaca guineensis* bark (UGB) (Table S6) at their MIC/2 and MIC/4 on selected strains of *Staphylococcus aureus*. (*Supplementary Materials*)

References

[1] V. G. Fowler Jr., L. L. Sanders, D. J. Sexton et al., "Outcome of *Staphylococcus aureus* bacteremia according to compliance with recommendations of infectious diseases specialists: experience with 244 patients," *Clinical Infectious Diseases*, vol. 27, no. 3, pp. 478–486, 1998.

[2] R. Jenkins, N. Burton, and R. Cooper, "Manuka honey inhibits cell division in methicillin-resistant *Staphylococcus aureus*," *Journal of Antimicrobial Chemotherapy*, vol. 66, no. 11, pp. 2536–2542, 2011.

[3] A. P. Johnson, "Methicillin-resistant *Staphylococcus aureus*: the European landscape," *Journal of Antimicrobial Chemotherapy*, vol. 66, no. S4, pp. iv43–iv48, 2011.

[4] Y. Taj, F. Essa, F. Aziz, and S. U. Kazmi, "Study on biofilm-forming properties of clinical isolates of *Staphylococcus aureus*," *Journal of Infection in Developing Countries*, vol. 6, no. 5, pp. 403–409, 2012.

[5] A. M. Ojeda-Sana, V. Repetto, and S. Moreno, "Carnosic acid is an efflux pumps modulator by dissipation of the membrane potential in *Enterococcus faecalis* and *Staphylococcus aureus*," *World Journal of Microbiology and Biotechnology*, vol. 29, no. 1, pp. 137–144, 2012.

[6] M. Mombeshora and S. Mukanganyama, "Development of an accumulation assay and evaluation of the effects of efflux pump inhibitors on the retention of chlorhexidine digluconate in *Pseudomonas aeruginosa* and *Staphylococcus aureus*," *BMC Research Notes*, vol. 10, no. 1, p. 328, 2017.

[7] T. Hirata, R. Wakatabe, J. Nielsen, T. Satoh, S. Nihira, and A. Yamaguchi, "Screening of an inhibitor of the tetracycline efflux pump in a tetracycline-resistant clinical-isolate of *Staphylococcus aureus* 743," *Biological & Pharmaceutical Bulletin*, vol. 21, no. 7, pp. 678–681, 1998.

[8] A. Paudel, H. Hamamoto, Y. Kobayashi, S. Yokoshima, T. Fukuyama, and K. Sekimizu, "Identification of novel deoxyribofuranosyl indole antimicrobial agents," *Journal of Antibiotics*, vol. 65, no. 2, pp. 53–57, 2011.

[9] J. P. Dzoyem, H. Hamamoto, B. Ngameni, B. T. Ngadjui, and K. Sekimizu, "Antimicrobial action mechanism of flavonoids from *Dorstenia* species," *Drug Discoveries & Therapeutics*, vol. 7, no. 2, pp. 66–72, 2013.

[10] M. M. Cowan, "Plant products as antimicrobial agents," *Clinical Microbiology Reviews*, vol. 12, no. 4, pp. 564–582, 1999.

[11] V. Kuete, "Medicinal plant research in Africa," in *Pharmacology and Chemistry*, V. Kuete, Ed., Elsevier, Oxford, UK, 2013.

[12] A. G. Fankam, V. Kuete, I. K. Voukeng, J. R. Kuiate, and J. M. Pages, "Antibacterial activities of selected Cameroonian spices and their synergistic effects with antibiotics against multidrug-resistant phenotypes," *BMC Complementary and Alternative Medicine*, vol. 11, p. 104, 2011.

[13] J. A. Noumedem, M. Mihasan, S. T. Lacmata, M. Stefan, J. R. Kuiate, and V. Kuete, "Antibacterial activities of the methanol extracts of ten Cameroonian vegetables against Gram-negative multidrug-resistant bacteria," *BMC Complementary and Alternative Medicine*, vol. 13, p. 26, 2013.

[14] D. E. Djeussi, J. A. Noumedem, J. A. Seukep et al., "Antibacterial activities of selected edible plants extracts against multidrug-resistant Gram-negative bacteria," *BMC Complementary and Alternative Medicine*, vol. 13, no. 1, p. 164, 2013.

[15] S. B. Tankeo, P. Tane, and V. Kuete, "*In vitro* antibacterial and antibiotic-potentiation activities of the methanol extracts from *Beilschmiedia acuta, Clausena anisata, Newbouldia laevis* and *Polyscias fulva* against multidrug-resistant Gram-negative bacteria," *BMC Complementary and Alternative Medicine*, vol. 15, no. 1, p. 412, 2015.

[16] H. A. Mohammed and A. F. A. Omer, "Antibacterial activity of *Azadirachta indica* (Neem) leaf extract against bacterial pathogens in Sudan," *American Journal of Research Communication*, vol. 3, pp. 246–251, 2015.

[17] S. Maragathavalli, S. Brindha, N. S. Kaviyarasi, B. Annadurai, and S. K. Gangwar, "Antimicrobial activity in leaf extract of neem (*Azadirachta indica* linn.)," *International Journal of Science and Nature*, vol. 3, pp. 110–113, 2012.

[18] A. G. Fankam, J. R. Kuiate, and V. Kuete, "Antibacterial and antibiotic resistance modulatory activities of leaves and bark extracts of *Recinodindron heudelotii* (Euphorbiaceae) against multidrug-resistant Gram-negative bacteria," *BMC Complementary and Alternative Medicine*, vol. 17, no. 1, p. 168, 2017.

[19] H. Kim, J. Y. Moon, A. Mosaddik, and S. K. Cho, "Induction of apoptosis in human cervical carcinoma HeLa cells by polymethoxylated flavone-rich *Citrus grandis* Osbeck (Dangyuja) leaf extract," *Food and Chemical Toxicology*, vol. 48, no. 8-9, pp. 2435–2442, 2010.

[20] Z. Zarina and S. Y. Tan, "Determination of flavonoids in *Citrus grandis* (Pomelo) peels and their inhibition activity on lipid peroxidation in fish tissue," *International Food Research Journal*, vol. 20, no. 1, pp. 313–317, 2013.

[21] A. Jabamalairaj, S. Dorairaj, S. A. Yadav, and C. Bathrachalam, "Detection of functional group and antimicrobial activity of leaf extracts of *Citrus grandis* (L.) against selected clinical pathogens," *Indo American Journal of Pharmaceutical Research*, vol. 5, pp. 1642–1648, 2015.

[22] P. Saha, U. K. Mazumder, and P. K. Haldar, "Acute and subchronic toxicity of *Cucurbita maxima* aerial parts," *International Journal of Research in Pharmaceutical and Biomedical Sciences*, vol. 2, pp. 634–639, 2011.

[23] P. Saha, U. K. Mazumder, P. K. Haldar et al., "Anticancer activity of methanol extract of *Cucurbita maxima* against Ehrlich as-cites carcinoma," *International Journal of Research in Pharmaceutical Sciences*, vol. 2, no. 1, pp. 52–59, 2011.

[24] D. K. Kushawaha, S. Chatterji, M. Yadav, and G. Watal, "Antimicrobial potential and preliminary phytochemical evaluation of *Cucurbita maxima* seeds," *International Journal of Pharma and Bio Sciences*, vol. 7, pp. 578–582, 2016.

[25] C. Y. Ragasa and K. Lim, "Sterols from *Cucurbita maxima*," *Philippine Journal of Science*, vol. 134, no. 2, pp. 83–87, 2005.

[26] N. Muruganantham, S. Solomon, and M. M. Senthamilselvi, "Antimicrobial activity of *Cucurbita maxima* flowers (Pumpkin)," *Journal of Pharmacognosy and Phytochemistry*, vol. 5, no. 1, pp. 15–18, 2016.

[27] K. K. Ajibesin, E. E. Essien, and S. A. Adesanya, "Antibacterial constituents of the leaves of *Dacryodes edulis*," *African Journal of Pharmacy and Pharmacology*, vol. 5, no. 15, pp. 1782–1786, 2011.

[28] A. Bouquet, *Witch Doctors and Traditional Medicine of Congo (Brazzaville)*, Mem Works and Documents of the Office of Scientific and Technical Research Overseas, Paris, France, 1969.

[29] P. T. Mpiana, D. S. Tshibangu, O. M. Shetonde, and K. N. Ngbolua, "*In vitro* antidrepanocytary actvity (anti-sickle cell anemia) of some congolese plants," *Phytomedicine*, vol. 14, no. 2-3, pp. 192–195, 2007.

[30] D. Daly, "Alternative medicinal course taught at the United States Medical School," *Journal of Alternative and Complementary Medicine*, vol. 3, no. 4, pp. 406–410, 1997.

[31] A. J. Uraku, P. M. Ajah, A. N.C. Okala, U. A. Ubiam, and P. N. Onu, "Effects of crude extract of *Abelmoshus esculentus* on albumin and total bilirubin of diabetic albino rats," *International Journal of Science and Nature*, vol. 1, no. 1, pp. 38–41, 2010.

[32] I. Dini, G. C. Tenore, and A. Dini, "Saponins in *Ipomoea batatas* tubers: Isolation, characterization, quantification and antioxidant properties," *Food Chemistry*, vol. 113, no. 2, pp. 411–419, 2009.

[33] M. Nagai, M. Tani, Y. Kishimoto et al., "Sweet potato (*Ipomoea batatas* L.) leaves suppressed oxidation of low density lipoprotein (LDL) in vitro and in human subjects," *Journal of

Clinical Biochemistry and Nutrition, vol. 48, no. 3, pp. 203–208, 2011.

[34] S. M. Hue, A. N. Boyce, and C. Somasundram, "Antioxidant activity, phenolic and flavonoid contents in the leaves of different varieties of sweet potato (*Ipomoea batatas*)," *Australian Journal of Crop Science*, vol. 6, no. 3, pp. 375–380, 2012.

[35] H. E. Etta, C. C. Olisaeke, and C. I. Iboh, "Effect of *Irvingia gabonensis* (AubryLecomte ex O'Rorke) seeds on the liver and gonads of male albino rats," *Journal of Biology, Agriculture and Healthcare*, vol. 4, pp. 10–15, 2014.

[36] O. Nworie, J. O. Orji, U. O. Ekuma, M. V. Agah, C. S. Okoli, and M. C. Nweke, "Antibacterial activity of the leaf and stem bark of *Irvingia gabonensis* (Bush Mango) against *Escherichia coli* and *Staphylococcus aureus*," *Global Journal of Pharmacology*, vol. 10, pp. 13–18, 2016.

[37] V. Kuete, "Potential of Cameroonian plants and derived products against microbial infections: a review," *Planta Medica*, vol. 76, no. 14, pp. 1479–1491, 2010.

[38] V. Kuete, G. F. Wabo, B. Ngameni et al., "Antimicrobial activity of the methanolic extract, fractions and compounds from the stem bark of *Irvingia gabonensis* (Ixonanthaceae)," *Journal of Ethnopharmacology*, vol. 114, no. 1, pp. 54–60, 2007.

[39] D. Heimler, P. Vignolini, M. G. Dini, and A. Romani, "Rapid tests to assess the antioxidant activity of *Phaseolus vulgaris* L. dry beans," *Journal of Agricultural and Food Chemistry*, vol. 53, no. 8, pp. 3053–3056, 2005.

[40] D. Madhu and S. Ramica, "Antidepressant activity of aqueous extract of *Phaseolus vulgaris* (black bean) in rodent models of depression," *International Journal of Nutrition, Pharmacology, Neurological Diseases*, vol. 4, no. 2, pp. 118–124, 2014.

[41] M. Pradeepkumar, S. D. Joshi, and V. H. Kulkarni, "Evaluation of antibacterial and antitubercular activity of *Phaseolus vulgaris* Linn, seeds," *Journal of Applied Pharmaceutical Science*, vol. 5, pp. 219–224, 2015.

[42] S. T. Milan, "Total Phenolic content, flavonoid concentration and antioxidant activity of *Marrubium peregrinum* L. extracts," *Kragujevac Journal of Science*, vol. 33, pp. 63–72, 2011.

[43] V. A. Oyono, C. Fokunang, J. P. A. Assam et al., "Acute toxicity studies, antioxidant and *in vitro* antibacterial activities of extract from the barks of *Ricinodendron heudoletti* (Euphorbiaceae)," *Journal of Pharmacognosy and Phytotherapy*, vol. 6, no. 4, pp. 47–53, 2014.

[44] J. Momeni, R. D. Djoulde, M. T. Akam, and S. F. Kimbu, "Chemical constituents and antibacterial activities of the stem bark extracts of *Ricinodendron heudelotii* (Euphorbiaceae)," *Indian Journal of Pharmaceutical Sciences*, vol. 67, pp. 386–399, 2005.

[45] C. A. Suresh-Kumar, R. Varadharajan, P. Muthumani, R. Meera, P. Devi, and B. Kameswari, "Psychopharmacological studies on the stem of *Saccharum spontaneum*," *International Journal of PharmTech Research*, vol. 2, no. 1, pp. 319–324, 2010.

[46] I. O. Williams, E. O. Onyenweaku, and I. J. Atangwho, "Nutritional and antimicrobial evaluation of *Saccharum officinarum* consumed in Calabar, Nigeria," *African Journal of Biotechnology*, vol. 15, no. 33, pp. 1789–1795, 2016.

[47] A. O. Ayoka, R. O. Akomolafe, O. S. Akinsomisoye, and O. E. Ukponmwan, "Medicinal and economic value of *Spondias mombin*," *African Journal of Biomedical Research*, vol. 11, no. 2, pp. 129–136, 2010.

[48] O. Aromolaran and O. K. Badejo, "Efficacy of fresh leaf extracts of *Spondias mombin* against some clinical bacterial isolates from typhoid patients," *Asian Pacific Journal of Tropical Disease*, vol. 4, no. 6, pp. 442–446, 2014.

[49] B. Trognitz, X. Scheldeman, K. Hansel-Hohl, A. Kuant, H. Grebe, and M. Hermann, "Genetic population structure of cacao plantings within a young production area in Nicaragua," *PLoS One*, vol. 6, no. 1, Article ID e16056, 2011.

[50] R. X. Santos, D. A. Oliveira, G. A. Sodre, G. Gosmann, M. Brendel, and C. Pungartnik, "Antimicrobial activity of fermented *Theobroma cacao* pod husk extract," *Genetics and Molecular Research*, vol. 13, no. 3, pp. 7725–7735, 2014.

[51] S. Nidhi, D. Shreyan, D. Abhirup, R. C. Akash, and A. Jayanthi, "Antimicrobial activity and cytotoxicity of *Theobroma cacao* extracts," *Der Pharmacia Lettre*, vol. 7, no. 7, pp. 287–294, 2015.

[52] E. J. Ndebia, B. N. Nkeh-Chungag, J. R. Temdie, Y. M. Fodjo, D. T. Ndinteh, and J. T. Mbafor, "Antinociceptive effects of the methanolic extract of *Uapaca guineensis* (Euphorbiaceae) stilt root bark," *Pharmacologyonline*, vol. 3, pp. 153–165, 2007.

[53] E. T. Ibibia, O. Peter, and O. S. Olugbemiga, "Evaluation of some Nigerian Savannah plants of antioxydent activity, and total phenols and flavonoid contents," *International Journal of Pharmaceutical Sciences Review and Research*, vol. 34, pp. 75–81, 2015.

[54] T. J. O. Ngalani, *Influence de la Multi-Résistance des Bactéries Entériques aux Antibiotiques Sur l'État Immunologique des Patients Séropositifs au VIH Venus en Consultation à l'Hôpital Adlucem Banka de Bafang*, Department of Biochemistry, University of Dschang, Dschang, Cameroon, 2015.

[55] J. Harbone, *Phytochemical Methods: a Guide to Modern Techniques of Plant Analysis*, Chapman & Hall, London, UK, 1973.

[56] J. N. Eloff, "A sensitive and quick microplate method to determine the minimal inhibitory concentration of plant extracts for bacteria," *Planta Medica*, vol. 64, no. 8, pp. 711–713, 1998.

[57] V. Kuete, F. Nana, B. Ngameni, A. T. Mbaveng, F. Keumedjio, and B. T. Ngadjui, "Antimicrobial activity of the crude extract, fractions and compounds from stem bark of *Ficus ovata* (Moraceae)," *Journal of Ethnopharmacology*, vol. 124, no. 3, pp. 556–561, 2009.

[58] V. Kuete, J. D. Wansi, A. T. Mbaveng et al., "Antimicrobial activity of the methanolic extract and compounds from *Teclea afzelii* (Rutaceae)," *South African Journal of Botany*, vol. 74, no. 4, pp. 572–576, 2008.

[59] J. Kovač, N. Gavari, F. Bucar, and S. Smole Možina, "Antimicrobial and resistance modulatory activity of *Alpinia katsumadai* seed extract, essential oil and post-distillation extract," *Food Technology and Biotechnology*, vol. 52, pp. 248–254, 2014.

[60] V. Kuete and T. Efferth, "Cameroonian medicinal plants: pharmacology and derived natural products," *Frontiers in Pharmacology*, vol. 1, p. 123, 2010.

[61] J. Bruneton, *Pharmacognosie: Phytochimie, Plantes medicinales*, Tec & Doc, Paris, France, 1999.

[62] M. Simões, R. N. Bennett, and E. A. Rosa, "Understanding antimicrobial activities of phytochemicals against multidrug resistant bacteria and biofilms," *Natural Product Reports*, vol. 26, no. 6, pp. 746–757, 2009.

[63] J. D. D. Tamokou, A. T. Mbaveng, and V. Kuete, "Chapter 8-antimicrobial activities of African medicinal spices and vegetables," in *Medicinal Spices and Vegetables from Africa*, pp. 207–237, Academic Press, Cambridge, MA, USA, 2017.

[64] T. J. Foster, "Antibiotic resistance in *Staphylococcus aureus*. Current status and future prospects," *FEMS Microbiology Reviews*, vol. 41, no. 3, pp. 430–449, 2017.

[65] M. C. Jennings, M. E. Forman, S. M. Duggan, K. P. C. Minbiole, and W. M. Wuest, "Efflux pumps might not be the major drivers of QAC resistance in methicillin-resistant *Staphylococcus aureus*," *ChemBioChem*, vol. 18, no. 16, pp. 1573–1577, 2017.

[66] P. W. Limaverde, F. F. Campina, F. A. da Cunha et al., "Inhibition of the TetK efflux-pump by the essential oil of *Chenopodium ambrosioides* L. and α-terpinene against *Staphylococcus aureus* IS-58," *Food and Chemical Toxicology*, vol. 109, pp. 957–961, 2017.

[67] M. M. Kristiansen, C. Leandro, D. Ordway et al., "Thioridazine reduces resistance of methicillin-resistant *Staphylococcus aureus* by inhibiting a reserpine-sensitive efflux pump," *In Vivo*, vol. 20, no. 3, pp. 361–366, 2006.

[68] A. Martins, L. Machado, S. Costa et al., "Role of calcium in the efflux system of *Escherichia coli*," *International Journal of Antimicrobial Agents*, vol. 37, no. 5, pp. 410–414, 2011.

[69] F. K. Touani, A. J. Seukep, D. E. Djeussi et al., "Antibiotic-potentiation activities of four Cameroonian dietary plants against multidrug-resistant Gram-negative bacteria expressing efflux pumps," *BMC Complementary and Alternative Medicine*, vol. 14, p. 258, 2014.

[70] L. C. Braga, A. A. Leite, K. G. Xavier et al., "Synergic interaction between pomegranate extract and antibiotics against *Staphylococcus aureus*," *Canadian Journal of Microbiology*, vol. 51, no. 7, pp. 541–547, 2005.

2

Bacterioplankton Activity in a Meso-eutrophic Subtropical Coastal Lagoon

Maria L. Schmitz Fontes[1,2] Heloísa Fernandes,[3] Manoela Brandão,[1] Mariana Coutinho Hennemann,[1] Raquel Aparecida Loss,[4] Valdelúcia Maria Alves de Souza Grinevicius,[5] Denise Tonetta,[1] Karina Cesca,[4] Mônica Hessel Silveira,[1] Mara Bedin,[1] Derce Recouvreux,[4] and Regina Vasconcellos Antônio[6]

[1]*Departamento de Ecologia e Zoologia, Centro de Ciências Biológicas, Universidade Federal de Santa Catarina, Trindade, 88040-970 Florianópolis, SC, Brazil*
[2]*PPGOCEANO, Departamento de Geociências, Centro de Filosofia e Ciências Humanas, Universidade Federal de Santa Catarina, Trindade, 88040-970 Florianópolis, SC, Brazil*
[3]*Departamento de Engenharia Sanitária, Centro Tecnológico, Universidade Federal de Santa Catarina, Trindade, 88040-970 Florianópolis, SC, Brazil*
[4]*Departamento de Engenharia Química e Engenharia de Alimentos, Centro Tecnológico, Universidade Federal de Santa Catarina, Trindade, 88040-970 Florianópolis, SC, Brazil*
[5]*Departamento deBioquímica, Centro de Ciências Biológicas, Universidade Federal de Santa Catarina, Trindade, 88040-970 Florianópolis, SC, Brazil*
[6]*Campus Araranguá, Universidade Federal de Santa Catarina, Trindade, 88000-000 Araranguá, SC, Brazil*

Correspondence should be addressed to Maria L. Schmitz Fontes; fontesml@me.com

Academic Editor: Hidetoshi Urakawa

The aim of this study was to investigate whether the bacterioplankton activity in the meso-eutrophic Conceição Lagoon would increase significantly under allochthonous inputs of inorganic nutrients and organic carbon. Abundance and biomass of bacterioplankton were evaluated under three treatments: light (14 h light/10 h dark), complete darkness (dark-control), and nutrient (C + N + P—dark, 100 : 10 : 1) enrichments during 72 h. Nutrient enrichments promoted a significant increase in abundance (maximum of 19.0×10^9 cells·L^{-1} in the first 32 hours) and biomass of the heterotrophic bacterioplankton, which induced the formation of large clusters. Bacterial biomass remained constant in the non-enriched incubations (dark-control and light). Bacterial growth rates were significantly higher after nutrient additions (1.35 d^{-1}), followed by control (0.79 d^{-1}), and light (0.63 d^{-1}) treatments, which were statistically equal ($p > 0.05$). Bacterial production rates were also significantly higher under nutrient additions (1.28 d^{-1}), compared to the control and light (0.50 d^{-1} and 0.44 d^{-1}, respectively), demonstrating that bacterial growth and production in this meso-eutrophic lagoon are under an immediate "bottom-up" regulation, followed by a potential top-down effect. These facts reinforce the urgency on improving the local wastewater management plan in order to prevent further expansion of anoxic waters.

1. Introduction

The comprehension of the pelagic bacterioplankton abundance and activity in coastal aquatic environments is of major importance in order to predict shifts in the microbial communities in response to frequent changes in such systems [1]. The inputs of organic material and inorganic nutrients to coastal lagoons mainly from watershed runoffs, infiltration, precipitation, air and sea exchanges can have a direct influence on activity and composition of microbial communities and, consequently, in the ecosystem function [2, 3]. Coastal lagoons are located in the transitional region between terrestrial and marine environments, and are considered highly dynamic environments, subjected to abrupt changes in physicochemical variables, residence times, and recipients of multiple sources of nutrients [3, 4].

Heterotrophic bacteria play important roles in the remineralization and transformation processes between dissolved organic carbon (DOC) and particulate organic carbon (POC) [5, 6], which will then be available to higher trophic levels through the microbial loop [7]. The functionality of bacterioplankton, predominantly net sink or link of carbon in a system, will depend on the chemical composition of the DOC pool [8], if predation is absent or minimized.

However, the cause-effect relationship of labile carbon forms and/or inorganic nutrients on the bacterial activity is not clear yet. Some studies have shown a positive effect of dissolved inorganic nitrogen and phosphate enrichment on marine bacterial growth [9–12], while others demonstrated no effect [13]. With the simultaneous addition of inorganic nutrients and glucose, several studies have shown a significant response in bacterial abundance and production [14–17]. The heterogeneity observed in the results has been related to the different trophic states of each environment [9].

In coastal ecosystems such as the Conceição Lagoon, a meso-eutrophic coastal lagoon [18], additional inputs of allochthonous material can stimulate heterotrophic microorganisms and consequently lead to oxygen depletion [19–21]. Anoxia occurrence has increased significantly in the last 15 years in the central area of the lagoon [18], and its trophic state has changed from oligo-mesotrophic to meso-eutrophic in the same period of time [18]. Thus, we hypothesized that bacterioplankton activity, determined here by bacterial production and growth, would increase in this meso-eutrophic system if exposed to simultaneous external inorganic nutrients' sources and organic matter.

2. Materials and Methods

2.1. Water Sampling and Study Site. A 20 L water sample was collected at subsurface in the central sector of Conceição Lagoon on the 2nd of November, 2011, and was kept at low temperature until transported to the laboratory. The lagoon is located between 27°30′17″ and 27°37′36″S and 48°25′30″ and 48°29′54″W (Figure 1).

Conceição Lagoon is connected to the Atlantic Ocean by a 2 km long channel, which drives a salinity variability in the central sector from 17.6 up to 30 [19, 22]. Water temperature during the experiment was maintained 24°C; usual dissolved inorganic nitrogen (ammonium + nitrate + nitrite) and dissolved inorganic phosphorus (phosphate) concentrations range between 1.0 and 13.2 μM and between 0.09 and 0.52 μM, respectively; with average DIN : DIP ratios above 25 : 1 [22]. Chlorophyll a concentrations average between 3.5 and 4.0 $\mu g \cdot L^{-1}$ [19, 20, 22]. Additionally, continuous runoffs rich in untreated (raw) domestic sewage flow directly into its water body [22, 23]. The average domestic sewage volume discharged into the system has been calculated to be around 160 m^3 of sewage $month^{-1}$, corresponding to 300 kg (biological oxygen demand, BOD_5 $day)^{-1}$ [24]. Due to this elevated sewage input, a considerable increase in dissolved inorganic nitrogen (ammonium, nitrite, and nitrate) concentrations has been observed in the lagoon water since 1980s [25].

2.2. Culture Preparation and Experimental Procedure. Water aliquots were collected from Conceição Lagoon and carefully filtered through a cellulose acetate membrane (Millipore 47 mm in diameter and 0.2 μm in pore size) in the laboratory. One liter of the filtered water or medium (dilution of the sample resulted in a reduction in the amount of grazers [14]) was transferred onto each of three sterile glass recipients resulting in 70% of filtered medium and 30% of inoculum (whole water—no previous filtration), equivalent to a volume of 1.43 L per replicate. Bacterioplankton cultures received the following labels according to the treatment: (1) light (*in situ* light intensity and photoperiod: 14 h light, 10 h dark, room temperature); (2) control (dark, room temperature); and (3) nutrients (C + N + P = glucose, NH_4Cl, and K_2HPO_4 additions at final concentrations of 100 μM C : 10 μM N : 1 μM P, dark, room temperature). All treatments were kept under the described conditions for 72 h. The reason why control and nutrients' treatments were kept in the dark was to prevent the accumulation of autotrophic biomass (presence of photoautotrophic bacteria and picoeukaryotes) throughout the experiment, which would interfere with the results by competition or release of extracellular DOC. On the other hand, light treatment was important for the effects of light/dark cycles on bacterial activity when compared to complete darkness.

2.3. Bacterial Counts and Biomass. Aliquots of 15 ml were taken from each treatment at 12 h intervals and fixed with PFA (paraformaldehyde) at 2% (final concentration), making 7 estimates for each treatment (t_0, t_{12}, t_{24}, t_{36}, t_{48}, t_{60}, and t_{72}). The aliquots were used to determine bacterial abundance, biomass, and size or biovolume. One ml of aliquots from each treatment was filtered through 0.2 μm dark polycarbonate membranes (diameter of 25 mm) (Millipore) and stained with 4′,6-diamidino-2-phenylindole (DAPI) for 15 min (1 $\mu g \cdot ml^{-1}$, final concentration).

Filters were mounted onto slides, and bacterial cells were counted using an epifluorescence microscope (Olympus BX-40) [26]. From each filter, ten random fields were counted and at least 200 bacterial cells were measured. Length, width, elongation, and area of each cell were measured using the freeware "UTHSCSA Image Tool" (University of Texas Health Science Center, San Antonio, Texas, US) [27]. Bacterial biovolume was calculated from length and width measurements and biomass as function of biovolume using the following algorithm: $B = 120\ V^{0.72}$, where B = biomass; V = biovolume (μm^3); 120 = conversion factor in $fg \cdot C \cdot \mu m^{-3}$ [28]. Total biomass was calculated by multiplying the average cellular biomass by bacterial abundance.

2.4. Data Analysis. Bacterial growth was calculated as the increase in cellular density from the beginning to the end of the exponential growth period. Bacterial production was calculated in the same manner using bacterial biomass data by the time.

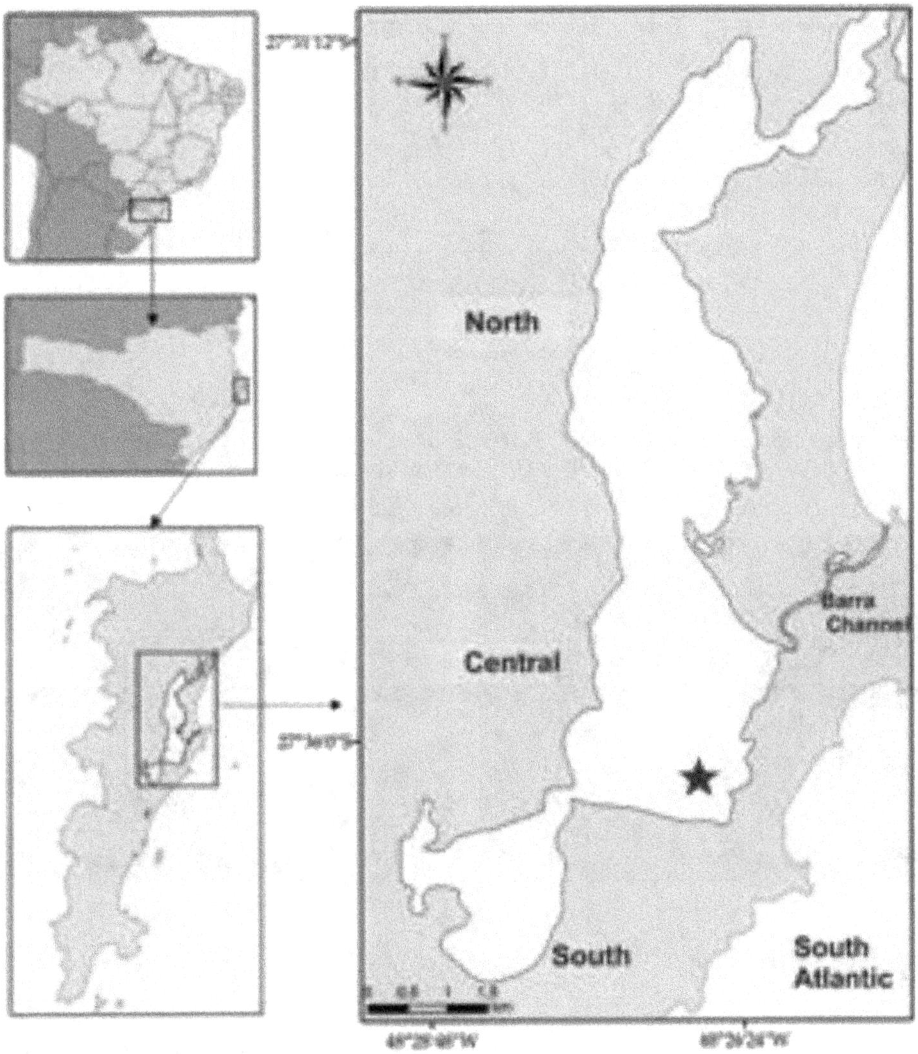

FIGURE 1: Location of the Conceição Lagoon (grey), Santa Catarina Island, Florianópolis, Brazil. Sampling site is represented by a star symbol (adapted from Fontes and Abreu [19]).

Bacterioplankton specific growth rates (μ) and biomass production rates were calculated as the slope of the linear regression of ln-transformed bacterial abundance or biomass with time during the exponential growth [29]. Daily growth rate (daily μ) and daily bacterial production rates were obtained by multiplying the hourly growth rate by 24. Doubling time was calculated by dividing 0.69 (ln 2) by the daily μ. Comparison between bacterial growth rates and production rates (linear regression angular coefficient) was made by analysis of covariance (ANCOVA), with subsequent multiple comparisons performed using Tukey's test with a statistical significance level of $p \leq 0.05$. Density, biomass, and biovolume means were compared by analysis of variance (ANOVA) or t-test, if the case, of $(\log + 1)$-transformed data, followed by Tukey's test (significance level of $p \leq 0.05$) [30, 31]. Statistical tests were conducted using Statistica 7 software.

3. Results

Samples showed different abundance and biomass values among treatments and time. The mean number of cells ($\text{cells}\cdot\text{L}^{-1}$) in each treatment after 72 h of incubation was higher in the nutrient treatment ($7.44 \times 10^9 \pm 2.55 \times 10^9\ \text{cells}\cdot\text{L}^{-1}$), followed by the dark-control ($3.00 \times 10^9 \pm 8.04 \times 10^8\ \text{cells}\cdot\text{L}^{-1}$) and light flasks ($1.77 \times 10^9 \pm 2.08 \times 10^8\ \text{cells}\cdot\text{L}^{-1}$). Bacterial abundance increased exponentially until 36 h (maximum $1.9 \times 10^{10}\ \text{cells}\cdot\text{L}^{-1}$), followed by a decrease and a second small increase between 60 and 72 h in the nutrient treatment. In the dark-control treatment, the number of cells increased until 48 h (maximum $7.18 \times 10^9\ \text{cells}\cdot\text{L}^{-1}$). In the light flasks, samples showed a lower abundance when compared to the other treatments and maintained approximately constant values throughout the experiment (Figure 2).

Considering cells arrangement in each treatment, there was an increase in the occurrence of cells clustering and size after 36 h of incubation only in the nutrient enrichment, with up to 200 cells clustered together (Figure 3). In addition, the abundance of filamentous bacteria increased between 12 and 24 h in the nutrients treatment, while in the dark-control, a smaller and delayed increase occurred between 36 and 48 h. In the light treatment, constant numbers of filamentous bacteria were observed throughout the

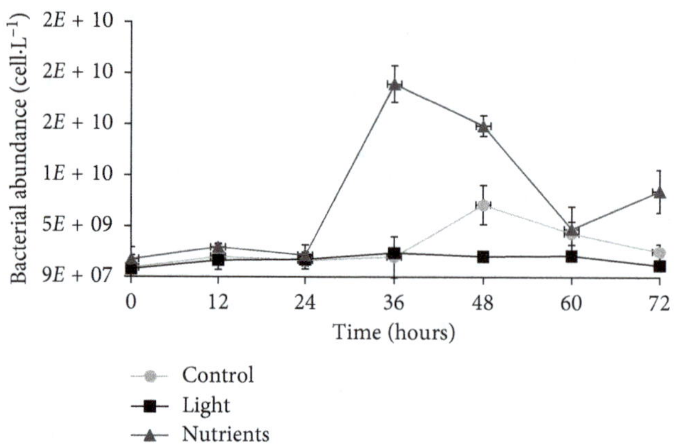

FIGURE 2: Bacterial abundance (mean ± standard error) in the three treatments during the 72 h incubation period. Treatments: dark-control = control, 14 h light/10 h dark cycle = light, and dark C + N + P enrichment = nutrients.

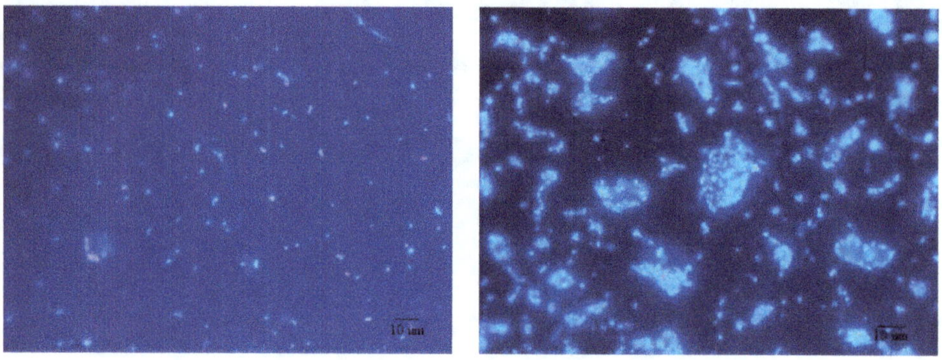
(a) (b)

FIGURE 3: Epifluorescence microphotographs (1000x) of bacteria stained with DAPI at 12 h (a) and 72 h (b) of the experiments in the C + N + P treatment.

experiment (Figure 4). However, no significant differences were observed concerning filamentous bacterial abundance among treatments (one-way ANOVA, $p > 0.05$). Considering the relative abundance of filamentous bacteria, a stronger contribution of this portion of the community was detected in the light and dark-control flasks (5.6% and 7.1%, respectively), against 2.5% contribution in the ones that received nutrients.

The bacterial size estimated as cellular biovolume (μm^3) has been used to calculate cellular biomass in function of C availability ($\mu g \cdot C \cdot cell^{-1}$). As a result, mean cellular biovolume ranged between 6.45 and 13.65 μm^3, and biomass, as a consequence, between 417.82 and 678.36 $\mu g \cdot C \cdot cell^{-1}$. The highest averages of biovolume were observed in the nutrients treatment at 36 h (13.65 μm^3) and in the light at 24 h (13.58 μm^3) (Figure 5). High oscillations in the mean biovolume can be attributed to different contributions of filamentous bacteria and to the presence of different bacteria morphotypes. However, biovolume was not significantly different among treatments (one-way ANOVA, $p > 0.05$) (Table 1). On the other hand, considering total biomass (carbon per volume unit ($mg \cdot C \cdot L^{-1}$)), the highest biomass was registered in the nutrients treatment, especially between 24 and 36 h (with a peak of 12.3 $mg \cdot C \cdot L^{-1}$) (Figure 6), similarly to what was shown for bacterial abundance (Figure 2).

Bacterial abundance and biomass were higher under nutrients additions when compared to the dark-control and light treatments (one-way ANOVA, $p < 0.05$) (Table 1). Bacterial growth rates were 0.79 d^{-1}, 1.34 d^{-1}, and 0.63 d^{-1} in the dark-control, nutrients, and light treatments, respectively. Bacterial production rates oscillated from 0.44 to 1.13 d^{-1} (Table 2). The ANCOVA result showed that bacterial growth and biomass production rates differed among treatments (ANCOVA: $F_{0.05;(1);2;9} = 9.00$ for growth and ANCOVA: $F_{0.05;(1);2;9} = 5.08$ for bacterial production). The difference among slopes was tested, and it was observed that bacterial growth and production were significantly higher in the nutrients treatment, while growth and production were similar in light and dark-control incubations (Table 2).

Doubling times (0.51 d^{-1}, 0.87 d^{-1}, and 1.10 d^{-1} for nutrients, dark-control, and light, respectively) showed that the heterotrophic bacterial community went through two duplications per day in the nutrients treatment, while in natural conditions (dark-control treatment), bacterioplankton duplicated once a day (Table 2).

When changes in cellular biovolume and carbon biomass are considered, slightly different results were observed, since both nutrients and light treatments showed small increases in the parameters at the end of the 72 h incubation period.

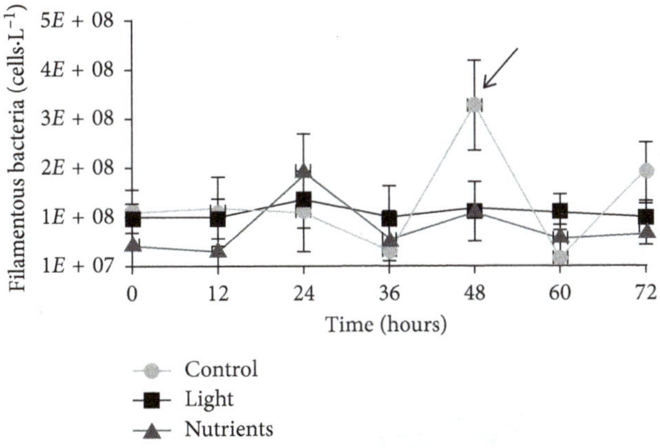

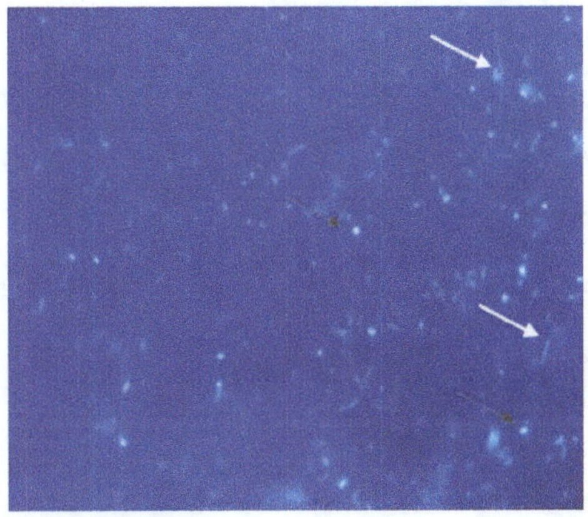

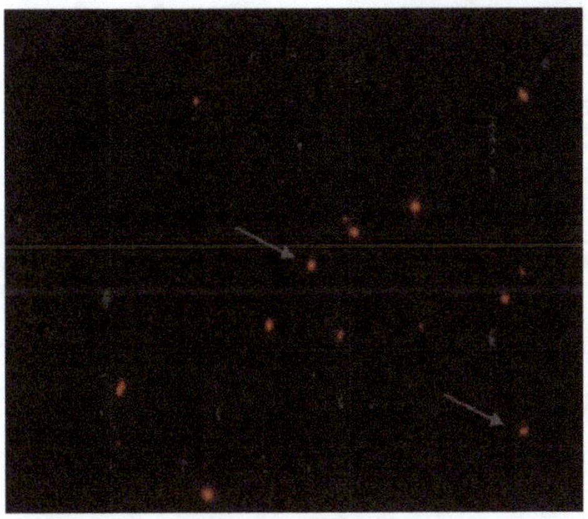

FIGURE 4: Filamentous bacterial abundance (mean ± standard error) (a) in the three treatments during the 72 h incubation period. Treatments: dark-control = control, 14 h light/10 h dark cycle = light, and dark C + N + P enrichment = nutrients. Examples of filamentous bacteria (b) and cyanobacteria (c) present in the dark-control treatment at the 48 h. (b) DAPI-stained cells, heterotrophic filamentous bacteria (indicated by white arrows). (c) Autofluorescence of cyanobacteria in red (indicated by grey arrows).

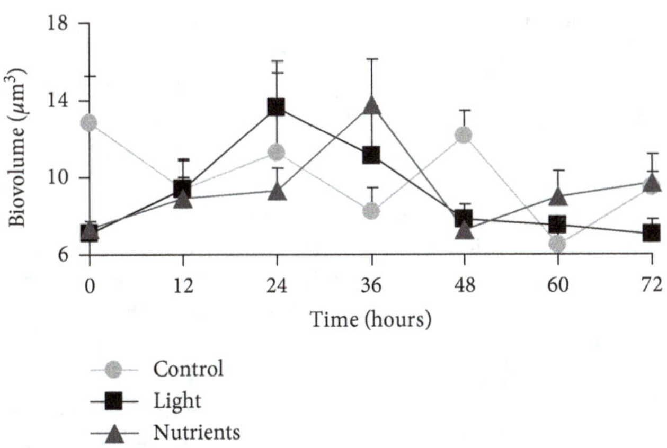

FIGURE 5: Cellular biovolume in μm^3 (mean ± standard error) in the three treatments during the 72 h incubation period. Treatments: dark-control = control, 14 h light/10 h dark cycle = light, and dark C + N + P enrichment = nutrients.

TABLE 1: Results of one-way ANOVA performed to test the effect of different treatments (dark-control, light, and dark C + N + P) on bacterial abundance, biovolume, and biomass.

Parameter		DF*	MS	F	p
Density (cells·L^{-1})	Treatment	2	0.4710	5.3	0.0154
	Error	18	0.0890		
Biovolume (μm^3)	Treatment	2	0.0031	0.3	0.7467
	Error	18	0.0105		
Biomass (mg·L^{-1})	Treatment	2	0.4734	4.6	0.0235
	Error	18	0.1017		

*DF = degrees of freedom.

4. Discussion

Bacterioplankton activity in this meso-eutrophic lagoon was significantly stimulated after the addition of dissolved organic carbon and inorganic nutrients, demonstrating the vulnerability of this system to external sources of organic and inorganic matter.

The average cellular size, 13 μm^3, did not vary significantly throughout the experiment; however, bacterial cells showed a great morphological variability. Naturally, isolated marine bacteria show low growth rates under environmental conditions [32]. These authors reported that cells with a biovolume of 0.008 μm^3 have growth rates of 0.3 d^{-1}, while 0.3 μm^3 cells show growth rates of 1.4 d^{-1}. The average biovolume found in our study was 13 μm^3, two orders of magnitude higher than usually found in marine environments. However, it is important to highlight the presence

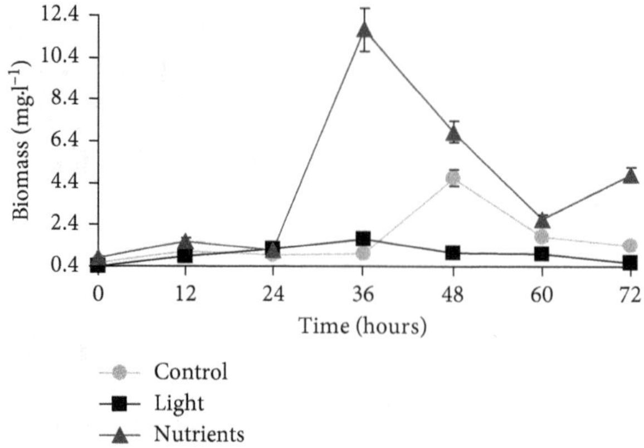

FIGURE 6: Bacterial biomass (mg·C·L^{-1} + standard error) in the three treatments during the 72 h incubation period. Treatments: dark-control = control, 14 h light/10 h dark cycle = light, and dark C + N + P enrichment = nutrients.

TABLE 2: Bacterial growth rates (hour^{-1} and day^{-1}) and coefficient of determination (R^2), production rates (hour^{-1} and day^{-1}) and coefficient of determination (R^2), and doubling time for light, dark-control, and dark C + N + P treatments. Bacterial growth and production rates were determined from the slope of the linear regression of ln-transformed bacterial abundance or biomass versus time (from time zero to the end of the exponential growth phase).

Treatment	Growth rate (h^{-1})	Growth rate (d^{-1})	R^2	Production rate (h^{-1})	Production rate (d^{-1})	R^2	Doubling time
Light	0.026^{b}	0.626	0.86	0.018^{b}	0.437	0.98	1.106
Dark-control	0.033^{b}	0.789	0.74	0.021^{a,b}	0.500	0.62	0.878
Dark C + N + P	0.056^{a}	1.346	0.74	0.047^{a}	1.128	0.67	0.515

Superscript letters indicate the result of ANCOVA and Tukey test for difference among slopes-q 0.05.

and the relative abundance of filamentous bacteria, especially around 48 h in the control treatment. The predominance of heterotrophic filamentous bacteria has already been described previously in the Conceição Lagoon as positively related to the coccoid cyanobacteria of genus *Synechococcus* [19], as shown in Figures 4(b) and 4(c). This has been described as a strategy to avoid predation, when bacteria alter their size towards filamentous forms when exposed to grazing pressure [33–35].

Some studies have suggested that viral lysis and predation by protists are the main regulators of bacterial growth and production, many times more important than resources availability [36–38]. However, in environments with N : P ratios above the Redfield ratio, where phosphorous is the limiting resource, as in Conceição Lagoon [22, 23], bottom-up control of bacterial growth and/or biomass has a much more important role than top-down control by bacterivory [39, 40]. Therefore, grazing cannot be ignored as a secondary regulator on the activity of bacterioplankton. The significant decrease of bacteria abundance between 36 and 60 h in the nutrients treatment was concurrent with a slight decrease in the average bacterial biovolume, which is illustrated by the formation of several clusters of bacterial cells resulting on smaller biovolume when attached to each other. There are a few hypotheses for this outcome: predation scape strategy, increase proximity, and signaling among cells to overcome resource limitation.

In Conceição Lagoon, bacterial growth rates and production doubled after the addition of nutrients, compared to the dark-control and light treatments. In a study carried out with samples from the oligotrophic Sargasso Sea, the additions of nitrogen, phosphorus, and carbon resulted in increasing bacterial production by 7 to 15 times when compared to control or inorganic nutrient additions alone [14]. This finding reinforces that bacterioplankton of oligotrophic environments (under constant starvation and competition pressure with phytoplankton) responds much faster after simultaneous additions of inorganic nutrients and carbon than those of meso-eutrophic environments.

Previous studies have shown contradicting results on the effect of nutrients and carbon enrichment on bacterial growth and secondary production (e.g., [40–46]). For example, a significant increase on bacterial production after the addition of nutrients and carbon are reported in a fjord in Norway [47], whereas no significant changes were described in the oligo-mesotrophic Uchiumi Bay, Japan [13]. The best explanation for such heterogeneity in growth and productivity of aquatic bacteria has been described to be the ecosystem's trophic state, as the primary response of bacteria to eutrophication is to increase their specific growth rates [48]. The ecosystem's trophic state is related to variables that directly or indirectly correlate with the system's primary production: total organic matter, dissolved oxygen concentration, total phosphorus, light penetration, and phytoplankton biomass indicated by chlorophyll a concentration [49, 50]. As Conceição Lagoon is under eutrophication, with persistent anoxic waters in the bottom of the central sector, the increase in aerobic respiration by microbes can be a major problem for the water quality of this ecosystem.

Thus, increasing nutrient loads (N and P) directly or indirectly (trophic state) will alter bacterial community composition by favouring certain bacterial populations that have the ability to rapidly consume these new available resources [51–55]. Our results show that bacterial activity remains unaltered when exposed to complete darkness (*t*-test - light x control - $p > 0.05$), and that under additions of inorganic nutrients and organic carbon, the total bacterial activity doubles in the first 24 hours. Apparently, there is an indicative of strong competition for nutrients in Conceição Lagoon.

The present study shows similar activities for total (auto + heterotrophic) bacterioplankton and strict heterotrophic bacteria when neither inorganic nutrient nor carbon is added onto the lagoon. Nevertheless, bacterioplankton growth and production were stimulated by carbon, nitrogen, and phosphorous additions, indicating that bacterioplankton biomass in this meso-eutrophic coastal lagoon is likely to be primarily bottom-up regulated. This study suggests that allochthonous carbon and inorganic nutrients via untreated sewage inputs or rainstorm runoffs into the lagoon have the potential to promote a quick response of bacterioplankton and, thus, increase the area and persistency of the "dead zones" within the lagoon. Our results reinforce the importance of an effective waste management plan for the area.

Conflicts of Interest

The authors declare that they have no conflicts of interest.

References

[1] B. Divya, S. O. Fernandes, G. Sheelu, N. Shanta, P. A. L. Bharathi, and D. Chandramohan, "Limno-tolerant bacteria govern nitrate concentration in Mandovi Estuary, India," *Estuarine, Coastal and Shelf Science*, vol. 82, no. 1, pp. 29–34, 2009.

[2] J. J. Barrera-Alba, S. M. F. Gianesella, G. A. O. Moser, and F. M. P. S. Correa, "Influence of allochthonous organic matter on bacterioplankton biomass and activity in a eutrophic, subtropical estuary," *Estuarine, Coastal and Shelf Science*, vol. 82, no. 1, pp. 84–94, 2009.

[3] F. A. Esteves, A. Caliman, J. M. Santangelo, R. D. Guariento, V. F. Farjalla, and R. L. Bozelli, "Neotropical coastal lagoons: an appraisal of their biodiversity, functionings, threats and conservation management," *Brazilian Journal of Biology*, vol. 68, no. 4S, pp. 967–981, 2008.

[4] B. Kjerfve, *Coastal Lagoon Processes*, Elsevier Science B.V., Amsterdam, Netherlands, 1994.

[5] D. Sigge, *Freshwater Microbiology: Biodiversity and Dynamic Interactions of Microorganisms in the Aquatic Environment*, Vol. 1, John Wiley & Sons Ltd., Chichester, UK, 2005.

[6] R. G. Wetzel, *Limnology*, Saunders Publishing, Philadelphia, PA, USA, 1982.

[7] F. Azam, T. Fenchel, J. G. Field, J. S. Gray, L. A. Meyer-Reil, and F. Thingstad, "The ecological role of water-column microbes in the sea," *Marine Ecology Progress Series*, vol. 10, pp. 257–263, 1983.

[8] M. Warkentin, H. M. Freese, and R. Schumann, "Bacterial activity and bacterioplankton diversity in the eutrophic river Warnow-direct measurement of bacteria l growth efficiency and its effect on carbon utilization," *Microbial Ecology*, vol. 61, no. 1, pp. 190–200, 2011.

[9] D. A. Caron, E. L. Lim, R. W. Sanders, M. R. Dennett, and U. G. Berninger, "Responses of bacterioplankton and phytoplankton to organic carbon and inorganic nutrient additions in contrasting oceanic ecosystems," *Aquatic Microbial Ecology*, vol. 22, pp. 175–184, 2000.

[10] M. P. Hoch and D. A. Bronk, "Bacterioplankton nutrient metabolism in the Eastern Tropical North Pacific," *Journal of Experimental Marine Biology and Ecology*, vol. 349, no. 2, pp. 390–404, 2007.

[11] L. R. Pomeroy and D. Deibel, "Temperature regulation of bacterial activity during the spring bloom in newfoundland coastal waters," *Science*, vol. 233, no. 4761, pp. 359–361, 1986.

[12] M. M. Sala, F. Peters, J. M. Gasol, C. Pedrós-Alió, C. Marrase, and D. Vaque, "Seasonal and spatial variations in the nutrient limitation of bacterioplankton growth in the northwestern Mediterranean," *Aquatic Microbial Ecology*, vol. 27, pp. 47–56, 2002.

[13] D. Ichinotsuka, T. Katan, H. Takeoka, and S. Nakano, "Effects of nutrient supplies on the growth rates of planktonic bacteria in Uchiumi Bay, Japan," *Aquatic Biology*, vol. 9, pp. 123–130, 2010.

[14] C. A. Carlson, S. J. Giovannoni, D. A. Hansel et al., "Effect of nutrient amendments on bacterioplankton production, community structure, and DOC utilization in the northwestern Sargasso Sea," *Aquatic Microbial Ecology*, vol. 30, pp. 19–36, 2002.

[15] M. J. Church, D. A. Hutchins, and H. W. Ducklow, "Limitation of bacterial growth by dissolved organic matter and iron in the Southern Ocean," *Applied and Environmental Microbiology*, vol. 66, no. 2, pp. 455–466, 2000.

[16] S. P. Donachie, R. James, J. R. Christian, and D. M. Karl, "Nutrient regulation of bacterial production and ectoenzyme activities in the subtropical North Pacific Ocean," *Deep Sea Research Part II: Topical Studies in Oceanography*, vol. 48, no. 8-9, pp. 1719–1732, 2001.

[17] L. R. Pomeroy, J. E. Sheldon, W. M. Sheldon Jr., and F. Peters, "Limits to growth and respiration of bacterioplankton in the Gulf of Mexico," *Marine Ecology Progress Series*, vol. 117, pp. 259–268, 1995.

[18] V. E. C. Silva, D. Franco, A. L. Fonseca, M. L. Schmitz Fontes, and A. R. Donnangelo, "Space time evolution of the trophic state of a subtropical lagoon: Lagoa da Conceição, Florianópolis Island of Santa Catarina, Brazil," *RBRH*, vol. 22, p. e10, 2017.

[19] M. L. S. Fontes and P. A. Abreu, "Spatiotemporal variation of bacterial assemblages in a shallow subtropical coastal lagoon in Southern Brazil," *Microbial Ecology*, vol. 58, no. 1, pp. 140–152, 2009.

[20] M. L. S. Fontes, M. T. Suzuki, M. T. Cottrell, and P. C. Abreu, "Primary production in a subtropical stratified coastal lagoon-contribution of anoxygenic phototrophic bacteria," *Microbial Ecology*, vol. 61, no. 1, pp. 223–237, 2011.

[21] H.-T. Ng, D. M. Marques, E. Jeppesen, and M. Søndergaard, "Bacterioplankton in the littoral and pelagic zones of subtropical shallow lakes," *Hidrobiologia*, vol. 646, no. 1, pp. 311–326, 2010.

[22] A. Fonseca, E. S. Braga, and B. B. Eichler, "Distribuição espacial dos nutrientes inorgânicos dissolvidos e da biomassa fitoplanctônica no sistema pelágico da Lagoa da Conceição, Santa Catarina, Brasil (Setembro, 2000)," *Atlântica*, vol. 24, pp. 69–83, 2002.

[23] M. L. S. Fontes, R. Cavellucci, A. Laurenti, E. C. Machado,

M. C. Camargo, and N. Brandini, "Detection of environmental impact on variations in dissolved nutrients and Chl-a in the Conceição Lagoon, Florianópolis, SC, Brazil," *Journal of Coastal Research*, vol. 39, pp. 1407–1412, 2006.

[24] I. S. Maia, "Avaliação de lagoas de maturação modificadas com a presença de suportes e macrófitas da família Lemnaceae no tratamento de esgotos domésticos," M.S. dissertation, Universidade Federal de Santa Catarina, Florianopólis, SC, Brazil, 2008.

[25] A. L. Fonseca, "Efeito da drenagem urbana nas características físico-químicas e biológicas da água superficial na Lagoa da Conceição (Florianópolis, SC, Brasil)," *Biotemas*, vol. 19, pp. 7–16, 2006.

[26] K. G. Porter and Y. S. Feig, "The use of DAPI for identifying and counting aquatic microflora," *Limnology and Oceanography*, vol. 25, no. 5, pp. 943–948, 1980.

[27] R. Massana, J. M. Gasol, P. K. Bjornsen et al., "Measurement of bacterial size via image analysis of epifluorescence preparations: description of an inexpensive system and solutions to some of the most common problems," *Scientia Marina*, vol. 61, no. 3, pp. 397–407, 1997.

[28] S. Norland, "The relationship between biomass and volume of bacteria," in *Handbook of Methods in Aquatic Microbial Ecology*, P. F. Kemp, B. F. Sherr, E. B. Sherr, and J. J. Cole, Eds., Lewis Publishers, Boca Raton, FL, USA, 1993.

[29] C. A. Carlson and H. W. Ducklow, "Growth of bacterioplankton and consumption of dissolved organic carbon in the Sargasso Sea," *Aquatic Microbial Ecology*, vol. 10, pp. 69–85, 1996.

[30] A. J. Underwood, *Experiments in Ecology: Their Logical Design and Interpretation Using Analysis of Variance*, Cambridge University Press, Cambridge, UK, 2006.

[31] J. H. Zar, *Biostatistical Analysis*, Prentice Hall, Englewood Cliffs, NJ, USA, 1996.

[32] J. M. Gasol, P. A. Giorgio, R. Massana, and C. M. Duarte, "Active versus inactive bacteria: size-dependence in a coastal marine plankton community," *Marine Ecology Progress Series*, vol. 128, pp. 91–97, 1995.

[33] G. Corno and K. Jürgens, "Direct and indirect effects of protist predation on population size structure of a bacterial strain with high phenotypic plasticity," *Applied and Environmental Microbiology*, vol. 72, no. 1, pp. 78–86, 2006.

[34] K. Jürgens and H. Gude, "The potential importance of grazing-resistant bacteria in planktonic systems," *Marine Ecology Progress Series*, vol. 112, pp. 169–188, 1994.

[35] K. Jürgens and C. Matz, "Predation as a shaping force for the phenotypic and genotypic composition of planktonic bacteria," *Antonie van Leeuwenhoek*, vol. 81, no. 1–4, pp. 413–434, 2002.

[36] L. Berdjeb, T. Pollet, I. Domaizon, and S. Jacquet, "Effect of grazers and viruses on bacterial community structure and production in two contrasting trophic lakes," *BMC Microbiology*, vol. 11, p. 88, 2011.

[37] M. G. Weinbauer and F. Rassoulzadegan, "Are viruses driving microbial diversification and diversity?," *Environmental Microbiology*, vol. 6, no. 1, pp. 1–11, 2004.

[38] M. G. Weinbauer, K. Horňák, J. Jezbera, J. Nedoma, J. R. Dolan, and K. Šimek, "Synergistic and antagonistic effects of viral lysis an protistan grazing on bacterial biomass, production and diversity," *Environmental Microbiology*, vol. 9, no. 3, pp. 777–788, 2007.

[39] D. J. Suggett, N. Stambler, O. Prásil et al., "Nitrogen and phosphorus limitation of oceanic microbial growth during spring in the Gulf of Aqaba," *Aquatic Microbial Ecology*, vol. 56, pp. 227–239, 2009.

[40] C. J. Williams, P. J. Lavrentyev, and F. J. Jochem, "Bottom-up and top-down control of heterotrophic bacterioplankton growth in a phosphorus-depleted subtropical estuary, Florida Bay, USA," *Marine Ecology Progress Series*, vol. 372, pp. 7–18, 2008.

[41] J. J. Cole, S. Findlay, and M. L. Pace, "Bacterial production in fresh and saltwater ecosystems: a cross-system overview," *Marine Ecology Progress Series*, vol. 43, pp. 1–10, 1988.

[42] A. L. Cuevas, J. K. Egge, F. T. Thingstad, and B. Töpper, "Organic carbon and mineral nutrient limitation of oxygen consumption, bacterial growth and efficiency in the Norwegian Sea," *Polar Biology*, vol. 34, no. 6, pp. 871–882, 2011.

[43] S. Drakare, P. Blomqvist, A. K. Bergstrom, and M. Jansson, "Primary production and phytoplankton composition in relation to DOC input and bacterioplankton production in humic Lake Ortrasket," *Freshwater Biology*, vol. 47, no. 1, pp. 41–52, 2002.

[44] J. Karlsson, M. Jansson, and A. Jonsson, "Similar relationships between pelagic primary and bacterial production in clearwater and humic lakes," *Ecology*, vol. 83, no. 10, pp. 2902–2910, 2002.

[45] D. L. Kirchman, "The uptake of inorganic nutrients by heterotrophic bacteria," *Microbial Ecology*, vol. 28, no. 2, pp. 255–271, 1994.

[46] M. M. Mills, C. M. Moore, R. Langlois et al., "Nitrogen and phosphorus co-limitation of bacterial productivity and growth in the oligotrophic subtropical North Atlantic," *Limnology and Oceanography*, vol. 53, no. 2, pp. 824–834, 2008.

[47] G. A. F. Flaten, T. Castberg, T. Tanaka, and F. Thingstad, "Interpretation of nutrient-enrichment bioassays by looking at sub-populations in a marine bacterial community," *Aquatic Microbial Ecology*, vol. 33, pp. 11–18, 2003.

[48] M. L. Pace and J. J. Cole, "Comparative and experimental approaches to top-down and bottom-up regulation of bacteria," *Microbial Ecology*, vol. 28, no. 2, pp. 181–193, 1994.

[49] R. E. Carlson, "A trophic state index for lakes," *Limnology and Oceanography*, vol. 22, no. 2, pp. 361–369, 1977.

[50] W. K. Dodds and J. J. Cole, "Expanding the concept of trophic state in aquatic ecosystems: it's not just the autotrophs," *Aquatic Sciences*, vol. 69, no. 4, pp. 427–439, 2007.

[51] B. Biddanda and J. B. Cotner, "Love handles in aquatic ecosystems: the role of dissolved organic drawdown, resuspended sediments, and terrigeous inputs in the carbon balance of Lake Michigan," *Ecosystems*, vol. 5, no. 5, pp. 431–445, 2002.

[52] B. J. Campbell, L. Yu, J. F. Heidelberg, and D. L. Kirchman, "Activity of abundant and rare bacteria in a coastal ocean," *Proceedings of the National Academy of Sciences*, vol. 108, no. 31, pp. 12776–12781, 2011.

[53] H. P. Grossart, J. Jezbera, K. Hornak, K. M. L. Hutalle, U. Buck, and K. Simek, "Top-down and bottom-up induced shifts in bacterial abundance, production and community composition in an experimentally divided humic lake," *Environmental Microbiology*, vol. 10, no. 3, pp. 635–652, 2008.

[54] K. Haukka, E. Kolmonen, R. Hyder et al., "Effect of nutrient loading on bacterioplankton community composition in Lake Mesocosms," *Microbial Ecology*, vol. 51, no. 2, pp. 137–146, 2006.

[55] R. J. Newton and K. D. McMahon, "Seasonal differences in bacterial community composition following nutrient additions in a eutrophic lake," *Environmental Microbiology*, vol. 13, no. 4, pp. 887–899, 2011.

3

Morphological and Molecular Identification of the Causal Agent of Anthracnose Disease of Avocado in Kenya

S. K. Kimaru ,[1] E. Monda,[2] R. C. Cheruiyot,[1] J. Mbaka,[3] and A. Alakonya[4]

[1]*Department of Plant Sciences, Kenyatta University, P.O. Box 43844, Nairobi, Kenya*
[2]*Department of Microbiology, Kenyatta University, P.O. Box 43844, Nairobi, Kenya*
[3]*Kenya Agricultural and Livestock Research Organisation, P.O. Box 220, Thika, Kenya*
[4]*Jomo Kenyatta University of Agriculture and Technology, P.O. Box 62000, Nairobi, Kenya*

Correspondence should be addressed to S. K. Kimaru; skimaru1@gmail.com

Academic Editor: Pierre Roques

Anthracnose disease of avocado contributes to a huge loss of avocado fruits due to postharvest rot in Kenya. The causal agent of this disease has not been clear but presumed to be *Colletotrichum gloeosporioides* as reported in other regions where avocado is grown. The fungus mainly infects fruits causing symptoms such as small blackish spots, "pepper spots," and black spots with raised margin which coalesce as infection progresses. Due to economic losses associated with the disease and emerging information of other species of fungi as causal agents of the disease, this study was aimed at identifying causal agent(s) of the disease. A total of 80 fungal isolates were collected from diseased avocado fruits in Murang'a County, the main avocado growing region in Kenya. Forty-six isolates were morphologically identified as *Colletotrichum* spp. based on their cultural characteristics, mainly whitish, greyish, and creamish colour and cottony/velvety mycelia on the top side of the culture and greyish cream with concentric zonation on the reverse side. Their spores were straight with rounded end and nonseptate. Thirty-four isolates were identified as *Pestalotiopsis* spp. based on their cultural characteristics: whitish grey mycelium with black fruiting structure on the upper side and greyish black one on the lower side and septate spores with 3-4 septa and 2 or 3 appendages at one end. Further molecular studies using ITS indicated *Colletotrichum gloeosporioides*, *Colletotrichum boninense*, and *Pestalotiopsis microspora* as the causal agents of anthracnose disease in avocado. However, with this being the first report, there is a need to conduct further studies to establish whether there is coinfection or any interaction thereof.

1. Introduction

The anthracnose disease is a common disease with wide host range causing severe economic loss. The disease has been reported on a wide variety of crops including avocado, almond, coffee, guava, apple, dragon fruit, cassava, mango, sorghum, and strawberry causing severe economic losses [1–4]. The causal agents of this disease are not clear. However, species of the genus *Colletotrichum* and *Pestalotiopsis* have been reported as causal agents of anthracnose in avocado [5, 6]. Such species includes *C. gloeosporioides*, *C. acutatum*, *C. boninense*, *C. karstii*, *C. godetiae*[7–11], and *Pestalotiopsis versicolor* [6].

Symptoms of anthracnose appear as pepper spot or speckle spot on immature fruit while still on tree and after fruit harvest during ripening as darkly, black coloured, sunken rounded spots with raised margins on fruit skins [12]. These lesions enlarge rapidly on the fruit skin and into the pulp leading to the death and rotting of the infected plant tissues [13]. The lesions may develop salmon-coloured, sticky spore masses typical of anthracnose diseases of this and many other plant species.

The avocado fruit has a high nutritional value since it contains vitamins (E, B, and C), minerals (potassium, iron, and phosphorus) and a great amount of oil [14]. In Kenya, avocado fruit is one of the most economically important fruits grown by both small and large scale farmers (HCDA, 2016). The fruit is mainly grown for fresh market but there is increasing demand from pharmaceutical, cosmetics, and vegetable oil industries (HCDA, 2016).

Anthracnose caused by *Colletotrichum gloeosporioides* has been associated with severe losses of avocado fruits both in the field and after harvest as compared to *Pestalotiopsis* spp. whose impact is not widely studied [5, 15–18]. Wasilwa et al. [19] reported that over 60% of the Kenyan avocado production cannot be marketed because of damage and low quality of fruits associated with anthracnose disease. Despite the huge losses associated with the anthracnose disease of avocado in Kenya, no cultural and molecular studies have been done to identify the causal agent(s). During the investigation, this paper aimed to identify the causal agent(s) of anthracnose of avocado in Kenya.

2. Materials and Methods

2.1. Fungal Isolation and Culturing. Samples of infected avocado fruits showing symptoms of anthracnose were collected from study area, Murang'a County, and brought to the laboratory for fungal isolation. The typical symptoms were small blackish spots "pepper spot" to larger blackish spots with raised margin [5]. The samples were cleaned using tap water and blotted to remove excess water. The fruits were surface sterilized using 0.5% sodium hypochlorite for 30 seconds. Small sections of the diseased area were cut aseptically and placed on hardened potato dextrose agar (PDA) in Petri dishes for fungal growth at room temperature (22–25°C). The emerging fungi were subcultured to obtain pure cultures. To obtain pure cultures, single spore isolation was done as follows: cultures were flooded with sterile distilled water and conidia were scraped off the plate using sterilized wire loop and suspended in 1 ml of sterile distilled water. A loopful of conidial suspension was spread evenly on 1.5% (wt/vol) water agar in a Petri dish and incubated at 25°C overnight. Using a sterilized glass needle germinated conidium was transferred onto hardened PDA in 9 cm diameter Petri dish and incubated at 25°C with a 12 h cycle of fluorescent light to induce growth and sporulation. To avoid bacterial contamination 0.5 g/l of streptomycin was added to PDA at molten state of about 50°C [20]. Single spore pure cultures of the pathogen were preserved in the slant universal bottle and stored in the fridge at 4°C for later use.

The fungi, *C. gloeosporioides* and *Pestalotiopsis* spp., were morphologically identified based on cultural and microscopical characteristics using published fungal key [21, 22].

2.2. Inoculation, Mycelial Growth, and Sporulation of C. gloeosporioides Isolates. Pure cultures of *C. gloeosporioides* preserved in universal bottles were used. Using a sterilized surgical scalpel, sections of mycelial plugs were cut aseptically and placed on hardened PDA on 9 cm diameter Petri dishes and incubated for 10 days for mycelial growth.

Five-millimetre mycelia plugs from the 10-day-old pure isolates of *C. gloeosporioides* were aseptically cut using five-millimetre diameter cork borer and placed individually at the centre of hardened PDA culture in 9 cm diameter Petri dishes. The cultures were incubated at room temperatures of ranges 22–25°C. Mycelia diameters of the isolates were measured at days 2, 4, 6, 8, and 10 after inoculation. On eleventh day, the cultures were flooded with distilled water and scrapped to bring the spores into suspension. The suspension was filtered through double layer cheese cloth to remove mycelia. The spore suspension was serially diluted to 10^{-6} for ease of counting the spores. The spore concentration was determined by use of haemocytometer.

2.3. Conidial Morphology and Size. Using a pipette, a drop of spore suspension (10^{-6}) was placed on a microscope slide, covered with a cover slip. The spores were stained with lactophenol cotton blue and observed under microscope. The shape of the spores from different isolates was noted and their sizes in terms of length and width were measured using a calibrated ocular slide and stage micrometer.

2.4. Determination of Genetic Diversity of Fungal Isolates

2.4.1. DNA Extraction. Pure fungal cultures in potato dextrose agar derived from a single spore from the original isolate were used. An improved fungal extraction protocol as described by Liu et al. [23] was used. About 40 mg of mycelia was placed in an Eppendorf tube containing 2 ml of extraction buffer (Tris-HCl, 100 mM; EDTA, 10 mM; NaCl, 1 M; SDS, 1%; proteinase K, 0.05 mg ml^{-1}; pH 8.0) and 10% (v/v) glass beads and ground into powder. The samples were vortexed and incubated at 65°C for 30 minutes. After incubation, the samples were centrifuged at 10,000 ×g for 15 min and supernatant was transferred to a fresh tube. To the supernatant, 150 μl of 3 M guanidine hydrochloride was added and incubated at −20°C for 10 minutes. The samples were centrifuged at 10,000 ×g for 10 minutes. After centrifugation the supernatant was transferred to a fresh tube, and an equal volume of isopropanol was added. Samples were incubated at −20°C for 3 h. The samples were centrifuged for 10 min at 10,000 ×g and thereafter 70% ethanol was added and more centrifugation done for 10 min at 10,000 ×g. The nucleic acid pellet obtained was air dried and dissolved in 50 μl of TE buffer (Tris-HCl, 10 mM, pH 8; EDTA, 1 mM). The nucleic acid dissolved in TE buffer was further treated with 3 μl of RNase (10 mg ml^{-1}), to precipitate RNA at 37°C, and the pure DNA obtained was stored at −20°C for later use. The quality of DNA was determined by loading 5 μl of DNA on 1% agarose before running it for 45 volts for 30 minutes and bands were noted by visualization under the UV gel imager.

2.4.2. Polymerase Chain Reactions and Gel Electrophoresis. The DNA extracted from isolates of the pathogens was used as template in polymerase chain reaction. Two sets of primers were used; the first set contained the universal primers ITS1 (5′-TCCGTAGGTGAACCTGCGG-3′) and ITS4 (5′-TCCTCCGCTTATTGATATGC-3′) targeting all fungal isolates while the other set was a primer CgInt (5′-GGGGAAGCCTCTCGCGG-3′) specific to *Colletotrichum gloeosporioides* combined with the universal primer ITS4 (TCCTCCGCTTATTGATATGC) for the identification of the isolates. The amplified regions were subsequently sequenced to obtain sufficient information for the identification of isolates.

TABLE 1: Cultural and morphological characteristics of the *Colletotrichum gloeosporioides* on PDA.

Number of isolates	Upper side		Lower side	Zonation	Conidial shape
	Colony colour	Texture	Colour		
24	Whitish grey	Cottony	Creamish grey	Concentric	Cylindrical and straight
12	Whitish cream	Velvety	Greyish orange	Concentric	Cylindrical and straight
10	Greyish pink	Velvety	Grey	Concentric	Cylindrical and straight

2.4.3. Agarose Gel Electrophoresis. A 1.5% agarose gel was prepared using 1x TBE-buffer and stained with 5 μl of SYBR Safe and poured into casting tray having a comb to solidify. The first well of the solid gel was loaded with 1.5 μl 1 Kb marker, followed by 2 μl of each amplified product and a control at the end. The gel was connected to electric voltage of 100 volts for 45 minutes to allow migration of amplified PCR products. The DNA bands formed were visualized under UV light and images photographed using a camera connected to a computer.

2.4.4. DNA Cleaning and Sequencing. The amplified products of the target fragments obtained above were cleaned using the Qiagen PCR cleaning kit according to the manufacturer instructions. The cleaned fragments were submitted for Sanger sequencing at Inqaba Africa Genomic platform in South Africa together with the primers used for amplification (ITS1 and ITS4 and ITS4 and CgInt).

2.5. Bioinformatics Analysis. The sequences obtained from Inqaba Africa Genomic platform at South Africa were trimmed before subjecting them to alignment with other Genbank sequences. Gene alignment was done using BioEdit software version while phylogenetic analysis was done by Mega Molecular Evolutionary Genetic Analysis version 7.1.8.

Alignment of sequences of the isolates and phylogenetic analyses was done using Mega 7.18 [24]. The maximum likelihood trees were obtained using the Close-Neighbour-Interchange algorithm while missing data and gaps were eliminated [25]. Clade stability of the resultant phylogenetic tree was based on bootstrap analysis with 1000 replicates [26]. Evolutionary distances in form of number of base substitution per site were computed using the Maximum Composite Likelihood method [24].

3. Results

3.1. Fungal Isolates. A total of 80 fungal isolates from diseased avocado fruits showing symptoms of anthracnose (Figure 1) collected from the study area were obtained. The isolates were identified based on cultural morphological characteristic on PDA and spore characteristics as observed under microscope [27, 28].

A total of 46 isolates had whitish, greyish, or creamish colour and cottony, velvety mycelium on the top side and greyish cream with circular orange-pinkish colour on the reverse side, Figure 2(a). Their spores were straight with rounded end (Figure 2(b)), typical of *C. gloeosporioides* [29]. The remaining 34 isolates had whitish grey mycelium with black fruiting structure on the upper side and greyish black one on the lower side (Figure 2(c)). Their spores had 3-4 septa and 2 or 3 appendages at one end characteristics of *Pestalotiopsis microspora* [30] (Figure 2(d)).

FIGURE 1: Anthracnose symptoms on avocado fruit.

The isolates were further confirmed through Koch's postulate where ten-day-old pure cultures of the isolates growing on PDA were used to inoculate healthy ripe avocado fruit, Fuerte variety. After two days, a characteristic black spot was formed by both *Colletotrichum* and *Pestalotiopsis* isolates. Each of the reisolated fungi from the diseased fruit showed similar morphological, cultural, and spore characteristics as initial isolates.

The *Colletotrichum* isolates were subjected to more detailed study while the *Pestalotiopsis* spp. was identified further at molecular level using universal primers ITS1 and ITS4.

3.2. The Mycelial Growth of Colletotrichum Isolates. The *Colletotrichum gloeosporioides* isolates from the study area grew rapidly on the PDA medium covering the whole surface of the Petri dish in 10–12 days after inoculation. The mycelial colour of the isolates varied between whitish grey, whitish cream, and greyish pink on the upper side of the culture (Table 1). Similarly, the lower side of the cultures had creamish grey, greyish orange, and grey (Table 1). In terms of mycelia structure, cottony one was observed in 24 isolates as compared to velvety one observed in 22 isolates (Table 1). There was significant difference (χ^2 = 23.455, df = 2, and $P < 0.001$) among the observed and expected frequencies for the cultural texture and colour of various *Colletotrichum gloeosporioides* isolates (Table 1).

3.3. Mycelial Growth of Colletotrichum gloeosporioides Isolates. The mycelial diameter of the isolates showed significant

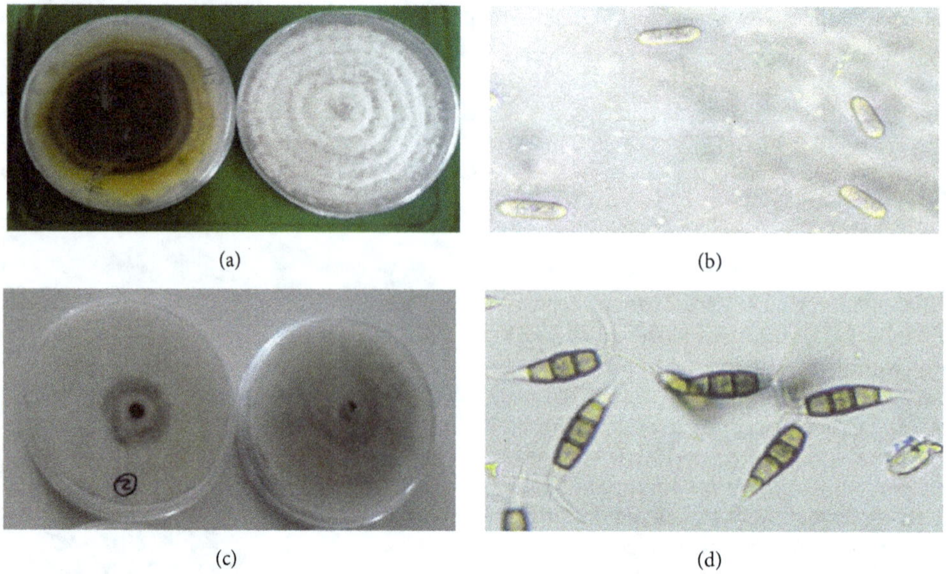

FIGURE 2: Mycelia of *Colletotrichum gloeosporioides* (a) and *Colletotrichum gloeosporioides spores* (b) (×400) and mycelia of *Pestalotiopsis microspora* (c) and *Pestalotiopsis microspora* spores (d) (×400).

differences ($P < 0.05$) throughout the growth period (Table 2). However, they exhibited similar trend in growth with day two having the least and day 10 having the largest diameter per isolate (Table 2). The radial diameter of all the isolates ranged from 0.3 to 0.93 cm and 2.37 to 4.5 cm in day 2 and day 10, respectively (Table 2).

3.4. Sporulation of the Colletotrichum Isolates. Sporulation of *Colletotrichum gloeosporioides* isolates exhibited a wide range of mean number of spores per isolate ranging from the lowest 0.65×10^6 to the highest 9.00×10^6 spores per ml (Table 2). These mean number of spores per ml differed significantly at $P < 0.05$ among isolates. Isolate s1 had the highest mean number of 9.0×10^6 per ml which was significantly different from the rest at $P < 0.05$ (Table 2). A total of 24 isolates had mean number of spores lower than the tabulated mean of all the isolates of 3.49×10^6 per isolate.

3.5. Conidial Morphology and Size. All the spores observed were cylindrical and straight with smooth round end. The spore size varied significantly at $P < 0.05$ among isolates ranging from 3.0 to 5.0 μm in width and 10.3 to 18.2 μm in length (Table 3). The spore widths of isolates s1, s40, and s44 were highest at 5.0 μm and they differed significantly from the rest at $P < 0.05$ (Table 3). Similarly, these isolates produced the longest spores of 18.2, 18.0, and 18.0 μm for isolate s1, s40, and s44, respectively. Isolates s37 and s46, however, produced the smallest spores having a mean of 3.0 μm in width and 10.3 μm in length. Thirty-one isolates produced spores having width within the range of 3.1–3.5 μm which did not differ significantly at $P \geq 0.05$ (Table 3). Overall, spore size in terms of width and length differed significantly at $P < 0.05$ among isolates.

3.6. Phylogenetic Studies of Colletotrichum gloeosporioides and Pestalotiopsis microspora. The molecular identification of *C. gloeosporioides* and *Pestalotiopsis* spp. was inferred from 13 sequences of *C. gloeosporioides* and 10 sequences of *Pestalotiopsis* isolates. A phylogenetic tree (Figure 1) was made of sequences from *C. gloeosporioides, Pestalotiopsis microspora*, and reference sequences from the Genbank (Figure 3). The identity of the isolates of both *Colletotrichum* and *Pestalotiopsis* to the Genbank isolates ranged from 98% to 100%.

The phylogenetic analysis of the isolates and references from the Genbank resulted into three clades: Clade 1 *(Colletotrichum gloeosporioides)*, Clade 2 *(Colletotrichum boninense)*, and Clade 3 *(Pestalotiopsis microspora)* (Table 4 and Figure 3).

4. Discussion and Conclusion

4.1. Cultural and Morphological Characteristics of Colletotrichum gloeosporioides and Pestalotiopsis microspora Isolates. The fungal isolates from diseased avocado fruits showing symptoms of anthracnose collected from the study area varied significantly in their cultural characteristics on PDA media in terms of texture and colour. A total of 46 isolates had whitish, greyish, or creamish colour and cottony, velvety mycelium on the top side and greyish cream with circular orange-pinkish colour on the reverse side. Similar cultural characteristics among *Colletotrichum gloeosporioides* isolates from avocado fruits were observed by [31]. Further, Chowdappa et al. [32] also noted the wide cultural variations among *C. gloeosporioides* isolates. The mycelial growth, however, had uniform radial growth characterised by circular ring-like patterns common to *C. gloeosporioides.* Sharma and Kulshrestha [4] reported similar mycelial growth

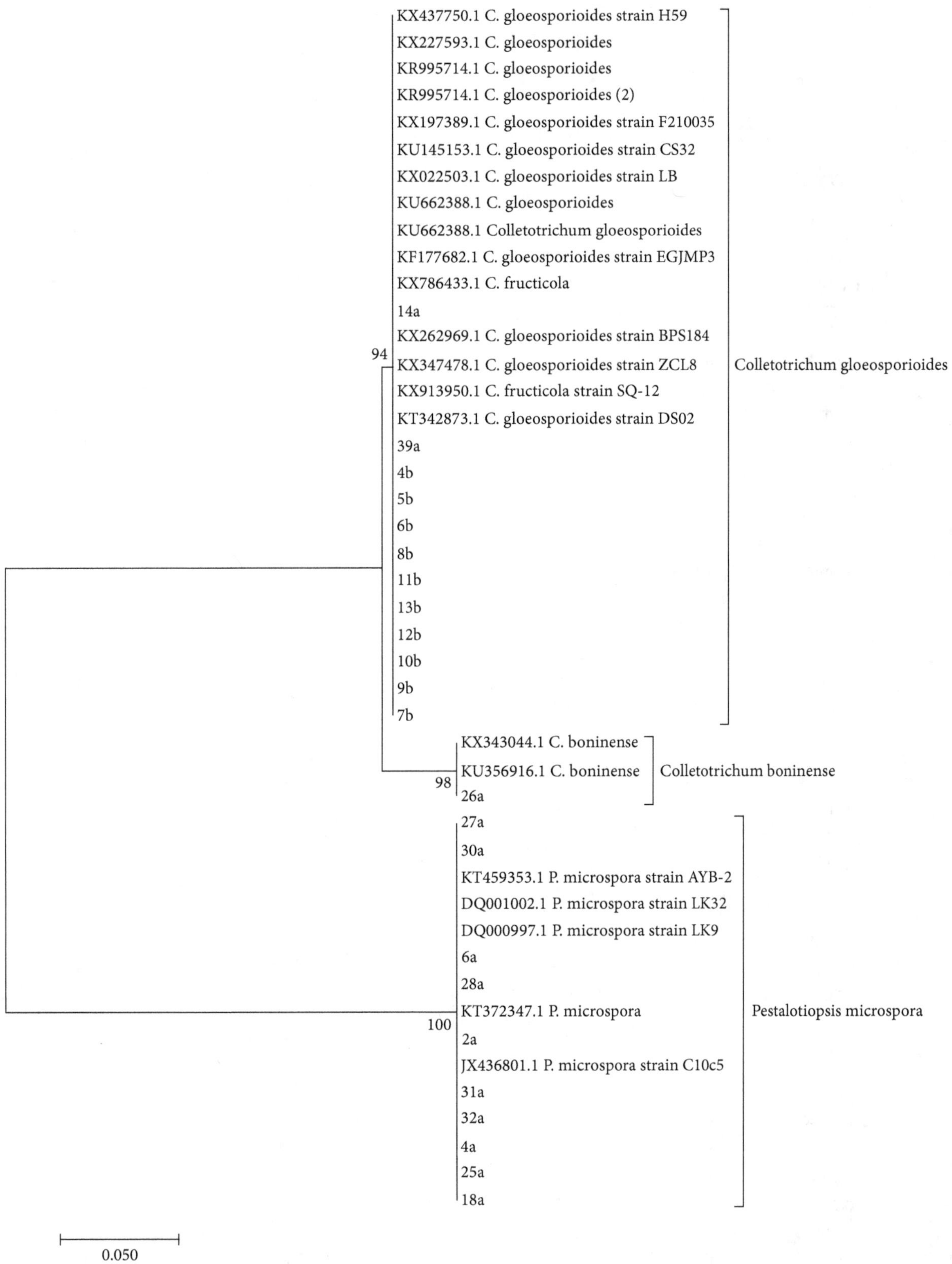

FIGURE 3: Maximum likelihood tree showing the relationship of *Colletotrichum gloeosporioides complex* and *Pestalotiopsis microspora* isolates based on ITS region.

TABLE 2: Daily mycelial growth and mean number of spores of *Colletotrichum gloeosporioides* isolates.

Isolate	Day 2	Day 4	Day 6	Day 8	Day 10	Mean number of spores (10^6 ml)
s1	0.77^{*bcd}	1.10^{jklmn}	1.33^{pq}	2.57^{l}	3.57^{m}	9.00^{*a}
s40	0.63^{defgh}	1.02^{mno}	1.60^{no}	3.05^{ijk}	4.00^{efghij}	6.00^{b}
s44	0.57^{fgh}	1.67^{d}	2.23^{def}	3.77^{ab}	4.03^{efghi}	5.67^{bc}
s3	0.63^{defgh}	1.02^{mno}	1.60^{no}	3.05^{ijk}	4.00^{efghij}	5.33^{bc}
s5	0.57^{fgh}	1.67^{d}	2.23^{def}	3.77^{ab}	4.03^{efghi}	5.00^{bc}
s6	0.65^{cdefgh}	1.43^{efg}	2.12^{efg}	$3.45^{bcdefgh}$	4.23^{bcde}	5.00^{bcd}
s19	0.63^{defgh}	1.47^{ef}	2.33^{de}	3.77^{ab}	4.17^{defg}	5.00^{bcd}
s21	0.53^{gh}	1.03^{lmno}	1.73^{lmn}	$3.30^{defghij}$	3.90^{hijkl}	5.00^{bcd}
s10	0.50^{h}	1.17^{ijklm}	1.83^{ijklm}	3.23^{fghij}	3.80^{ijklm}	5.00^{bcd}
s18	0.63^{defgh}	1.10^{jklmn}	1.97^{ghijk}	3.40^{cdefgh}	4.07^{efgh}	4.67^{bcde}
s25	0.63^{defgh}	1.10^{jklmn}	1.97^{ghijk}	3.40^{cdefgh}	4.07^{efgh}	4.67^{bcde}
s28	0.60^{efgh}	0.93^{no}	1.20^{q}	1.47^{n}	2.60^{n}	4.67^{bcde}
s14	0.60^{efgh}	0.93^{no}	1.20^{q}	1.47^{n}	2.60^{n}	4.33^{bcde}
s17	0.60^{efgh}	0.87^{o}	1.73^{lmn}	3.63^{abcd}	4.17^{defg}	4.33^{bcde}
s2	0.60^{efgh}	0.87^{o}	1.73^{lmn}	3.63^{abcd}	4.17^{defg}	4.00^{cdef}
s22	0.50^{h}	1.43^{efg}	2.03^{fghi}	3.67^{abc}	4.07^{efgh}	4.00^{cdef}
s24	0.70^{cdef}	1.27^{ghij}	1.97^{ghijk}	3.63^{abcd}	4.10^{efgh}	4.00^{cdef}
s29	0.60^{efgh}	1.20^{hijkl}	1.70^{lmno}	3.70^{abc}	4.00^{efghij}	3.67^{def}
s30	0.73^{bcde}	1.37^{fgh}	2.03^{fghi}	3.60^{abcde}	4.20^{cdef}	3.33^{defg}
s43	0.87^{ab}	1.57^{de}	2.80^{b}	3.70^{abc}	4.03^{efghi}	3.33^{defg}
s13	0.60^{efgh}	1.10^{jklmn}	1.65^{mno}	2.82^{kl}	4.37^{abcd}	3.33^{defg}
s15	0.60^{efgh}	1.00^{mno}	1.60^{no}	2.97^{jk}	4.03^{efghi}	3.17^{efg}
s4	0.67^{cdefg}	1.20^{hijkl}	1.97^{ghijk}	$3.43^{bcdefgh}$	4.13^{Defgh}	3.00^{efg}
s8	0.93^{a}	1.90^{c}	3.23^{a}	3.57^{bcdef}	4.10^{Efgh}	3.00^{efg}
s38	0.50^{h}	1.43^{efg}	2.03^{fghi}	$3.30^{defghij}$	4.50^{a}	3.00^{efg}
s16	0.63^{defgh}	2.18^{b}	2.37^{cd}	3.28^{efghij}	3.65^{Lm}	3.00^{efg}
s26	0.80^{abc}	1.13^{jklm}	1.88^{hijkl}	3.18^{ghij}	3.90^{Hijkl}	3.00^{efg}
s45	0.80^{abc}	1.90^{c}	2.58^{bc}	3.15^{hijk}	3.93^{Ghijk}	3.00^{efg}
s9	0.67^{cdefg}	1.32^{fghi}	2.08^{fgh}	$3.43^{bcdefgh}$	4.02^{Efghij}	2.67^{fg}
s12	0.70^{cdef}	1.37^{fgh}	2.10^{fgh}	$3.37^{cdefghi}$	4.43^{Abc}	2.67^{fg}
s20	0.77^{bcd}	1.23^{hijk}	1.60^{no}	3.13^{hijk}	3.97^{Fghij}	2.33^{fgh}
s27	0.80^{abc}	2.63^{a}	3.40^{a}	3.93^{a}	4.13^{Defgh}	2.33^{fgh}
s31	0.60^{efgh}	1.03^{lmno}	1.50^{op}	2.50^{l}	3.77^{Jklm}	2.33^{fgh}
s32	0.57^{fgh}	0.93^{no}	1.33^{pq}	2.03^{m}	2.37^{n}	2.33^{fgh}
s33	0.30^{i}	0.40^{p}	0.97^{r}	1.43^{n}	2.60^{n}	2.33^{fgh}
s35	0.70^{cdef}	1.03^{lmno}	1.70^{lmno}	3.25^{fghij}	4.07^{Efgh}	2.00^{gh}
s36	0.70^{cdef}	1.17^{ijklm}	1.97^{ghijk}	3.63^{abcd}	4.10^{Efgh}	2.00^{gh}
s41	0.63^{defgh}	1.10^{jklmn}	1.77^{klmn}	2.83^{kl}	4.00^{Efghij}	2.00^{gh}
s7	0.60^{efgh}	1.13^{jklm}	1.33^{pq}	2.53^{l}	3.80^{Ijklm}	1.00^{h}
s11	0.57^{fgh}	1.07^{klmn}	2.00^{ghij}	3.50^{bcdefg}	4.47^{Ab}	1.00^{h}
s23	0.70^{cdef}	1.13^{jklm}	1.77^{klmn}	3.50^{bcdefg}	4.47^{Ab}	0.67^{h}
s34	0.77^{bcd}	1.07^{klmn}	1.78^{jklmn}	3.27^{efghij}	4.03^{Efghi}	0.67^{h}
s39	0.77^{bcd}	1.07^{klmn}	1.67^{lmno}	1.87^{m}	4.03^{Efghi}	0.66^{h}
s42	0.77^{bcd}	1.07^{klmn}	1.67^{lmno}	1.87^{m}	4.03^{Efghi}	0.66^{h}
s37	0.77^{b-d}	1.13^{jklm}	1.63^{mno}	2.97^{jk}	3.70^{Klm}	0.65^{hi}
s46	0.70^{c-f}	1.58^{de}	2.67^{b}	3.55^{bcdef}	4.22^{Bcdef}	0.65^{hi}
LSD (0.05)	0.16	0.17	0.23	0.34	0.25	1.673
P-value	<0.001	<0.001	<0.001	<0.001	<0.001	<0.001

*Data are means of three replicates. Means on the same column followed by similar letter(s) are not significantly different at $P \geq 0.05$ according to Fisher's protected LSD test.

Table 3: The mean width and length of spores in micron (μm) produced by 10-day-old *Colletotrichum gloeosporioides* isolates.

Isolate	Width	Length
s1	5.0^{*a}	18.2^{a}
s40	5.0^{a}	18.0^{a}
s44	5.0^{a}	18.0^{a}
s3	4.7^{a}	17.1^{ab}
s5	4.7^{a}	17.1^{ab}
s6	4.7^{a}	16.2^{bc}
s19	3.7^{b}	16.1^{bc}
s21	3.7^{b}	16.1^{bc}
s10	3.6^{bc}	16.0^{bc}
s18	3.6^{bc}	15.4^{cd}
s25	3.6^{bc}	15.3^{cd}
s28	3.6^{bc}	15.2^{cde}
s14	3.5^{bcd}	15.2^{cde}
s17	3.5^{bcd}	15.2^{cde}
s2	3.5^{bcd}	15.1^{cde}
s22	3.5^{bcd}	15.0^{cde}
s24	3.5^{bcd}	15.0^{cde}
s29	3.5^{bcd}	15.0^{cde}
s30	3.5^{bcd}	15.0^{cde}
s43	3.5^{bcd}	14.5^{def}
s13	3.4^{bcde}	14.4^{defg}
s15	3.4^{bcde}	14.2^{defgh}
s4	3.4^{bcde}	14.0^{efgh}
s8	3.4^{bcde}	14.0^{efgh}
s38	3.3^{bcde}	13.5^{fgh}
s16	3.3^{bcde}	13.2^{ghi}
s26	3.3^{bcde}	13.2^{ghi}
s45	3.3^{bcde}	13.2^{ghi}
s9	3.3^{bcde}	13.1^{hi}
s12	3.2^{cde}	13.1^{hi}
s20	3.2^{cde}	13.0^{hi}
s27	3.2^{cde}	12.2^{ij}
s31	3.2^{cde}	12.2^{ij}
s32	3.2^{cde}	12.2^{ij}
s33	3.2^{cde}	12.1^{ij}
s35	3.2^{cde}	12.1^{ij}
s36	3.2^{cde}	12.1^{ij}
s41	3.2^{cde}	12.1^{ij}
s7	3.2^{cde}	12.0^{ij}
s11	3.1^{de}	12.0^{ij}
s23	3.1^{de}	11.1^{jk}
s34	3.1^{de}	11.0^{jk}
s39	3.1^{de}	11.0^{jk}
s42	3.1^{de}	10.4^{k}
s37	3.0^{e}	10.3^{k}
s46	3.0^{e}	10.3^{k}
LSD	0.4031	1.2096
P value	<0.001	<0.001

*Data are means of three replicates. Means on the same column followed by similar letter(s) are not significantly different at $P \geq 0.05$ according to Fisher's protected LSD test.

Table 4: *Colletotrichum* and *Pestalotiopsis* isolates with their Genbank accession number.

Species	Culture	Location	Genbank number
Colletotrichum gloeosporioides	14^{a}	Murang'a County	MG013524
	39^{a}		MG013525
	4^{b}		MG013527
	5^{b}		MG013528
	6^{b}		MG013529
	8^{b}		MG013530
	10^{b}		MG013531
	11^{b}		MG013532
	12^{b}		MG013533
	13^{b}		MG013534
	9^{b}		MG013535
	7^{b}		MG013536
Colletotrichum boninense	26^{a}	Murang'a County	MG013526
Pestalotiopsis microspora	27^{a}	Murang'a County	MG013537
	30^{a}		MG013538
	6^{a}		MG013539
	28^{a}		MG013540
	2^{a}		MG013541
	31^{a}		MG013542
	32^{a}		MG013543
	4^{a}		MG013544
	25^{a}		MG013545
	18^{a}		MG013546

characteristic of *C. gloeosporioides* in vitro. The cultures produced spores which were straight with rounded end, ranging within 3.0–5.0 μm in width and 10.3–18.2 μm in length, characteristic of *Colletotrichum gloeosporioides* as also reported by Chowdappa et al. [32].

The remaining 34 isolates had whitish cream mycelium with black fruiting structure, acervuli on the upper side, and had light orange to orange at lower side. These isolates produced spores having 3-4 septa and 2 or 3 appendages, characteristics of *Pestalotiopsis microspora* as reported by El-argawy [33]. Further molecular characteristic of these isolates using universal primers ITS1 and ITS4 confirmed the species as *Pestalotiopsis microspora* as discussed below in molecular characterisation section.

The differences observed in cultural and morphological characters of the isolates could be associated with their genetic variations and repeated subculturing [34, 35]. Further, the isolates had significantly different growth rate at $P < 0.05$ (Table 2). Such growth rate among *C. gloeosporioides* isolates was also reported by Zakaria and Bailey [31]. Overall the proportion in percentage (%) of the variance in mycelia radial diameter that was predictable from days for the isolates varied from the lowest ($R2 = 0.812$) to highest value ($R2 = 0.993$).

In this study, sporulation by *Colletotrichum gloeosporioides* isolates exhibited a wide range of mean number of spores per isolate from lowest 0.67×10^{6} to the highest 9.00×10^{6} spores per ml (Table 3) similar to observation made by

Peres et al. [9]. These mean numbers of spores per ml differed significantly at $P < 0.05$ among isolates (Table 3).

The spores observed in this study were cylindrical and straight with smooth round end. Similar spores of *C. gloeosporioides* were observed by Chowdappa et al. [32]. The spore size varied significantly at $P < 0.05$ among isolates ranging within 3.0–5.0 μm in width and 10.3–18.2 μm in length (Table 3). Overall, spore size in terms of width and length differed significantly at $P < 0.05$ among isolates.

4.2. Phylogenetic Studies of Colletotrichum gloeosporioides and Pestalotiopsis microspora. Phylogenetic results showed that ribosomal internal transcribed spacers (ITS) DNA can be used to indicate the relationships within *Colletotrichum* and *Pestalotiopsis* species. The study identified *Colletotrichum gloeosporioides* and *Colletotrichum boninense* based on a randomly selected sample of 13 sequences of *Colletotrichum gloeosporioides* isolates using ITS4 (universal) and CgInt (*C. gloeosporioides* specific) primers yielding single band of 450 bp and 98–100% homology with nucleotide sequence of ITS region of DNA with *C. gloeosporioides* isolates in the Genbank (Figure 1). This was in agreement with the nucleotide sequence of ITS region of ribosomal DNA of *Colletotrichum gloeosporioides* isolates from orchids amplified using specific primers CgInt and ITS4 as reported by Chowdappa et al. [32]. Further, molecular identification of *Pestalotiopsis microspora* was based on 10 randomly selected sequences of *Pestalotiopsis microspora* isolates using universal ITS1 and ITS4 primers for DNA sequencing.

Twelve isolates of *C. gloeosporioides*, one isolate of *C. boninense*, and ten isolates of *P. microspora* from the study area gave identical sequences of the published sequences in the Genbank for the same species with 94%, 98%, and 100% bootstraps value, respectively (Figure 1). *Colletotrichum gloeosporioides* identified in this study has been reported as the most common and wide spread pathogen in all avocado growing region worldwide [36–38]. Further it has been associated with infection of other hosts such as almond, coffee, guava, apple, dragon fruit, cassava, mango, sorghum, and strawberry [2, 15, 39]. Though *C. boninense* identified is not very common in the study area, its sequence was identical to published sequence (KX 343044.1 and KU356916.1) in the Genbank. This species, among others like *Colletotrichum acutatum*, *Colletotrichum godetiae*, *C. fioriniae*, *C. aenigma*, and *Colletotrichum gigasporum*, has been reported to cause anthracnose of avocado [11, 16, 40, 41].

The ten randomly sampled isolates of *P. microspora* identified in this study showed 100% identity with the published sequences in the Genbank thereby confirming their identity (Figure 1). *Pestalotiopsis* species are widespread in both tropical and temperate region [42, 43]. Individual species of *Pestalotiopsis* are known to cause infection on a wide range of hosts [44, 45]. *Pestalotiopsis clavispora* is known to cause stem end rot of avocado [6]; *P. versicolor* has been reported as a causal agent of anthracnose in avocado [46]. *P. palmarum* known to cause leaf spot and fruit canker in avocado [47]. Though *P. microspora* is prevalent in both tropics and subtropics, its association with host plants is not well researched [48]. It has been regarded both as an endophyte and as a pathogen causing postharvest diseases [49]. It has been reported to cause scab disease of guava fruits in Hawaii [48]. In this study, the fungus was isolated from diseased avocado fruits showing symptoms associated with anthracnose disease of avocado. However, despite its prevalence, *P. microspora* and its role in plant ecology are poorly understood [49].

Conclusively, *Colletotrichum gloeosporioides, Colletotrichum boninense,* and *Pestalotiopsis microspora* were identified for the first time as the causal agents of anthracnose of avocado in Kenya through cultural, morphological, and molecular techniques.

Conflicts of Interest

The authors declare that there are no conflicts of interest regarding the publication of this article.

Authors' Contributions

S. K. Kimaru conceived, designed, and performed experiments and was responsible for collection of diseased avocado fruits and isolation of fungi, morphological identification of the fungi, DNA extraction, running gel electrophoresis, analysis of sequence and construction of phylogenetic tree, data analysis, and drafting manuscript. R. C. Cheruiyot participated in the creation of the concept of the manuscript, revision of the manuscript, and approval of the manuscript. J. Mbaka was responsible for confirmation of the identity of fungi morphologically, methodical consultation in revision of the manuscript, and approval of manuscript. A. Alakonya was responsible for revision of the manuscript mainly regarding molecular section and approval of manuscript. E. Monda participated in the conception and design of the experiment, identification of fungi morphologically, sequence analysis, and methodical consultation in verification and approval of the manuscript.

Acknowledgments

The authors acknowledge Kenyatta University for granting study leave, Kenya Agricultural Research and Livestock Organisation (KARLO), Kandara and Kenya Plant Health Inspectorate Services (KEPHIS) where morphological and molecular studies were conducted, respectively, and Inqaba Biotec, South Africa, for DNA sequencing.

References

[1] A. T. Djami-Tchatchou, C. J. Straker, and F. Allie, "454 Sequencing for the identification of genes differentially expressed in avocado fruit (cv. Fuerte) infected by Colletotrichum gloeosporioides," *Journal of Phytopathology*, vol. 160, no. 9, pp. 449–460, 2012.

[2] J. E. Erpelding, "Field assessment of anthracnose disease response for the Sorghum germplasm collection from the Mopti region," *American Journal of Agricultural and Biological Sciences*, vol. 5, no. 3, pp. 363–369, 2010.

[3] C. A. Onyeani and N. A. Amusa, "Incidence and severity of anthracnose in mango fruits and its control with plant extracts

in south west nigeria," *International Journal of Agricultural Research*, vol. 10, no. 1, pp. 33–43, 2015.

[4] M. Sharma and S. Kulshrestha, "Colletotrichum gloeosporioides: an anthracnose causing pathogen of fruits and vegetables," *Biosciences, Biotechnology Research Asia*, vol. 12, no. 2, pp. 1233–1246, 2015.

[5] H. V. Silva-Rojas and G. D. Ávila-Quezada, "Phylogenetic and morphological identification of Colletotrichum boninense: a novel causal agent of anthracnose in avocado," *Plant Pathology*, vol. 60, no. 5, pp. 899–908, 2011.

[6] A. Valencia, R. Torres, and B. Latorre, "First report of Pestalotiopsis clavispora and Pestalotiopsis spp. causing postharvest stem end rot of avocado in Chile," *Plant Disease*, vol. 95, no. 4, pp. 492-493, 2011.

[7] H. J. Boesewinkel, "A list of 142 new plant disease recordings from New Zealand and short notes on three diseases," *Australasian Plant Pathology*, vol. 11, no. 4, pp. 40–43, 1982.

[8] F. Liu, M. Wang, U. Damm, P. W. Crous, and L. Cai, "Species boundaries in plant pathogenic fungi: a Colletotrichum case study," *BMC Evolutionary Biology*, vol. 16, no. 1, article 649, 2016.

[9] N. A. R. Peres, E. E. Kuramae, M. S. C. Dias, and N. L. De Souza, "Identification and characterization of Colletotrichum spp. affecting fruit after harvest in Brazil," *Journal of Phytopathology*, vol. 150, no. 3, pp. 128–134, 2002.

[10] D. G. Rodriguez-Sanchez, M. Flores-García, C. Silva-Platas et al., "Isolation and chemical identification of lipid derivatives from avocado (Persea americana) pulp with antiplatelet and antithrombotic activities," *Food & Function*, vol. 6, no. 1, pp. 193–203, 2015.

[11] M. V. Valle and A. Campos-Martínez, "First report of avocado anthracnose caused by Colletotrichum karstii in Mexico," *Plant Disease*, vol. 100, no. 2, p. 534, 2016.

[12] S. L. Willingham, A. W. Cooke, L. M. Coates, and K. G. Pegg, "Pepper spot: a new preharvest Colletotrichum disease of avocado cv. Hass," *Australasian Plant Pathology*, vol. 29, no. 2, p. 151, 2000.

[13] D. F. Farr, M. C. Aime, A. Y. Rossman, and M. E. Palm, "Species of colletotrichum on agavaceae," *Mycological Research*, vol. 110, no. 12, pp. 1395–1408, 2006.

[14] B. Schaffer, B. Wolstenholme, and A. Whiley, *The Avocado: Botany, Production and Uses*, CABI, 2013.

[15] G. N. Agrios, *Plant Pathology*, Elsevier Academic Press, 2005.

[16] E. Dann, R. Ploetz, L. Coates, and K. Pegg, "Foliar, fruit and soilborne diseases," in *The Avocado: Botany, Production and Uses*, pp. 380–422, 2013.

[17] A. Gautam, "Colletotrichum gloeosporioides: biology, pathogenicity and management in India," *Plant Physiology & Pathology*, vol. 2, no. 2, article 2, 2014.

[18] K. Pernezny and R. Marlatt, "Diseases of avocado in Florida," Plant Pathol Fact Sheet 21, 2000.

[19] L. Wasilwa, J. Njuguna, and E. Okoko, *Status of Avocado Production in Kenya*, Kenya Agricultural Research Institute, Nairobi, Kenya, 2004.

[20] Y. Choi, K. Hyde, and W. Ho, *Fungal Diversity*, Fungal Diversity Press, 1999.

[21] S. Freeman, T. Katan, and E. Shabi, "Characterization of *Colletotrichum* species responsible for anthracnose diseases of various fruits," *Plant Disease*, vol. 82, no. 6, pp. 596–605, 1998.

[22] A. Nagamani and I. K. Kunwar, *Handbook of Soil Fungi*, Edited by K. Indra and C. Manoharachary, I.K. International, 2009.

[23] B. Liu, F. J. Louws, T. B. Sutton, and J. C. Correll, "A rapid qualitative molecular method for the identification of Colletotrichum acutatum and C. gloeosporioides," *European Journal of Plant Pathology*, vol. 132, no. 4, pp. 593–607, 2012.

[24] S. Kumar, G. Stecher, and K. Tamura, "MEGA7: Molecular Evolutionary Genetics Analysis version 7.0 for bigger datasets," *Molecular Biology and Evolution*, vol. 33, no. 7, pp. 1870–1874, 2016.

[25] K. Takahashi and M. Nei, "Efficiencies of fast algorithms of phylogenetic inference under the criteria of maximum parsimony, minimum evolution, and maximum likelihood when a large number of sequences are used," *Molecular Biology and Evolution*, vol. 17, no. 8, pp. 1251–1258, 2000.

[26] H. Kishino and M. Hasegawa, "Evaluation of the maximum likelihood estimate of the evolutionary tree topologies from DNA sequence data, and the branching order in hominoidea," *Journal of Molecular Evolution*, vol. 29, no. 2, pp. 170–179, 1989.

[27] K. Domsch, W. Gams, and T. Anderson, *Compendium of Soil Fungi*, vol. 1, Academic Press, 1980.

[28] A. J. Rabha, A. Naglot, G. D. Sharma et al., "Morphological and molecular diversity of endophytic Colletotrichum gloeosporioides from tea plant, Camellia sinensis (L.) O. Kuntze of Assam, India," *Journal of Genetic Engineering and Biotechnology*, vol. 14, no. 1, pp. 181–187, 2016.

[29] C. F. B. Silva and S. J. Michereff, "Biology of Colletotrichum spp. and epidemiology of the anthracnose in tropical fruit trees," *Revista Caatinga*, vol. 26, no. 4, pp. 130–138, 2014.

[30] K. D. Hyde, R. H. Nilsson, S. A. Alias et al., "One stop shop: backbones trees for important phytopathogenic genera: I (2014)," *Fungal Diversity*, vol. 67, no. 1, pp. 21–125, 2014.

[31] M. Zakaria and J. A. Bailey, "Morphology and cultural variation among *Colletotrichum* isolates obtained from tropical forest nurseries," *Journal of Tropical Forest Science*, vol. 12, no. 1, pp. 1–20, 2000.

[32] P. Chowdappa, C. S. Chethana, R. Bharghavi, H. Sandhya, and R. P. Pant, "Morphological and molecular characterization of Colletotrichum gloeosporioides (Penz) Sac. isolates causing anthracnose of orchids in India," *Biotechnology, Bioinformatics and Bioengineering*, vol. 2, no. 1, pp. 567–572, 2012.

[33] E. El-Argawy, "haracterization and control of Pestalotiopsis spp. the causal fungus of guava scabby canker in el-beheira governorate, Egypt," *International Journal of Phytopathology*, vol. 4, no. 3, pp. 121–136, 2016.

[34] A. Vidyalakshmi and C. V. Divya, "New report of Colletotrichum gloeosporioides causing anthracnose of Pisonia alba in India," *Archives of Phytopathology and Plant Protection*, vol. 46, no. 2, pp. 201–204, 2013.

[35] W. V. Parker, P. R. Johnston, and U. Damm, "*Colletotrichum gloeosporioides* species complex," *Studies in Mycology*, vol. 73, pp. 115–180, 2012.

[36] J. O. Honger, S. K. Offei, K. A. Oduro, G. T. Odamtten, and S. T. Nyaku, "Identification and molecular characterisation of Colletotrichum species from avocado, citrus and pawpaw in Ghana," *South African Journal of Plant and Soil*, vol. 33, no. 3, pp. 177–185, 2016.

[37] Y. Siddiqui and A. Ali, "Chapter 11—*Colletotrichum gloeosporioides* (Anthracnose)," in *Postharvest Decay*, 2014.

[38] M. Twizeyimana, H. Förster, V. McDonald, D. H. Wang, J. E. Adaskaveg, and A. Eskalen, "Identification and pathogenicity of fungal pathogens associated with stem-end rot of avocado in California," *Plant Disease*, vol. 97, no. 12, pp. 1580–1584, 2013.

[39] R. Dean, J. A. L. Van Kan, Z. A. Pretorius et al., "The Top 10 fungal pathogens in molecular plant pathology," *Molecular Plant Pathology*, vol. 13, no. 4, pp. 414–430, 2012.

[40] F. R. Giblin, L. M. Coates, and J. A. G. Irwin, "Pathogenic diversity of avocado and mango isolates of Colletotrichum gloeosporioides causing anthracnose and pepper spot in Australia," *Australasian Plant Pathology*, vol. 39, no. 1, pp. 50–62, 2010.

[41] A. Hernandez-Lauzardo and A. Campos-Martínez, "First report of Colletotrichum godetiae causing anthracnose on avocado in Mexico," *Plant Disease*, vol. 99, no. 10, p. 555, 2015.

[42] G. Ding, Z. Zheng, S. Liu, H. Zhang, L. Guo, and Y. Che, "Photinides A–F, cytotoxic benzofuranone-derived γ-lactones from the plant endophytic fungus Pestalotiopsis photiniae," *Journal of Natural Products*, vol. 72, no. 5, pp. 942–945, 2009.

[43] L. Liu, S. Liu, X. Chen, L. Guo, and Y. Che, "Pestalofones A–E, bioactive cyclohexanone derivatives from the plant endophytic fungus Pestalotiopsis fici," *Bioorganic & Medicinal Chemistry*, vol. 17, no. 2, pp. 606–613, 2009.

[44] Y. Suto and T. Kobayashi, "Taxonomic studies on the species of *Pestalotiopsis*, parasitic on conifers in Japan," *Transactions of the Mycological Society of Japan*, 1993.

[45] A. M. Ismail, G. Cirvilleri, and G. Polizzi, "Characterisation and pathogenicity of Pestalotiopsis uvicola and Pestalotiopsis clavispora causing grey leaf spot of mango (Mangifera indica L.) in Italy," *European Journal of Plant Pathology*, vol. 135, no. 4, pp. 619–625, 2013.

[46] J. Darvas, "Fungi associated with pre-and postharvest diseases of avocado fruit at Westfalia Estate, South Africa," *Phytophylactica*, vol. 19, no. 1, pp. 83–86, 1987.

[47] M. Kamhawy, M. Hassan, and S. Sharkawy, "Morphological and phylogenetic characterization of pestalotiopsis in relation to host association," *Egyptian Journal of Agricultural Research*, vol. 89, no. 1, pp. 1–16, 2011.

[48] L. M. Keith, M. E. Velasquez, and F. T. Zee, "Identification and characterization of Pestalotiopsis spp. causing scab disease of guava, Psidium guajava, in Hawaii," *Plant Disease*, vol. 90, no. 1, pp. 16–23, 2006.

[49] A. M. Metz, A. Haddad, J. Worapong et al., "Induction of the sexual stage of Pestalotiopsis microspora, a taxol-producing fungus," *Microbiology*, vol. 146, no. 8, pp. 2079–2089, 2000.

4

Illumina MiSeq Sequencing for Preliminary Analysis of Microbiome Causing Primary Endodontic Infections in Egypt

Sally Ali Tawfik,[1] **Marwa Mohamed Azab**,[1]
Ali Abdellah Abdelrahman Ahmed,[1] **and Dalia Mukhtar Fayyad**[2]

[1] *Department of Microbiology & Immunology, Faculty of Pharmacy, Suez Canal University, Ismailia, Egypt*
[2] *Department of Endodontics, Faculty of Dentistry, Suez Canal University, Ismailia, Egypt*

Correspondence should be addressed to Sally Ali Tawfik; lovelysally_17@hotmail.com

Academic Editor: Joseph Falkinham

The use of high throughput next generation technologies has allowed more comprehensive analysis than traditional Sanger sequencing. The specific aim of this study was to investigate the microbial diversity of primary endodontic infections using Illumina MiSeq sequencing platform in Egyptian patients. Samples were collected from 19 patients in Suez Canal University Hospital (Endodontic Department) using sterile # 15K file and paper points. DNA was extracted using Mo Bio power soil DNA isolation extraction kit followed by PCR amplification and agarose gel electrophoresis. The microbiome was characterized on the basis of the V3 and V4 hypervariable region of the 16S rRNA gene by using paired-end sequencing on Illumina MiSeq device. MOTHUR software was used in sequence filtration and analysis of sequenced data. A total of 1858 operational taxonomic units at 97% similarity were assigned to 26 phyla, 245 families, and 705 genera. Four main phyla Firmicutes, Bacteroidetes, Proteobacteria, and Synergistetes were predominant in all samples. At genus level, *Prevotella*, *Bacillus*, *Porphyromonas*, *Streptococcus*, and *Bacteroides* were the most abundant. Illumina MiSeq platform sequencing can be used to investigate oral microbiome composition of endodontic infections. Elucidating the ecology of endodontic infections is a necessary step in developing effective intracanal antimicrobials.

1. Introduction

The human microbiome is determined as the ecological community of commensal, symbiotic, and pathogenic microorganisms that actually share our body space [1, 2]. The oral microbiome is associated with the development of oral diseases such as dental caries, periodontal disease, and endodontic infection [3, 4]. Microorganisms that initially invade and colonize the pulp can lead to primary endodontic infection which is characterized by a polymicrobial infection dominated by anaerobic bacteria [5, 6]. On the other hand, secondary (persistent) infection refers to microorganisms that were members of a primary or secondary infection and that, in some way, resisted intracanal antimicrobial procedures and were able to survive during periods of nutrient deprivation in treated canals [7]. It has been reported that the bacterial communities in primary endodontic infections were more diverse than those in persistent infections [7, 8]. Apical periodontitis refers to inflammatory disease of apical periradicular tissues caused by microbial infection within the root canal system of the implicated tooth. Apical periodontitis is mainly the consequence of dental caries when the root canal system is infected by oral microbiota [9, 10]. Since the etiology and pathogenesis of apical periodontitis have not been finally elucidated, additional research should be conducted in this field.

Previously, bacterial diversity of the infected root canal system was determined by broad-range polymerase chain reaction followed by cloning and Sanger sequencing as well as molecular fingerprinting techniques such as denaturing gradient gel and terminal restriction fragment length polymorphism analysis [5, 11]. These techniques offered initial insights into the bacterial diversity but despite their high sensitivity these methods could detect only the most prevalent bacterial community members. Low abundant species may occupy critical niches within a complex microbial community and thus are potentially important in maintaining the stability and virulence of a microbial community [12].

Microbial diversity related to the human body is much greater than previously thought [13, 14]. Molecular methods have demonstrated a higher complexity of the endodontic microbiota than previously reported by cultivation approaches. Recent studies have moved away from the concept that a single pathogen causes a disease to the concept that the whole community is responsible for the pathogenicity [10, 15]. Molecular biology methods have facilitated the identification of specific bacterial species with apical periradicular disease which led to the discovery of novel endodontic pathogens [6]. Next generation sequencing has been widely used for bacterial diversity analyses using 16S ribosomal RNA gene [16]. The use of high throughput technology enables a large number of reads in a single run, providing increased sampling depth compared with other techniques [17]. The major advantage of next generation sequencing is the detection of low abundant genera [17, 18].

The aim of this study was to investigate bacterial diversity in primary endodontic infections through taxonomic classification of 16S rRNA gene by Illumine MiSeq sequencing platform. Therefore, the recognition of community profiles involved in endodontic disease with periapical lesion may represent an important step toward a better understanding of the pathogenesis of the disease and establishment of more effective therapeutic protocols.

2. Material and Methods

2.1. Subjects and Patients. Samples were obtained from 19 adult patients (15 females and 4 males; aged 18 to 51 years; mean age 31 years) who visited Suez Canal University Dental Hospital for the endodontic treatment of primary endodontic infections (untreated root). They were recruited as study subjects (10 with upper incisor, 1 with lower incisor, 6 with upper premolar, and 2 with upper molar). Patients who participated in this study were volunteers. All patients were aware about the nature of the study. They were included into the study after signing an ethics committee-approved informed consent form. This study was carried out in accordance with the guidelines of, and after approval by, the Ethic Committee of Human Research at Faculty of Pharmacy, Suez Canal University, Egypt (reference number: 201611MH1). Only teeth from adult patients with carious lesions, necrotic pulps, and radiographic evidence of apical periradicular disease were involved in this study. We excluded patients who had systemic disease, cancer, diabetes, immunodeficiency disorder, and a history of using antibiotics or fluoride in the previous 3 months [19, 20].

2.2. Microbiome Sample Collection. At the beginning, the tooth crown was cleansed with pumice; then local anesthesia was applied and a rubber dam was placed. 30% hydrogen peroxide was used to clean the tooth and the surrounding field which were then decontaminated with a 2.5% sodium hypochlorite solution (NaOCl) for 30 seconds each. Complete access preparations were carried out using sterile burs without water spray. 2.5% NaOCl was again used to swab the operative field including the pulp chamber. The NaOCl solution was inactivated with 5% sodium thiosulfate [21]. Little amount of sterile saline solution was inserted into the canal in case the root canal was dry. Samples were firstly collected by means of #15 K-type file with the handle cut-off. The file was inserted to a level approximately 1 mm short of the tooth apex with a gentle filing motion that depended on the diagnostic radiographs. Then, two sequential paper points were introduced to the same level and used to soak up the fluid in the canal. Each paper point was kept in position for 1 minute. The two paper points and the file were both transferred to 1.5 ml microcentrifuge tube containing 1 ml of lysis Mo Bio buffer. Afterwards, the samples were transferred to the laboratory for DNA extraction in frozen ice packs. A negative control (sample code 19) using sterile file and paper points which were not applied to the root canal was used in parallel to test for the existence of contaminants in the file and paper points [19, 22, 23].

2.3. DNA Extraction. Microcentrifuge tubes containing endodontic samples were vortexed at low speed to disperse the trapped bacteria. Genomic DNA from samples and the negative control were extracted immediately after sample collection using Mo Bio power soil DNA isolation kit with some modifications (Mo Bio Laboratories, Carlsbad, CA, USA, cat. number 12888-50). The file and paper points were removed and the solution was transferred to bead tubes. Then, 60 μL of solution C1 had been added to the bead tubes. Using water bath, tubes were incubated at 65°C for 10 minutes and then vortexed horizontally at maximum speed for 2 minutes using a flat-bed vortex pad with tape at maximum speed for 10 minutes. Afterwards, instructions were followed as directed by the manufacturer protocol except eluting DNA with 30 μL instead of 100 μL of PCR DNase-free water. DNA samples were measured using a Nanodrop ND-1000 spectrophotometer (ND-1000; Thermo Scientific, Waltham, MA, USA), by measuring absorbance values at 260 and 280 nm.

2.4. PCR Amplification. PCR amplification was carried out immediately after DNA extraction, using primers targeting the V3 and V4 regions of the 16S rRNA gene which used the extracted DNA as a template. Hypervariable regions V3-V4 of 16S rRNA gene were amplified using the following primers with Illumina adapters (underlined): forward primer: 5′ TCGTCGGCAGCGTCAGATGTGTATAAGAGACAGCCTACGGGNGGCWGCAG; reverse primer: 5 GTCTCGTGGGCTCGGAGATGTGTATAAGAGACAGGACTACHVGGGTATCTAATCC [24].

The PCR mixture was composed of 0.8 μL for each forward and reverse primer (10 μM, Metabion, Germany), 3 μL of template DNA for the samples, and 12.5 μL of 1x of Hot Master Mix (Promega GoTaq® Green Master Mix) to a final volume of 25 μL. For negative control, 3 μL of elution solution was used. The amplifications were performed under the following conditions: initial denaturation at 95°C for 2 minutes, followed by 30 cycles of denaturation at 95°C for 30 seconds, primer annealing at 60°C for 30 seconds, and extension at 72°C for 30 seconds, with a final elongation at 72°C for 5 minutes. The presence of PCR products was confirmed by electrophoresis in a 2% agarose gel conducted at 80 V/cm in Tris-Borate-EDTA (TBE) buffer. Ethidium bromide (0.5 lg/ml) was used to stain the gel for 15 min which is then visualized under 300 nm ultraviolet light. The resulting image was then captured using a computer software program (AlphaEase™, Alpha lnnotech). PCR amplicons that were positive on the agarose gel electrophoresis images were subjected to purification by means of the MinElute kit (QIAGEN). The purified DNA products were pooled together according to equal concentrations where short fragments were removed using Ampure beads (Agencourt Bioscience, MA, USA). The eluted DNA product obtained after purification was quantified using Qubit Kit Assays (Invitrogen, Life Technologies). Bioanalyzer 2100 with the DNA 1000 Chip kit (Agilent, Palo Alto, CA, USA) was used to assess the quality of the final products for each sample individually [19, 24]. There was no amplification product observed in the negative extraction controls.

2.5. Illumina Sequencing of 16S rRNA Gene. Sequencing was carried out at IGA Technology Services (Udine, Italy) using paired-end Illumina MiSeq sequencing on an Illumina MiSeq device (Illumina Inc., San Diego, CA, USA) with 600 cycles (300 cycles for each paired read and 12 cycles for the barcode sequence) according to the manufacturer's instructions. To artificially increase the genetic diversity, it has been common to mix in a control library of genomic DNA from the phage phix to prevent focusing and phasing problems due to the sequencing of "low diversity" libraries. Sequence analysis was conducted using the 16S-based metagenomics workflow of MiSeq Reporter v2.3 (Illumina). 16S rRNA gene is the widely used target to identify microbes without the need to sequence entire genome. Illumina workflow started with purified genomic DNA. Primers were tailed with sequences to include indexing barcodes. Samples were gathered into a single library for sequencing on Illumina MiSeq sequencing system which generated paired 300 bp reads. Sequences were then demultiplexed based on index sequences. FASTQ files with Quality Score Encoding were created. OTUs clustering and classification at several taxonomic levels, kingdom, phylum, class, order, family, genus, and species, were performed. Illumina-curated version of the Greengenes database was used as a taxonomy database for the metagenomics workflow (http://greengenes.secondgenome.com/downloads/database/13_5).

2.6. Analysis of Microbial Community with MOTHUR. Analysis of sequenced data was done by using MOTHUR software (v.1.38.1) and the pipeline adapted from standard operating procedure (SOP) from Schloss et al. [25]. The two sets of reads for each sample (forward and reverse reads) were combined using (make.contigs) command. Sequences that failed to fulfil any one of the following criteria were excluded: maximum length of 600 bases, the presence of any ambiguities, maximum homopolymer length of 8 nt, and more than 1 nucleotide mismatch to the primer using (screen.seqs) command. Duplicated sequences were removed using (unique.seqs) command [25, 26]. SILVA database was customized to the targeted V3-V4 region of 16S rRNA gene using the (pcr.seqs) command which corresponds to the similar region in *Escherichia coli* starting from 6426 and ending at 27645. Unique sequences were aligned in MOTHUR using SILVA reference database [27]. Columns in the alignment that only contain gap characters were pulled out without losing any information using (filter.seqs) command. Then, sequences were preclustered using (pre.cluster) command which could permit at most two differences between sequences. Chimeras were detected using UCHIME algorithm and then removed using (chimera.uchime) command [28]. Sequences were taxonomically classified using the Naive Bayesian classification with 80% confidence threshold by using (classify.seqs) command [29]. Sequences that were not classified to any one of the domains (unknown) or classified in Chloroplast, Mitochondria, Eukaryota, and Archaea were eliminated using (remove.lineage) command. Sequences were analyzed from more than one point of view. Firstly, we considered each sample as a separate community. Secondly, samples were combined into two groups according to sex (males and females). Thirdly, samples were grouped into four groups according to age. Groups 1, 2, 3, and 4 ranged within 18–20, 21–30, 31–40, and 41–50 years, respectively. Lastly, based on 16S rRNA dendrogram three separate groups were constructed according to community structure and population compositions (Group A, Group B, and Group C) using (tree.shared) command which was visualized using FigTree program (v1.4.3).

2.7. Phylotypes Based Analysis. Sequences were analyzed into phylotypes according to their taxonomic classification using (phylotype) command. A distance matrix was created and the sequences were clustered into operational taxonomic units (OTUs) at 3% dissimilarity cut-off (97% similarity) using (dist.seqs) and (cluster) command, respectively. The cut off numbering of the phylotypes equals 4 which corresponded to order level used in (cluster.split) command. The number of sequences in each OTU was determined using (make.shared) command. The taxonomy for each OTU was specified using (classify.otu) command. To equalize read sizes of the samples for the comparison of read sizes among the samples, random subsampling was performed (i.e., 2,000) using (sub.sample) command [25].

2.8. Operation Taxonomic Unit (OTU) Based Analysis

2.8.1. Alpha Diversity. Rarefaction curves describe the number of OTUs observed as a function of sampling effort which were performed using (rarefaction.single) command. Community richness indices, number of observed OTUs, Chao 1 richness estimate, abundance based coverage richness estimate, and Jackknife estimator, were calculated. Community diversity indices, Simpson diversity index, inverse Simpson index, Shannon diversity index, nonparametric estimate of classical Shannon diversity index, Q statistic index, and Berger-Parker index, were also conducted [30]. Community richness and diversity indices were performed in MOTHUR software using (summary.single) command.

2.8.2. Beta Diversity. Shared community membership and community structure were analyzed using Jaccard and Theta index, respectively. And (dist.shared) command was used to calculate beta diversity [31]. Shared community richness, shared Sobs, shared Chao 1, and shared ACE, was conducted using (summary.shared) command. The UniFrac based metrics were used to assess the similarity between two communities' membership (unifrac.unweighted) and structure (unifrac.weighted). The relation between samples was identified using principal coordinate analysis (PCoA) that was visualized to compare our samples using graphing calculator 3D program (v.6.6.2) [32, 33]. A visualized Venn diagram was illustrated using MOTHUR software using (Venn) command [25]. Dendrogram was created which described similarities between the samples using (tree.shared) command [25].

2.9. Statistical Analysis. For statistical analyses, PAST software (v.3.16) was used to calculate Mann–Whitney (MW) U test and Kruskal-Wallis (KW) rank sum test. The difference between the 2 groups was analyzed using Mann–Whitney U test whereas for 3 groups Kruskal-Wallis analysis of variance by ranks was used, respectively. Statistical significance was assumed at $p < 0.05$ [34]. The correlation between OTUs was assessed using Spearman correlation coefficient (SpCC) with $p < 0.001$. The command (otu.association) in MOTHUR software was used for this assessment [25]. At genus level, hierarchical dendrogram based on the Bray-Curtis distances was calculated using Vegan package (v.2.4–4) [35] and gplots package (v.3.0.1) [36] in the R programing language version (3.4.0) 2016 [37].

3. Results

Paired-end sequencing on Illumina MiSeq of root canal content samples showed that all 19 cases were positive for the presence of bacterial DNA which yielded data of 2083824 raw sequences, Table 1. 1313300 unique sequences were produced from MOTHUR software analysis. Final data of 383401 sequence reads (>300 base pairs) was obtained after excluding low-quality sequence reads, preclustering, and the chimeras removal. A total of 1858 operational taxonomic units (OTUs) at 97% similarity were assigned to 26 phyla, 245 families, and 705 genera. Phyla with a representation of 0.5% and higher (relative abundance) are presented in Figure 1: Firmicutes (40.7%), Bacteroidetes (34.7%), Proteobacteria (6.1%), Synergistetes (3.9%), Fusobacteria (3.5%), Actinobacteria (3.4%), Spirochetes (1.2%), and Cyanobacteria (0.5%). Tenericutes, Chloroflexi, and Verrucomicrobia phyla were found in relatively low proportions less than 0.5%. These phyla were the most predominant, accounting for approximately 90% of the distinct OTUs found in the root canal content samples. The 19 samples were divided into three groups according to 16S rRNA dendrogram as illustrated in Figure 2. Samples 1, 2, 5, 6, 8, 9, 10, 13, and 18 (Group A) showed similarities as did samples 3, 4, 7, 11, 12, 14, and

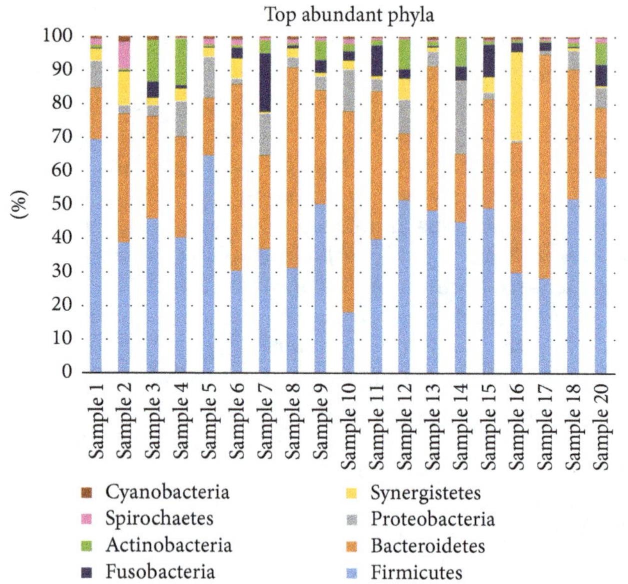

FIGURE 1: Relative abundance of the top most predominant bacterial phyla in each one of the 19 samples from the root canal content of endodontic infections. Data presented in average proportion (%) of all sequences.

20 (Group B). Also samples 15, 16, and 17 showed similar characteristics (Group C). Since sample code number 19 was the control sample, therefore, it was not involved in our analysis. When samples were divided into 3 groups based on 16S rRNA dendrogram, Firmicutes and Bacteroidetes were predominately found in Groups A, B, and C. The relative abundance of Firmicutes was equal in both Groups A and B, Figure 3. At genus level, the predominant genera ranked by abundance (over 1% of the microbiome) were *Prevotella* genus which accounted for 17.2% of the sequences, followed by *Bacillus* (5.1%), *Porphyromonas* (3.6%), *Streptococcus* (3.5%), and *Bacteroides* (3.2%) as shown in Figure 4. *Prevotella, Bacillus, Bacteroides, Staphylococcus,* and *Porphyromonas* were abundant in Group A, whereas *Prevotella, Streptococcus, Veillonella, Leptotrichia,* and *Lactobacillus* were abundant in Group B. *Prevotella, Atopobium, Pyramidobacter, Dialister,* and *Fusobacterium* were relatively abundant in Group C Figure 5.

The *Prevotella* genus was found with the highest proportions in all samples ranging from 13 to 53% of reads except for samples 1, 5, and 13. It was found that in samples 1, 5, and 13 *Bacillus* genus was greater than *Prevotella* genus in their abundance. The species of the top three genera with their relative abundance were mentioned below. For *Prevotella* genus the main species detected were *Prevotella tannarae* (3.2%), *Prevotella intermedia* (2.3%), and *Prevotella oris* (2.1%) of all sequences. On the other hand, the major species of *Bacillus* were *Bacillus firmus* (2.4%), *Bacillus siralis* (1%), and *Bacillus horneckiae* (0.4%). *Porphyromonas endodontalis* (2%) and *Porphyromonas gingivalis* (0.8%) were the major species in *Porphyromonas* genus. Classification of sequences by alignment with reference database SILVA at 97% similarity revealed that there were sequences that were not aligned

TABLE 1: Sequence analysis for the generated reads and the remaining reads after preprocessing, filtration, preclustering, and chimeras removal.

Sample code	Total reads	Reads passing quality filtering	Number of unique reads	Total number of reads before preclustering	Total number of reads after preclustering	Number of chimeras	Number of reads after chimeras removal	Percentage of chimeras sequence
1	130378	102374	69001	52390	22726	4334	18392	19.10%
2	99713	78230	52970	43726	20319	6216	14103	30.60%
3	148927	117143	83994	65789	30793	7323	23470	23.80%
4	136015	107981	73508	52409	24382	5070	19312	20.80%
5	91410	72590	54671	43282	22740	4037	18702	17.80%
6	159563	124396	81290	65593	29018	7503	21515	25.90%
7	179597	140847	94637	72983	32522	6942	25580	21.30%
8	153991	121551	76195	59673	26540	8457	18083	31.90%
9	140290	108819	77616	59922	27164	5045	22119	18.60%
10	93517	75671	50873	40472	17933	4568	13365	25.50%
11	153126	120768	91730	75798	41305	18169	23136	44%
12	159655	129432	95261	67017	31948	6394	25554	20%
13	72461	57647	40291	32138	16499	5210	11289	31.60%
14	98370	77375	56593	46254	22690	5828	16862	25.70%
15	126414	99959	65761	52345	24373	7679	16694	31.50%
16	128433	101274	65627	52214	24255	9275	14980	38.20%
17	202643	155799	107509	84438	42083	18983	23100	45.10%
18	183073	144386	96114	73998	33752	8485	25267	25.10%
20	188766	147582	107487	84994	41383	9505	31878	23%
Average	139281.2	109674.9	75848.84	59233.42	28022.37	7843.32	20179	27%

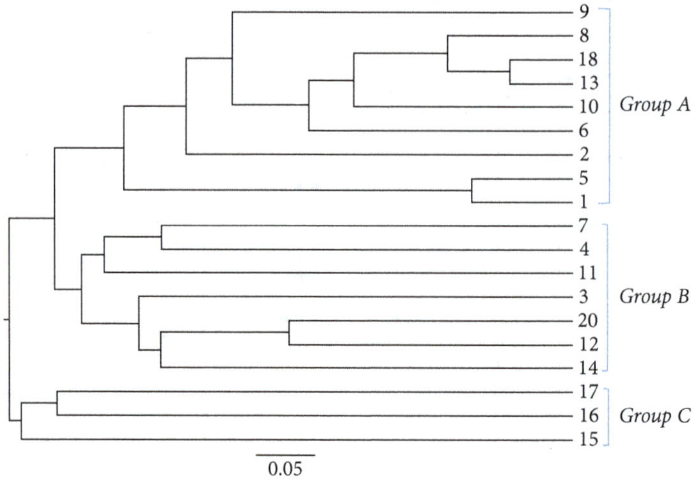

FIGURE 2: Phylogenetic tree dendrogram based on 16S rRNA gene from 19 primary endodontic patients for comparison of bacterial communities which demonstrated three separate community structure and population compositions (Groups A, B, and C).

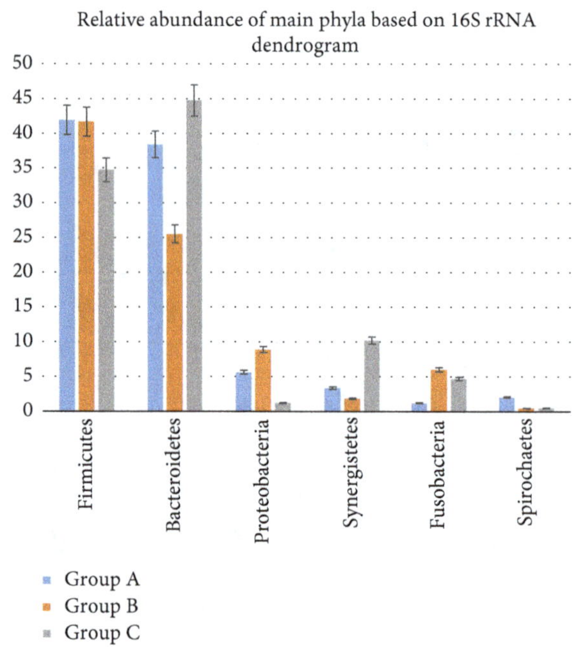

FIGURE 3: Comparison of the groups (A, B, and C) at the phylum level with error bars.

to reference sequence at different taxonomic levels. At the genus or species levels, the number of unclassified sequences increased by aligning to higher taxonomic levels. At the phylum level the mean percentage of sequences not assigned to any phyla was 5.7% while at the class, order, family, genus, and species level the mean percentage of unclassified sequences was 5.93%, 7.32%, 15.88%, 16.98%, and 36.55% respectively.

3.1. Alpha Diversity. Table 2 depicts data from community richness calculations. A predicted mean of 6161 and 16966 distinct OTUs per root canal content sample from Chao and ACE nonparametric measures of richness had been revealed, respectively. In addition, Table 3 depicts data from community diversity calculations. The good estimator detected > 90% coverage for the overall data, meaning that only 10 OTUs would be found if 100 additional sequences have been analyzed. Moreover, the rarefaction curve shape showed that bacterial richness in the root canal content samples was completely revealed by the number of sequences analyzed, Figure 6. Alpha diversity indices at 97% similarity were calculated using random resampling (2000 sequences).

3.2. Beta Diversity. The mean value of pairwise comparisons between different samples was 0.96 for Jaccard and 0.86 for Theta. These results revealed more similarity between samples. Table 4 displayed data on shared observed species for OTUS (at genus level) and shared Chao index.

3.3. Grouping of Samples and Its Statistical Results. Samples were classified into three categories A, B, and C to assist a comparative analysis. The results of using Kruskal-Wallis (KW) rank sum test to compare between 3 groups were presented in Table 5. Mann–Whitney U test was used to compare between 2 groups. Comparison between male and female groups revealed significant differences only for *Prevotella* ($p = 0.01$) and *Bacteroides* ($p = 0.005$) genera. The relative abundance of *Prevotella* genus in males (14.3%) was higher than females (3.8%) whereas, in the *Bacteroides* genus, the relative abundance was greater in females (4%) than males (0.6%). On the other hand, there was no significant difference between the four age groups using Kruskal-Wallis (KW) rank sum test. Illustrated Venn diagram was represented for sex and age groups in Figures 7 and 8.

Relationships between various samples were evaluated using UniFrac based principal component analysis (PCoA). Principal coordinate analysis based on UniFrac revealed that clustering of samples was according to grouping of the 16S rRNA dendrogram rather than samples, Figure 9. Correlations between various members of the endodontic

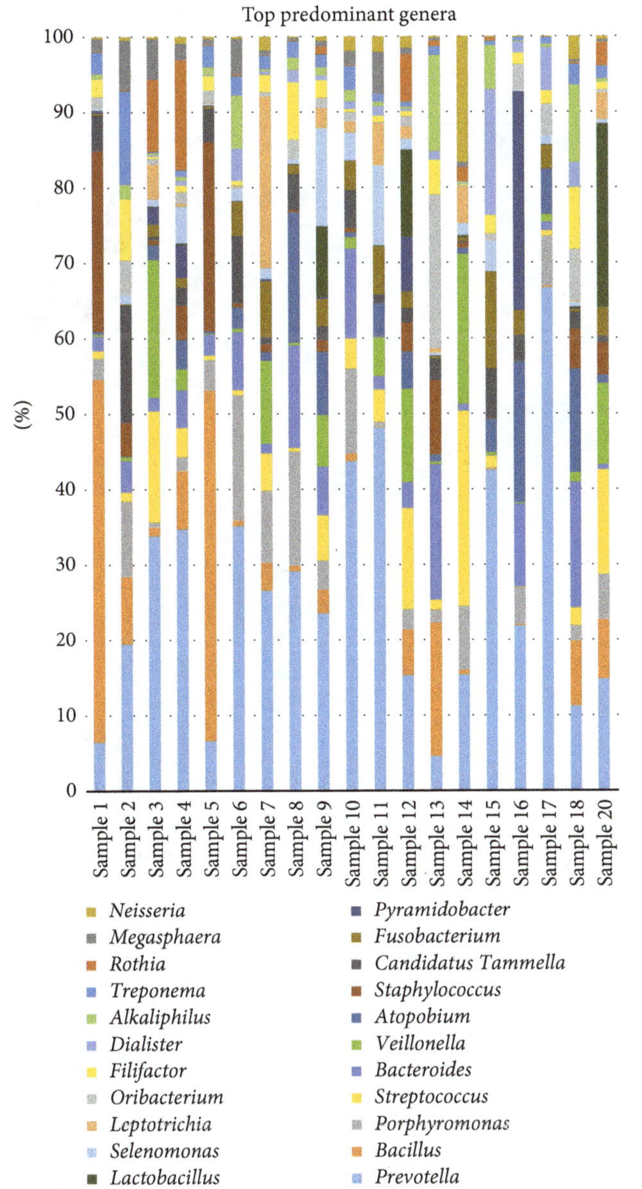

FIGURE 4: Average relative abundance of the top bacterial genera from root canal content of 19 patients with primary endodontic infections.

microbiome at different taxonomic levels were assessed using Spearman correlation coefficient (SpCC). There were 8829 (r => 0.6, p value < 0.001) significant positive correlations which expressed as a positive relationship between the two variables. The strongest correlations ($r = 0.82$) were detected between the most abundant phyla in all samples: Firmicutes, Bacteroidetes, Actinobacteria, Synergistetes, and Fusobacteria. The majority of samples showed positive SpCC mainly between OTUs which belonged to Lachnospiraceae family. While at genus level, the most significant positive SpCC was detected between OTUs assigned to *Oribacterium* and *Atopobium* genera. In parallel, minor samples resulting with negative SpCC (r =< -0.6, p value < 0.05) between OTUs were related to Fusobacteriaceae and Spirochaetaceae families. The significant negative SpCC showed between OTUs was related to *Fusobacteria, Treponema, Bacillus,* and *Prevotella* genera.

TABLE 2: Community richness indices: Sobs, Chao, ACE, and Jackknife.

Sample	Sobs	Chao	ACE	Jackknife
1	382	4815.077	12727.76	2609.788
2	258	3575.143	7171.851	15878.39
3	414	3142.682	8772.472	2691.427
4	425	3297.435	7720.03	2891.89
5	320	4077.091	16280.43	3722.471
6	480	7129.714	22638.01	3919.938
7	593	8777	21779.88	4134.895
8	546	10491.08	35406.91	4547.667
9	494	4541.652	11181.97	3302.81
10	254	1773.4	5302.989	1317.182
11	498	4931.143	10405.96	3422.119
12	565	5020.37	13945.89	5031.037
13	406	5131.067	23696.3	3136.874
14	339	3190.867	8166.598	2103.402
15	439	10414	23551.97	3477.703
16	466	10524.33	19000.8	4022.645
17	735	11259.55	37465.82	5812.206
18	370	3902.5	9036.737	2307.078
20	800	11067.44	28107.83	6655.509

Sobs: number of observed OTUs (at genus level); Chao: Chao 1 richness estimate; ACE: abundance based coverage richness estimate; Jackknife: Jackknife estimator.

4. Discussion

Primary endodontic infection is caused by microorganisms that invade and colonize the necrotic pulp tissue. It is distinguished by a mixed society prominently dominated by anaerobic bacteria [38, 39]. There were variations at the bacterial profiles of the endodontic microbiota from person to person [39]. This means that each individual is hosted by unique endodontic microbiota in terms of species richness and abundance. This suggests that primary endodontic infection has a heterogeneous etiology where no single species can be considered a major endodontic pathogen [19]. Primary endodontic infection is caused by multiple bacterial combinations which were evident by high numbers of different taxa that have been detected [22].

Culture-independent molecular biology methods had overcome many limitations of culture techniques and have been used recently for microbial characterization in endodontic research [38, 40–42]. Root canal microbiome was investigated by several methods. Many studies used extracted teeth subjected to cryogenic grinding approach where PCR products from extracted DNA were separated by denaturing gradient gel electrophoresis to obtain fingerprinting of bacterial communities [8, 43]. Other studies also used extracted teeth but extracted DNA was subjected to pyrosequencing

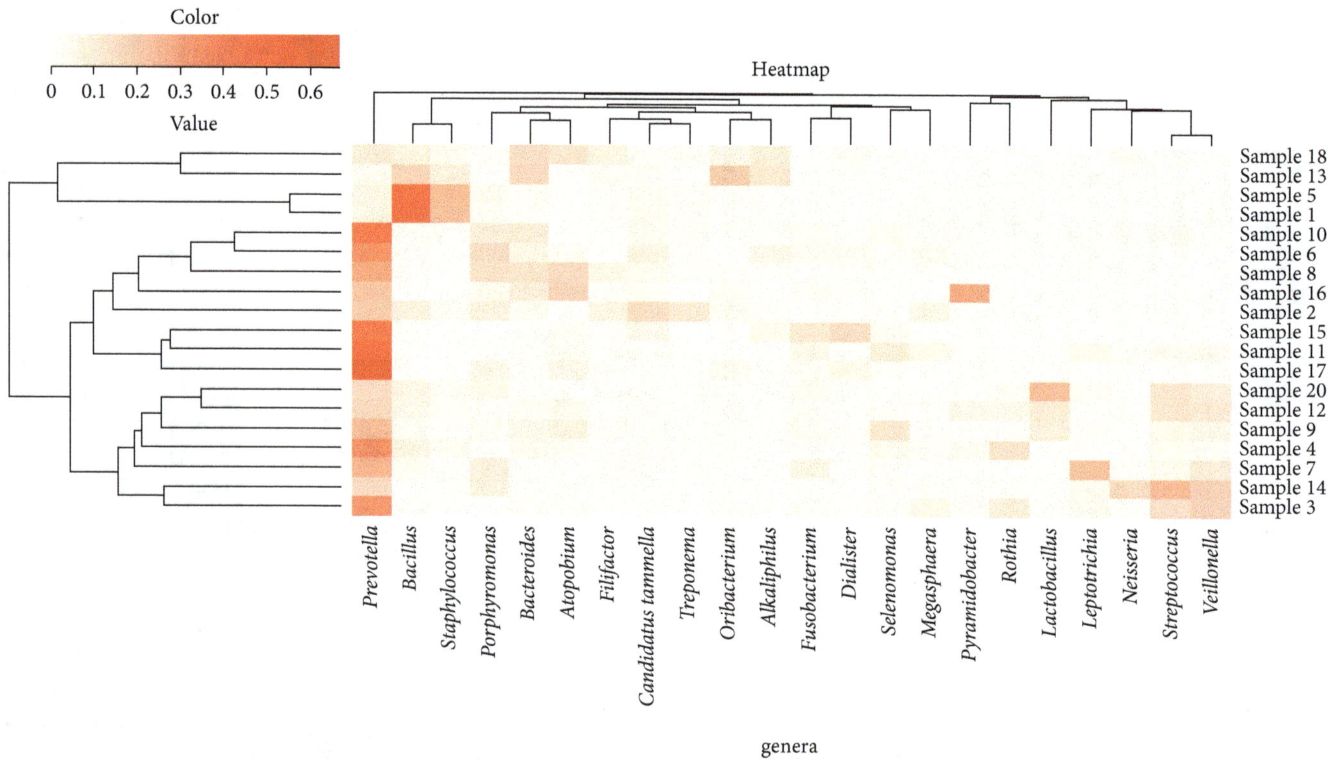

FIGURE 5: Heatmap with dendrogram at the genus level using a gradient heatmap (over 1% of the microbiome). The 22 most abundant genera were used in hierarchical clustering to evaluate the relationships between 19 samples using weighted pair clustering based on Barry-Curtis measurements. The darker the red color was the more predominant the genus was.

TABLE 3: Community diversity indices: Coverage, Simpson, Invsimpson, Shannon, Npshannon, Qstat, and Berger-Parker.

Group	*N*seqs	Coverage	Simpson	Invsimpson	Shannon	Npshannon	Qstat	Berger-Parker
1	2000	0.99	0.05	20.46	4.64	5.29	253.91	0.19
2	2000	0.90	0.03	38.10	4.52	4.90	160.14	0.09
3	2000	0.94	0.02	65.38	5.07	5.51	265.46	0.05
4	2000	0.96	0.02	59.33	5.21	5.78	278.44	0.08
5	2000	0.97	0.03	30.90	4.69	5.32	214.96	0.13
6	2000	0.94	0.03	29.92	4.80	5.33	321.00	0.14
7	2000	0.99	0.02	60.74	5.32	5.85	392.41	0.08
8	2000	0.91	0.07	14.44	4.53	5.28	375.82	0.24
9	2000	0.91	0.02	49.64	5.12	5.65	327.49	0.09
10	2000	0.92	0.08	12.21	3.90	4.44	164.47	0.23
11	2000	0.99	0.02	56.74	5.23	5.77	326.05	0.07
12	2000	0.92	0.01	119.61	5.65	6.20	372.94	0.04
13	2000	0.95	0.06	17.85	4.90	5.85	282.05	0.23
14	2000	0.99	0.02	43.05	4.88	5.44	221.45	0.09
15	2000	0.99	0.04	28.08	4.83	5.45	293.59	0.15
16	2000	0.97	0.03	31.07	4.91	5.56	313.06	0.12
17	2000	0.95	0.02	45.51	5.46	6.20	506.39	0.11
18	2000	0.96	0.08	13.15	4.35	4.88	236.60	0.26
20	2000	0.95	0.01	114.42	5.86	6.44	534.52	0.05

*N*seqs: number of sequences in the sample; Simpson: Simpson diversity index; Invsimpson: inverse Simpson index ($1/D$); Shannon: Shannon diversity index; Npshannon: nonparametric estimate of classical Shannon diversity index; Qstat: Q statistic index; Berger-Parker: Berger-Parker index.

Table 4: Shared OTUs (bold) and shared Chao index (italicized) between each pair of the dental root canal content samples.

	Sample 1	Sample 2	Sample 3	Sample 4	Sample 5	Sample 6	Sample 7	Sample 8	Sample 9	Sample 10	Sample 11	Sample 12	Sample 13	Sample 14	Sample 15	Sample 16	Sample 17	Sample 18	Sample 20
Sample 1	-	41	34	36	41	30	38	19	39	25	18	42	23	17	14	13	15	36	43
Sample 2	57.04	-	29	34	42	30	33	25	34	47.50	110.00	90.38	26.50	86.50	66.50	36.50	36.65	80.50	101.11
Sample 3	68.33	112.64	-	44	32	24	43	23	46	40.58	92.50	92.01	35.77	110.10	54.75	33.00	31.25	62.71	77.54
Sample 4	57.85	84.22	77.33	-	37	32	48	25	49	42.10	57.42	80.74	50.06	88.07	49.75	53.13	54.00	74.97	85.3125
Sample 5	63.27	80.57	72.85	83.75	-		31	26	37	43.90	78.00	62.87	40.64	73.00	72.00	42.00	47.00	72.28	84.68
Sample 6	46.68	56.58	49.23	41.33	67.50	-	34	29	41	46.89	49.58	49.50	30.90	70.75	33.61	44.50	39.90	57.08	53.71
Sample 7	89.63	63.42	76.10	96.48	55.50	39.26	-	26	47	45.64	77.17	91.47	69.33	71.33	43.50	38.00	36.86	58.58	77.67
Sample 8	29.25	46.17	39.42	43.48	64.00	41.60	36.50	-	35	40.88	66.33	49.42	42.38	31.83	35.83	29.00	39.23	53.76	38.79
Sample 9	63.79	80.93	63.71	105.64	76.76	47.14	65.66	73.64	-	39.82	85.21	101.58	71.64	79.50	57.08	40.00	36.05	125.85	80.58
Sample 10	48.79	25	27	30	23	26	34	22	30	-	32	42	23	26	20	19	19	29	39
Sample 11	33.00	25	46	34	24	38	39	31	48	44.95	-	49	21	41	34	22	31	37	49
Sample 12	67.14	36	65	56	35	40	56	27	60	77.94	115.39	-	32	52	31	35	32	47	68
Sample 13	38.17	20	22	31	25	20	25	22	32	44.07	34.04	60.35	-	18	19	18	14	37	28
Sample 14	46.00	21	46	34	20	25	31	16	37	64.75	68.48	124.23	33.70	-	20	12	16	29	50
Sample 15	28	17	26	27	21	29	23	25	35	50.70	57.58	54.11	42.14	36.83	-	18	23	29	26
Sample 16	16	16	18	28	15	23	23	20	23	29.25	41.50	63.43	22.67	33.50	26.50	-	25	25	23
Sample 17	21	18	21	25	19	30	27	28	30	20.20	37.92	45.65	16.15	34.83	25.89	40.70	-	28	27
Sample 18	84.1	37	37	43	38	39	35	35	54	45.00	71.04	63.49	74.19	61.00	50.42	31.00	39.75	-	49
Sample 20	70.4	51	48	51	45	34	48	27	54	59.00	81.24	114.24	42.02	111.21	102.14	48.70	38.42	92.80	-

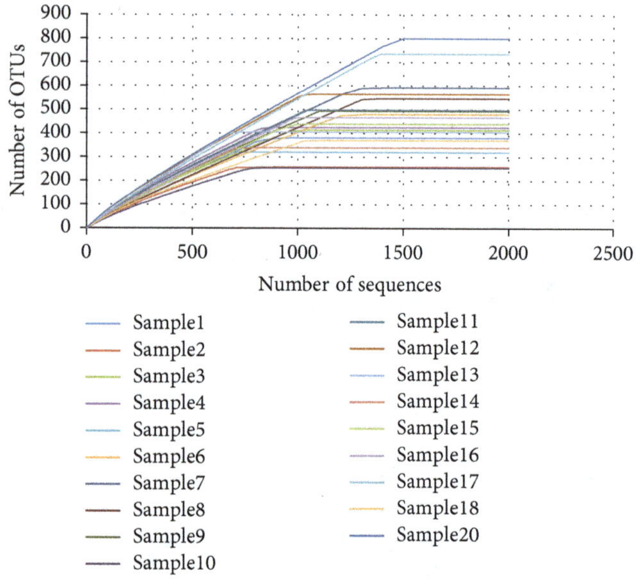

FIGURE 6: Rarefaction curves of 16S rRNA gene sequences for each sample calculated for OTUs at 97% similarity. Vertical axis shows operational taxonomic units, and horizontal axis shows the number of samples sequenced. OTUs = operational taxonomic units.

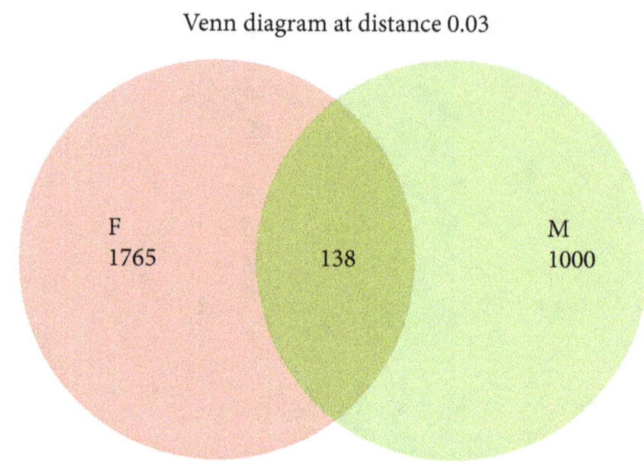

The number of species in group F is 1765

The number of species in group M is 1000

The number of species shared between groups F and M is 138

Percentage of species that are shared in group F and M is 4.75370

The total richness for all groups is 2903

FIGURE 7: Venn diagram illustrated the number of shared OTUs between males and females at 97% similarity. Colored circles represented each sex and intersection part between circles represented the number of shared OTUs.

TABLE 5: *Derived from Kruskal-Wallis sum rank test. Bolded numbers are statistically significant ($p < 0.05$).

Genus	Relative abundance (%)			p value (KW)*
	Group A	Group B	Group c	
Prevotella	11.37	17.74	33.60	0.05
Bacillus	8.81	2.42	0.12	**0.04**
Porphyromonas	4.40	2.74	3.05	0.4
Streptococcus	1.05	7.94	0.60	**0.004**
Bacteroides	4.75	1.31	3.27	**0.04**
Veillonella	0.70	7.75	0.38	**0.002**
Atopobium	2.58	1.67	7.63	0.12
Staphylococcus	4.61	1.35	0.04	**0.04**
Candidatus Tammella	3.35	0.69	2.50	**0.02**
Fusobacterium	0.93	2.17	4.69	**0.04**
Pyramidobacter	0.06	1.22	7.71	0.3
Lactobacillus	0.61	3.76	0.04	0.08
Selenomonas	1.39	2.05	1.33	0.5
Leptotrichia	0.31	3.90	0.03	**0.002**
Oribacterium	2.15	0.41	2.23	0.08
Filifactor	2.19	0.55	1.47	**0.04**
Dialister	0.73	0.29	5.85	**0.03**
Alkaliphilus	2.10	0.23	1.53	**0.004**
Treponema	2.10	0.58	0.50	**0.02**
Rothia	0.16	3.12	0.12	0.1

analysis [44, 45] or Illumina sequencing [46]. Moreover, apical periradicular lesions obtained during apical surgery have been used for sampling root canal [47]. Our study used files and paper points for sampling as other studies did, because it is a simple and easy method [9, 17, 19, 22, 23, 48, 49]. In addition, another advantage of this method of collection is that it does not require extracted tooth to be processed or teeth subject to root-end resection during apical periradicular surgery. The collection method however had limitations for the characterization of the total endodontic microbiome due to anatomical variations between the patients such as fins, lateral canals, dentinal tubules, and isthmuses [19]. There is no ideal approach for endodontic microbiological sampling without limitations. The method of sample collection should be adequate to the purpose of the study in the research design. The results of the Illumina MiSeq sequencing have showed that the root canal can harbor a highly diverse microbiome [19]. These previous studies used pyrosequencing technique [9, 17, 19, 20, 44, 45, 49] and their sample sizes were 20, 18, 7, 20, 17, 10, and 10, respectively. On the other hand, these two studies [22, 46] with sample sizes of 10 and 12 patients were utilized in Illumina sequencing platform. Therefore, the use of 19 participants was suitable for our study aim.

The analysis of the 16S ribosomal RNA gene sequence is commonly used in metagenomic studies. The 16S gene is universal in all bacteria as it gathers the advantage of being long enough to provide high information but, also at the same time, short in length that it can be easily sequenced. The 16S rRNA gene is composed of around 1500 base pairs. It contains nine variable and conserved regions in such a manner that conserved region is followed by variable region. Phylogenetic classification is performed using variable regions of 16S rRNA. To debate, the reason of choosing variable region depends on various factors such as objectives of the experiment, experimental design, and sample type [50]. Since the

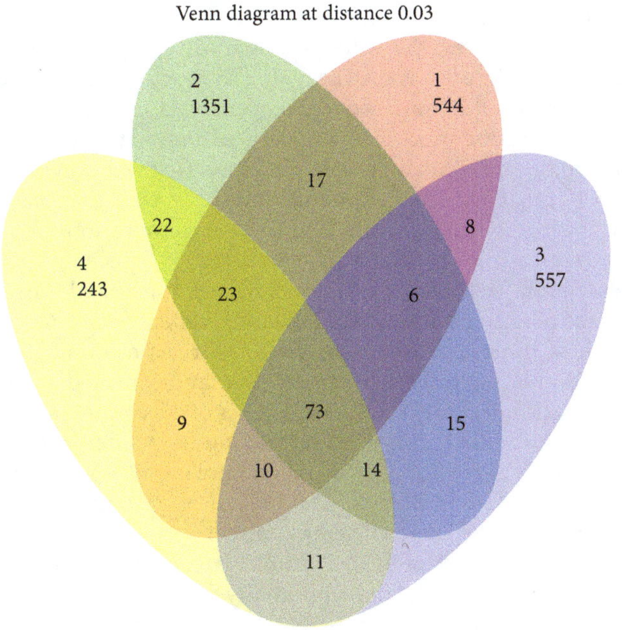

The number of species in group 1 is 544
The number of species in group 2 is 2351
The number of species in group 3 is 557
The number of species in group 4 is 243
The number of species shared between groups 1 and 2 is 119
The number of species shared between groups 1 and 3 is 97
The number of species shared between groups 1 and 4 is 115
The number of species shared between groups 2 and 3 is 108
The number of species shared between groups 2 and 4 is 132
The number of species shared between groups 3 and 4 is 108
The number of species shared between groups 1, 2, and 3 is 79
The number of species shared between groups 1, 2, and 4 is 96
The number of species shared between groups 1, 3, and 4 is 83
The number of species shared between groups 2, 3, and 4 is 87
The total richness for all groups is 2903

FIGURE 8: Venn diagram illustrated the number of shared OTUs according to four age groups at 97% similarity. Colored circles represented each age group and intersection part between circles represented the number of shared OTUs.

bacterial diversity is largely wide, it may be recommended that more than one variable region be targeted. This increases the specificity, accuracy, and sensitivity of the 16S gene study [20]. Our study used V3 and V4 regions. An experimental study revealed that targeting the V3 and V4 regions produced high quality of sequenced data [51]. In addition, these regions were recommended by Illumina protocol manual [52]. To ensure that microbial classification is of high quality, our study targeted two variable regions in order to achieve pair-ended reads of approximately 490 bp.

The main bacterial phyla detected in the primary endodontic infection were Firmicutes and Bacteroidetes, which together represented about three-quarters of all sequences obtained. The next most abundant phyla were Proteobacteria

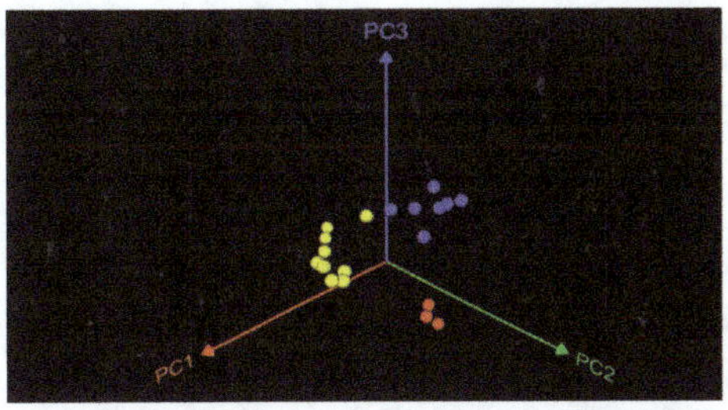

FIGURE 9: A three-dimensional plot of weighted UniFrac based principal coordinate analysis (PCoA). Colors represented the three groups based on 16S rRNA dendrogram. Yellow, blue, and red balls resembled Groups A, B, and C respectively. Plot was created using the pairwise weighted UniFrac distances (where PC1 is variability at 25.57%, PC2 is variability at 14.69%, and PC3 is variability at 11.44%). Samples from the same group cluster more closely together.

and Synergistetes. In previous studies using culturing and molecular methods, Firmicutes was the most abundant phylum in endodontic infection [5, 6]. In other previous studies which used pyrosequencing, there was a difference in the most dominated phylum. Some studies showed Bacteroidetes [9, 17, 23] as the major phylum whereas other studies demonstrated Firmicutes [19, 22, 45, 49] and Proteobacteria [44, 46, 47], respectively. The difference in the phylum results may be due to different sampling method, different clinical expressions and interventions, analytical artifacts during PCR, and filtering sequences or identification. It is important to take into consideration the fact that most previous studies used pyrosequencing whereas few studies used Illumina sequencing and our study was one of these studies. Our data was in agreement with those provided by Vengerfeldt et al. [22] which revealed Firmicutes and Bacteroidetes as the top two main phyla. We would like to focus on the fact that this study was the first study which characterized endodontic infection using next generation sequencing (MiSeq Illumina) in the Middle East, Egypt. As a result, this study gains the benefit of simplicity of sampling procedure and millions of reads were generated per instrument run with less cost and more depth coverage in comparison to pyrosequencing.

Due to variations in the abundance of OTUs in the groups, the groups appeared to show different stages of endodontic infections. The *Lactobacillus* genus was abundant in Group B, which indicated the initial stage of endodontic infection [53, 54]. Bacteroidetes and *Porphyromonas* were abundant in Group A. Group A might appear in another stage of infection. In a previous study of Gomes et al. [55], *Tannerella forsythia* was found to be a member of the *Bacteroides* genus which is most likely recovered from an acute dental abscess. *Tannerella forsythia* species was associated with tenderness to percussion [55]. Also *Porphyromonas* was present in high levels in Group A where in a previous study *P. endodontalis* and *P. gingivalis* were associated with the presence of a sinus tract and abscess formation [56]. Gomes et al. [57, 58] recognized that the microbiota found in teeth with a sinus was predominantly mixed. Group C included *Fusobacterium* species with high abundance in comparison to the other two groups. *Fusobacterium* was recognized with more severe endodontic infection with pain or history of pain and was linked to polymicrobial infections due to bacterial synergism [48, 59, 60]. *Prevotella* genus was present in high levels in all the three groups but in higher proportion in Group C followed by Group B and then Group A. *Prevotella* species are strictly anaerobes and highly virulent species which was the reason for their high abundance [61]. This might also give indication that higher abundance of *Prevotella* is related to the degree and stage of infection. Jung et al. [62] indicated that the degree of severity of an endodontic infection is related not only to the presence of pathogens but also to the numbers of those organisms in the infected site. Our study was in agreement with the previous studies results which showed that the species *Porphyromonas gingivalis, Porphyromonas endodontalis, Prevotella intermedia,* and *Prevotella nigrescens* were abundant in endodontic infection [61, 63, 64]. *Treponema* genus was from the top twenty genera found in this study. Our research is in line with the findings of Siqueira Jr. and Rôças [39] which revealed that *Treponema denticola, Treponema socranskii, Treponema maltophilum, Treponema amylovorum,* and *Treponema medium* were from the most prevalent species in *Treponema.*

This study identified 1858 different OTUs with >3% dissimilarity, belonging to 705 genera. When identifying each sample for its highest dominated genera, it revealed that 9 samples were dominated by *Prevotella* and only 2 samples were dominated by *Bacillus* genus. There were 8 samples dominated by bacteria that could not be identified for the genus level (unclassified). This was the number of genera that were found by these studies: Hong et al. [9] specified 133, Tzanetakis et al. [23] 347, and Anderson et al. [26] 525. The differences between these studies are linked to the number of samples examined, the depth of sequencing, the analytical methods for identification, and the selected region sampled (whole main canal in the paper point approach versus apical canal in the cryopulverization approach) [46].

Acidobacteria, Chloroflexi, and Cyanobacteria are phyla which exist in soil, wastewater plants, and water. When next generation sequencing instruments were invented, these phyla were detected in endodontic infections which were not previously known to be found in endodontic infections [17]. Most of these phyla were found in relatively low abundance and were unnoticed in previous studies characterizing endodontic communities.

Regarding composition level, heatmap examination revealed a remarkable interindividual variability in the bacterial communities' composition. Each patient harbored unique endodontic microbiome for abundance and richness of species. In fact, microbiota composition varied between individuals having the same disease which means that endodontic infections were caused by heterogeneous etiology. In other words, similar disease resulted from multiple communities. Regardless of this interindividual variance, according to the pairwise weighted UniFrac distances, samples tended to cluster based on 16S rRNA dendrogram.

Alpha diversity analysis indicated that the diversity is almost investigated as illustrated in the rarefaction curves but we must take into consideration the fact that using paper points and file in our sample as a collection method may not be the most favorable method to determine the whole microbiome in lateral canals, dentinal tubules, and isthmus. This suggests that the whole bacterial diversity in primary endodontic infections may be higher than actually identified.

Our results from sequencing analyses indicated that primary endodontic infections were dominated by both anaerobic and facultative anaerobic species. It was previously thought that primary endodontic infections were dominated by anaerobic species while facultative anaerobes were more predominant in secondary infections [42, 65]. Recently, next generation sequencing studies have demonstrated mixed results of anaerobic and facultative anaerobic species [66].

It is important to characterize the composition of the endodontic microflora because this may be related to the different clinical presentations or stages of development of an endodontic infection as well as its responses to different treatments [19]. Large numbers of OTUs were found at low abundance at our study. This ensures the power of Illumina analysis to detect the bacterial communities in primary endodontic infection [22]. However, it is difficult to determine the role of each bacterium in the community [18]. But it is clearly seen that even low abundant species play an important role in the endodontic oral microbiome [18]. Shifts in environmental conditions may result in the fact that members with low abundance may become dominant as a response to these changes [49]. Our study demonstrated that Illumina MiSeq platform sequencing technology allowed us to identify low abundant bacteria in infected root canals and also detect bacterial diversity which was not previously known in endodontic microbiota. As a result, more accurate estimation of species abundance and prevalence can be determined in the endodontic microbial community [12]. Endodontic microflora found in low proportions can even occupy critical niches within a complex microbial community, and so it is important to detect these members to maintain the stability and virulence of a microbial community [12]. The utilization of MiSeq Illumina sequencing can broaden our understanding of the pathogenesis of endodontic infections and has the power to improve treatment outcomes.

Next generation sequencing instruments are unable to differentiate between living and dead microorganisms and so all genetic material is measured [15]. This may be considered as an overestimate of bacterial count because even after cell death DNA can persist for a period of one year [67]. There is an argument that assessing living and dead microorganisms is essential because these microorganisms may have been dominant in the early stages of disease [47]. In addition, there has been an argument concerning the quality of taxonomic identification since that next generation sequencing used short sequencing reads [50]. But recently, NGS compared to early NGS are able to sequence different read lengths ranging from 50 bp up to 700 bp [66]. There are other factors which can affect the sequencing results such as sequencing errors, primer selection bias, chimera formation, interpretation of the huge data produced by bioinformatics programs, and PCR conditions [68, 69].

5. Conclusion

This study revealed that microbiota of endodontic infection with periapical lesions had high polymicrobial communities.

To fully understand the etiology of endodontic disease, a further and deeper host-microbiome analyses should be performed. Since that the results of our community largely varied from case to case, therefore this suggests that this disease is characterized by multispecies bacterial communities having a heterogeneous etiology.

Abbreviations

NGS:	Next generation sequencing
OTUs:	Operational taxonomic units
MWU test:	Mann–Whitney U test
KW rank sum test:	Kruskal-Wallis rank sum test
PCR:	Polymerase chain reaction
PCoA:	Principal coordinate analysis
NaOCl:	Sodium hypochlorite
UniFrac:	Unique fraction
rRNA:	Ribosomal RNA
DNA:	Deoxyribonucleic acid.

Conflicts of Interest

The authors declare that they have no conflicts of interest.

Authors' Contributions

Ali Abdellah Abdelrahman and Marwa Mohamed Azab proposed the original idea. Sally Ali Tawfik conducted the experiments, analyzed the data, and wrote the manuscript. Marwa Mohamed Azab did critical revision of the article. All authors read and approved the manuscript.

Acknowledgments

Funding for all aspects of this study including the collection of samples, analysis, interpretation of data, and writing and editing of manuscript was provided by the author Sally Ali Tawfik (personal funding). The authors are truly indebted and thankful to Mohammed Ramadan Gomma for his efforts, guidance, and unlimited support throughout the research.

References

[1] J. Peterson, S. Garges, M. Giovanni et al., "The NIH human microbiome project," *Genome Research*, vol. 19, no. 12, pp. 2317–2323, 2009.

[2] S. R. Gill, M. Pop, R. T. DeBoy et al., "Metagenomic analysis of the human distal gut microbiome," *Science*, vol. 312, no. 5778, pp. 1355–1359, 2006.

[3] I. Struzycka, "The oral microbiome in dental caries," *Polish Journal of Microbiology*, vol. 63, no. 2, pp. 127–135, 2014.

[4] B. Liu, L. L. Faller, N. Klitgord et al., "Deep sequencing of the oral microbiome reveals signatures of periodontal disease," *PLoS ONE*, vol. 7, no. 6, Article ID e37919, 2012.

[5] M. A. Munson, T. Pitt-Ford, B. Chong, A. Weightman, and W. G. Wade, "Molecular and cultural analysis of the microflora associated with endodontic infections," *Journal of Dental Research*, vol. 81, no. 11, pp. 761–766, 2002.

[6] J. F. Siqueira Jr. and I. N. Rôçac, "Diversity of endodontic microbiota revisited," *Journal of Dental Research*, vol. 88, no. 11, pp. 969–981, 2009.

[7] J. F. Siqueira Jr. and I. N. RôÇas, "Distinctive features of the microbiota associated with different forms of apical periodontitis," *Journal of Oral Microbiology*, vol. 1, 2009.

[8] N. Chugal, J.-K. Wang, R. Wang et al., "Molecular characterization of the microbial flora residing at the apical portion of infected root canals of human teeth," *Journal of Endodontics*, vol. 37, no. 10, pp. 1359–1364, 2011.

[9] B.-Y. Hong, T.-K. Lee, S.-M. Lim et al., "Microbial analysis in primary and persistent endodontic infections by using pyrosequencing," *Journal of Endodontics*, vol. 39, no. 9, pp. 1136–1140, 2013.

[10] J. F. Siqueira and I. N. Rôças, "Microbiology and treatment of acute apical abscesses," *Clinical Microbiology Reviews*, vol. 26, no. 2, pp. 255–273, 2013.

[11] I. N. Rôças, J. F. Siqueira Jr., M. C. R. Aboim, and A. S. Rosado, "Denaturing gradient gel electrophoresis analysis of bacterial communities associated with failed endodontic treatment," *Oral Surgery, Oral Medicine, Oral Pathology, Oral Radiology, and Endodontology*, vol. 98, no. 6, pp. 741–749, 2004.

[12] J. F. Siqueira Jr. and I. N. Rôças, "Community as the unit of pathogenicity: An emerging concept as to the microbial pathogenesis of apical periodontitis," *Oral Surgery, Oral Medicine, Oral Pathology, Oral Radiology, and Endodontology*, vol. 107, no. 6, pp. 870–878, 2009.

[13] P. J. Turnbaugh, R. E. Ley, M. Hamady, C. M. Fraser-Liggett, R. Knight, and J. I. Gordon, "The human microbiome project," *Nature*, vol. 449, no. 7164, pp. 804–810, 2007.

[14] W. W. L. Hsiao and C. M. Fraser-Liggett, "Human Microbiome Project-paving the way to a better understanding of ourselves and our microbes," *Drug Discovery Therapy*, vol. 14, no. 7-8, pp. 331–333, 2009.

[15] J. F. Siqueira Jr. and I. N. Rôças, "Uncultivated phylotypes and newly named species associated with primary and persistent endodontic infections," *Journal of Clinical Microbiology*, vol. 43, no. 7, pp. 3314–3319, 2005.

[16] J. F. Siqueira Jr., A. F. Fouad, and I. N. Rôças, "Pyrosequencing as a tool for better understanding of human microbiomes," *Journal of Oral Microbiology*, vol. 4, 2012.

[17] L. Li, W. W. L. Hsiao, R. Nandakumar et al., "Analyzing endodontic infections by deep coverage pyrosequencing," *Journal of Dental Research*, vol. 89, no. 9, pp. 980–984, 2010.

[18] M. L. Sogin, H. G. Morrison, J. A. Huber et al., "Microbial diversity in the deep sea and the underexplored 'rare biosphere'," *Proceedings of the National Acadamy of Sciences of the United States of America*, vol. 103, no. 32, pp. 12115–12120, 2006.

[19] K. H. Yun, H.-S. Lee, O. H. Nam, C. Y. Moon, J.-H. Lee, and S. C. Choi, "Analysis of bacterial community profiles of endodontically infected primary teeth using pyrosequencing," *International Journal of Paediatric Dentistry*, vol. 27, no. 1, pp. 56–65, 2017.

[20] H. E. H. M. Alenezi, *Microbiological Analysis of Root Canal Infections Using High Throughput Sequencing on the Illumina MiSeq Platform*, University of Leeds (School of Dentistry, 2015.

[21] M. Sakamoto, J. F. Siqueira Jr., I. N. Rôças, and Y. Benno, "Molecular analysis of the root canal microbiota associated with endodontic treatment failures," *Oral Microbiology and Immunology*, vol. 23, no. 4, pp. 275–281, 2008.

[22] V. Vengerfeldt, K. Špilka, M. Saag et al., "Highly diverse microbiota in dental root canals in cases of apical periodontitis (data of Illumina sequencing)," *Journal of Endodontics*, vol. 40, no. 11, pp. 1778–1783, 2014.

[23] G. N. Tzanetakis, M. A. Azcarate-Peril, S. Zachaki et al., "Comparison of bacterial community composition of primary and persistent endodontic infections using pyrosequencing," *Journal of Endodontics*, vol. 41, no. 8, pp. 1226–1233, 2015.

[24] M. Ramadan, S. Solyman, M. Taha, and A. Hanora, "Preliminary characterization of human skin microbiome in healthy Egyptian individuals," *Cellular and molecular biology (Noisy-le-Grand, France)*, vol. 62, no. 8, pp. 21–27, 2016.

[25] P. D. Schloss, S. L. Westcott, T. Ryabin et al., "Introducing mothur: open-source, platform-independent, community-supported software for describing and comparing microbial communities," *Applied and Environmental Microbiology*, vol. 75, no. 23, pp. 7537–7541, 2009.

[26] A. C. Anderson, A. Al-Ahmad, F. Elamin et al., "Comparison of the bacterial composition and structure in symptomatic and asymptomatic endodontic infections associated with root-filled teeth using pyrosequencing," *PLoS ONE*, vol. 8, no. 12, Article ID e84960, 2013.

[27] C. Quast, E. Pruesse, P. Yilmaz et al., "The SILVA ribosomal RNA gene database project: improved data processing and web-based tools," *Nucleic Acids Research*, vol. 41, no. 1, pp. D590–D596, 2013.

[28] R. C. Edgar, B. J. Haas, J. C. Clemente, C. Quince, and R. Knight, "UCHIME improves sensitivity and speed of chimera detection," *Bioinformatics*, vol. 27, no. 16, pp. 2194–2200, 2011.

[29] Q. Wang, G. M. Garrity, J. M. Tiedje, and J. R. Cole, "Naïve Bayesian classifier for rapid assignment of rRNA sequences into the new bacterial taxonomy," *Applied and Environmental Microbiology*, vol. 73, no. 16, pp. 5261–5267, 2007.

[30] C. E. Shannon, "A mathematical theory of communication," *Bell System Technical Journal*, vol. 27, no. 4, pp. 623–656, 1948.

[31] J. C. Yue and M. K. Clayton, "A similarity measure based on species proportions," *Communications in Statistics—Theory and Methods*, vol. 34, no. 11, pp. 2123–2131, 2005.

[32] J. Chen, K. Bittinger, E. S. Charlson et al., "Associating microbiome composition with environmental covariates using generalized UniFrac distances," *Bioinformatics*, vol. 28, no. 16, pp. 2106–2113, 2012.

[33] C. Lozupone and R. Knight, "UniFrac: a new phylogenetic method for comparing microbial communities," *Applied and Environmental Microbiology*, vol. 71, no. 12, pp. 8228–8235, 2005.

[34] Ø. Hammer, D. A. T. Harper, and P. D. Ryan, "PAST: paleontological statistics software package for education and data analysis," *Palaeontologia Electronica*, vol. 4, no. 1, pp. 1–9, 2001.

[35] J. Oksanen, F. Blanchet, and R. Kindt, "Vegan: community ecology package. 2013," *R Package Version*, pp. 2.0–7, 2013.

[36] G. R. Warnes, B. Bolker, L. Bonebakker et al., "gplots: various R programming tools for plotting data. R package version 2.17.0," Computer software, 2015, https://cran.r-project.org/web/packages/gplots/index.html.

[37] R. C. Team, *R: A Language and Environment for Statistical Computing*, R Foundation for Statistical Computing, Vienna, Austria, 2014.

[38] J. F. Siqueira Jr. and I. N. Rôças, "Exploiting molecular methods to explore endodontic infections: Part 2 - Redefining the endodontic microbiota," *Journal of Endodontics*, vol. 31, no. 7, pp. 488–498, 2005.

[39] J. F. Siqueira Jr. and I. N. Rôças, "Treponema species associated with abscesses of endodontic origin," *Oral microbiology and immunology*, vol. 19, no. 5, pp. 336–339, 2004.

[40] J. F. Siqueira Jr. and I. N. Rôças, "Exploiting molecular methods to explore endodontic infections: Part 1 - Current molecular technologies for microbiological diagnosis," *Journal of Endodontics*, vol. 31, no. 6, pp. 411–423, 2005.

[41] D. A. Spratt, "Significance of bacterial identification by molecular biology methods," *Endodontic Topics*, vol. 9, no. 1, pp. 5–14, 2004.

[42] J. C. Baumgartner, "Microbiological and molecular analysis of endodontic infections," *Endodontic Topics*, vol. 7, no. 1, pp. 35–51, 2004.

[43] F. R. F. Alves, J. F. Siqueira Jr., F. L. Carmo et al., "Bacterial community profiling of cryogenically ground samples from the apical and coronal root segments of teeth with apical periodontitis," *Journal of Endodontics*, vol. 35, no. 4, pp. 486–492, 2009.

[44] J. F. Siqueira Jr., F. R. F. Alves, and I. N. Rôças, "Pyrosequencing analysis of the apical root canal microbiota," *Journal of Endodontics*, vol. 37, no. 11, pp. 1499–1503, 2011.

[45] Y. Ping, J. Wang, and J. Liang, "Pyrosequencing analysis of apical microbiota of teeth with persistent apical periodontitis," *Journal of Dental Sciences*, vol. 10, no. 4, pp. 365–371, 2015.

[46] J. F. Siqueira, H. S. Antunes, I. N. Rôç, C. T. C. C. Rachid, and F. R. F. Alves, "Microbiome in the apical root canal system of teeth with post-treatment apical periodontitis," *PLoS ONE*, vol. 11, no. 9, Article ID e0162887, 2016.

[47] M. H. Saber, K. Schwarzberg, F. A. Alonaizan et al., "Bacterial flora of dental periradicular lesions analyzed by the 454-pyrosequencing technology," *Journal of Endodontics*, vol. 38, no. 11, pp. 1484–1488, 2012.

[48] E. T. Pinheiro, B. P. F. A. Gomes, C. C. R. Ferraz, E. L. R. Sousa, F. B. Teixeira, and F. J. Souza-Filho, "Microorganisms from canals of root-filled teeth with periapical lesions," *International Endodontic Journal*, vol. 36, no. 1, pp. 1–11, 2003.

[49] A. L. Santos, J. F. Siqueira Jr., I. N. Rôças, E. C. Jesus, A. S. Rosado, and J. M. Tiedje, "Comparing the bacterial diversity of acute and chronic dental root canal infections," *PLoS ONE*, vol. 6, no. 11, Article ID e28088, 2011.

[50] J. Kuczynski, C. L. Lauber, W. A. Walters et al., "Experimental and analytical tools for studying the human microbiome," *Nature Reviews Genetics*, vol. 13, no. 1, pp. 47–58, 2012.

[51] D. W. Fadrosh, B. Ma, P. Gajer et al., "An improved dual-indexing approach for multiplexed 16S rRNA gene sequencing on the Illumina MiSeq platform," *Microbiome*, vol. 2, no. 1, article no. 6, 2014.

[52] Illumina, *16S MetagenOmic Sequencing Library Preparation*, 2013.

[53] Z. Chen, M. R. Wilkins, N. Hunter, and M. A. Nadkarni, "Draft genome sequences of two clinical isolates of Lactobacillus rhamnosus from initial stages of dental pulp infection," *Genome Announcements*, vol. 1, no. 1, 2013.

[54] M. A. Nadkarni, M. R. Simonian, D. W. S. Harty, H. Zoellner, N. A. Jacques, and N. Hunter, "Lactobacilli are prominent in the initial stages of polymicrobial infection of dental pulp," *Journal of Clinical Microbiology*, vol. 48, no. 5, pp. 1732–1740, 2010.

[55] B. P. F. A. Gomes, R. C. Jacinto, E. T. Pinheiro et al., "Molecular analysis of filifactor alocis, tannerella forsythia, and treponema denticola associated with primary endodontic infections and failed endodontic treatment," *Journal of Endodontics*, vol. 32, no. 10, pp. 937–940, 2006.

[56] H. Cao, Z. Qi, H. Jiang, J. Zhao, Z. Liu, and Z. Tang, "Detection of Porphyromonas endodontalis, Porphyromonas gingivalis and Prevotella intermedia in primary endodontic infections in a Chinese population," *International Endodontic Journal*, vol. 45, no. 8, pp. 773–781, 2012.

[57] B. P. Gomes, D. B. Drucker, and J. D. Lilley, "Associations of specific bacteria with some endodontic signs and symptoms," *International Endodontic Journal*, vol. 27, no. 6, pp. 291–298, 1994.

[58] B. P. F. A. Gomes, J. D. Lilley, and D. B. Drucker, "Clinical significance of dental root canal microflora," *Journal of Dentistry*, vol. 24, no. 1-2, pp. 47–55, 1996.

[59] L. J. Peterson, "Contemporary management of deep infections of the neck," *Journal of Oral and Maxillofacial Surgery*, vol. 51, no. 3, pp. 226–231, 1993.

[60] L. M. Sassone, R. A. Fidel, M. Faveri et al., "A microbiological profile of symptomatic teeth with primary endodontic infections," *Journal of Endodontics*, vol. 34, no. 5, pp. 541–545, 2008.

[61] L. F. Tomazinho and M. J. Avila-Campos, "Detection of Porphyromonas gingivalis, Porphyromonas endodontalis, Prevotella intermedia, and Prevotella nigrescens in chronic endodontic infection," *Oral Surgery, Oral Medicine, Oral Pathology, Oral Radiology, and Endodontology*, vol. 103, no. 2, pp. 285–288, 2007.

[62] I.-Y. Jung, B.-K. Choi, K.-Y. Kum et al., "Identification of oral spirochetes at the species level and their association with other bacteria in endodontic infections," *Oral Surgery, Oral Medicine, Oral Pathology, Oral Radiology, and Endodontology*, vol. 92, no. 3, pp. 329–334, 2001.

[63] S. E. Gharbia, M. Haapasalo, H. N. Shah et al., "Characterization of Prevotella intermedia and Prevotella nigrescens isolates from periodontic and endodontic infections.," *Journal of Periodontology*, vol. 65, no. 1, pp. 56–61, 1994.

[64] K.-S. Bae, J. C. Baumgartner, T. R. Shearer, and L. L. David, "Occurrence of prevotella nigrescens and prevotella intermedia in infections of endodontic origin," *Journal of Endodontics*, vol. 23, no. 10, pp. 620–623, 1997.

[65] D. Figdor and G. Sundqvist, "A big role for the very small—understanding the endodontic microbial flora," *Australian Dental Journal*, vol. 52, no. 1, pp. S38–S51, 2007.

[66] J. M. Di Bella, Y. Bao, G. B. Gloor, J. P. Burton, and G. Reid, "High throughput sequencing methods and analysis for microbiome research," *Journal of Microbiological Methods*, vol. 95, no. 3, pp. 401–414, 2013.

[67] G. Young, S. Turner, J. K. Davies, G. Sundqvist, and D. Figdor, "Bacterial DNA persists for extended periods after cell death," *Journal of Endodontics*, vol. 33, no. 12, pp. 1417–1420, 2007.

[68] J. Reeder and R. Knight, "The 'rare biosphere': a reality check," *Nature Methods*, vol. 6, no. 9, pp. 636-637, 2009.

[69] C. Quince, A. Lanzén, T. P. Curtis et al., "Accurate determination of microbial diversity from 454 pyrosequencing data," *Nature Methods*, vol. 6, no. 9, pp. 639–641, 2009.

5

Predicted Cold Shock Proteins from the Extremophilic Bacterium *Deinococcus maricopensis* and Related *Deinococcus* Species

Michael J. LaGier

Department of Biology, Grand View University, Des Moines, IA, USA

Correspondence should be addressed to Michael J. LaGier; mlagier@grandview.edu

Academic Editor: Todd R. Callaway

While many studies have examined the mechanisms by which extremophilic Deinococci survive exposure to ionizing radiation, very few publications have characterized the cold shock adaptations of this group, despite many species being found in persistent cold environments and environments prone to significant daily temperature fluctuations. Bacterial cold shock proteins (Csps) are a family of conserved, RNA chaperone proteins that commonly play a role in cold temperature adaptation, including a downward shift in temperature (i.e., cold shock). The primary aim of this study was to test whether a representative, desert-dwelling *Deinococcus*, *Deinococcus maricopensis*, encodes Csps as part of its genome. Bioinformatic approaches were used to identify a Csp from *D. maricopensis* LB-34. The Csp, termed Dm-Csp1, contains sequence features of Csps including a conserved cold shock domain and nucleic acid binding motifs. A tertiary model of Dm-Csp1 revealed an anticipated Csp structure containing five anti-parallel beta-strands, and ligand prediction experiments identified N-terminally located residues capable of binding single-stranded nucleic acids. Putative Csps were identified from 100% of (27 of 27) Deinococci species for which genome information is available; and the Deinococci-encoded Csps identified contain a C-terminally located region that appears to be limited to members of the class Deinococci.

1. Introduction

The Genus *Deinococcus* currently contains 59 recognized species (http://www.bacterio.net), including 27 species for which genomic information is available (https://www.ncbi.nlm.nih.gov/genome/browse/). Members of *Deinococcus* demonstrate a capacity to resist environmental stress including exposure to significant levels of ionizing radiation and desiccation [1]. Species of Deinococci have been isolated from a wide range of environments including air, desert soil, fresh and marine waters, alpine environments, Antarctica, hot springs, and radioactive sites [2]. One species, *Deinococcus maricopensis*, was originally isolated from an arid soil sample collected from the Sonoran Desert in Arizona (USA), and the genome of the type strain of this species (LB-34) has been recently sequenced as part of the Genomic Encyclopedia of Bacteria and Archaea project [3, 4]. *D. maricopensis* LB-34 cells are mesophilic, Gram-positive, rod-shaped, and nonmotile, with an optimal growth temperature of 40°C and an observed growth range of 10°C to 45°C [3]. The relatively wide growth range of *D. maricopensis* correlates with observed diurnal swings of 15°C or more during a typical day within the Sonoran Desert (https://science.nature.nps.gov/im/units/sodn/sonoran.cfm). Hence, *D. maricopensis* has adapted to an environment prone to rapid temperature changes, including a shift to colder temperatures.

Although many studies have examined the mechanisms by which Deinococci survive exposure to significant levels of ionizing radiation, few studies have looked at the cold shock adaptations of this group, despite many species being found in persistent cold environments (e.g., Antarctica, cold water oceans) and environments prone to significant daily temperature fluctuations such as deserts [5–7]. Using the model species of the group, *Deinococcus radiodurans,* data from a previous proteomic study showed changes in global protein expression patterns following a cold shock [8]. Although this study was useful in demonstrating that gene expression changes do occur in response to cold shock, studies are lacking that examine additional Deinococci genomes for putative cold shock proteins.

Bacterial cold shock proteins (Csps) represent a family of small proteins (smaller than 100 amino acids in length) with common structural features and a capacity to bind to single-stranded nucleic acids by way of conserved, N-terminally located nucleic acid-binding motifs [9]. Csp expression levels are typically increased following an ambient temperature decrease, but not always [9, 10]. Within the Deinococci species, the best characterized protein that shares significant sequence homology to known Csps is from *D. radiodurans* [11]. The protein, termed PprM, appears to play a role in the radiation and oxidative stress resistance phenotypes of *D. radiodurans* [11]. Interestingly, while one study showed PprM to be induced [8] after heat shock (and apparently not induced after cold shock), a second study failed to demonstrate induction of PprM after heat shock [12]. Hence, our understanding of the potential role of Csps within the Deinococci species as a group is limited.

Outside the Deinococci species, the best studied Csps come from the mesophile *Escherichia coli,* although putative Csps have been identified from psychrophiles, mesophiles, and thermophiles [13]. In *E. coli,* nine members of the Csp family have been identified: CspA, B, C, D, E, F, G, H, and I. Although a significant number of the *E. coli* Csps are induced by cold (A, B, G, and I), E and C are constitutively expressed at physiological temperature, and D is induced via nutrient stress [9]. Of the nine, CspA is best characterized and appears to play a role in bacterial cold shock by acting as an RNA chaperone, destabilizing secondary structures within mRNAs that are more likely to form as extracellular temperatures decrease, and thus can reduce transcriptional and translational efficiency [9, 13, 14].

The current study identifies and characterizes, using bioinformatics, a potential Csp homolog from *D. maricopensis,* a desert-dwelling species that is naturally exposed to significant temperature fluctuations including daily temperature downshifts [3]. The *D. maricopensis* Csp, termed Dm-Csp1, contains expected nucleic acid binding motifs, shares significant sequence identity and similarity to known Csp proteins, contains five conserved anti-parallel beta-strands according to structural modeling, and is predicted to have a capacity to bind single-stranded nucleic acids. In addition, the presence of Csps was identified in all Deinococci genomes queried (27 of 27). Interestingly, the Csps identified all contain a C-terminal region that is unique among members of the class deinococci; and for which a potential function is unknown.

2. Materials and Methods

2.1. Sequence Retrieval. Protein data available for *Deinococcus maricopensis* LB-34 at the National Center for Biotechnology Information (NCBI) was queried via a text search using "cold shock" as the input. An 86 amino acid long protein, NCBI Accession #WP 013556416.1 was then selected for further study by storing the protein sequence locally as a FASTA file. This protein was termed Dm-Csp1.

2.2. Analysis of Primary Features of Dm-Csp1. A tool within the ExPASy site, ProtParam (http://web.expasy.org/protparam/), was used to predict the basic physiological and chemical features of Dm-Csp1. For comparison, a known Csp protein (CspA) sequence from *E. coli* K12 (NCBI Accession #P0A9X9) was subjected to ProtParam analysis in parallel. Default settings were used for ProtParam analyses. ProtParam predicts properties of input protein sequences including molecular weight, isoelectric pH, aliphatic index, and extinction coefficients.

2.3. Sequence Similarity Analyses of Dm-Csp1. The FASTA protein sequence of Dm-Csp1 was used as a query in standard BLASTp searches against the nonredundant database (nr) as well as the PDB protein database, which includes only sequences for which a solved protein structure is available (http://www.rcsb.org/pdb/home/home.do). To identify potential Csp homologs in the Deinococci as a group, a standard BLASTp search was run (using Dm-Csp1 as the query) for which the nr database was limited to sequences from members of the Genus *Deinococcus* (27 species total, including *D. maricopensis*) containing permanent draft or finished genomic data available at the Integrated Microbial Genomes and Microbiomes (IMG/M) site (https://img.jgi.doe.gov/). A combination of Pfam and ScanProsite was used to identify conserved protein domains within the sequence of Dm-Csp1, under default settings [15, 16]. To determine the number and type of Csps in *D. maricopensis,* CspA, B, C, D, E, F, G, H, and I protein sequences from *E. coli* K12 were used via BLASTp to query the genome of *D. maricopensis* LB-34. Unless stated otherwise, BLASTp was run according to default parameters.

2.4. Multiple Sequence Alignment. BLASTp, with Dm-Csp1 as a query, was used to identify closely related proteins from within the Genus Deinococcus. Dm-Csp1, along with 10 putative Csp proteins from 10 representative Deinococci, was retrieved as FASTA files from NCBI and aligned using standard parameters of the BioEdit Sequence Alignment Editor (http://www.mbio.ncsu.edu/bioedit/bioedit.html). Amino acids 100% conserved (Figure 1) or 50% conserved (Figure 4) among the 11 aligned proteins were highlighted in blue, and functional motifs were identified by visual inspection of the output alignment.

2.5. Protein to Protein Interactions. The STRING tool was used to identify proteins that Dm-Csp1 may interact with *in vivo.* The STRING database contains information from experimental data and public text collections to predict protein-protein interactions [17]. The basic interaction unit in STRING is the functional association, likely contributing to a common biological purpose. Predicted protein-protein interactions are derived from multiple sources

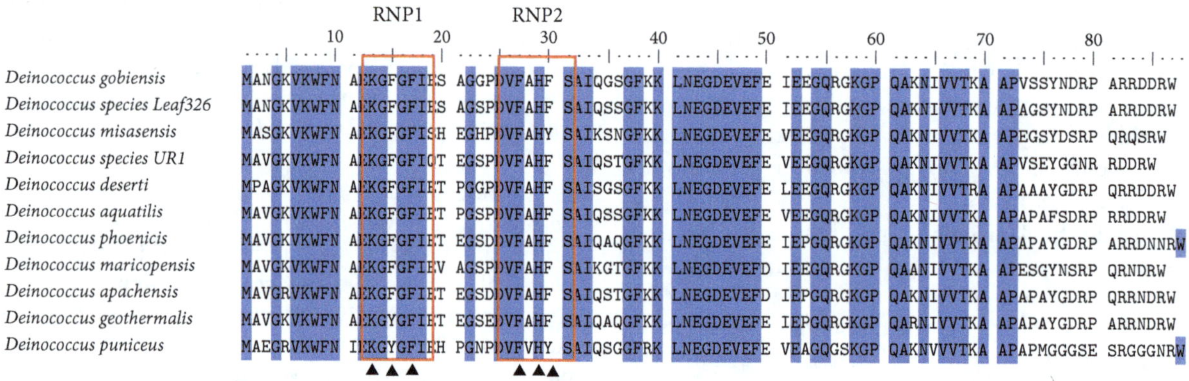

FIGURE 1: *Multiple sequence alignment of Dm-Csp1 and related, representative Deinococci-encoded cold shock proteins.* The species names are indicated. Areas of 100% conservation are highlighted in blue. The position of conserved RNA-binding motifs (RNP1 and RNP2) is enclosed within red rectangles. Conserved, basic, and aromatic residues within RNP1 and RNP2 that form a nucleic acid binding surface are indicated by arrowheads.

including known experimental interactions (coexpression, experiments), genome functional pathway knowledge from manually curated databases (databases), automated text-mining searches (text-mining), shared gene neighborhood locations across multiple genomes (neighborhood), and interactions observed in one organism and systematically transferred to other organisms, via precomputed orthology relations (cooccurrence and gene fusion). Default parameters of STRING were used.

2.6. Dm-Csp1 Structure Prediction and Model Quality Assessment. The secondary structure of Dm-Csp1 was predicted by the PSIPRED server (http://bioinf.cs.ucl.ac.uk/psipred/), under default settings. The tertiary structure of Dm-Csp1 was modeled using the $(PS)^2$-v2 server [18]. In $(PS)^2$-v2, autodetect functionality was employed to select a suitable template structure to model against. The output model of Dm-Csp1 was visualized by Chimera software [19]. The quality of the predicted structure was assessed by PROCHECK [20], QMEAN6 [21], and verify3D [22]. PROCHECK and QMEAN6 are programs within the SWISS-MODEL workspace (https://swissmodel.expasy.org), and verify3D is found at the UCLA-DOE Structure Evaluation Server (http://services.mbi.ucla.edu/Verify_3D/).

2.7. Dm-Csp1 Ligand Binding Site Prediction. The model file generated from $(PS)^2$-v2 was used as an input into COACH [23] to predict potential ligand binding regions within Dm-Csp1. COACH is a metaserver approach to protein ligand binding site prediction. COACH generates predictions using TM-SITE and S-SITE, which recognize ligand binding templates from the BioLiP protein function database by binding specific substructure and sequence profile comparisons. These predictions are combined with predictions from COFACTOR, FINDSITE, and ConCavity to generate final ligand binding site predictions [23].

3. Results and Discussion

3.1. Identification of Csps from D. maricopensis and Related Deinococci. Initial text queries of *D. maricopensis* protein sequences available at NCBI identified a potential protein of 86 amino acids in length denoted as a cold shock protein (Csp). The protein, designated in this study as Dm-Csp1 (NCBI Accession #WP 013556416.1), was then used as a query sequence in additional BLASTp searches. Specifically, BLASTp analysis against the nonredundant (nr) and PDB protein database showed that Dm-Csp1 shares significant sequence identity and similarity to bacterial proteins annotated as Csps, including those for which a protein structure has been solved (Table 1). In addition, predicted physiological and chemical features of Dm-Csp1 were similar to those obtained when compared to a known Csp from the model bacterium *E. coli*. As shown in Table 2, Dm-Csp1 is predicted, according to ProtParam, to have a molecular mass of 7.4 Kilodaltons and an isoelectric pH of 5.58.

To better understand the conservation of Csps among the Genus *Deinococcus*, the sequence of Dm-Csp1 was used as a BLASTp query to scan the genomes of related Deinococci species, for which genomic sequence information is available (27 total species, including *D. maricopensis*). As summarized in Table 3, at least one protein sharing significant sequence identity and similarity (as well as significant *E*-values), was identified in 100% of the Deinococci species queried. Although the number of Csps per bacterial genome can vary according to species, the reasoning behind such variation is unknown [10, 13]. The absolute conservation of Csps among the Deinococci sampled implies that Csps may be a significant feature of this groups' capacity in adapting to cold shock conditions and/or adapting or responding to other stress conditions.

While *E. coli* has been shown to contain a total of 9 Csps per genome, the *D. maricopensis* LB-34 genome appears to contain 2 Csps, according to BLAST testing (Section 2). Specifically, in addition to Dm-Csp1, the LB-34 genome

TABLE 1: BLASTp analysis of Dm-Cspl (NCBI Accession #WP 013556416.1).

Database identifier[a]	BLASTp hit species	Name of hit	GenBank hit number	E-value	Identity (%)	Similarity (%)	Hit query coverage (%)
nr	*Deinococcus maricopensis*	Cold shock protein	WP 013556476.1[b]	$5.00E-56$	95	96	100
nr	*Deinococcus peraridilitoris*	Cold shock protein	WP 015234625.1	$2.00E-48$	84	90	100
nr	*Deinococcus aqualitis*	Cold shock protein	WP 019008855.1	$7.00E-48$	80	91	100
nr	*Deinococcus misasensis*	Cold shock protein	WP 034335902.1	$1.00E-47$	80	88	100
nr	*Deinococcus pimensis*	Cold shock protein	WP 027481785.1	$2.00E-47$	83	87	100
PDB	*Bacillus caldolyticus*	Cold shock protein	115F_A[c]	$6.00E-24$	64	75	76
PDB	*Bacillus caldolyticus*	Cold shock protein	1C90_A	$6.00E-24$	66	75	76
PDB	*Bacillus subtilis*	Cold shock protein	2ES2_A	$6.00E-24$	64	73	82
PBD	*Bacillus caldolyticus*	Cold shock protein	1HZB_A	$6.00E-24$	64	75	76
PDB	*Bacillus subtilis*	Cold shock protein	2I5L_X	$8.00E-24$	66	73	74

[a]Nonredundant database (nr) or protein database (PDB), acquired via independent BLASTp searches using Dm-Cspl as a query. [b]Referred to as Dm-Csp2 within text. [c]Protein database identifier code.

TABLE 2: Physiological and chemical properties of Dm-Cspl and *E. coli* K12 CspA[a].

Protein identifier	Amino acid length	Molecular weight	Isoelectric point	Aliphatic index	Estimated half-life	Positively charged residues (%)	Negatively charged residues (%)
Dm-Cspl	86	9472.54	5.79	57.91	>10 hours	13.95	15.12
E. coli CspA	70	7403.28	5.58	61.29	>10 hours	10	11.42

[a]Calculated using default settings of ProtParam (http://web.expasy.org/protparam/).

appears to contain a second Csp, termed in this study as Dm-Csp2 (NCBI Accession #WP 013556476.1), also 86 amino acids in total length. In comparison to *E. coli* Csps (A–I), Dm-Cspl shares the greatest homology with *E. coli* CspA (BLASTp E-value of $9.0E-23$) and Dm-Csp2 shares greatest homology with CspB (BLASTp E-value of $2.0E-21$). Dm-Cspl and Dm-Csp2 share 95% sequence identity and 96% sequence similarity. Dm-Cspl and Dm-Csp2 differ in a total of 4 amino acids, none of which are found within key, conserved regions of known Csps (data not shown). Hence, Dm-Cspl and Dm-Csp2 may serve similar functions in *D. maricopensis*. Within a single genome, it is anticipated that bacterial Csps share significant levels of sequence identity and similarity, as is the case with Csps from *E. coli* and *Yersinia pseudotuberculosis* [9]. In *E. coli* and *Y. pseudotuberculosis*, having more than one Csp sharing significant sequence homology is thought to represent functional redundancy, as mutagenesis of one or more Csps can be functionally compensated by remaining, nonmutated homologs [24, 25].

In addition to sequence homology testing, the primary sequence of Dm-Cspl was compared to a sample of Deinococci-encoded Csps by multiple sequence alignments and was also used as a query to identify conserved protein domains. Pfam searching identified a conserved cold shock DNA-binding domain (Clan #CL0021) within the N-terminal region of Dm-Cspl with a significant E-value score of $2.9E-25$. Consistent with Pfam, ScanProsite results indicated an N-terminally located cold shock domain signature (#PS00352) from amino acid position 15 to 33 with good confidence. A multiple sequence of Dm-Cspl with 10 related Deinococci-encoded Csps revealed the presence of an additional feature of Csps, two nonspecific RNA-binding sequence motifs [9], ribonucleotide motifs 1 and 2 (Figure 1, RNP1 and RNP2). In bacterial Csps, the canonical sequence for RNP1 is (K/S-G-F/K/Y-G-F/L-I) and for RNP2 is (L/I/V-F/Q-V/A/L-H-X-S/T/R). In the aligned representative Deinococci (Figure 1) Csps, the sequences of RNP1 and RNP2 are very similar to expected, with RNP1 being

TABLE 3: BLASTp analysis of Dm-Csp1 (NCBI Accession #WP 013556416.1) versus deinococci group.

BLASTp hit species	Name of hit	Gene ID number[a]	*E*-value	Identity (%)	Similarity (%)
Deinococcus maricopensis	Cold shock protein	649901717	5.00E − 56	95	96
D. reticulitermitis	Cold shock protein	2617968483	6.00E − 45	76	87
D. peraridilitoris	Cold shock protein	2509591723	2.00E − 48	84	91
D. deserti	Cold shock protein	643799117	7.00E − 45	78	85
D. xibeiensis	Cold shock protein	2586308013	5.00E − 45	77	86
D. marmoris	Cold shock protein	2562219581	1.00E − 33	72	85
D. wulumuqiensis	Cold shock protein	2553983725	5.00E − 45	77	86
D. species 2009	Cold shock protein	2516938626	1.00E − 44	78	87
D. pimensis	Cold shock protein	2509082093	1.00E − 47	83	87
D. misasensis	Cold shock protein	2571066521	6.00E − 47	79	87
D. phoenicis	Cold shock protein	2579632034	1.00E − 40	78	84
D. geothermalis	Cold shock protein	641229653	2.00E − 45	80	87
D. apachensis	Cold shock protein	2519136834	3.00E − 47	84	87
D. gobiensis	Cold shock protein	25128201432	3.00E − 47	81	88
D. proteolyticus	Cold shock protein	649963862	5.00E − 33	62	77
D. hopiensis	Cold shock protein	2509586483	2.00E − 46	83	86
D. grandis	Cold shock protein	2691382520	1.00E − 45	80	90
D. species RL	Cold shock protein	2611619696	3.00E − 47	81	88
D. radiodurans	Cold shock protein	2558116098	5.00E − 45	77	86
D. murrayi	Cold shock protein	2528309711	3.00E − 47	81	88
D. ficus	Cold shock protein	2526061465	1.00E − 33	78	87
D. fringens	Cold shock protein	2562223450	1.00E − 33	72	85
D. humi	Cold shock protein	2597538219	2.00E − 47	79	92
D. soli	Cold shock protein	2637699313	1.00E − 45	80	90
D. species Leaf 326	Cold shock protein	2645444050	2.00E − 46	82	90
D. aquatilis	Cold shock protein	2520600036	7.00E − 48	80	92
D. species YIM 77859	Cold shock protein	2582523845	8.00E − 47	82	91

[a]According to the Integrated Microbial Genomes database (https://img.jgi.doe.gov/).

(K-G-F/Y-G-F-I-X-X) and RNP2 being (V-F-V/A-H-F/Y-S). Within the known tertiary structures of Csps, the basic and aromatic residues within RNP1 and RNP2, collectively, form a nucleic acid-binding surface (Figure 1).

3.2. Structural Analyses of Dm-Csp1. To better understand the potential biological function of Dm-Csp1, a structural model of Dm-Csp1 was constructed. The secondary structure of Dm-Csp1 was predicted using the PSIPRED server. The result of PSIPRED, as shown in Figure 2, has good confidence and contains 5 beta-strands, which are expected among Csps [9, 26].

The modeling of the likely tertiary structure of Dm-Csp1, as predicted by $(PS)^2$-v2, used a known Csp from *Bacillus caldolyticus* (PDB #1C9O) as a template. The *B. caldolyticus* Csp crystal structure used as a template is 66 amino acids in length and shares 61.76% sequence identity with Dm-Csp1 [27]. The model of Dm-Csp1 was found to contain five antiparallel beta-strands that form a barrel structure (Figure 3). These features are known characteristics of Csps. The quality of the output model structure was assessed using a combination of PROCHECK, QMEAN6, and verify3D (Section 2). PROCHECK reveals that all amino acids residues within the model (68 of 68) are within the

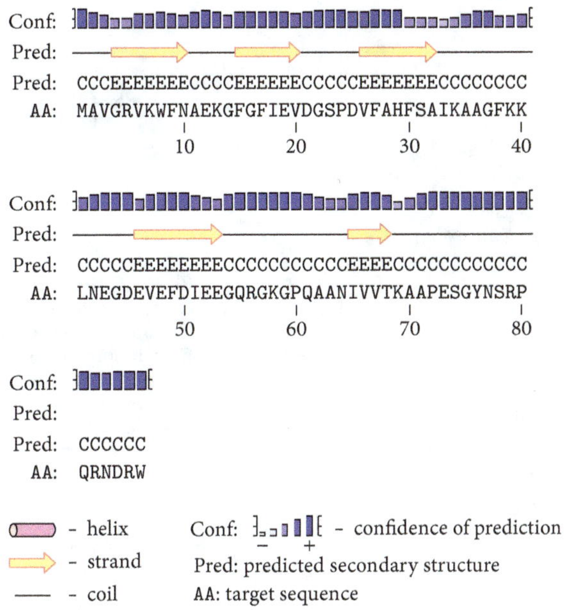

FIGURE 2: *Secondary protein structure of Dm-Csp1 as predicted by PSIPRED.* The relative height of blue bars indicates confidence of prediction.

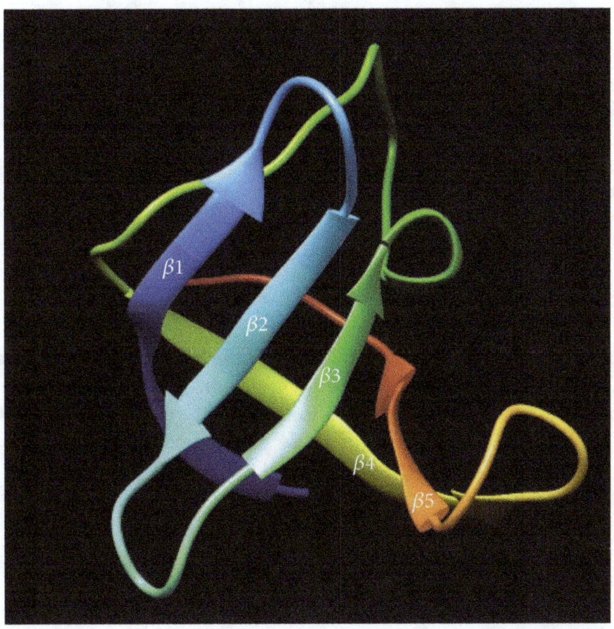

FIGURE 3: *Modeled tertiary structure of Dm-Csp1.* The five anti-parallel strands that collectively form a barrel structure are indicated.

limits of a Ramachandran plot, with 96.4% of residues in the most favored regions and only a single residue (1.8%) in a disallowed region (Figure 4(a)). A good model is expectedvspace.5pt to have over 90% of residues in most favored regions [20]. QMEAN6, which estimates the global quality of structural models on the basis of 6 structural descriptors, places the model of Dm-Csp1 in the dark region of the estimated absolute model quality graph with a significant QMEAN6 score of 0.781 and a significant Z-score of 0.413 (Figure 4(b)). Using verify3D, the environmental profile of the model was acceptable (Figure 4(c)), yielding a high score of 0.57 [22].

The C-terminal region of Dm-Csp1 could not be modeled, specifically from amino acids 69 to 86. The reason

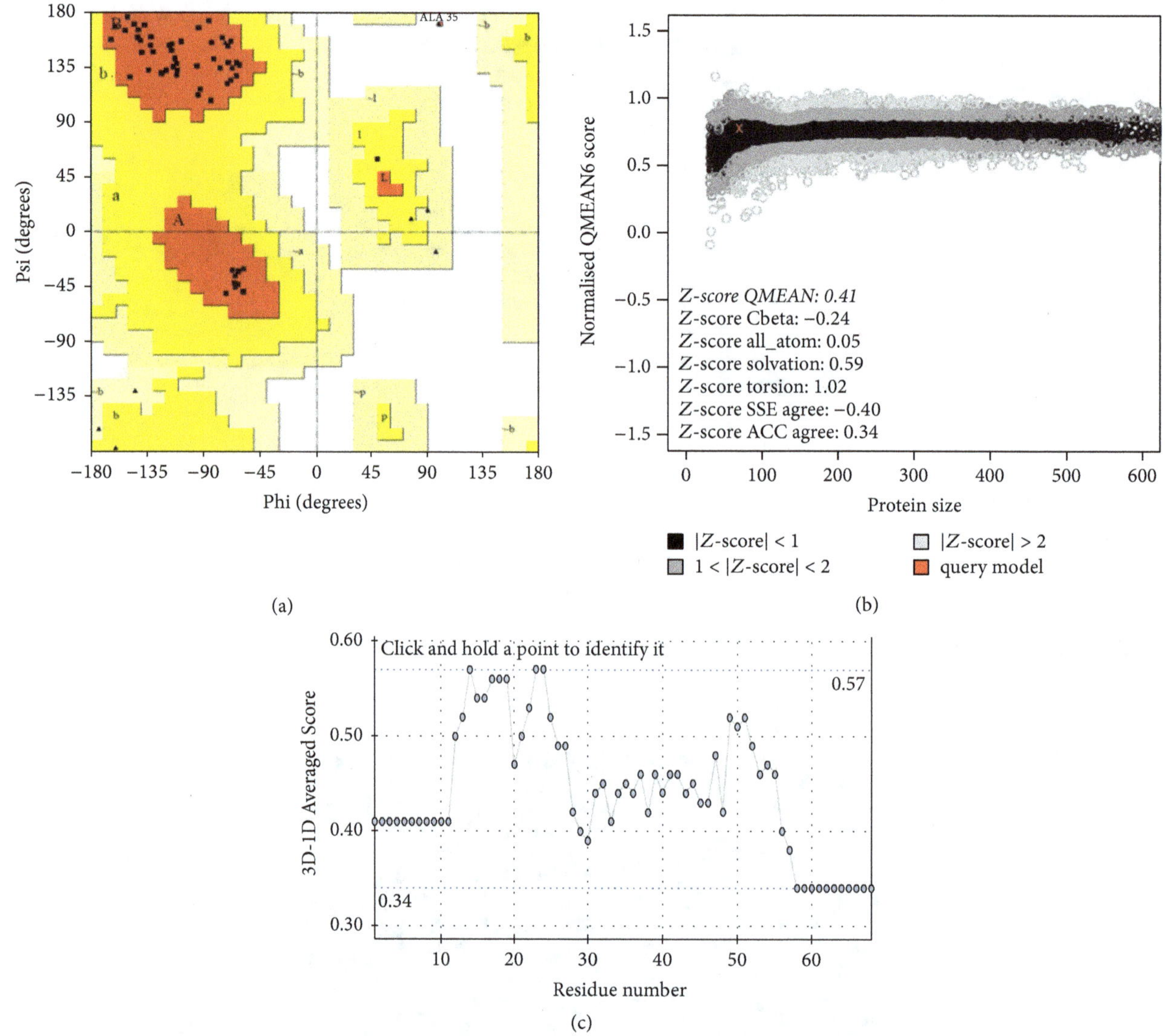

FIGURE 4: *Tertiary structural model assessment.* (a) Ramachandran plot of protein model generated by PROCHECK. The position of 96.4% amino acids within the generated structural model are within most favored regions. The position of 1 amino acid (ALA-35, red text) is found within a disallowed region. (b) Graphical output of estimation of absolute quality of model via QMEAN6. The position of the Dm-Csp1 tertiary model is indicated by a red X. (c) Verify3D output data for Dm-Csp1 predicted tertiary structure, yielding an acceptably high score of 0.57.

for this inability to model is due to the overall length of Dm-Csp1 being 86 amino acids. Bacterial Csps identified thus far are approximately 70 amino acids in length, with the N-terminal region containing conserved nucleic acid binding regions [9, 13]. The extended C-terminal region of Dm-Csp1, according to BLASTp analysis, appears to be unique to the class Deinococci. For example, in addition to species of Deinococci, some members (for which genomic information is available) of the class Deinococci encode at least one putative Csp with the C-terminal extension including *Truepera radiovictrix* (data not shown, NCBI Accession #WP_013178602). As shown in Figure 5, all the Deinococci Csps included in the Figure 1 alignment contain an extended C-terminal region that appears to be rich in positively charged arginine residues and end in the combination arginine-tryptophan. The potential function of this conserved region within Deinococci Csps is unknown but warrants further investigation.

3.3. Biological Function Prediction. The synthesized, and quality assessed, structural model of Dm-Csp1 was used as a template to predict the potential of the protein to bind ligands. Using the COACH server, the N-terminal region of Dm-Csp1 appears able to bind single-stranded nucleic acids, with a significant support score (C-score) of 0.2 [23]. A total of 8 residues are predicted to bind nucleic acids, 4 of which are located with the previously identified within

Species	10	
Deinococcus gobiensis	APVSSYNDRP	ARRDD-RW
Deinococcus species Leaf326	APAGSYNDRP	ARRDD-RW
Deinococcus misasensis	APEGSYDSRP	-QRQS-RW
Deinococcus species UR1	APVSEYGGN-	-RRDD-RW
Deinococcus deserti	APAAAYGDRP	QRRDD-RW
Deinococcus aquatilis	APAPAFSDRP	-RRDD-RW
Deinococcus phoenicis	APAPAYGDRP	ARRDNNRW
Deinococcus maricopensis	APESGYNSRP	-QRND-RW
Deinococcus apachensis	APAPAYGDRP	QRRND-RW
Deinococcus geothermalis	APAPAYGDRP	ARRND-RW
Deinococcus puniceus	APAPMGGGSE	SRGGGNRW
	▲	▲▲ ▲

Figure 5: *Multiple sequence alignment of C-terminal region of Dm-Csp1 and related, representative Deinococci-encoded cold shock proteins.* Species names are indicated, regions of 50% or greater conservation are marked blue, and the positions of arginine-rich areas are denoted by arrowheads.

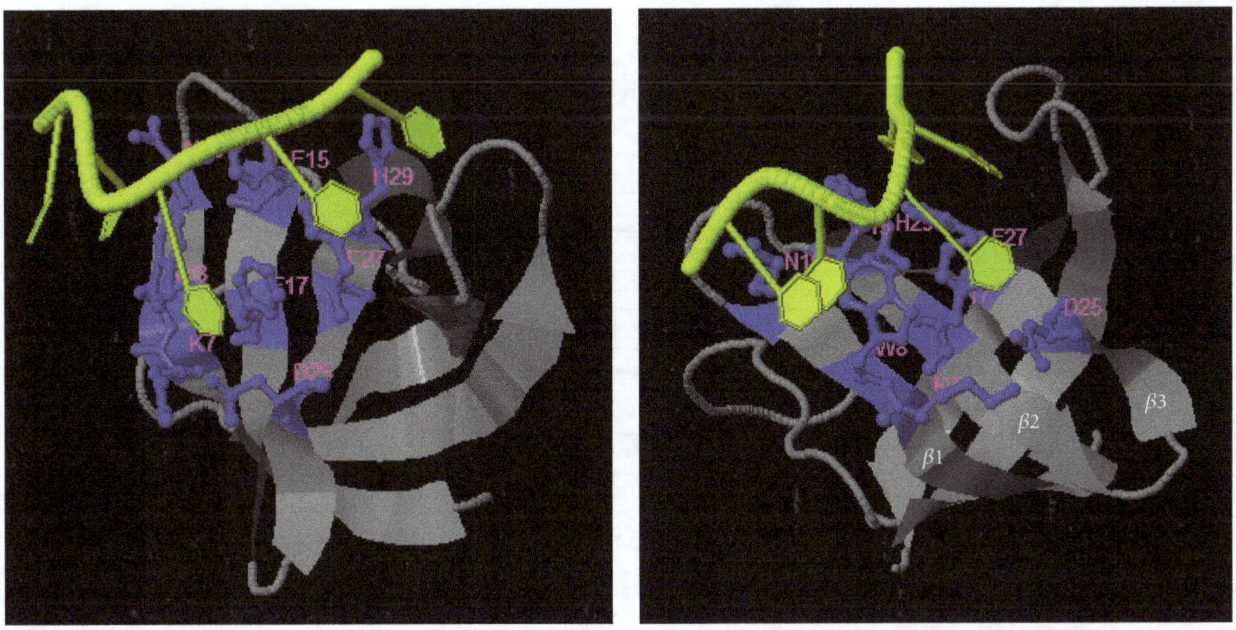

(a) (b)

Figure 6: *Functional site prediction of Dm-Csp1.* COACH output identifying N-terminally located residues predicted to bind single-stranded nucleic acids ((a), nucleic acid colored green, binding residues colored blue), and the same output repositioned to show the predicted nucleic acid binding residues is found within three anti-parallel beta-strands that collectively form a nucleic acid binding surface (b).

the RNP1 and RNP2 motifs including F15, F17, F27, and H29 (Figure 6(a)). The binding of nucleic acids using a combination of three sequential beta-strands (Figure 6(b)) is consistent with previously solved structures of Csps [9, 13, 14, 26, 28]. According to STRING analysis, Dm-Csp1 is predicted to functionally interact with a variety of proteins including those (rpoB, rpoC, and rpoA, and rpoZ, RNA polymerase subunits) involved in transcription and a polyribonucleotide nucleotidyltransferase (pnp) potentially playing a role in mRNA degradation (Figure 7). The predicted interactions of Dm-Csp1 with transcription-related proteins make sense when considering previous studies demonstrating the role of Csps in modulating transcription, including acting as a RNA chaperone [9, 13, 29, 30].

4. Conclusions

The findings here suggest that one adaptation by which *D. maricopensis* deals with significant, diurnal temperature shifts is by utilizing genome-encoded cold shock proteins (Csps). Significantly, all additional members of the Genus Deinococcus, for which genomic information is available, also encode Csps. Interestingly, the identified Deinococci Csps contain a C-terminal extension not observed in bacteria

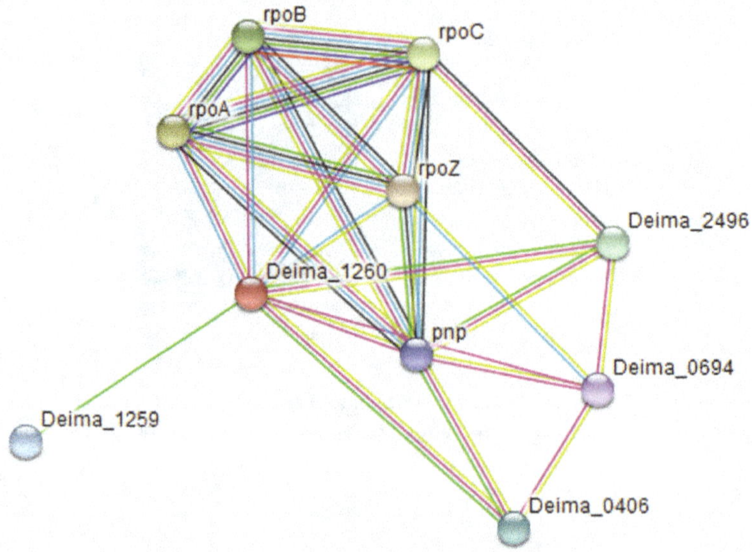

(a)

Your Input:

●	Deima_1260	*cold-shock DNA-binding domain-containing protein (86 aa)*

Predicted Functional Partners:

			Neighborhood	*Gene fusion*	*Cooccurence*	*Coexpression*	*Experiments*	*Databases*	*Text-mining*	*[Homology]*	*Score*
●	rpoZ	*DNA-directed RNA polymerase subunit omega; Promotes RNA polymerase assembly. Latches the N- and C- terminal regio...*						●	●		0.770
●	rpoA	*DNA-directed RNA polymerase subunit alpha; DNA-dependent RNA polymerase catalyzes the transcription of DNA into RN...*					●	●	●		0.723
●	rpoC	*DNA-directed RNA polymerase subunit beta′; DNA-dependent RNA polymerase catalyzes the transcription of DNA into RN...*					●	●	●		0.641
●	rpoB	*DNA-directed RNA polymerase subunit beta; DNA-dependent RNA polymerase catalyzes the transcription of DNA into RN...*					●	●			0.625
●	Deima_2496	*DEAD/DEAH box helicase (604 aa)*	●				●		●		0.609
●	Deima_0406	*DEAD/DEAH box helicase; The UvrABC repair system catalyzes the recognition and processing of DNA lesions. A damage...*	●				●		●		0.597
●	Deima_1259	*hypothetical protein (299 aa)*	●								0.483
●	pnp	*polyribonucleotide nucleotidyltransferase; Involved in mRNA degradation. Hydrolyzes single-stranded polyribonucleotides...*					●		●		0.481
●	Deima_0694	*ATP-dependent helicase HrpB (831 aa)*					●				0.406

(b)

FIGURE 7: *STRING protein-protein interactions of Dm-Csp1.* STRING network map (a) and summary table of predicted interactions (b). The category of interaction is shown in the table. Color saturation of the edges represents the confidence score of a given functional or physical association. The position of Dm-Csp1 is labeled Deima_1260. Shown interactions are considered of high confidence when yielding scores of greater than 0.7.

outside the class Deinococci, and for which a function is undefined. Future studies will seek to further characterize the biological roles of Dm-Csp1, and related Csps, via *in vitro* approaches.

Conflicts of Interest

The author does not have any conflicts of interest to declare.

Acknowledgments

The author thanks all members of the Biology Department at Grand View University for encouragement and project support.

References

[1] K. S. Makarova, L. Aravind, Y. I. Wolf et al., "Genome of the extremely radiation-resistant bacterium *Deinococcus radiodurans* viewed from the perspective of comparative genomics," *Microbiology and Molecular Biology Reviews*, vol. 65, no. 1, pp. 44–79, 2001.

[2] E. Gerber, R. Bernard, S. Castang et al., "Deinococcus as new chassis for industrial biotechnology: biology, physiology and tools," *Journal of Applied Microbiology*, vol. 119, no. 1, pp. 1–10, 2015.

[3] F. A. Rainey, K. Ray, M. Ferreira et al., "Extensive diversity of ionizing-radiation-resistant bacteria recovered from Sonoran Desert soil and description of nine new species of the genus Deinococcus obtained from a single soil sample," *Applied and Environmental Microbiology*, vol. 71, no. 9, pp. 5225–5235, 2005.

[4] R. Pukall, A. Zeytun, S. Lucas et al., "Complete genome sequence of *Deinococcus maricopensis* type strain (LB-34 T)," *Standards in Genomic Sciences*, vol. 4, no. 2, pp. 163–172, 2011.

[5] A. de Groot, V. Chapon, P. Servant et al., "*Deinococcus deserti* sp. nov., a gamma-radiation-tolerant bacterium isolated from the Sahara Desert," *International Journal of Systematic and Evolutionary Microbiology*, vol. 55, no. 6, pp. 2441–2446, 2005.

[6] S. Srinivasan, S. Lim, J.-H. Lim, H.-Y. Jung, and M. K. Kim, "*Deinococcus rubrus* sp. Nov., a bacterium isolated from Antarctic coastal sea water," *Journal of Microbiology and Biotechnology*, vol. 27, no. 3, pp. 535–541, 2017.

[7] M. Yuan, W. Zhang, S. Dai et al., "*Deinococcus gobiensis* sp. nov., an extremely radiation-resistant bacterium," *International Journal of Systematic and Evolutionary Microbiology*, vol. 59, part 6, pp. 1513–1517, 2009.

[8] A. Airo, S. L. Chan, Z. Martinez, M. O. Platt, and J. D. Trent, "Heat shock and cold shock in *Deinococcus radiodurans*," *Cell Biochemistry and Biophysics*, vol. 40, no. 3, pp. 277–288, 2004.

[9] G. Horn, R. Hofweber, W. Kremer, and H. R. Kalbitzer, "Structure and function of bacterial cold shock proteins," *Cellular and Molecular Life Sciences*, vol. 64, no. 12, pp. 1457–1470, 2007.

[10] D. N. Ermolenko and G. I. Makhatadze, "Bacterial cold-shock proteins," *Cellular and Molecular Life Sciences*, vol. 59, no. 11, pp. 1902–1913, 2002.

[11] S.-H. Park, H. Singh, D. Appukuttan et al., "PprM, a cold shock domain-containing protein from *Deinococcus radiodurans*, confers oxidative stress tolerance to *Escherichia coli*," *Frontiers in Microbiology*, vol. 7, article no. 2124, 2017.

[12] A. K. Schmid, M. S. Lipton, H. Mottaz, M. E. Monroe, R. D. Smith, and M. E. Lidstrom, "Global whole-cell FTICR mass spectrometric proteomics analysis of the heat shock response in the radioresistant bacterium *Deinococcus radiodurans*," *Journal of Proteome Research*, vol. 4, no. 3, pp. 709–718, 2005.

[13] R. Keto-Timonen, N. Hietala, E. Palonen, A. Hakakorpi, M. Lindström, and H. Korkeala, "Cold shock proteins: a minireview with special emphasis on Csp-family of enteropathogenic Yersinia," *Frontiers in Microbiology*, vol. 7, p. 1151, 2016.

[14] K. Yamanaka, "Cold shock response in Escherichia coli," *J. Mol. Microbiol. Biotechnol*, vol. 1, no. 2, pp. 193–202, 1999.

[15] E. de Castro, C. J. A. Sigrist, A. Gattiker et al., "ScanProsite: detection of PROSITE signature matches and ProRule-associated functional and structural residues in proteins," *Nucleic Acids Research*, vol. 34, pp. W362–W365, 2006.

[16] R. Finn, S. Griffiths-Jones, and A. Bateman, "Identifying protein domains with the Pfam database," in *Current Protocols in Bioinformatics*, chapter 2, Unit 2.5, 2003.

[17] D. Szklarczyk, A. Franceschini, S. Wyder et al., "STRING v10: protein-protein interaction networks, integrated over the tree of life," *Nucleic Acids Research*, vol. 43, pp. D447–D452, 2015.

[18] C.-C. Chen, J.-K. Hwang, and J.-M. Yang, "$(PS)^2$-v2: template-based protein structure prediction server," *BMC Bioinformatics*, vol. 10, article 366, 2009.

[19] E. F. Pettersen, T. D. Goddard, C. C. Huang et al., "UCSF Chimera—a visualization system for exploratory research and analysis," *Journal of Computational Chemistry*, vol. 25, no. 13, pp. 1605–1612, 2004.

[20] A. Bhattacharya, Z. Wunderlich, D. Monleon, R. Tejero, and G. T. Montelione, "Assessing model accuracy using the homology modeling automatically software," *Proteins*, vol. 70, no. 1, pp. 105–118, 2008.

[21] P. Benkert, S. C. E. Tosatto, and D. Schomburg, "QMEAN: a comprehensive scoring function for model quality assessment," *Proteins*, vol. 71, no. 1, pp. 261–277, 2008.

[22] R. Luthy, J. U. Bowie, and D. Eisenberg, "Assesment of protein models with three-dimensional profiles," *Nature*, vol. 356, no. 6364, pp. 83–85, 1992.

[23] J. Yang, A. Roy, and Y. Zhang, "Protein-ligand binding site recognition using complementary binding-specific substructure comparison and sequence profile alignment," *Bioinformatics*, vol. 29, no. 20, pp. 2588–2595, 2013.

[24] K. Neuhaus, S. Rapposch, K. P. Francis, and S. Scherer, "Restart of exponential growth of cold-shocked Yersinia enterocolitica occurs after down-regulation of cspA1/A2 mRNA," *Journal of Bacteriology*, vol. 182, no. 11, pp. 3285–3288, 2000.

[25] B. Xia, H. Ke, and M. Inouye, "Acquirement of cold sensitivity by quadruple deletion of the cspA family and its suppression by PNPase S1 domain in Escherichia coli," *Molecular Microbiology*, vol. 40, no. 1, pp. 179–188, 2001.

[26] G. D'Auria, C. Esposito, L. Falcigno et al., "Dynamical properties of cold shock protein A from Mycobacterium tuberculosis," *Biochemical and Biophysical Research Communications*, vol. 402, no. 4, pp. 693–698, 2010.

[27] U. Mueller, D. Perl, F. X. Schmid, and U. Heinemann, "Thermal stability and atomic-resolution crystal structure of the Bacillus caldolyticus cold shock protein," *Journal of Molecular Biology*, vol. 297, no. 4, pp. 975–988, 2000.

[28] J. Lee, K.-W. Jeong, B. Jin et al., "Structural and dynamic features of cold-shock proteins of Listeria monocytogenes, a psychrophilic bacterium," *Biochemistry*, vol. 52, no. 14, pp. 2492–2504, 2013.

[29] V. Chaikam and D. T. Karlson, "Comparison of structure, function and regulation of plant cold shock domain proteins to bacterial and animal cold shock domain proteins," *BMB Reports*, vol. 43, no. 1, pp. 1–8, 2010.

[30] S. Phadtare, V. Tadigotla, W.-H. Shin, A. Sengupta, and K. Severinov, "Analysis of Escherichia coli global gene expression profiles in response to overexpression and deletion of CspC and CspE," *Journal of Bacteriology*, vol. 188, no. 7, pp. 2521–2527, 2006.

6

Current Progress in the Structural and Biochemical Characterization of Proteins Involved in the Assembly of Lipopolysaccharide

Thomas E. Bohl and Hideki Aihara

Department of Biochemistry, Molecular Biology, & Biophysics, University of Minnesota, 321 Church Street, 6-155 Jackson Hall, Minneapolis, MN 55455, USA

Correspondence should be addressed to Thomas E. Bohl; bohlx031@umn.edu

Academic Editor: David M. Iovannisci

The lipid component of the outer leaflet of the outer membrane of Gram-negative bacteria is primarily composed of the glycolipid lipopolysaccharide (LPS), which serves to form a protective barrier against hydrophobic toxins and many antibiotics. LPS is comprised of three regions: the lipid A membrane anchor, the nonrepeating core oligosaccharide, and the repeating O-antigen polysaccharide. The lipid A portion is also referred to as endotoxin as its overstimulation of the toll-like receptor 4 during systemic infection precipitates potentially fatal septic shock. Because of the importance of LPS for the viability and virulence of human pathogens, understanding how LPS is synthesized and transported to the outer leaflet of the outer membrane is important for developing novel antibiotics to combat resistant Gram-negative strains. The following review describes the current state of our understanding of the proteins responsible for the synthesis and transport of LPS with an emphasis on the contribution of protein structures to our understanding of their functions. Because the lipid A portion of LPS is relatively well conserved, a detailed description of the biosynthetic enzymes in the Raetz pathway of lipid A synthesis is provided. Conversely, less well-conserved biosynthetic enzymes later in LPS synthesis are described primarily to demonstrate conserved principles of LPS synthesis. Finally, the conserved LPS transport systems are described in detail.

1. Introduction

Gram-negative bacteria are distinguished from Gram-positive bacteria by the secondary lipid bilayer that surrounds their peptidoglycan cell wall [1]. In the majority of Gram-negative bacteria, this outer membrane is an asymmetric bilayer with phospholipids on the inner leaflet and lipopolysaccharide (LPS) on the outer leaflet [1, 2]. LPS is a glycolipid composed of a lipid A membrane anchor, a core oligosaccharide, and a repeating O antigen polysaccharide (Figure 1) [1, 5]. This layer of LPS provides a permeability barrier to environmental toxins. When this layer of LPS is disrupted by invasion of phospholipids [2], loss of LPS synthesis or transport [6], or loss of glycosylation [7], the bacteria show increased sensitivity to hydrophobic toxins, such as detergent or bile salts or antibiotics. Furthermore, a complete loss of LPS by mutation of lipid A synthesis genes made bacteria less virulent to *Caenorhabditis elegans* and human epithelial cells [8].

The role of LPS in disease is related not only to its protective function in Gram-negative bacteria but also to its interaction with the host's innate immune system. Humans possess a cohort of proteins that allow rapid response to the presence of LPS [9]. LPS binding protein is thought to extract LPS from bacterial membranes and present an LPS molecule to CD14 [9, 10]. CD14, which may be secreted or lipid-linked to the extracellular face of cells, transfers this LPS to the complex of myeloid differentiation factor 2 and toll-like receptor 4 (MD2/TLR4) [9, 11–13]. Activation of TLR4 by LPS leads to the production and secretion of proinflammatory cytokines and type I interferon [9, 14]. While this response is beneficial for clearing small bacterial

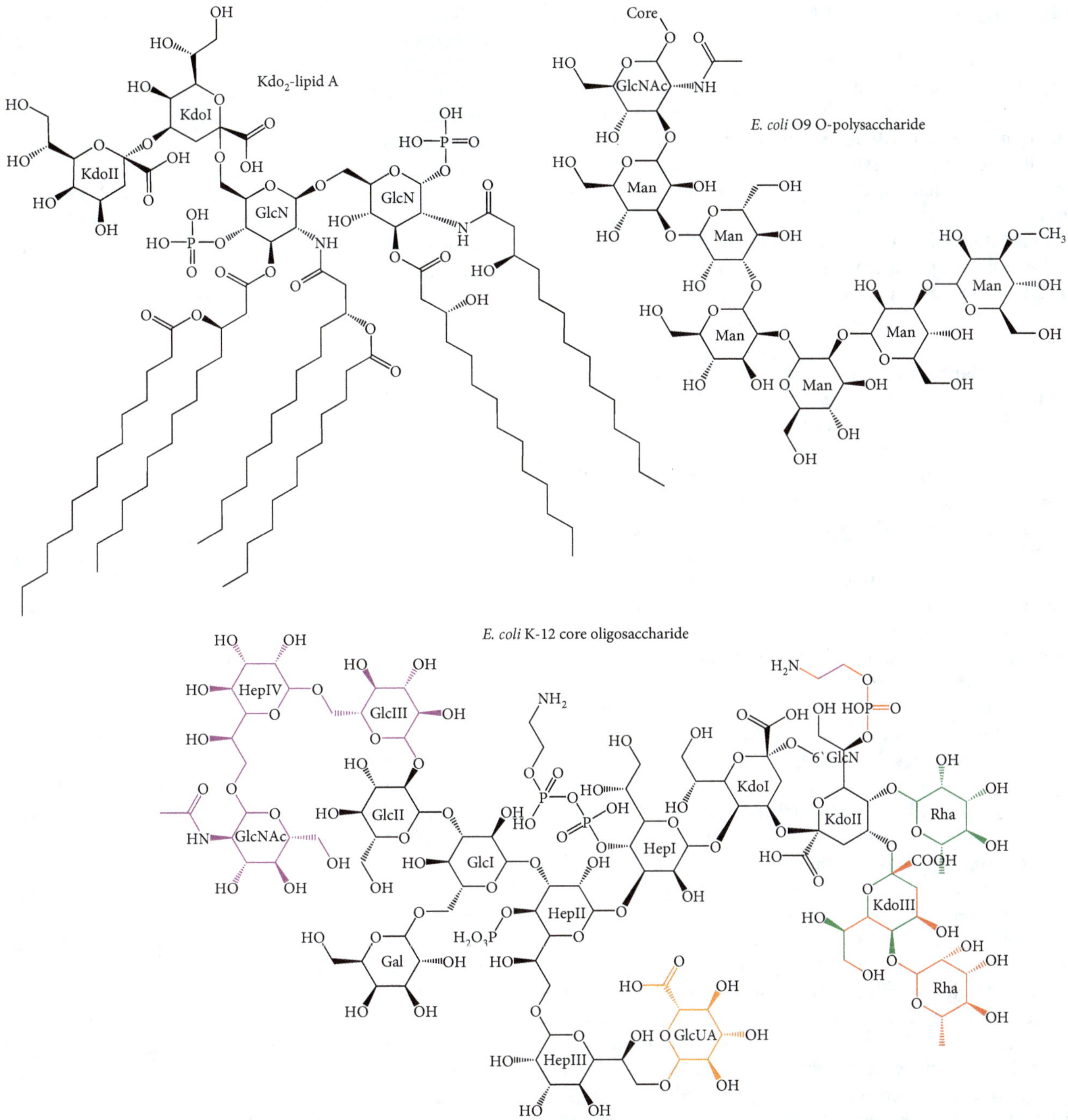

FIGURE 1: LPS chemical structure. Chemical structures of the canonical lipid A of *E. coli*, the core oligosaccharide of *E. coli* K-12, and the *E. coli* O9 O-antigen (showing one repeat unit) [1]. The text for the distal glucosamine of lipid A attached to the core oligosaccharide is colored blue. The text for the residues at each end of the core oligosaccharide is colored red. The text for the ends of the O-antigen is colored green, and that for the repeat region is colored purple. In the core oligosaccharide, atom and bond coloring indicates different core structures [3]. Black indicates the conserved structure, and purple, red, and green indicate moieties present in three different variants of the K-12 core. Glucuronic acid (GlcUA, orange) may be attached to HepIII in the absence of the phosphate on HepII [4]. Dual coloring of moieties indicates their presence in both variants. Chemical structure figures were generated with ChemBioDraw Ultra 14.0.

infections, overstimulation of the inflammatory response during systemic infection, such as sepsis (presence of bacteria in the blood), is potentially fatal and can cause permanent organ damage or neurological problems [15, 16].

Because of the severity of this condition, rapid treatment with broad-spectrum antibiotics before the causative bacteria can be identified is recommended for patients with sepsis [15]. The highly conserved and (in most cases) essential enzymes involved in the synthesis of the lipid A portion of LPS are promising targets for the development of new broad-spectrum antibiotics to treat sepsis and infections by antimicrobial-resistant Gram-negative strains [1, 7, 17, 18].

2. Main Text

2.1. Raetz Pathway. The canonical Raetz pathway of lipid A synthesis is a 9 enzyme pathway that produces Kdo_2-lipid A (Figure 2) [1]. While some variation exists between species, particularly with regards to acylation, the Raetz pathway, especially the 4 of the first 5 enzymes, is well conserved in Gram-negative bacteria, excluding specialized species that do not produce LPS [17, 19]. Most of these enzymes have been characterized at the biochemical and structural levels and the following summarizes what is known about the enzymes of the Raetz pathway.

2.1.1. LpxA. Early experiments with radiolabeled substrates indicated that UDP-*N*-acetylglucosamine is incorporated into the lipid A synthesis pathway and suggested that acylation of the 3-hydroxyl of glucosamine precedes the removal of acetate [20]. The acyltransferase responsible for the addition of this acyl chain was denoted LpxA and was found to be specific for UDP-*N*-acetylglucosamine and R-3-hydroxymyristoyl-acyl carrier protein (ACP), with little activity toward the S-enantiomer or the coenzyme A adduct of β-hydroxymyristate or toward palmitoyl-ACP, myristoyl-ACP, or acyl chain acceptors with chains more than 3 carbons long at the 2-amine of glucosamine or with uracil substituted with any base other than thymine [21, 22].

Furthermore, radiolabeled LpxA product was converted to other intermediates of lipid A synthesis [21].

Crystal structures of LpxA revealed that LpxA is composed of an unusual *N*-terminal left-handed parallel β-helix domain and a C-terminal α-helical domain and forms a trimer with 3-fold symmetry (Figure 3) [25]. Each β-helix approximates an equilateral triangular prism, and these helices come together in the trimer to form a large cleft between each pair of subunits [25]. Extended loops within helical repeats 4 and 5 form additional contacts between adjacent subunits [26]. Crystal structures with UDP-*N*-acetylglucosamine or the product bound show that substrates bind in this cleft between subunits with the uridine and *N*-acetylglucosamine moieties contacting adjacent subunits and with the R-3-hydroxymyristoyl chain extending up the cleft toward the C-terminal domain (Figure 3) [23, 26]. The extended loop within helical repeat 4 provides part of the contacts for *N*-acetylglucosamine [23, 26].

Furthermore, these crystal structures provide insights into the mechanism and selectivity of LpxA [23, 26]. LpxA is thought to catalyze the nucleophilic attack of the glucosamine 3-hydroxyl on the thioester of R-3-hydroxymyristoyl-ACP, and a conserved His (H125 in *E. coli* LpxA) is positioned to act as the catalytic base and accept the proton from this hydroxyl, while the backbone amide of G143 is positioned to stabilize the oxyanion intermediate [23, 26]. The role of H125 was supported by activity assays that showed it to be critical for acyl transfer [27]. D126 hydrogen bonds with and stabilizes the position of H125 forming a catalytic dyad similar to that observed in other acyltransferases [26]. In addition, the product-bound structure suggests that H191 of *E. coli* LpxA limits the length of the acyl chain by capping the channel, wherein the chain binds and mutation of another residue that lines this channel (G173) to larger residues decreased activity toward the natural acyl donor substrate and increased activity toward the 10 carbon chain analogue [23, 28]. Moreover, selection of the R-3-hydroxylated fatty acid may be controlled by hydrogen-bonding interactions between this functional group and H122, Q73, and an ordered water bound to H99 [23]. Finally, UDP-*N*-acetylglucosamine selectivity is governed by extensive hydrogen-bonding as well as the stereospecific architecture of the *N*-acetylglucosamine binding pocket that would interfere with the binding of a sugar with axial substituents (Figure 3) [26].

2.1.2. LpxC. LpxC catalyzes the second step in the Raetz pathway, the deacetylation of UDP-3-O-(R3-hydroxymyristoyl)-*N*-acetylglucosamine [29, 30]. Early experiments showed that the LpxA product is deacetylated when incubated with *E. coli* extracts and that the deacetylated product can be converted into other intermediates in the pathway [29]. Furthermore, LpxC was found to have poor activity toward UDP-*N*-acetylglucosamine [31]. Because the reaction catalyzed by LpxA is freely reversible and, surprisingly, thermodynamically unfavorable, LpxC catalyzes the committed step of lipid A synthesis [22]. In accordance with the thermodynamic importance of this step, LpxC is also an important point of regulation for lipid A synthesis [32, 33]. LpxC concentration is regulated by proteolytic degradation by FtsH, which may be controlled by the levels of acyl-ACP or Raetz pathway intermediates [32, 33]. In addition, the heat shock-induced protein LapB was found to be essential for this function of FtsH [34]. LapB likely acts as a scaffold or chaperone for inner membrane proteins involved in LPS synthesis and transport including FtsH, LpxM, and WaaC [34].

LpxC is a Zn^{2+}-dependent metalloenzyme though it displays higher activity in the presence of Ni^{2+} or Co^{2+}, which may be due to inhibition by Zn^{2+} binding at a second site [31]. Consistent with these enzymatic assays, the crystal structure of *Aquifex aeolicus* LpxC was solved with two Zn^{2+} in the active site, representing the zinc-inhibited state, and the crystal structure of *E. coli* LpxC in complex with the lipidic product, which was crystallized in a high concentration of phosphate, was solved with a single high-affinity Zn^{2+} in the active site [35, 36]. Native mass spectrometry was also consistent with the binding of one high-affinity Zn^{2+} [36].

Solution NMR and crystal structures of LpxC revealed that the enzyme is composed of two α/β-domains with a conserved topology that form a five-stranded β-sheet and two α-helices [35, 37]. The α-helices of the conserved fold are packed together giving LpxC an overall pseudo-twofold symmetry (Figure 4) [35, 37]. The domains contain divergent insertions between the fourth β-strand and first α-helix: the *N*-terminal domain's insertion forms a three-stranded β-sheet, and the C-terminal domain's insertion contributes a short helix to the helical core and forms an α/β-subdomain [35–37].

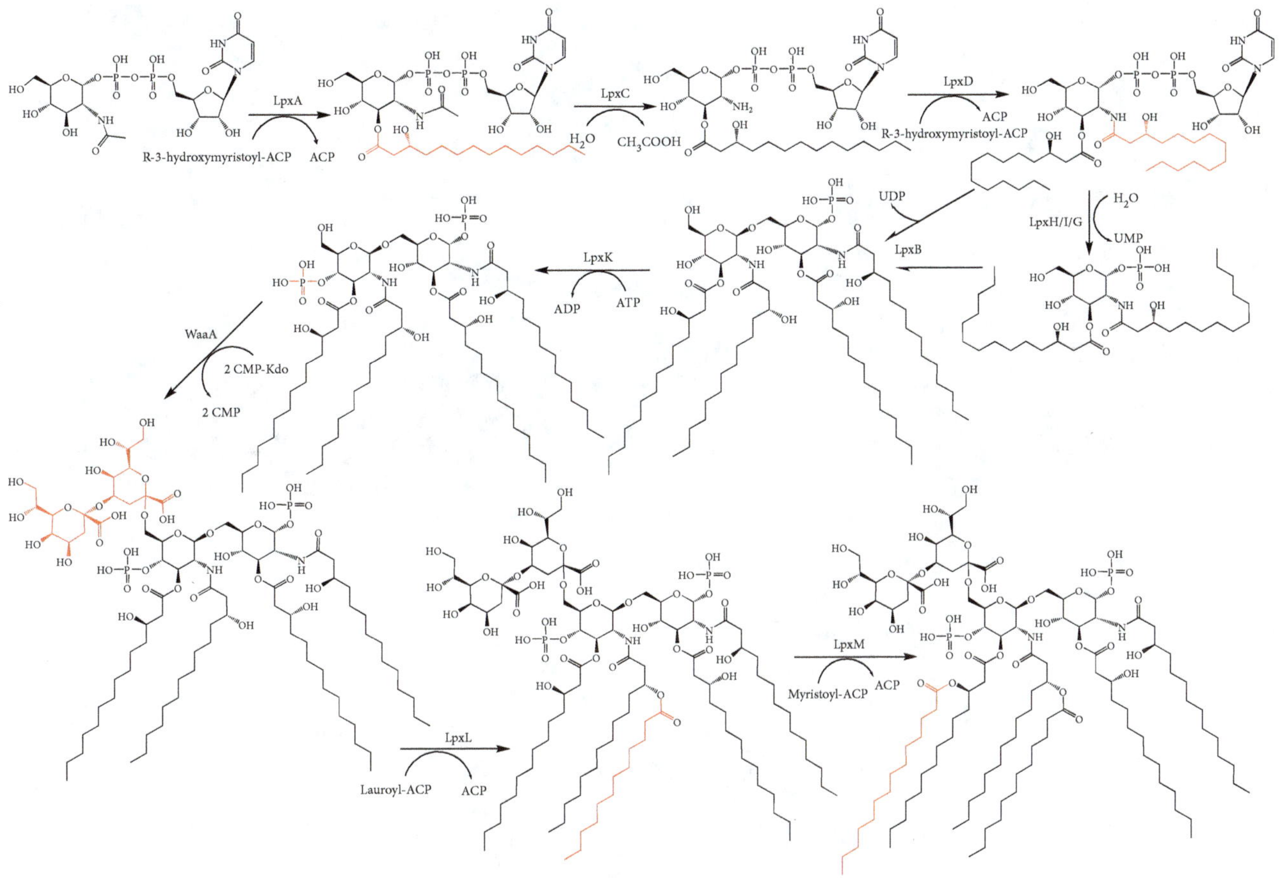

FIGURE 2: Raetz pathway of lipid A synthesis. The moieties added in each step are shown in red [1].

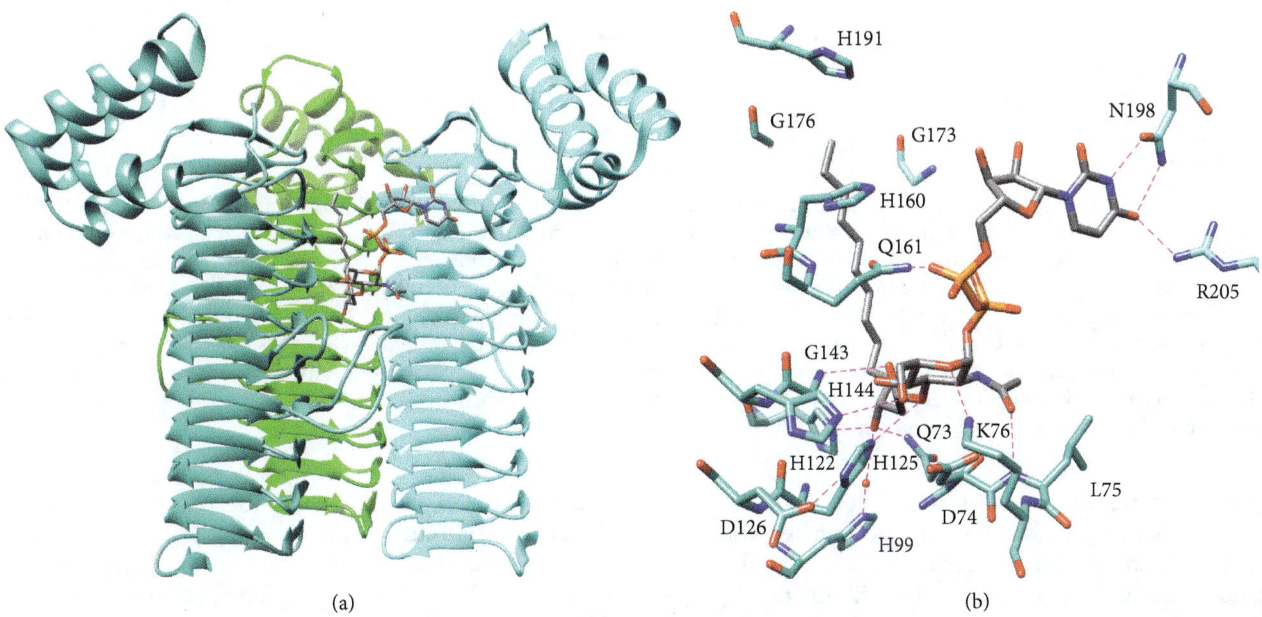

FIGURE 3: LpxA: (a) ribbon diagram of the LpxA trimer bound to its UDP-3-O-(R3-hydroxymyristoyl)-*N*-acetylglucosamine product (PDB: 2QIA) [23]; (b) residues involved in product binding from (a). Subscripts are used below to distinguish subunits in the trimer. Hydrogen-bonding interactions are shown. Distances of interactions in Ångstroms are as follows: 3.3 from 3′-acyl carbonyl O to $G143_{0.3}$ N 3.2 from uracil O4 to $N198_{0.1}$ Nδ, 3.0 from uracil O4 to $R205_{0.1}$ Nη2 2.7 from uracil N3 to $N198_{0.1}$ Oδ, 2.9 from water to $H99_{0.3}$ Nδ, 2.8 from same water to β-hydroxyl, 2.8 from β-hydroxyl to $H122_{0.3}$ Nε, 3.0 from β-hydroxyl to $Q73_{0.3}$ Nε, 3.3 from glucosamine 3′O to $H125_{0.3}$, 2.8 from $H125_{0.3}$ Nδ to $D126_{0.3}$, 2.9 from glucosamine 6′-hydroxyl to $H144_{0.3}$ Nε, 3.0 from the glucosamine ring O to $K76_{0.3}$ Nζ, 3.0 from 2′-acetyl carbonyl to $L75_{0.3}$ N, and 2.6 from α-phosphate to $Q161_{0.3}$ Nε. Protein structure figures were generated using UCSF Chimera [24].

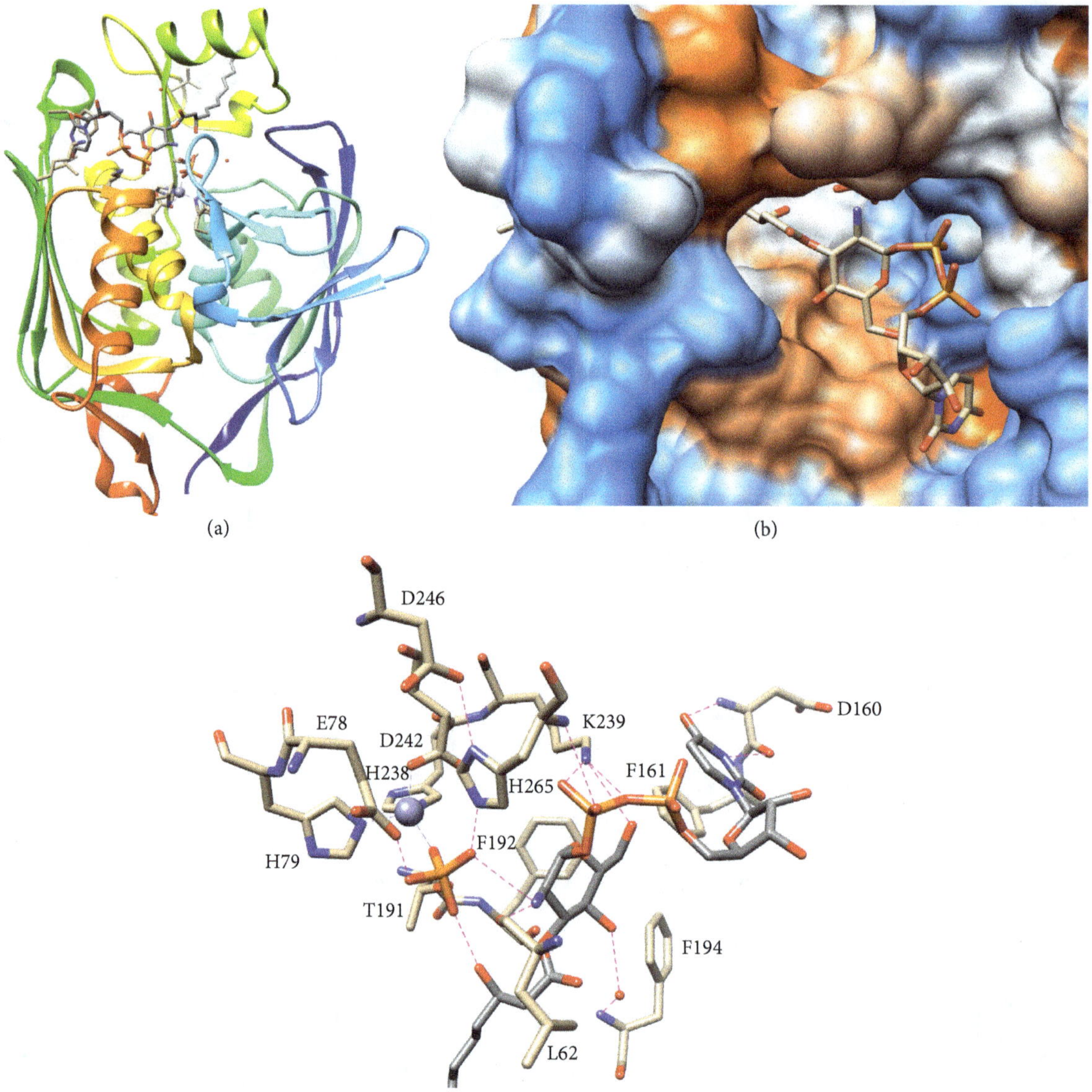

FIGURE 4: LpxC. (a) Ribbon structure of LpxC bound to its UDP-3-O-(R3-hydroxymyristoyl)-glucosamine product (PDB: 4MDT) [36]. Spectrum coloring begins with blue at the *N*-terminus. (b) Hydrophobicity surface of the structure in (a) with a orange-blue scale. Orange is more hydrophobic. (c) Residues involved in product binding from (a). Zinc coordination is shown in purple, and hydrogen-bond and salt-bridge interactions are shown in pink. Distances of interactions in Ångstroms are as follows: 1.9 from Zn^{2+} to H238 Nε, 2.0 from Zn^{2+} to H79 Nε, 2.0 from Zn^{2+} to D242 Oδ1, 1.9 from Zn^{2+} to phosphate O2, 2.5 from water to β-phosphate and 3.2 to α-phosphate and 3.0 to A266 N 3.2 from pyrophosphate bridging O and 2.8 from β-phosphate and 3.5 from glucosamine 6′-hydroxyl to K239 Nζ, 2.8 from β-hydroxyl to phosphate O1, 3.2 from water to F194 N and 2.9 to glucosamine 4′-hydroxyl, 3.1 from glucosamine 2′-amine to phosphate O3 and 2.7 to L62 O 2.6 from E78 Oε2 to phosphate O4, 2.4 from H265 Nε to phosphate O3, 3.5 from uracil O4 to D160 N 2.7 from D246 Oδ2 to H265 Nδ, and 3.5 from uracil N3 to D160 O.

The structure of *E. coli* LpxC with UDP-3-O-(R-3-hydroxymyristoyl)-glucosamine and phosphate bound in the active site shows that the product primarily binds to the insertions and the C-terminal domain with the first helix of the *N*-terminal domain contributing one Zn^{2+}-coordinating residue and the putative catalytic base [36]. Zn^{2+} is coordinated with tetrahedral geometry by two His side chains: one Asp side chain (H79, H238, and D242 in *E. coli* LpxC) and a molecule from the crystallization solution [35–37]. The importance of these residues was supported by enzymatic assays that showed decreased activity in mutants of any one of these residues as well as greatly decreased Zn^{2+}-binding in the His to Ala mutants [38].

In the product-bound structure of *E. coli* LpxC, the fourth ligand of Zn^{2+} was modeled as phosphate, and it may approximate the tetrahedral transition state of the leaving group and hydrogen bonding with the glucosamine amine [36]. The phosphate forms hydrogen bonds with E78, T191, and H265 implicating these residues in catalysis, and mutants of these residues were indeed found to have decreased activity [36, 38, 39]. Specifically, E78 is positioned to accept a proton from a Zn^{2+}-coordinated hydrolytic water, and this role was supported by the greater reduction in activity of E78Q in comparison to E78 A as E78Q cannot act as a catalytic base but likely blocks the compensatory positioning of another residue or solvent molecule [36, 38]. On the contrary, H265 is positioned to donate a proton to the glucosamine amine, and D246, which hydrogen-bonds with the opposite nitrogen of the H265 azole thus tuning pK_a of H265, was also critical for activity [36, 38].

As stated above, the product primarily interacts with the insertions and the C-terminal domain [36]. The UDP moiety binds to the C-terminal domain: uracil hydrogen-bonds with the backbone of D160 and forms a π-stacking interaction with F161, and the pyrophosphate interacts with K239 and the backbone of H265 (Figure 4) [36]. Recognition of the glucosamine moiety involves all three subdomains [36]. The 6′-hydroxyl also hydrogen-bonds with K239, and the 2′-amine hydrogen-bonds with the backbone of L62 (Figure 4). The importance of K239 was confirmed by enzymatic assays showing a slightly increased K_M and greatly decreased k_{cat} of the K239 A mutant [39]. In addition, the 4′-hydroxyl hydrogen-bonds to an ordered water bound to the F194 backbone [36]. Finally, the acyl chain binds in a hydrophobic channel formed by the C-terminal domain insertion with the chain terminus extending into solvent, suggesting that, while presence of the chain is critical for activity, LpxC has little specificity for chain length (Figure 4) [31, 35, 36].

2.1.3. LpxD. Early experiments indicated that addition of the second R3-hydroxymyristoyl group at the glucosamine 2-amino group precedes the hydrolysis of the phosphoanhydride of UDP [20]. This structural gene for the UDP-3-O-(R3-hydroxymyristoyl)-glucosamine-*N*-acyltransferase (*lpxD*) was identified by homology to *lpxA* and its position in the same operon, and enzymatic assays showed LpxD is highly specific for *β*-hydroxymyristoyl-ACP, showing no activity for myristoyl-ACP and 10-fold reduced activity for *β*-hydroxylauroyl-ACP [40, 41]. Furthermore, steady-state kinetic experiments showed that LpxD follows an ordered sequential mechanism, wherein R3-hydroxymyristoyl-ACP binds first and ACP dissociates last [41].

Crystal structures of LpxD reveal many similar features shared with LpxA [42, 43]. In particular, LpxD has a very similar left-handed *β*-helix domain that includes two elongated loops (within helical repeats 5 and 6) that contribute additional interactions with the adjacent subunit in the trimer [42]. Moreover, LpxD also has an *α*-helical C-terminal domain, although in LpxD, the C-terminal helices of the subunits come together to form a trimeric helix bundle [42].

However, LpxD has an additional *N*-terminal domain composed of a five-stranded *β*-sheet surrounded by *α*-helices and a two-stranded *β*-sheet (Figure 5) [42]. Furthermore, crystal structures of LpxD bound to substrates and substrate analogues have revealed conserved catalytic residues and mechanisms of substrate specificity [42, 44]. Specifically, a conserved His (H239 in *E. coli*) is positioned to accept a proton from the glucosamine amine, while the G257 amide is positioned to stabilize the oxyanion intermediate [42–44]. As in LpxA, the substrates bind at the cleft between subunits [42, 44].

In the structure of *Chlamydia trachomatis* LpxD, UDP-*N*-acetylglucosamine binds to the *N*-terminal domain and the left-handed *β*-helix domain of the adjacent subunit with the *N*-terminal domain primarily binding the uridine moiety [42]. At the active site formed by chains A and B, uracil is sandwiched in π-stacking interactions with $F43_B$ and $Y49_B$ (F41 and Y47 in *E. coli*) and forms hydrogen-bonds with the backbone of residues 33_B and 43_B, and 45_B (Figure 5) [42]. Ribose is hydrogen-bonded to $E32_B$ and $E34_B$ (S30 and Q32 in *E. coli*) and $Q248_A$ (Q240 in *E. coli*), while the pyrophosphate is engaged by $Y49_B$, $N46_B$, and $H284_A$ (N44 and H276 in *E. coli*), and *N*-acetylglucosamine binds to $N46_B$ and $H247_A$ (H239 in *E. coli*, the predicted catalytic base) [42]. Kinetic assays in *E. coli* LpxD confirmed that H239 is critical for catalysis: H239 A decreased k_{cat} almost 3 orders of magnitude with a moderate decrease of the acyl acceptor K_M [41]. In addition, F41 was confirmed to be critical for acyl acceptor binding: F41 A increased the acceptor K_M 29-fold with only a small decrease in k_{cat} [41]. Conversely, Y47 A had little effect, and H276 A decreased k_{cat} 32-fold [41].

E. coli LpxD crystallized in complex with ACP yielded structures with R3-hydroxymyristoyl-ACP and with holo-ACP and two *β*-hydroxymyristate molecules bound to LpxD [44]. ACP primarily binds to the C-terminal domain with the pantetheine arm on ACP S36 extending up the active site cleft 0(Figure 5) [44]. Association with ACP is largely electrostatic with an acidic patch on ACP contacting a basic patch on LpxD [44]. For example, the Ala mutant of R193, which forms an ion pair with E41 of ACP, was found to cause a 23-fold increase of the R3-hydroxymyristoyl-ACP K_M [41, 44]. In the intact substrate structure, the R3-hydroxymyristoyl chain turns 180° and extends back toward the C-terminus of LpxD deeper in the cleft [44]. Specificity for the length of the acyl chain attached to ACP is conferred by the length of the hydrophobic pocket that accommodates the chain [41, 42, 44]. In *E. coli* LpxD, the ~13 Å deep pocket is terminated by M290, and M290 A allowed incorporation of 16 and 18 carbon acyl chains *in vivo* [43]. Moreover, in *C. trachomatis* LpxD, which transfers a *β*-hydroxyarachidoyl chain, the corresponding residue is G298, and the hydrophobic pocket is ~18 Å deep [42]. In the active site formed by chains B and C, the *β*-hydroxyl group of the transferred chain is recognized by hydrogen-bonds to $D232_B$ and $Q236_C$ of *E. coli* LpxD, and the acyl chain carboxylate hydrogen-bonds to the $G257_C$ amide, consistent with its proposed stabilization of the oxyanion (Figure 5) [41, 43, 44]. Finally, the structure of *E. coli* LpxD bound to holo-ACP and two *β*-hydroxymyristates reveals the binding pocket for the 3′-acyl chain of the acceptor substrate: this pocket is formed by the left-handed *β*-helix of the subunit that contributes the *N*-terminal domain, the elongated loop of repeat 5 from the adjacent subunit, and the pantetheine group (Figure 5), consistent with kinetic experiments that showed R3-hydroxymyristoyl-ACP binds first [41, 44].

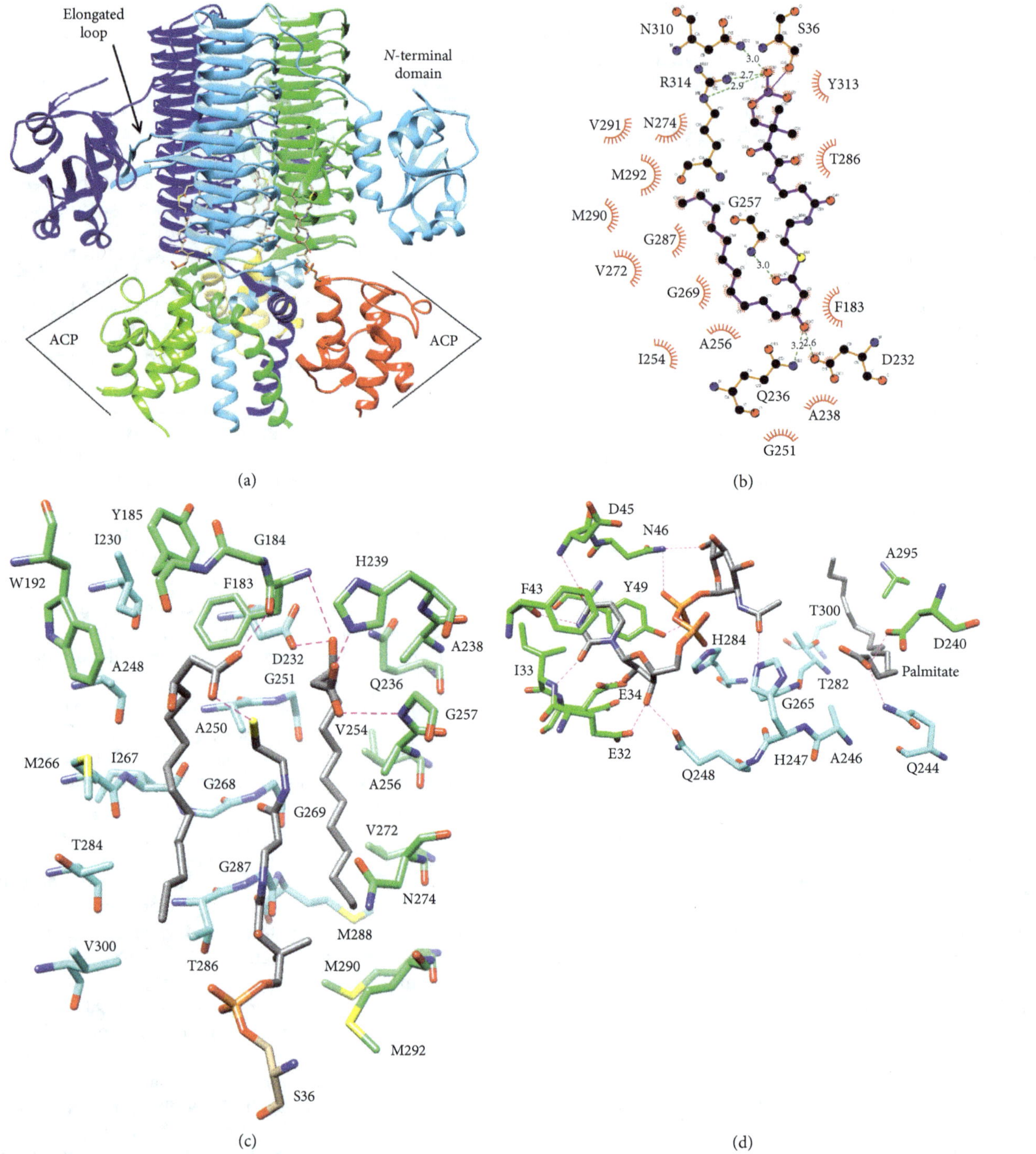

Figure 5: LpxD. (a) Heterohexamer of LpxD and R3-hydroxymyristoyl-ACP (PDB: 4IHF) [44]. ACP chains are colored red, yellow, and chartreuse, and LpxD chains are colored blue, cyan, and sea green. (b) LigPlot+ diagram of acyl donor substrate from (a) [45]. Distances of hydrogen-bond and salt-bridge interactions shown in Ångstroms are as follows: 3.2 from acyl chain β-hydroxyl to Q236 Nε and 2.6 to D232 Oδ2, 3.0 from acyl chain carbonyl to G257 N 3.0 from phosphate to N310 Nδ and 2.7 and 2.9, respectively, to R314 Nη and Nε. (c) Acyl chain binding in LpxD/ACP heterohexamer (PDB: 4IHG) [44]. Subscripts are used below to distinguish protein chains in the complex. Hydrogen-bonds are shown, and corresponding distances in Ångstroms are as follows: 2.9 from putative 2′-R3-hydroxymyristate carboxylate to $G257_C$ N 2.1 from putative 2′-R3-hydroxymyristate carboxylate to $H239_C$ Nε, 3.0 from putative 2′-R3-hydroxymyristate β-hydroxyl to $D232_B$ Oδ2 and 3.6 to $F183_C$ N and 2.8 from putative 3′-R3-hydroxymyristate carboxylate to $F183_C$ O and 3.2 to S of phosphopantetheine attached to ACP $S36_H$. (d) UDP-*N*-acetylglucosamine and palmitate binding to LpxD trimer (PDB: 2IU9) [42]. Hydrogen-bonds are shown with the following distances in Ångstroms: 2.8 from palmitate carboxylate to $Q244_A$ Nε and 2.9 to $D240_B$ Oδ2, 2.6 from 2′-acetyl carbonyl to $H247_A$ Nε, 2.8 from uracil O2 to $I33_B$ N 2.7 from uracil N3 to $F43_B$ O 2.8 from uracil O4 to $D45_B$ N 2.6 from ribose 3′-hydroxyl to $E32_B$ Oε2 and 2.9 to $Q248_A$ Oε, 3.0 from ribose 2′-hydroxyl to $E34_B$ Oε1, 2.4 from β-phosphate to $Y49_B$ Oη and 3.2 to $N46_B$ Nδ, 2.7 from α-phosphate to $H284_A$ Nε, and 3.1 from glucosamine 6′-hydroxyl to $N46_B$ Nδ.

2.1.4. LpxH. Early experiments indicated that hydrolysis of the phosphoanhydride of UDP-2,3-bis[O-(3R)-3-hydroxymyristoyl]-glucosamine (UDP-DAG) is the fourth step in the Raetz pathway [20, 46]. The responsible pyrophosphatase in *E. coli* (LpxH) was identified as an essential Mn^{2+}-dependent hydrolase that catalyzes the attack of water on the α-phosphate of UDP-DAG [47–49]. Electron paramagnetic resonance was utilized to further characterize Mn^{2+} binding, and these data were consistent with the presence of a binuclear manganese center [49].

Crystal structures of LpxH from *Haemophilus influenzae* and *Pseudomonas aeruginosa* revealed that LpxH has a calcineurin-like metal-dependent phosphoesterase fold with the addition of a unique helical cap, comprising 3 major α-helices, that covers the active site and binds the lipid substrates (Figure 6) [50, 51]. The core domain is composed of 2 five-stranded β-sheets surrounded by 4 α-helices, and the helical cap domain is inserted between β-strand 6 and the final α-helix [50, 51]. The crystal structures confirm that LpxH has a binuclear manganese center in its active site [50, 51]. In the *P. aeruginosa* structure (PDB: 5B49), the first Mn^{2+} is coordinated by D8, H10, D41 (bridges the Mn^{2+}), H197, a bridging water, and a second water that completes the octahedral geometry and likely represents the attacking water [50]. The second Mn^{2+} is coordinated by D41, the bridging water, N79, H114, and H195, which leaves one open coordination site that may be filled by the α-phosphate in the substrate-bound structure (Figure 6) [50]. The importance of several of the Mn^{2+}-coordination residues was confirmed by enzymatic assays in *H. influenzae* LpxH (HiLpxH) that showed large (3–5 orders of magnitude) drops in specific activity when individual residues were mutated to Ala [49].

Structures of LpxH bound to its product, 2,3-bis[(3R)-3-hydroxymyristoyl]-glucosaminyl-1-phosphate (lipid X), show that the lipid X head group is recognized by extensive hydrogen-bonding interactions, and the acyl chains bind in the hydrophobic pocket formed by the capping domain and the interface with the core domain: the amide-linked chain is buried in the hydrophobic pocket while the ester-linked chain exits between the helices and extends across the top of the lid or into solvent [50, 51]. In the *P. aeruginosa* LpxH (PaLpxH) structure (PDB: 5B49), the lipid X phosphate is bound by N79 and R80, which is a conserved residue particular to LpxH and is usually His in other metal-dependent phosphoesterases of the calcineurin-like superfamily [50].

Consistent with the importance of this Arg, the Ala mutation in HiLpxH decreased activity 7000-fold [49]. In addition, glucosamine is hydrogen-bonded to S160, T164, K167, and H195, and the β-hydroxyl groups of the acyl chains are hydrogen-bonded to R157 and K167 (Figure 6) [50]. Finally, Y125 packs on top of the glucosamine ring [50]. All of these residues noted from PaLpxH are conserved in *E. coli* LpxH (same numbering) and HiLpxH (numbering +1) with the exception of T164, which is Asn and Lys, respectively.

Furthermore, the structural dynamics of the substrate-binding cap domain have been explored by hydrogen-deuterium exchange and molecular dynamics simulations [52]. The relatively rapid hydrogen-deuterium exchange rate of the capping helices in a solubilized form of *E. coli* LpxH supported the idea that the capping domain is highly flexible in the absence of the substrate (Figure 7) [52]. Molecular dynamics simulations were consistent with the highly flexible nature of the capping domain in the absence of the substrate and predicted that capping helices of apo PaLpxH could undergo an opening motion, wherein the helices spread apart to expose the active site [52] (Figure 7(c)). In addition, molecular dynamics simulations predicted that a loop of the core pyrophosphatase containing F82 and L83 acts like a wedge to open the capping helices and that mutation of these residues to Gly would result in the collapse of the capping domain into a fully closed state (Figure 7(a)) [52]. Activity assays with solubilized *E. coli* LpxH confirmed that the F82 G/L83 G mutations significantly decreased activity [52].

2.1.5. LpxG and LpxH2. LpxG was identified as a gene from *Chlamydia trachomatis* that could complement a *δlpxH* strain of *E. coli* [53]. Activity assays showed that, like LpxH, LpxG is a Mn^{2+}-dependent pyrophosphatase that catalyzes the nucleophilic attack of water on the α-phosphate of UDP-DAG [49, 53]. Consistently, mutation of the predicted metal-coordinating residue D59 to Ala significantly decreased activity [53]. Also like LpxH, homology suggests that LpxG is a member of the calcineurin-like metal-dependent phosphoesterase family [53]. However, LpxG is predicted to have an additional *N*-terminal transmembrane helix, and the sequence identity with LpxH is very low (11%) [53]. Furthermore, while sequence alignments suggest that LpxG has a similarly located insertion in the calcineurin-like metal-dependent phosphoesterase fold, and the length and sequence of the insertions in LpxH and LpxG are different, suggesting that these proteins either diverged early in their evolution or arose by convergent evolution from different phylogenetic clades within the same enzyme family [50]. The structure of LpxG has not been determined.

Another gene called *lpxH2* because of its sequence homology to *lpxH* is present in some Gram-negative bacteria without *lpxH* but can also be present in addition to *lpxH* [17, 47]. Expression of LpxH2 could not complement the *δlpxH* strain of *E. coli*, and the sequence of the substrate-binding helical cap domain of LpxH is not conserved in LpxH2 [47, 50]. Thus, the function of *lpxH2* and its relevance to lipid A synthesis remains unknown.

2.1.6. LpxI. LpxI is a nonhomologous alternative to LpxH present in some Gram-negative bacteria that lack LpxH and LpxG, particularly α-proteobacteria [17, 54]. *LpxI* was discovered as a gene of unknown function present in the same operon as *lpxA*, *lpxD*, and *lpxB* in bacteria lacking *lpxH* [54]. Unlike LpxH, LpxI does not utilize a catalytic metal center, although it did require divalent cations, specifically Mg^{2+}, Mn^{2+}, or Co^{2+}, for optimal activity [54, 55]. Furthermore, ^{18}O incorporation from $H_2{}^{18}O$ indicated that LpxI catalyzes

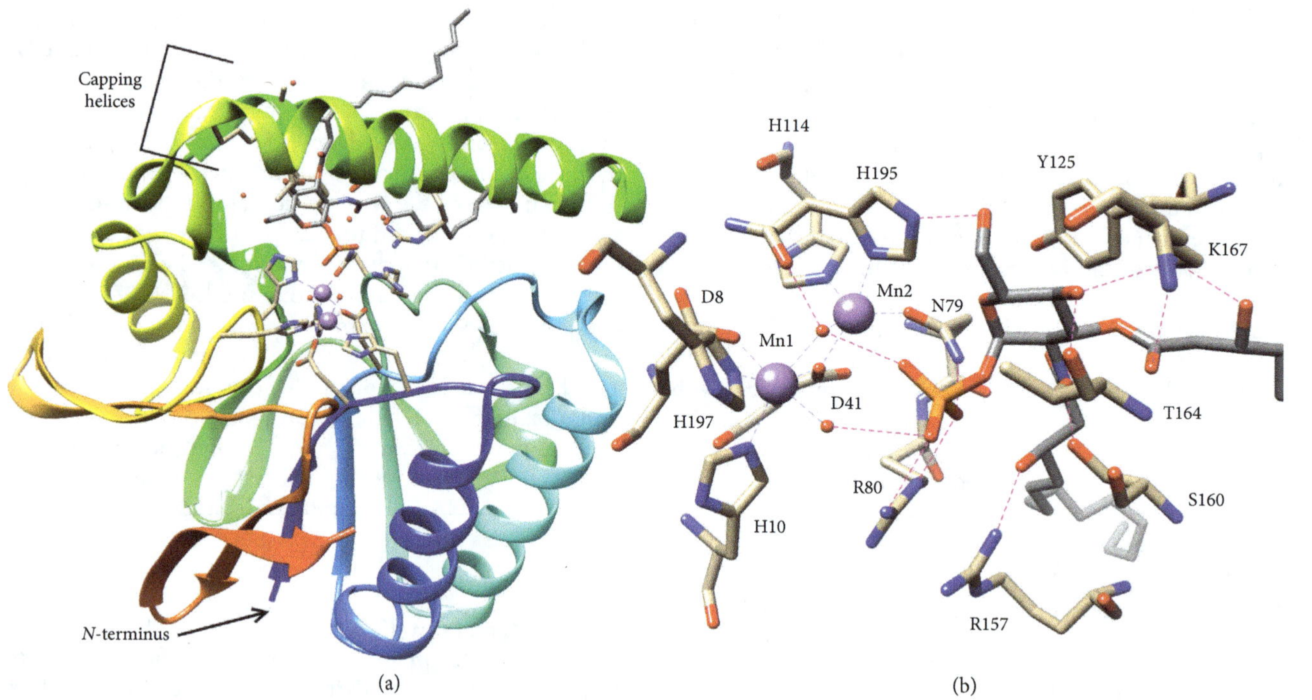

Figure 6: Pseudomonas LpxH bound to the lipid X product. (a) Ribbon structure of PaLpxH bound to lipid X (PDB: 5B49) [50]. Lipid X carbons are colored grey, and protein carbons are colored tan. Mn^{2+} ions are colored purple. Spectrum coloring begins with blue at the *N*-terminus. (b) Residues involved in lipid X binding from (a). Metal coordination is shown in purple with the following distances in Ångstroms: from Mn301 to H10 Nε (2.3), to H197 Nε (2.3), to D8 Oδ1 (2.2), to D41 Oδ2 (2.3), to bridging water (2.3), and to adjacent water (2.2) and from Mn302 to D41 Oδ2 (2.2), to N79 Oδ (2.1), to H195 Nδ (2.3), to H114 Nε (2.2), and to bridging water (2.2). Hydrogen-bond and salt-bridge interactions are shown in pink with distances in Ångstroms of 2.7 from H195 O to bridging water, 2.9 from 2-acyl chain β-hydroxyl to R157 Nη1, 2.8 from glucosamine 4-hydroxyl to T164 Oγ and 2.9 to K167 Nζ, 3.3 from K167 Nζ to 3-acyl chain carbonyl and 2.8 to its β-hydroxyl, 2.8 from glucosamine 6-hydroxyl to H195 Nε, 2.8 from glucosamine 2-amino to S160 Oγ, and 2.7 from 1-phosphate to bridging water and 3.4 to adjacent water and 2.6 to N79 Nδ and 2.8 and 3.1, respectively, to R80 Nε and Nη2.

the nucleophilic attack of water on the β-phosphate of UDP-DAG [54].

Crystal structures of *Caulobacter crescentus* LpxI and LpxI-D225 A bound, respectively, to 2,3-bis(3R-hydroxymyristoyl)-glucosaminyl-1-phosphate (lipid X) and UDP-DAG revealed that LpxI has a novel two-domain fold wherein the *N*-terminal domain forms the binding pocket for the majority of the lipid substrate and the C-terminal domain forms the pyrophosphatase active site [55]. The *N*-terminal domain is composed of a parallel β-sheet surrounded by 4 α-helices, and the C-terminal domain is composed of a six-stranded β-sheet surrounded by 4 α-helices and a small two-stranded β-sheet [55]. The two domains are connected by a flexible linker that undergoes a secondary structure rearrangement to allow the domains to come together and bring the substrate into the active site (Figure 8) [55]. In the crystal structures, LpxI forms a dimer with two-fold symmetry (Figure 8), and LpxI also appeared to form a dimer in solution [55]. However, the functional relevance of dimerization is unclear [55].

In the structure of LpxI-D225 A in complex with UDP-DAG, uracil is bound by the backbone amides of residues 12 and 232, and ribose is hydrogen-bonded to D188 and T233 [55]. The pyrophosphate forms an ion pair with K214 and hydrogen-bonds with Q169 and T187 (Figure 8) [55].

Furthermore, the importance of T187 was supported by activity assays that showed T187 A had 175-fold less activity than wild-type LpxI [55]. Glucosamine is hydrogen-bonded to N74 and T288, and the carbonyls of the acyl chains are hydrogen-bonded to the backbone amides of residues 75 and 226, while the hydrocarbon chains bind in the complementary hydrophobic pocket of the *N*-terminal domain [55]. Although LpxI activity was stimulated by divalent cations, none are bound to the pyrophosphatase domain of either the crystal structure or are putative nucleophilic waters apparent [54, 55]. Therefore, the exact mechanism of catalysis remains unclear. However, positioning of A225 suggests that D225 may position the nucleophilic water for attack on the β-phosphate and/or act as the catalytic base, and in this case, the divalent cation may simply coordinate the unengaged side of the pyrophosphate to balance the negative charge to allow nucleophilic attack [55].

2.1.7. LpxB. LpxB is the glycosyltransferase that forms the base glucosamine disaccharide of lipid A, catalyzing the formation of the glycosidic bond between the anomeric carbon of the UDP-DAG glucosamine and the 6-hydroxyl of lipid X to form lipid A disaccharide [56]. However, *lpxB* was first identified by a mutation that (in the presence of a

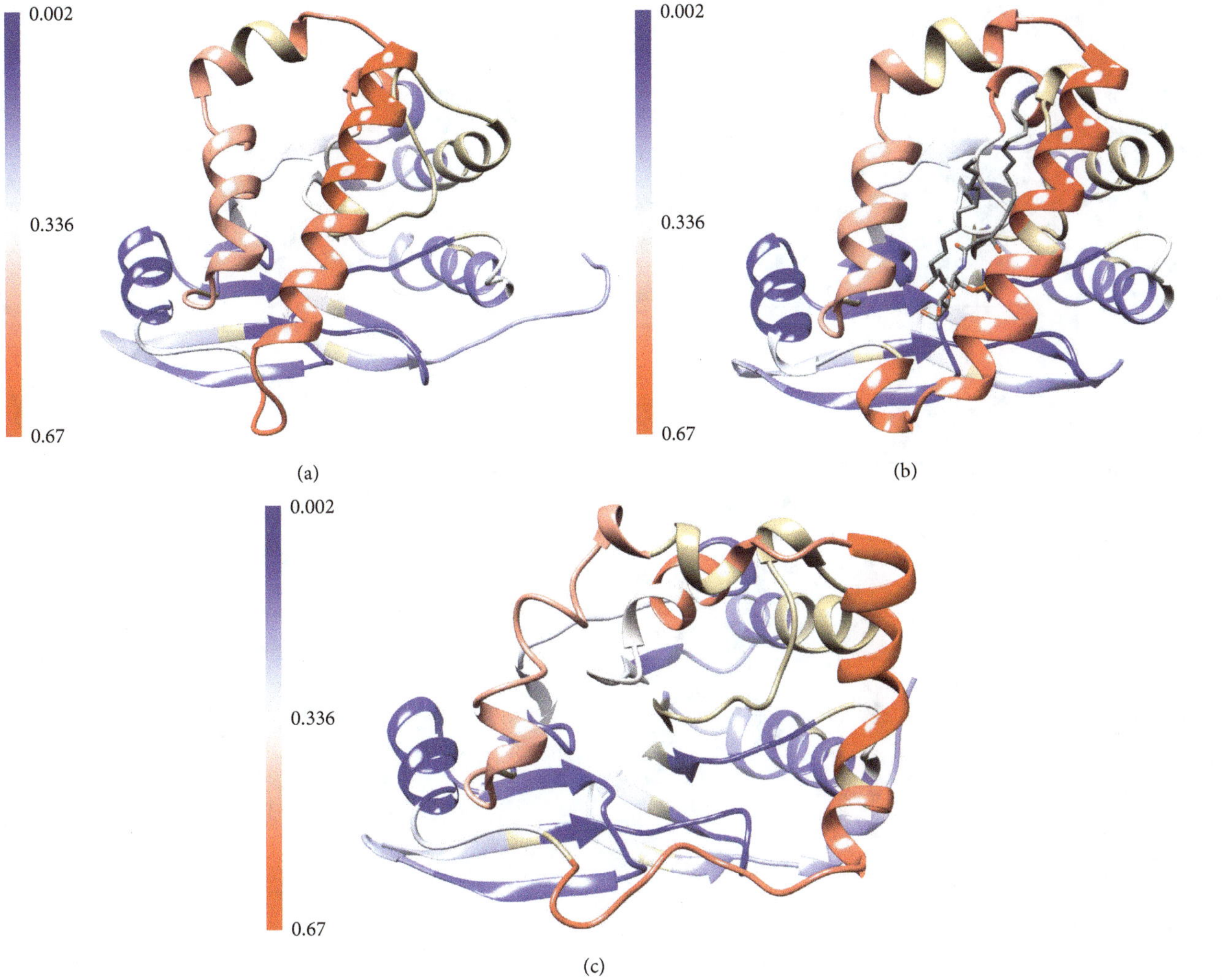

FIGURE 7: Structural dynamics of LpxH cap. PaLpxH ribbons are colored by fractional deuterium uptake at 10 min observed in hydrogen-deuterium exchange experiments of solubilized *E. coli* LpxH [52]. Blue indicates less deuterium uptake. (a) Closed/collapsed cap model observed in molecular dynamics simulations of PaLpxH-F82 G/L83 G [52]. (b) Lipid X-bound PaLpxH crystal structure (PDB: 5B49) [50]. (c). Open cap model observed in molecular dynamics simulations of PaLpxH without lipid X bound [52].

mutation in *pgsA*) decreased synthesis of phosphatidylglycerol at 42°C [57]. *LpxB* mutants were also found to accumulate lipid X, a palmitoylated form of lipid X (lipid Y), and UDP-DAG, suggesting a role in lipid A synthesis, and furthermore, experiments with overexpressed and purified LpxB showed that LpxB forms the glucosamine disaccharide of lipid A [46, 56–61]. In addition, enzymatic assays showed that LpxB is diluted by increasing detergent concentration above the critical micelle concentration, which suggested that LpxB is a surface-active enzyme [61]. Despite being the founding member of the Raetz pathway, LpxB remained structurally uncharacterized until recently.

The crystal structure of a solubilized form of *E. coli* LpxB, which was generated by the mutation of a surface-exposed hydrophobic patch, showed that LpxB forms a glycosyltransferase B (GT-B) dimer with a unique C-terminal swap wherein the last 87 residues of one subunit complete the fold of the opposite subunit (Figure 9) [62–64]. In total, the dimer is composed of four Rossmann-like domains with β-sheets surrounded by α-helices that are connected by α-helical linkers (Figure 9) [62]. The mutated hydrophobic patch appears to be part of a conserved membrane-association motif in GT-B enzymes that often consists of a hydrophobic loop followed by an amphipathic helix containing basic residues in the *N*-terminal domain [63]. This motif has been most well characterized in the GT-B PimA: the hydrophobic loop and basic residues were both found to be required for activity, and a Trp in the amphipathic helix was shown to associate with the membrane [65, 66]. Similarly, decreasing hydrophobicity of the surface-exposed patch in LpxB as hydrophobic residues was mutated to Ser correlated with decreasing activity [62].

The UDP-bound structure of LpxB was also solved, showing that the nucleotide-charged sugar donor substrate (UDP-DAG) binds to the C-terminal domain at the cleft between the domains as is typical for GT-B enzymes (Figure 9) [62–64]. The sugar acceptor substrate (lipid X) likely binds on the *N*-terminal side of the active site cleft near the

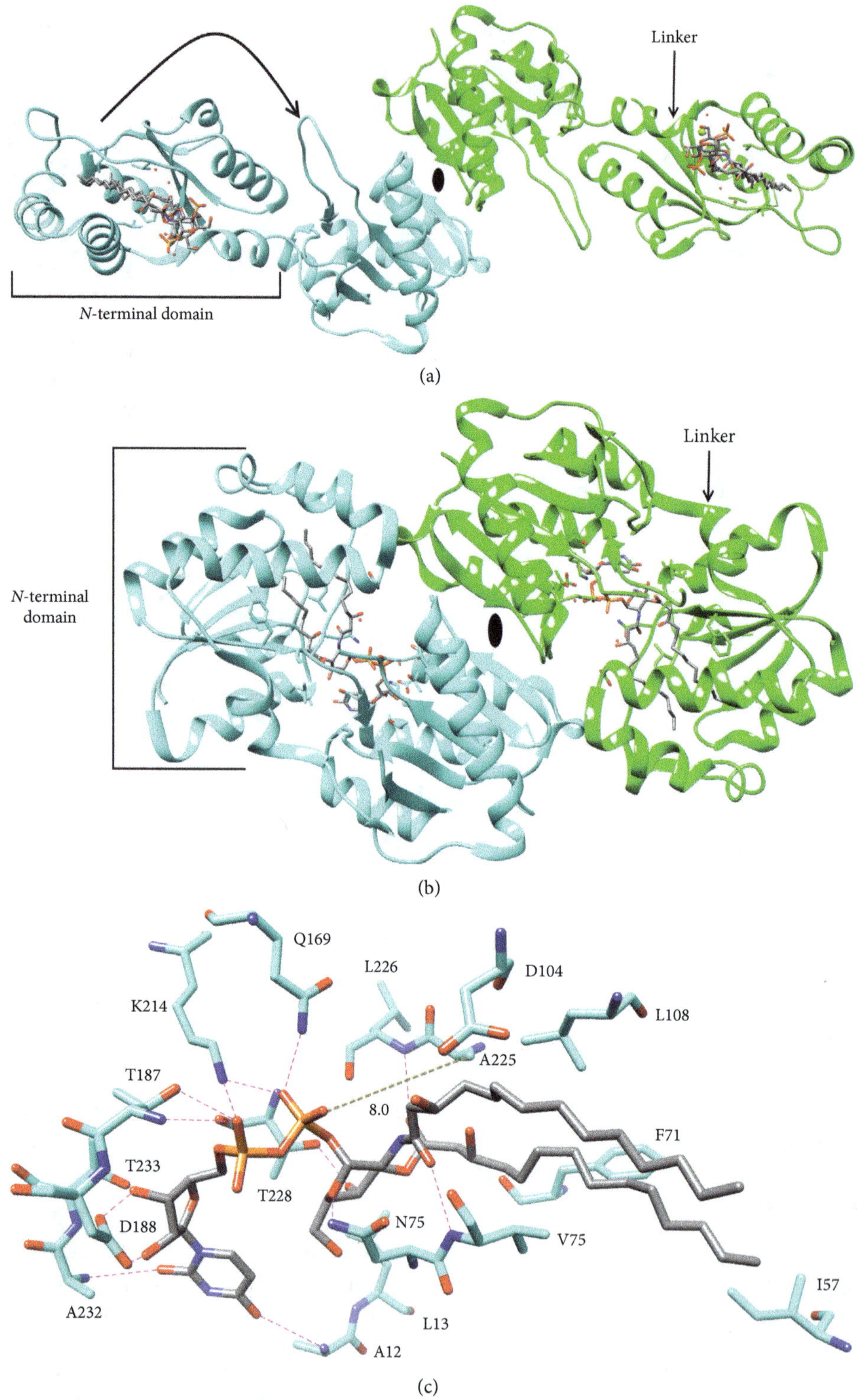

FIGURE 8: LpxI. (a) LpxI dimer bound to lipid X (PDB: 4GGM) [55]. (b) LpxI-D225 A dimer bound to UDP-DAG (PDB: 4J6E) [55]. (c) Residues involved in UDP-DAG binding to LpxI-D225 A from (b). Hydrogen-bond and salt-bridge interactions are shown, and the distance between A225 and the β-phosphate is highlighted. Distances in Ångstroms are as follows: 3.3 from 3′-acyl carbonyl to L226 N 2.5 from ribose 2′-hydroxyl to D188 Oδ2, 2.3 from ribose 3′-hydroxyl to D188 Oδ1 and 3.0 to T233 Oγ, 3.3 from α-phosphate to T187 N and 2.8 to T187 Oγ and 3.3 to K214 Nζ, 2.6 from β-phosphate to K214 Nζ and 2.9 to Q169 Nε, 2.5 from glucosamine 4′-hydroxyl to T228 Oγ, 3.3 from uracil O2 to A232 N 2.9 from the glucosamine ring O to N74 Nδ, 3.4 from uracil O4 to A12 N, and 3.1 from 2′-acyl carbonyl to V75 N.

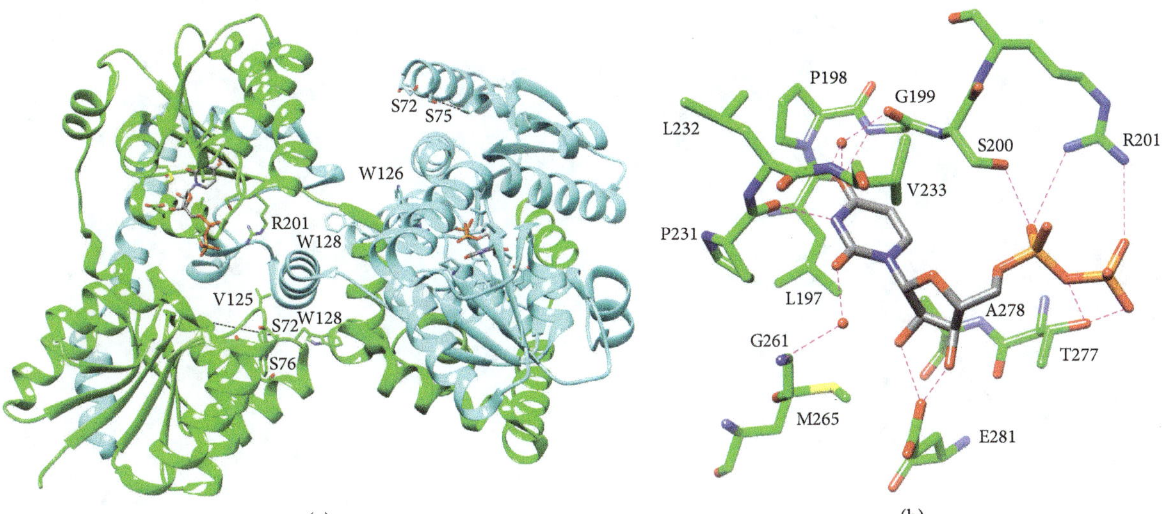

(a) (b)

FIGURE 9: LpxB. (a) Ribbon structure of LpxB dimer bound to UDP (PDB: 5W8X) [62]. Three Ser side chains of six in the hydrophobic patch are shown; S66, S68, and S69 are part of preceding disordered loop and were not visible in the electron density [62]. V125, W126, and W128, which may line the hydrophobic groove that binds the substrates' acyl chains, are also shown [62]. (b) Residues involved in UDP binding from (a). Hydrogen-bond and salt-bridge interactions are shown. Distances of these interactions in Ångstroms are as follows: 2.5 from E281 Oε2 to ribose O2$'$ and 2.4 to ribose O3$'$, 2.9 from T277 Oγ to β-phosphate O3 and 2.9 to pyrophosphate bridging O 3.8 from R201 Nη1 to α-phosphate O2 and 3.8 from Nη2 to β-phosphate O1, 2.9 from G261 N to water and 2.9 from this water to uracil O2, 2.5 from P231 O to uracil N3, 2.7 from G199 N to uracil O4, 3.0 and 2.8 from V233 N and G199 O (respectively) to water, and 3.2 from this second water to uracil O4.

predicted catalytic base (D98), and a hinge-like movement of the *N*-terminal domain, similar to conformational changes observed in other GT-B enzymes, may be required to bring the substrates into alignment for catalysis [61–64].

Uracil is recognized by hydrogen-bonding interactions with the backbone of residues 199 and 231 and via two water molecules to the backbone of residues 199, 233, and 261 (Figure 9) [62]. Ribose is recognized by E281, and the pyrophosphate is bound by hydrogen-bonds to S200 and T277 and a salt-bridge with R201 (Figure 9) [62]. R201 may serve to stabilize the negative charge of the UDP leaving group, and activity assays confirmed that this residue is critical for activity [61, 62]. Molecular docking of UDP-DAG correctly identified the UDP binding pocket and predicted that the acyl chains extend up a hydrophobic groove lined by V125, W126, and W128, and activity assays showed that W126 is indeed important for activity [61, 62]. Finally, the molecular docking model suggested that lipid X binds on top of UDP-DAG, which implies that LpxB follows an ordered sequential mechanism [62].

2.1.8. LpxK. LpxK is the kinase that phosphorylates the 4$'$-hydroxyl of the distal glucosamine of lipid A disaccharide to form lipid IV_A [67, 68]. LpxK has been proposed to be the second regulatory point of the Raetz pathway: LpxK may be stimulated by unsaturated fatty acids, and its lipid A disaccharide substrate may stimulate the proteolysis of LpxC by FtsH, which requires the critical scaffold/chaperone protein LapB [33, 34]. This regulation allows Gram-negative bacteria to balance phospholipid and LPS synthesis as both pathways compete for β-hydroxymyristoyl-ACP [32]. High flux through the Raetz pathway depletes β-hydroxymyristoyl-ACP, limiting the production of unsaturated fatty acids, and decreased unsaturated fatty acid production leads to decreased LpxK activity, which causes a buildup of lipid A disaccharide [32, 33]. Buildup of this intermediate then leads to the proteolysis of LpxC, decreasing the consumption of β-hydroxymyristoyl-ACP by the Raetz pathway [32, 33]. In addition, LpxK is the final absolutely required step of the Raetz pathway as *E. coli* could transport lipid IV_A to the outer membrane and remain viable [7, 69].

LpxK was most active with ATP and an equimolar concentration of the divalent cation, showing little activity when divalent cations were removed with EDTA [67, 70]. For the lipid substrate, LpxK was specific for the glucosamine disaccharide head-group, showing no activity for lipid X, UDP-DAG, or substrates already containing Kdo core oligosaccharide sugars [68]. However, LpxK had little specificity for the number of acyl chains attached to the glucosamine disaccharide: LpxK could phosphorylate substrates with 2–6 acyl chains and substrates wherein the ester-linked chains were replaced with amide-linked chains [67, 68].

Apo, substrate analogue-bound, and product-bound crystal structures of *Aquifex aeolicus* LpxK have been determined [70–72]. These structures reveal that LpxK is composed of two Rossmann-like domains with β-sheets surrounded by α-helices, and these domains are connected by two β-strands that are coextensive with the β-sheets of both domains (Figure 10) [71]. The larger *N*-terminal domain contains the P-loop/Walker A and Walker B motifs thus

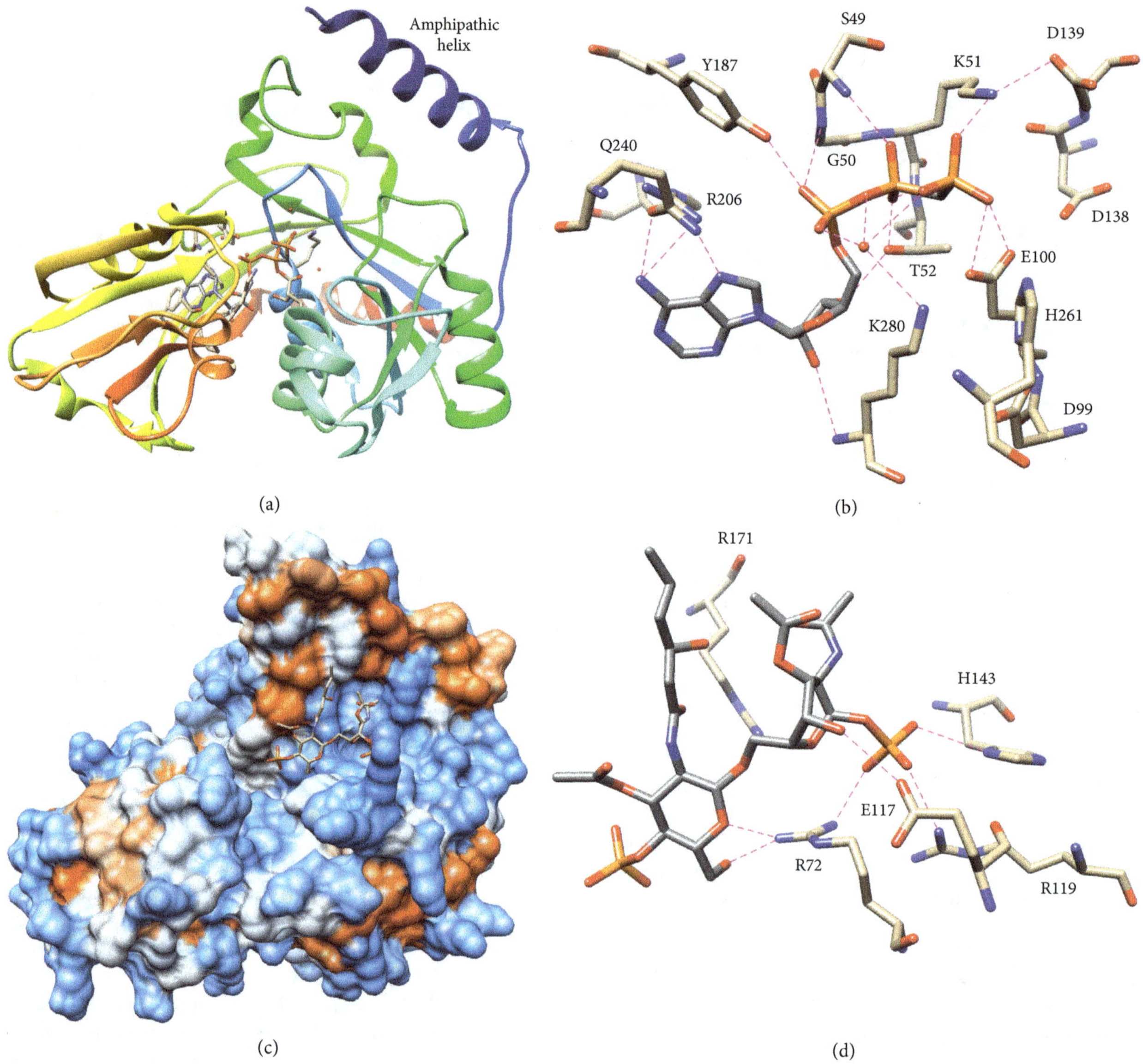

FIGURE 10: LpxK. (a) LpxK bound to AMP-PCP ATP analogue (PDB: 4ITL) [70]. Spectrum coloring begins with blue at the *N*-terminus. (b) Residues involved in AMP-PCP binding from (a). Distances for hydrogen-bond and salt-bridge interactions are shown with distances in Ångstroms as follows: 3.0 from adenine N6 to Q240 Oε and 3.5 to R206 Nη1, 3.0 from adenine N7 to Q240 Nε, 2.8 from ribose 3$'$-hydroxyl to T52 Oγ, 2.9 from β-phosphate to S49 N and 3.1 to K51 N and 2.7 to T52 N and 2.6 to T52 Oγ, 3.0 from α-phosphate to G50 N and 2.6 to Y187 Oη, 3.4 from ribose 2$'$-hydroxyl to K280 N 2.8 from D99 Oδ1 to H261 Nδ, 3.0 from K51 Nζ to D139 Oδ1 and 2.7 to γ-phosphate, 2.9 and 2.8 from γ-phosphate to E100 Oε2/1, 3.2 from K280 Nζ to water, and 2.2, 3.0, and 3.0 from this water to the phosphates. (c) Hydrophobicity surface of LpxK bound to lipid IV$_A$ (PDB: 4LKV) [72]. Hydrophobicity is shown in an orange-blue scale with blue representing the hydrophilic end. (d) Residues involved in binding lipid IV$_A$ from (c). Distances in Ångstroms for hydrogen-bonds and salt-bridges are as follows: 2.4 from 6$'$-hydroxyl to R72 Nη2 and 3.2 from the distal glucosamine ring O 3.4 from 1-phosphate to R119 Nη1 and 3.2 to H143 Nδ and 2.9 from R72 Nη1, and 3.3 from 4-hydroxyl to E117 Oε.

binding ATP, and the C-terminal domain binds the lipid substrate [70–72]. A hinge-like movement of the domains closes the distance between the binding pockets observed in the apo structure to bring the substrates into position for phosphoryl transfer [71]. In addition, LpxK has an *N*-terminal amphipathic helix and adjacent basic patch, which likely mediate membrane association; consistent with the presence of these structural elements, LpxK was optimally active at the critical micelle concentration of Triton X-100, and surface dilution of LpxK activity was observed at higher concentrations [70, 71].

Moreover, the amphipathic helix was required for activity [72]. LpxK was the first membrane active kinase of the P-loop-containing NTPase superfamily to be identified [71].

The AMP-PCP-bound structure of *A. aeolicus* LpxK shows that adenine is recognized by hydrogen-bonding to Q240 and R206, while ribose is bound by T52 and the backbone amide of K280 [70]. Furthermore, the α- and β-phosphates are engaged in extensive hydrogen-bonding interactions with T52, Y187, the backbone amides of residues 49–52, and an ordered water bound to K280 [70]. The

γ-phosphate is positioned for transfer by a salt-bridge with K51 and hydrogen bonds to E100 and the ordered water (Figure 10) [70]. Homology with other P-loop-containing ATPases suggests that T52, E100, and D138 will coordinate Mg^{2+} in the complex with ATP-Mg [70, 71]. The role of E100 in Mg^{2+} coordination rather than simply in hydrogen-bonding is supported by E100 A/D/Q mutants, wherein E100Q showed the greatest decrease in k_{cat}, and E100D showed the least as Asp can also coordinate Mg^{2+}, and Ala may allow binding of a compensatory water [70]. In addition, H261 is positioned to accept a proton from the attacking 4′-hydroxyl of lipid A disaccharide, and D99, which is hydrogen-bonded to H261, likely serves to increase its pK_a [70]. Consistent with this role, D99 A/E/N mutants showed the same trend as the E100 mutants [70]. Kinetic experiments also confirmed the importance of K51, T52, and H261, showing 3000-fold decreases in k_{cat} for K51 A and T52 A and an 850-fold decrease for H261 A, and the Walker B motif residues D138 and D139 were observed to be critical for catalysis and ATP-Mg binding, decreasing k_{cat} 4700- and 8100-fold and increasing K_M 3.3- and 4-fold, respectively [70].

The structure of LpxK bound to its lipid IV_A product shows that LpxK binds the lipid IV_A glucosamines with the acyl chains extending into solvent and is largely disordered, which explains the minimal specificity for the extent of acylation [67, 68, 72]. The lipidic substrate binding site is on the same face as the *N*-terminal amphipathic helix, which suggests that LpxK may only partially extract lipid A disaccharide from the membrane during catalysis [72]. Even with glucosamine disaccharide, there are relatively few interactions [72]. The 1-phosphate is recognized by R72, R119, and H143; however, the proximal glucosamine is only hydrogen-bonded to E117, and the distal glucosamine is only hydrogen-bonded to R72 (Figure 10) [72]. Kinetic experiments confirmed the importance for lipid A disaccharide binding of R72 and H143, as Ala mutants increased K_M 4- and 7-fold, respectively, but R119 was largely dispensable, showing no change in K_M and only a 67-fold decrease in k_{cat} [72]. Finally, R171 A increased K_M 20-fold, but R171 was not directly bound to lipid IV_A, which suggests that the lipid IV_A-bound structure may not fully recapitulate productive lipid A disaccharide binding [72].

2.1.9. WaaA (KdtA). WaaA, also called KdtA, is the glycosyltransferase that, in *E. coli*, transfers two Kdo residues from CMP-Kdo: first to the 6′-hydroxyl of the distal glucosamine and then to the 4′-hydroxyl of the first Kdo [73, 74]. While *E. coli* WaaA is thus a bifunctional glycosyltransferase, WaaA variants from different species transfer 1–4 Kdo residues [1, 75, 76]. Activity assays also showed that, like LpxK, WaaA had little specificity for the extent of acylation, accepting substrates with 3–6 acyl chains, nor was 1-phosphate required [73, 74, 76]. However, 4′-phosphate was important as lipid A disaccharide was a poor substrate for *E. coli* WaaA although even lipid A disaccharide could be glycosylated to a lesser extent by *Aquifex aeolicus* WaaA [73, 74, 76]. Conversely, WaaA was highly specific for the donor substrate, showing activity and specific binding for CMP but no other nucleoside monophosphates [74, 76]. Interestingly, WaaA is another point of regulation in the Raetz pathway as WaaA was also degraded by the FtsH protease that targets LpxC [77].

Crystal structures of the monofunctional WaaA from *A. aeolicus* revealed that WaaA has a typical glycosyltransferase B superfamily (GT-B) fold with two Rossmann-like domains (Figure 11) [78]. Moreover, the *N*-terminal domain of WaaA has a basic and hydrophobic surface as observed for other membrane surface-active GT-B enzymes (including LpxB as discussed above), which is thought to be required for productive membrane association [62, 63, 65, 66, 78, 79]. As expected for a GT-B enzyme, CMP was observed to bind to the C-terminal domain, and a putative lipid IV_A binding pocket was identified on the *N*-terminal domain [63, 78]. WaaA appears to have been crystallized in an open conformation with a wide cleft separating the domains; therefore, a hinge-like movement of the domains may be required to bring the substrates into position for nucleophilic attack [63, 65, 78–81].

The CMP-bound structure of *A. aeolicus* WaaA shows that cytosine binds in a hydrophobic pocket formed by F247 and L250 and is recognized by hydrogen-bonds with the backbone carbonyl of P211 and amide of G248 [78]. Ribose is hydrogen-bonded to E276, and the phosphate is engaged by R212 and N273 (Figure 11) [78]. Activity assays confirmed the importance of R212, which may stabilize the negative charge of the CMP leaving group during catalysis [78]. In addition, activity assays of Ala mutants confirmed the importance of residues G30, E31, E98, and K162 in the putative lipid IV_A binding pocket [78]. E31 appears well positioned to act as the catalytic base by accepting a proton from the 6′-hydroxyl of the distal glucosamine, and G30 could prevent steric clash from residue 30 that could interfere with proper alignment for nucleophilic attack [78]. If this role is correct for E31, then E98 and K162 may be important for positioning Kdo for transfer, possibly by contacting its 1′-carboxylate, and residues S28, S54, and R56 may bind the 1-phosphate of lipid IV_A [78]. Mutation of all 3 of these putative 1-phosphate binding residues to Ala significantly reduced activity [78]. Finally, increase in Trp-fluorescence upon lipid IV_A binding in the wild-type but not W102 A WaaA suggested that the acyl chains bind on the top of W102, shielding it from solvent quenching [78].

2.1.10. LpxL and LpxP. In *E. coli*, LpxL is the acyltransferase that adds a lauroyl group to the β-hydroxyl of the 2′-β-hydroxymyristate on the distal glucosamine of Kdo_2-lipid IV_A [82, 83]. Enzymatic assays showed that LpxL was a membrane surface active enzyme, and a predicted *N*-terminal transmembrane helix was required for activity *in vivo* [83]. In addition, enzymatic assays showed that LpxL was selective for the presence of the Kdo residues, as its activity for lipid IV_A was 6000-fold lower [83]. Furthermore, LpxL was selective for lauroyl-ACP: LpxL was 5% as active with lauroyl-coenzyme A (CoA), showed further

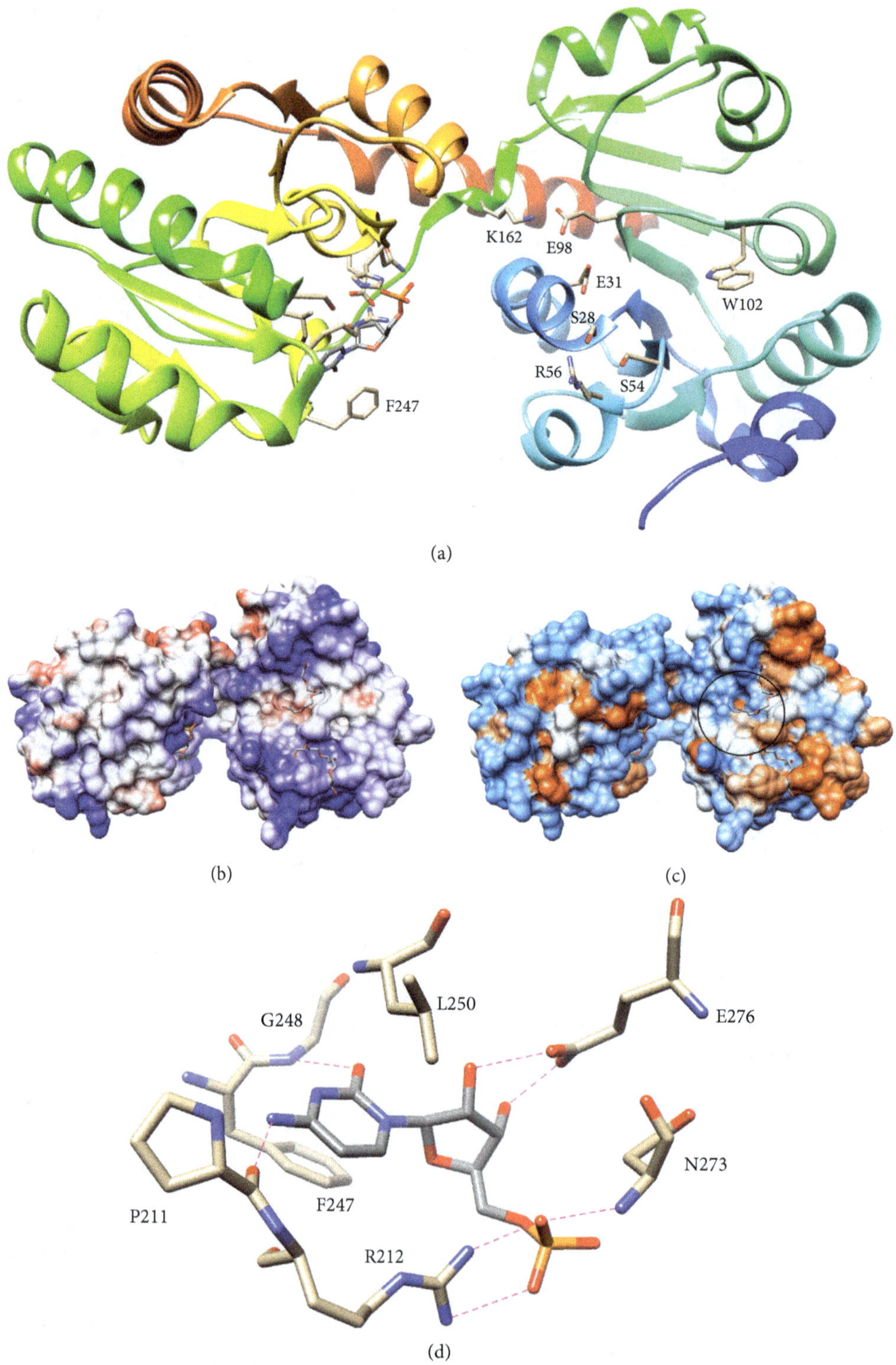

FIGURE 11: WaaA. (a) WaaA bound to the CMP product (PDB: 2XCU) [78]. Spectrum coloring ends with red at the C-terminus. (b) Coulombic surface of WaaA from (a). Charge is shown with a blue-red scale with blue at the positive end. Bound polyethylene glycol molecules are shown. (c) Hydrophobicity surface of WaaA from (a). Hydrophobicity is shown on an orange-blue scale with orange as the hydrophobic end. The putative lipid IV_A-binding pocket is circled. (d) Residues involved in CMP binding from (a). Hydrogen-bond and salt-bridge interactions are shown with the following distances in Ångstroms: 3.0 from phosphate to N273 N and 2.6 and 3.2 to R212 Nη1/2, 2.5 from E276 Oε1 to ribose 3′-hydroxyl and 2.8 to the 2′-hydroxyl, 2.9 from cytosine O2 to G248 N, and 3.1 from cytosine N4 to P211 O.

reduced activity with decanoyl-CoA and myristoyl-CoA, and incorporated little or no acyl chains 16 carbons or longer from acyl-CoA [83].

Conservation of catalytic residues suggested that LpxL is related to the glycerol-3-phosphate acyltransferase (GPAT) family [83]. Consistent with this hypothesis, mutation to Ala of H132 and E137 of the predicted catalytic dyad in *E. coli* LpxL decreased activity 1000- and 3000-fold, respectively [83]. Moreover, Ala mutations of conserved R169 and D200 residues decreased activity 170- and 15-fold [83]. However, P238 A had little effect even though this Pro is important in the GPAT family, and this Pro was not conserved in the related enzyme LpxP [83].

E. coli also have a cold-shock inducible alternative to *lpxL* (*lpxP*) that replaces LpxL when cells are grown at 12°C [84]. LpxP transfers a palmitoleoyl chain from ACP instead of a lauroyl chain [84]. Like LpxL, LpxP was specific for the presence of the Kdo groups of Kdo_2-lipid IV_A as LpxP failed to acylate lipid IV_A [84]. However, *E. coli* waaA mutants could produce penta- and hexa-acylated forms of lipid A containing laurate, myristate, and palmitoleate when grown on minimal media at 21°C, indicating that LpxL, LpxP, and LpxM have some activity for lipid IV_A [69]. In addition, LpxP was highly specific for (16 : 1) palmitoleoyl-ACP, showing little activity with shorter or saturated acyl chains or with palmitoleoyl-CoA [84]. While LpxL could substitute for LpxP at 12°C in a *δlpxP E. coli* strain, this strain showed increased sensitivity to rifampicin and vancomycin at 12°C, which suggests that incorporation of unsaturated acyl chains into the outer leaflet of the outer membrane by LpxP maintains the selective barrier at low temperature [85]. Neither LpxL nor LpxP have been structurally characterized, but the related enzyme LpxM has been [86].

2.1.11. LpxM. In *E. coli*, LpxM is the acyltransferase that attaches a myristoyl group to the β-hydroxyl of the 3′-β-hydroxymyristate on the distal glucosamine of the LpxL product (Kdo_2-lipid V_A) to produce the end product of the Raetz pathway, hexa-acylated Kdo_2-lipid A [87]. Enzymatic assays suggested that *E. coli* LpxM was specific for the penta-acylated substrate as LpxM acylated Kdo_2-lipid IV_A and lipid IV_A much more slowly [87]. However, while LpxM showed poor activity for decanoyl-ACP, β-hydroxymyristoyl-ACP, palmitoyl-ACP, and palmitoleoyl-ACP, LpxM appeared to only have a slight preference for myristoyl-ACP over lauroyl-ACP [87].

The crystal structure of LpxM from *Acinetobacter baumannii* reveals that LpxM has a unique fold with a twisted, 7-stranded β-sheet surrounded by 10 α-helices that form a deep hydrophobic pocket and with an *N*-terminal transmembrane/membrane-association helix [86].

Unlike *E. coli* LpxM, *A. baumannii* LpxM transfers lauroyl chains to the β-hydroxyls of the acyl chains at the 2- and 3′-positions of the glucosamine disaccharide; therefore, *A. baumannii* LpxM is a bifunctional acyltransferase that transfers secondary lauroyl groups to both glucosamines to produce hepta-acylated lipid A [86]. Kinetic experiments showed substrate inhibition by lauroyl-ACP, suggesting that LpxM utilizes an ordered sequential mechanism in which the acyl acceptor must bind first [86]. Consistent with the bifunctionality of *A. baumannii* LpxM, substrate inhibition was best fit by a two-site binding model wherein only one lauroyl-ACP is inhibitory, and substrate inhibition was alleviated in the K282 E/K285 E double mutant of a putative ACP-binding site at the top of the lipid-binding pocket [86].

Like LpxL, LpxM has a conserved His and Glu (H122 and E127 in *A. baumannii*) that, in distantly related acyltransferases, forms the catalytic dyad with the His acting as the catalytic base that accepts a proton from the nucleophilic hydroxyl [86, 88]. Kinetic experiments confirmed that E127 was critical for activity [86]. In addition, LpxM has a conserved Arg and Asp (R159 and D192); R159 was also confirmed to be critical for activity [86]. The crystal structure of *A. baumannii* LpxM reveals the location of these residues in the hydrophobic pocket, and the positions of these residues suggest that, at least in this bifunctional LpxM, the roles of the conserved HX_4E/D motif have been separated to form two catalytic dyads that catalyze acyl transfer to the 2- and 3′-β-hydroxymyristoyl chains (Figure 12) [86]. In the glycerol-3-phosphate acyltransferase from *Cucurbita moschata*, the His and Asp of the HX_4E/D catalytic dyad are hydrogen-bonded, suggesting that Asp increases the pK_a of His [89]. However, in *A. baumannii* LpxM, H122 and E127 are separated by 7.6 Å, and H122 is hydrogen-bonded to D192 [86]. Moreover, R159 forms an ion pair with E127 [86]. Therefore, H122 and D192 may serve as the catalytic dyad for the transfer of one lauroyl group, and E127 and R159 may act form the catalytic dyad for the transfer of the second lauroyl group. Finally, the structure of the *A. baumannii* LpxM E127 A mutant was resolved with an acyl chain (modeled as undecanoic acid) bound in one of two particularly deep, narrow channels in the active site (Figure 12) [86]. These channels may act as hydrocarbon rulers to control the length of the transferred acyl chains similar to those in LpxA and LpxD as discussed above [86]. Consistent with the hypothesis that H122/D192 and R159/E127 form two catalytic dyads for separate acyl transfer reactions, one of these putative hydrocarbon-ruler channels is near H122 and D192, and the other channel is near R159 and E127 (Figure 12).

2.1.12. LpxJ and LpxN. Some bacteria encoding LpxL but not LpxM instead encode a distant homologue of LpxL-, LpxM-, and LpxP-designated LpxJ [90]. LpxJ enzymes from three ε-proteobacteria species: *Helicobacter pylori*, *Campylobacter jejuni*, and *Wolinella succinogenes*, which were characterized, showing a range of substrate specificities [90]. Ultraviolet photon dissociation mass spectrometry combined with activity assays confirmed that *H. pylori* LpxJ acylated the β-hydroxyl of the 3′-acyl chain, like *E. coli* LpxM, and not the acyl chains of the proximal glucosamine [90]. However, none of these enzymes required the presence of the Kdo groups, and *H. pylori* LpxJ was not specific for the presence of the 2′-secondary acyl chain added by LpxL, while the other LpxJ variants were only active on Kdo_2-lipid IV_A or lipid IV_A, not the LpxL product [90]. *H. pylori* and *W. succinogenes* LpxJ were most active with lauroyl-ACP, but *H. pylori* LpxJ also utilized myristoyl-ACP [90]. On the contrary, *C. jejuni* LpxJ was most active with palmitoyl- or stearoyl-ACP [90]. In addition, *Vibrio cholerae* utilizes another alternative to

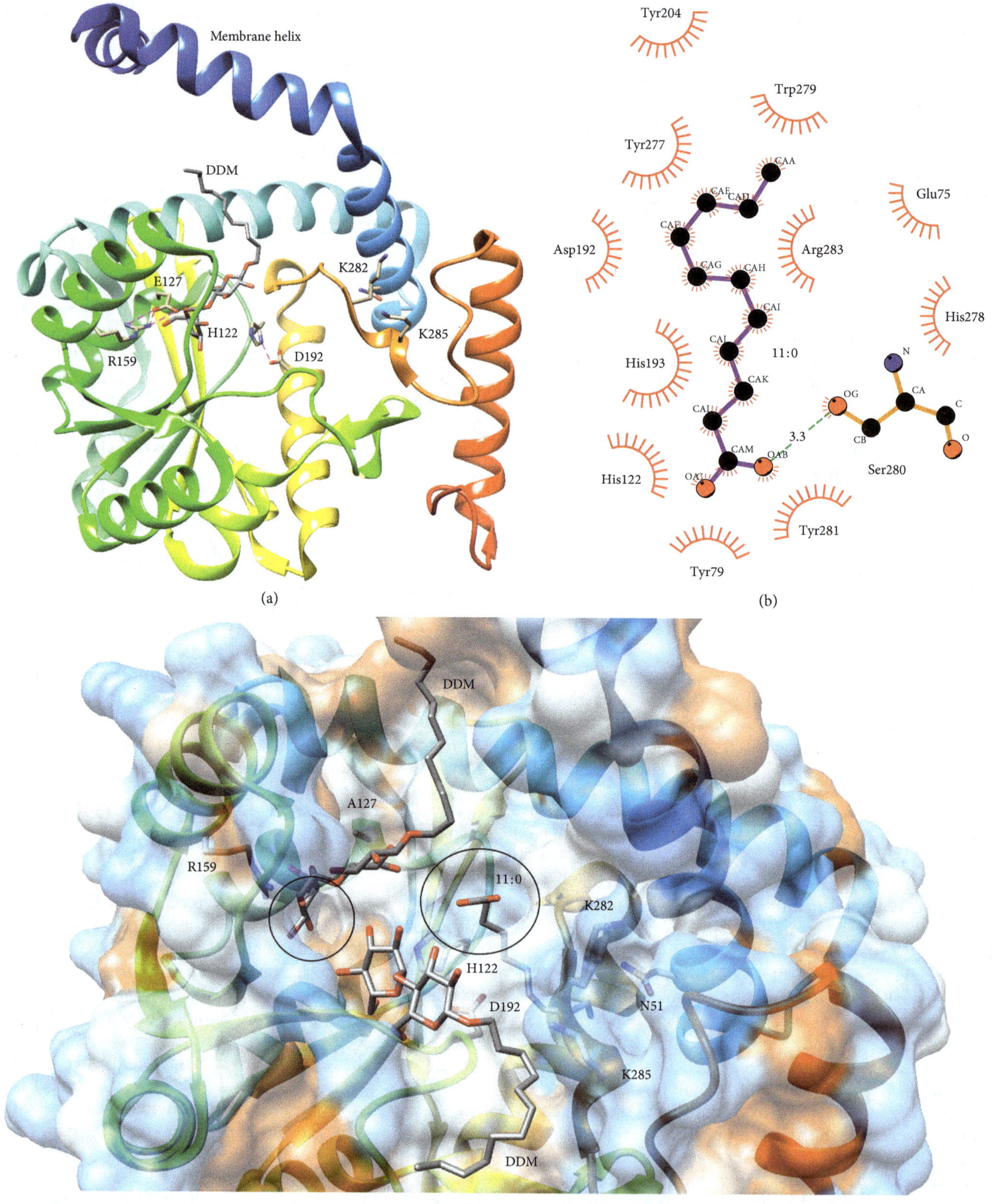

FIGURE 12: LpxM. (a) Ribbon diagram of LpxM (PDB: 5KN7) [86]. Spectrum coloring ends with red at the C-terminus. Interactions between putative catalytic residues are shown. Distances in Ångstroms are 2.6 from H122 Nδ to D192 Oδ1, 2.4 from E127 Oε1 to R159 Nη2 and 3.4 to Nη1, and 3.2 from E127 Oε2 to R159 Nη2. (b) Ligplot+ diagram of fatty acid binding to LpxM (PDB: 5KNK) with distance shown in Ångstroms [45, 86]. Hydrophobicity surface of LpxM (PDB: 5KNK). Hydrophobicity is shown on an orange-blue scale with blue as the hydrophilic end. The entrances to putative hydrocarbon-ruler channels are circled.

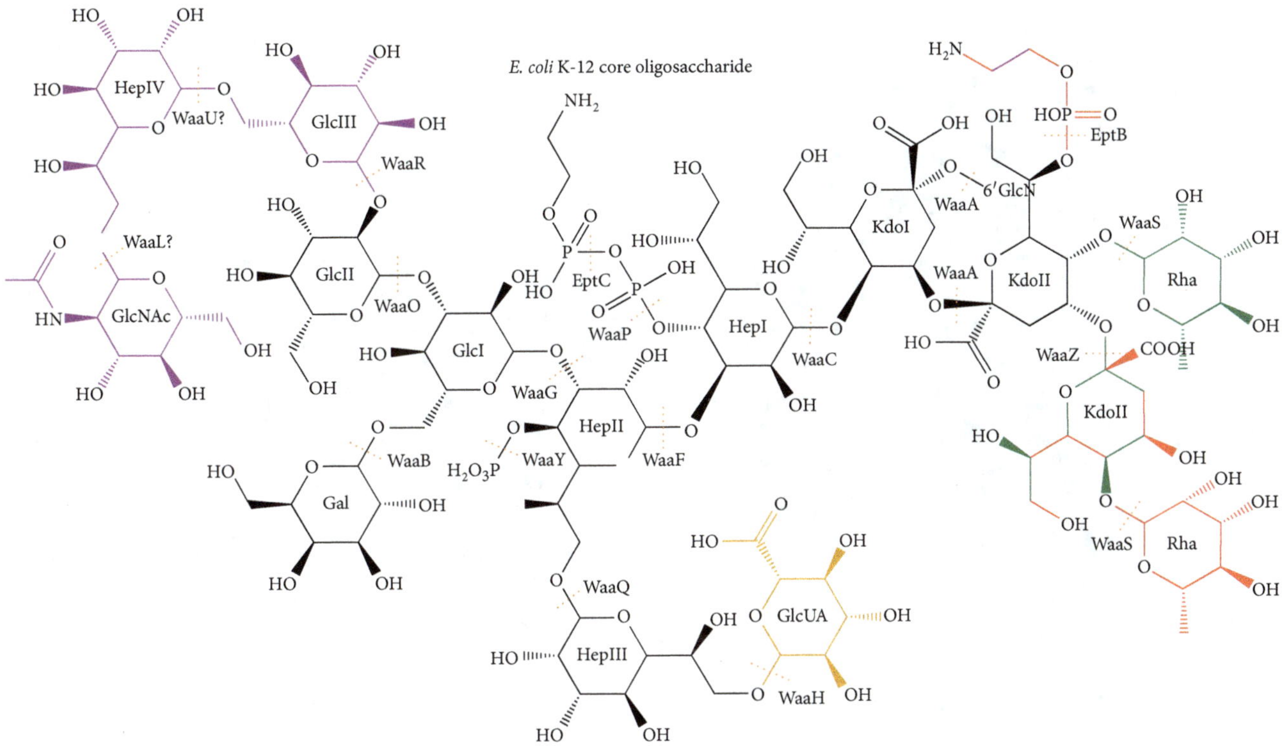

FIGURE 13: Core oligosaccharide synthesis. Coloring of core oligosaccharide is the same as in Figure 1. Transferases responsible for the formation of bonds linking residues in the K-12 core oligosaccharide are marked in orange [1, 3, 4, 95–100].

LpxM, designated LpxN, that transfers a β-hydroxylauroyl chain to the β-hydroxyl of the $3'$-acyl chain [91]. Thus, the different activities and substrate specificities of the LpxL/M homologues encoded by different bacterial species result in diversity in the length, extent, and position of acylation in the final lipid A product of the Raetz pathway in different species [1].

2.2. LPS Synthesis and Transport

2.2.1. Core Oligosaccharide Synthesis. Following the completion of the Raetz pathway, the core oligosaccharide is completed at the cytoplasmic side of the inner membrane [1, 92]. The core oligosaccharide can be divided into inner and outer cores, and the inner core tends to be conserved within a species or genus, while species typically have a small number of different outer cores synthesized in different strains [1, 92]. Because of the variability of core oligosaccharides, a discussion of the biochemical and structural details of biosynthetic enzymes that synthesize the many different core oligosaccharides is beyond the scope of this review. For an overview of the structure and biosynthesis of different core oligosaccharides, the reader is referred to the following reviews [1, 93, 94]. However, the biosynthetic pathway of the core oligosaccharide from *E. coli* K-12 is briefly covered as an example. As described above, the first two Kdo residues are added during the Raetz pathway, and these sugar residues are the most conserved moiety of the inner core with all known core oligosaccharides containing Kdo [1, 73, 74]. In addition, the inner core typically contains L-glycero-D-mannoheptose (Hep) or, less frequently, the D-glycero-D-mannoheptose isomer [1]. In *E. coli*, the first and second Hep residues (Figure 13) are added sequentially to the first Kdo by WaaC and WaaF, respectively, and a branching Hep is added to the second Hep by WaaQ [1, 95]. The first and second Hep residues are phosphorylated by WaaP and WaaY, respectively, and the activity of these kinases is dependent on the addition of the first outer core residue by WaaG though WaaP retained some activity in δ*waaG* strains [1, 96, 101].

These phosphorylated inner core residues are thought to be important for crosslinking of LPS molecules by divalent cations, and loss of phosphorylation increases sensitivity of *E. coli* to detergent and some antibiotics and strongly induces RpoE-regulated envelope stress response via increased expression of this σ factor [1, 96, 101]. Moreover, the second Hep is the attachment site of this first outer core residue, which is glucose in *E. coli* [1, 97]. In *E. coli* K-12, the first glucose is the site of attachment of a branching galactose by WaaB and a second glucose residue by WaaO [97]. This WaaO product is a branch point in core synthesis in *E. coli* K-12: either WaaR adds a third glucose to the second glucose, or WaaZ adds a third Kdo to the second Kdo [3]. Competition between these pathways appears to be controlled by the relative levels of WaaR and WaaZ, which are dependent on growth conditions with phosphate limitation favoring WaaZ and optimal growth conditions favoring WaaR [3]. Following addition of the third Kdo, WaaS can attach L-rhamnose (Rha) to the second Kdo or, if the second

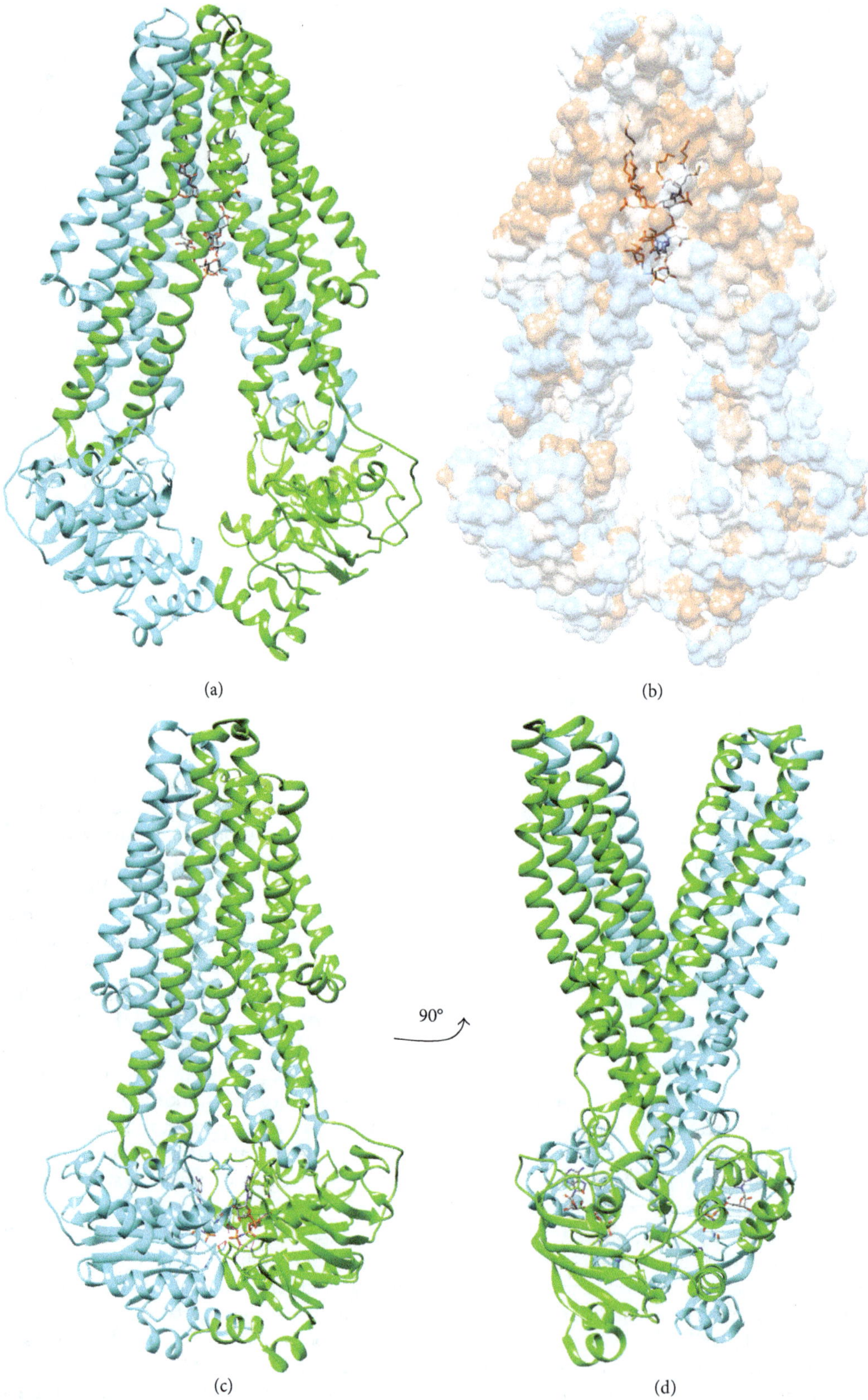

Figure 14: MsbA. (a) Cytoplasmic-open conformation of MsbA bound to LPS (PDB: 5TV4) [104]. (b) Hydrophobicity surface of MsbA in (a). Orange is the hydrophobic end of the scale, and blue is the hydrophilic end. The mostly orange region roughly indicates the coverage of the nanodisc membrane in the electron density [104]. (c) Periplasmic-open conformation of MsbA bound to AMP-PNP ATP analogue (PDB: 3B60) [103].

Kdo has already been modified with phosphoethanolamine by EptB, to the third Kdo [3]. In the WaaR pathway, a fourth heptose is attached to the third glucose to complete the outer core (Figure 13) [1]. Sequence similarity with other heptosyltransferases suggests that this reaction is catalyzed by WaaU [99].

2.2.2. Transport of Core-lipid across the Inner Membrane by MsbA. Following the completion of the core oligosaccharide, the core-lipid A molecule is flipped across the inner membrane by the ATP-binding cassette (ABC) transporter MsbA [7, 92, 102]. Decreased activity of MsbA for earlier intermediates of LPS synthesis lacking core sugars or secondary acyl chains likely helps to prevent premature transport [7]. Structures of MsbA obtained by cryogenic electron microscopy (cryo-EM) and X-ray crystallography have provided insights into the mechanism of transport and selectivity [103–105]. These structures show that MsbA forms a functional dimer wherein the transmembrane domains are comprised of 4 transmembrane helices from one subunit and 2 transmembrane helices from the opposite subunit (Figure 14) [103, 104]. A cytoplasmic P-loop containing the ATPase domain is appended to TM6, and cytoplasmic helices between TM2 and TM3 and between TM4 and TM5 contact the ATPase domains [103, 104]. The helix between TM2 and TM3 contacts its own ATPase domain at the ATP-binding site, and the helix between TM4 and TM5 contacts the ATPase domain of the opposite subunit (Figure 14) [103, 104].

The ATPase domains also act as a functional dimer: the 3.7 Å crystal structure of *Salmonella* MsbA bound to a nonhydrolyzable ATP analogue (AMP-PNP) shows that ATP binds between the ATPase domains (Figure 14(c)) [103]. One half of the ATPase active site formed by the interstrand loops of two β-sheets and an α-helix sandwiched between them from one ATPase domain, and the other half is contributed by the α-helical side of the other ATPase domain [103]. In this structure, the transmembrane domain assumes a periplasmic open conformation (Figure 14(c)) [103].

A 4.2 Å cryo-EM structure of *E. coli* MsbA in the cytoplasmic open conformation shows core-lipid A bound inside the transmembrane domains with partially resolved acyl chains and core oligosaccharide resolved out to the three inner core Hep residues (Figure 14) [104]. The core-lipid A binding pocket is separated into hydrophobic lipid A-binding and hydrophilic oligosaccharide-binding regions by a ring of primarily basic residues that recognize the phosphorylated glucosamine disaccharide of lipid A [104]. Additionally, a recent 2.9 Å crystal structure of MsbA bound to core-lipid A and a small molecule inhibitor fully resolved the acyl chains, and the tight packing of the chains into the hydrophobic pocket suggests that MsbA selectivity for completed core-lipid A may be partially contributed by packing of acyl chains of the correct length and number [105].

The acyl chains are directed toward the periplasm suggesting that flipping occurs after a conformational change to open the periplasmic side of MsbA; however, examination of the hydrophobicity of MsbA (Figure 14(b)) suggests that the requisite translation has already occurred with the acyl chains already extending to the periplasmic leaflet of the inner membrane [104]. A second 4.8 Å cryo-EM structure, determined in the presence of ADP and vanadate, shows MsbA in an occluded conformation with parallel transmembrane helices [104]. Comparison of the structures shows that transition of the transmembrane domains from cytoplasmic-open to periplasmic-open involves a shuffling of the TM helices that may help flip core-lipid A [104].

Ergo, the structures of MsbA suggest a mechanism, wherein core-lipid A binds the cytoplasmic open conformation, and ATP binding brings the ATPase domains together, which is coupled to a conformational change in the transmembrane domains from cytoplasmic open to periplasmic open [103, 104]. Core-lipid A flips and diffuses into the periplasmic leaflet [103, 104]. Next, ATP hydrolysis is coupled to a periplasmic open to occluded conformational change, which prevents the transporter from acting in reverse [104]. Finally, release of phosphate allows MsbA to return to the starting cytoplasmic open conformation [103, 104].

Recently, a set of small molecules that selectively inhibit MsbA core-lipid A transport and ATP hydrolysis were identified, and the binding of two were characterized by X-ray crystallography [105]. These inhibitors were found to bind simultaneously with core-lipid A in an adjacent membrane-exposed cleft formed by TM4-6 of each subunit in the cytoplasmic open conformation [105]. Because this pocket will be deformed by the transition to the periplasmic open conformation, these inhibitors may lock the MsbA transmembrane domains into the cytoplasmic open conformation [105]. Moreover, inhibitor binding appeared to strain TM4 leading to a movement of the ATPase domains via the coupling cytoplasmic helix between TM4 and TM5 [105]. The ultimate result of this inhibitor-induced conformational change was the partial detachment of one ATPase domain from its corresponding coupling cytoplasmic helices and the loss of the symmetrical arrangement of the ATPase domains required for ATP hydrolysis, which suggests that these inhibitors act by locking MsbA in a catalytically inactive state [105].

2.2.3. O-Antigen Polysaccharide Synthesis and Transport. The O-antigen is a repeating polysaccharide that varies between bacterial strains [1]. Different O-antigens are synthesized by one of three divergent pathways: Wzy dependent, ABC transporter dependent, and synthase dependent [1, 92]. However, O-antigen synthesis has a few conserved features: O-antigen is built on undecaprenol phosphate (und-P), and sugar residues are transferred to this lipid from nucleotide-charged sugar donors at the cytoplasmic face of the inner membrane [1, 92]. Furthermore, a sugar-1-phosphate [1-phospho-*N*-acetyl-glucosamine (P-GlcNAc) in *E. coli*] is transferred to und-P to form a phosphoanhydride product (GlcNAc-PP-und in *E. coli*) [1]. Finally, the completed O-antigen is ligated to core-lipid A at the periplasmic face of the inner membrane to form

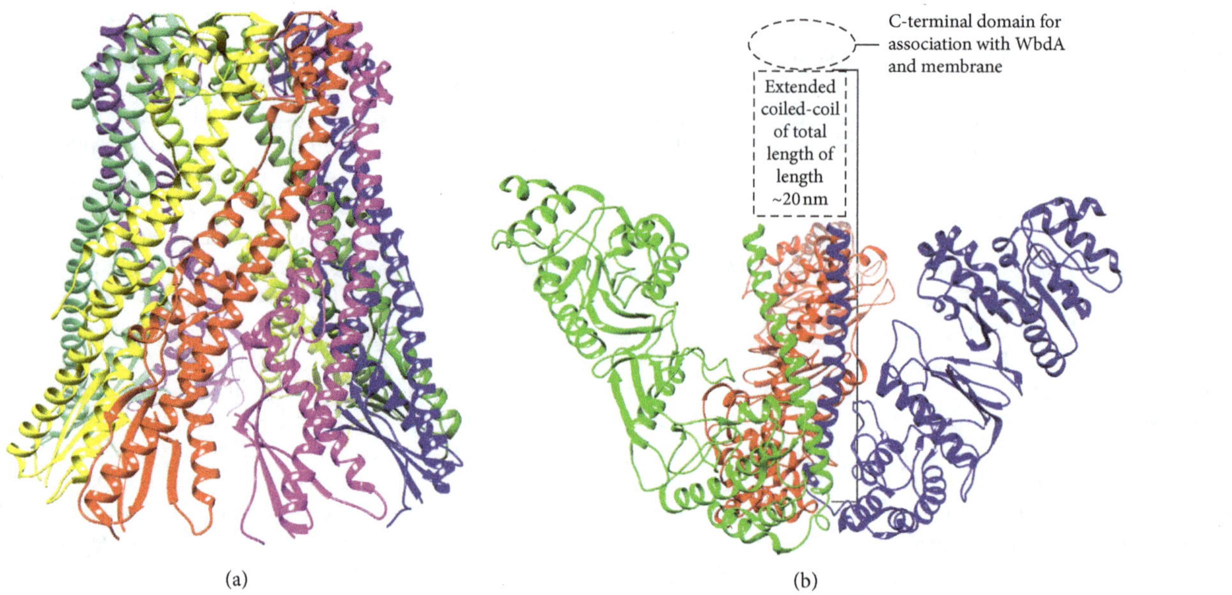

FIGURE 15: Proteins that control O-antigen modal length. (a) Octamer of WzzB periplasmic domains (PDB: 4E29) [113]. (b) WbdD trimer (PDB: 4UW0) with the location of the unresolved extended coiled-coil domain and uncharacterized C-terminal domain marked by a rectangle and oval, respectively [114, 115].

completed LPS by WaaL, which is specific for core oligosaccharides but not O-antigens [1]. The lack of polysaccharide specificity of WaaL also allows it to attach other capsular polysaccharides (K-antigens and colonic acid in *E. coli*) to core-lipid A [100, 106]. Moreover, while *E. coli* K-12 does not synthesize O-antigen, conservation of the machinery shared with other polysaccharide synthesis pathways still enables this strain to synthesize GlcNAc-PP-und and transport it across the inner membrane [99, 106, 107]. Therefore, WaaL may also transfer *N*-acetyl-glucosamine to the terminal Hep of the K-12 outer core (Figure 13) [99, 100]. Studies of the topology of WaaL indicated that the ligase contained 12 transmembrane segments and 2 major periplasmic segments, and enzymatic assays showed that a conserved Arg in a short periplasmic segment and a conserved His in the largest periplasmic segment were essential for activity [108, 109]. Examination of the reactants and products of the WaaL reaction suggest that WaaL catalyzes an S_N2 substitution reaction, wherein the terminal residue of the core oligosaccharide (7′-hydroxyl of the fourth Hep in *E. coli* K-12) attacks the anomeric carbon of the basal residue of the O-antigen with und-PP acting as the leaving group [109]. The positions of these residues and the metal-independent activity of WaaL make it tempting to speculate that the conserved Arg and His could bind the pyrophosphate of O-antigen-PP-und and act as the catalytic base for the attacking hydroxyl, respectively [108, 109].

The Wzy dependent pathway is unique from the other pathways, and in that single repeat units are synthesized and transported across the inner membrane [1]. Once the repeat unit has been synthesized by its specific glycosyltransferases, the repeat unit-PP-und is flipped across the inner membrane by a Wzx flippase [110]. Little is known about the mechanism of Wzx flippases; even whether they require energy is a matter of contention though *in vitro* transport of a soluble analogue of GlcNAc-PP-und was not affected by the addition of ATP, NADH, or ionophores [107, 110]. However, Wzx flippases are thought to have 12 transmembrane segments, and they tend to be specific for a particular O-antigen repeat unit with the first sugar attached to und-PP playing the most important role in substrate specificity [108, 110]. Next, the O-antigen repeat units are polymerized at the periplasmic face of the inner membrane on a single und-PP by Wzy, which catalyzes the transfer of the growing polysaccharide to the free end of the new repeat unit-PP-und [1]. Topology studies of Wzy from *Pseudomonas aeruginosa* PAO1 suggested that the polymerase contained 14 transmembrane segments and 4 major periplasmic segments [108]. Two of these periplasmic segments were found to contain RX_{10} G motifs with highly conserved Arg residues that were essential for activity [111, 112].

The presence of two similar motifs in Wzy corresponding with two similar substrates and the conservation of Arg suggests that the Arg residues could be involved in binding a common motif in Wzy substrates such as the pyrophosphate of oligo-PP-und [108, 111, 112]. In addition to a specific sugar composition, O-antigens have a strain specific modal length, and this is controlled by Wzz [1, 113, 116]. While crystal and cryo-EM structures of Wzz length regulators from multiple species have been determined, the mechanism of length regulation has remained obscure [113, 116, 117]. Wzz is primarily a periplasmic protein with *N*- and C-terminal transmembrane helices, and the periplasmic region forms bell-shaped oligomers with 5–12 subunits though open trimers have also been observed [113, 116, 117]. The periplasmic region has an α/β-domain that forms the base of Wzz adjacent to the membrane and an extended α-helical domain that forms a twisted helix bundle

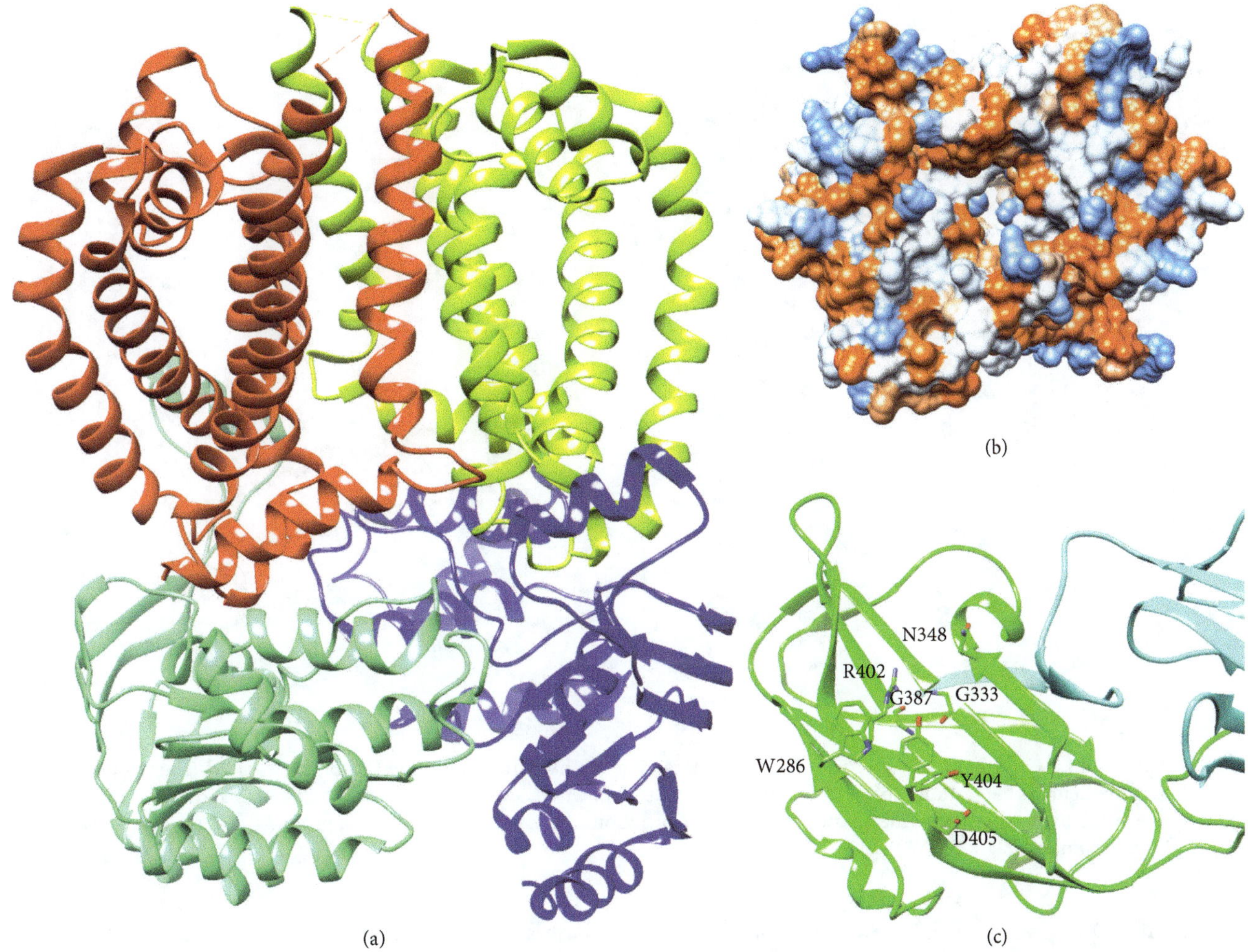

Figure 16: O-antigen ABC transporter. (a) Wzm-Wzt heterotetramer ABC transporter from *Aquifex aeolicus* (PDB: 6AN7) [120]. The Wzm subunits are shown in red and yellow, and the ATPase domain of the Wzt subunits are shown in cyan and blue. (b) Hydrophobicity surface of Wzm-Wzt heterotetramer from (a) showing periplasmic end of the continuous channel. Hydrophobicity is shown on an orange-blue scale with orange as the hydrophobic end. The minimum radius of the channel is 3.5 Å [120]. (c) C-terminal carbohydrate-binding domain of Wzt from *E. coli* O9a (PDB: 2R5O) [119]. The domain was crystallized as a C-terminally swapped dimer, but this C terminal swap may not occur in the full Wzm-Wzt complex. Residues found to be involved in specific recognition of terminally modified O9a O-antigen are shown [119].

in oligomers (Figure 15) [113, 116, 117]. *In vivo* assays of Wzz mutants and chimeras indicated that surface-exposed residues at the top of the α-helical domain controlled O-antigen length, but the role of oligomerization in length regulation (if any) has not been established [113, 116].

In the ABC transporter dependent pathway, the O-antigen is synthesized continuously on a single und-P carrier at the cytosolic face of the inner membrane [1, 118]. Consequently, the initial 1-phospho-sugar and second adaptor residue are not repeated, forming the adaptor region before the repeat region of the O-antigen [1, 118]. Moreover, ABC transporter-dependent O-antigens can contain terminal modifications to the free end of the O-antigen that act as a form of quality control to ensure only completed O-antigen is transported across the inner membrane [1, 118, 119]. In terminally modified O-antigens synthesized by the ABC transporter dependent pathway, the strain specific modal length of the O-antigen is controlled by the enzyme that makes the terminal modifications, which halt the activity of the glycosyltransferases that add the repeat units [114, 118]. Crystal structures and small angle x-ray scattering (SAXS) data of terminating enzyme of *E. coli* O9a (WbdD) revealed that the catalytic domain of WbdD is separated from a structurally uncharacterized C-terminal domain by an extended α-helical domain that forms a ~200 Å coiled-coil in WbdD trimers (Figure 15) [114]. The C-terminal domain mediated association with the membrane and the glycosyltransferase that synthesizes the O9a repeat units (WbdA) [115]. Deletions that decreased the length of the coiled-coil domain decreased the modal length of the O-antigen *in vivo*, and the opposite effect was observed for insertions designed to increase the length of the coiled-coil [114]. Therefore, these data support a model wherein O-antigen is lengthened by WbdA until long enough to reach from the membrane to the catalytic domain of WbdD at which point termination depends on competition between WbdA and WbdD for the

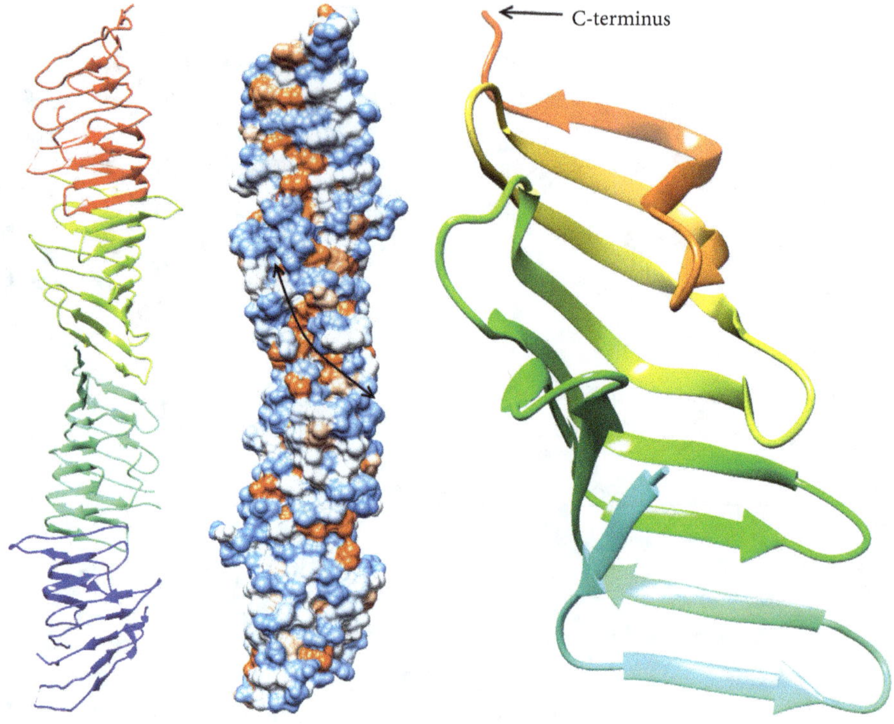

FIGURE 17: LptA and LptC. (a) LptA tetramer with subunits interacting in a head-to-tail fashion (PDB: 2R1A) [129]. The tetramer is shown with the *N*-terminal end at the bottom. (b) Hydrophobicity surface of LptA tetramer in (a). Hydrophobicity is shown on an orange-blue scale with blue at the hydrophilic end. The arrow indicates the continuous, helical, and hydrophobic groove [129]. (c) Ribbon structure of LptC periplasmic domain (PDB: 3MY2) [130].

free end of O-antigen [114]. However, some ABC transporter-dependent O-antigens lack terminal modifications, and in this case, the modal length is controlled by the stoichiometry of the glycosyltransferases that synthesize the repeat units and of the ABC transporter [118]. Finally, the completed O-antigen is transported across the membrane by the ABC transporter [1, 118].

The structure of the heterotetramer complex of the Wzm permease and the *N*-terminal ATPase domain of Wzt from *Aquifex aeolicus* and the structure of the C-terminal carbohydrate-binding domain of Wzt from *E. coli* O9a have been solved by X-ray crystallography [119, 120]. The Wzm-Wzt ABC transporter complex was solved in an open conformation and forms a continuous channel with a minimum radius of 3.5 Å, suggesting that the polysaccharide can occupy the channel through multiple rounds of ATP hydrolysis (Figure 16) [120]. The Wzm channel is lined with several aromatic residues that may form stacking interactions with the sugar rings [120]. The Wzt C-terminal domain forms an immunoglobulin-like β-sandwich with a groove containing several residues that were shown to be involved in recognition of the terminally modified O9a polysaccharide *in vivo* and *in vitro* (Figure 16) [119].

As in the ABC transporter-dependent pathway, synthase-dependent O-antigen is synthesized on a single und-P carrier and thus contains an unrepeated adaptor region [1]. The synthase (WbbF) is thought to simultaneously add repeat units and transport the growing O-antigen across the membrane [1]. However, only O:54 O-antigen of *Salmonella enterica* is known to be synthesized by this pathway, and it has thus not been studied extensively [1].

2.2.4. Transport of LPS to the Outer Membrane. Completed LPS is transported from the periplasmic leaflet of the inner membrane to the outer leaflet of the outer membrane by the Lpt complex, which is comprised of proteins LptA-G and spans from the cytoplasmic face of the inner membrane to the extracellular face of the outer membrane [121–124]. This complex, which is formed by β-jellyroll domains in LptF, LptG, LptC, LptA, and LptD, forms a continuous hydrophobic slide for LPS across the periplasm, and transport is driven by the ATPase activity of LptB [123–132]. Extensive structural and functional data have been obtained to support this transenvelope protein bridge mechanism.

Some of the earliest functional evidence for the transport of LPS via a transenvelope bridge was obtained by the study of lipid transport from inner membrane spheroplasts [133]. These spheroplasts were found to contain pieces of the outer membrane that could be separated by centrifugation [133]. Newly synthesized (^{14}C-labeled) LPS was transported to the outer membrane in an MsbA-dependent, periplasmic extract-independent manner, and this contrasted with the transport of lipoproteins, which could be released from spheroplasts by the addition of the soluble periplasmic carrier protein LolA, and with the transport of newly synthesized phospholipids,

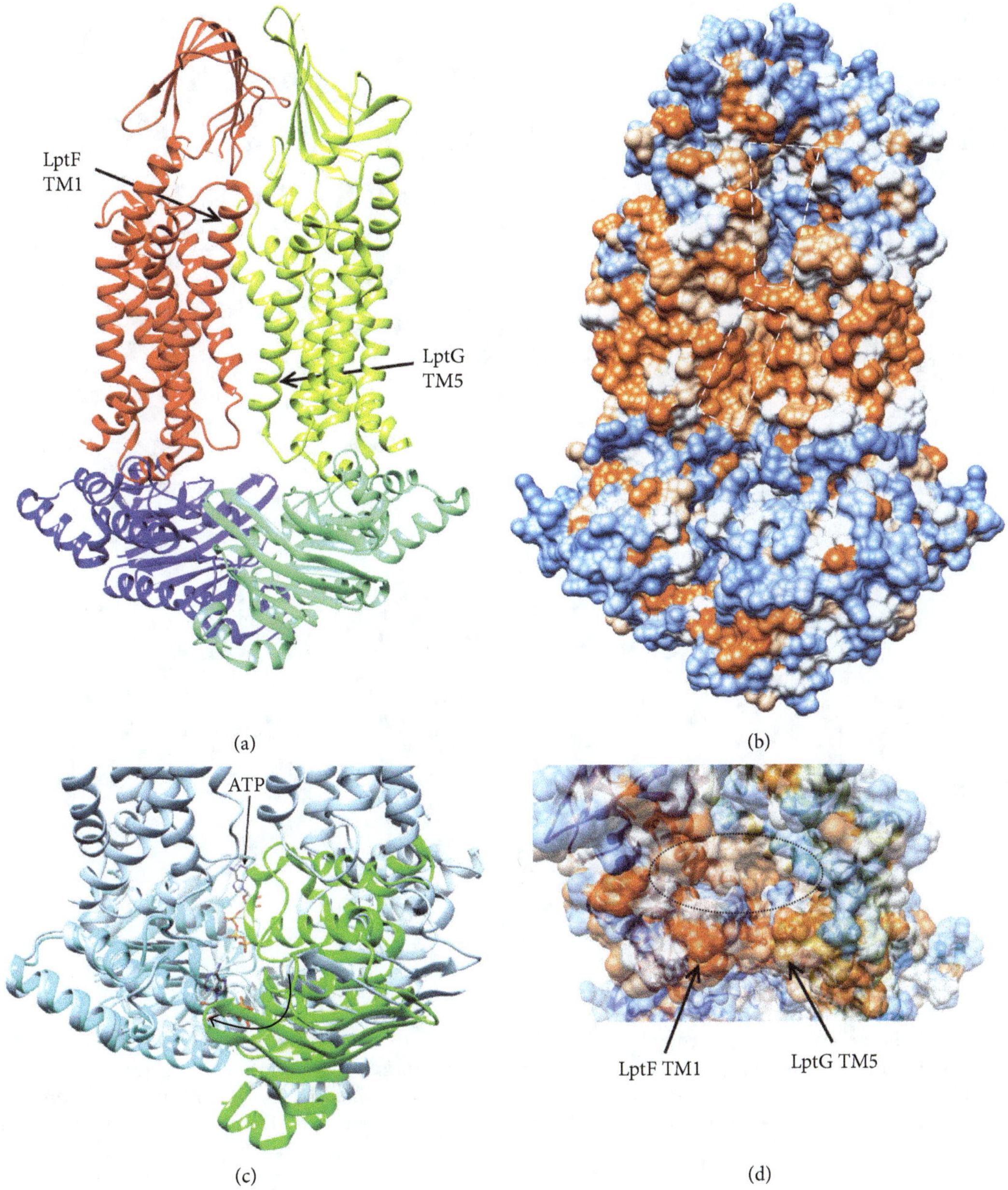

FIGURE 18: LptB2FG ABC transporter. (a) Ribbon structure $LptB_2FG$ ABC transporter (PDB: 5X5Y) [128]. LptF is shown in red, while LptG is shown in yellow, and LptB subunits are shown in cyan and blue. (b) Hydrophobicity surface of (a). Hydrophobicity is shown on an orange-blue scale with orange as the most hydrophobic. Quadrilaterals mark spaces between LptF TM1 and LptG TM5. (c) Comparison of ATP-bound LptB dimer (PDB: 4P33) [134] with LptB subunits in nucleotide-free $LptB_2FG$ complex from (a). Subunits from ATP-bound LptB-E163Q dimer are colored cyan and green, and the $LptB_2FG$ complex is colored light blue. Cyan LptB-E163Q was overlaid with LptB chain B of complex [24]. Curved arrow shows movement of other LptB subunits from nucleotide-free conformation in $LptB_2FG$ complex to ATP-bound conformation in the LptB-E163Q dimer. (d) Rotated view of (b) showing opening to V-shaped pocket on top/periplasmic side of $LptB_2FG$ complex. Ellipse marks the opening.

which could not be transported to the outer membrane outside of intact cells [133]. Further studies utilized lipid A modifying enzymes to demonstrate that LPS accumulated at the periplasmic face of the inner membrane when the soluble periplasmic protein LptA was inactivated thus identifying a putative LPS carrier for this process [121]. The idea of a protein bridge was further supported by crystal structures of LptA [129]. The crystal structures revealed that LptA is primarily comprised of two 8-stranded β-sheets that fold into a β-jellyroll with a hydrophobic cleft at one edge (Figure 17). Moreover, in the presence of LPS (though LPS is not visible in the electron density), LptA crystallized as a chain of head-to-tail interacting subunits that formed two continuous β-sheets and a continuous hydrophobic groove suggestive of a hydrophobic slide for LPS [129].

Next, the crystal structure of the LptC periplasmic domain showed a very similar structure to that of LptA (Figure 17) supporting the model of a continuous β-jellyroll slide for LPS [130]. Moreover, LptA could abstract LPS from LptC but not vice versa thus supporting affinity-based

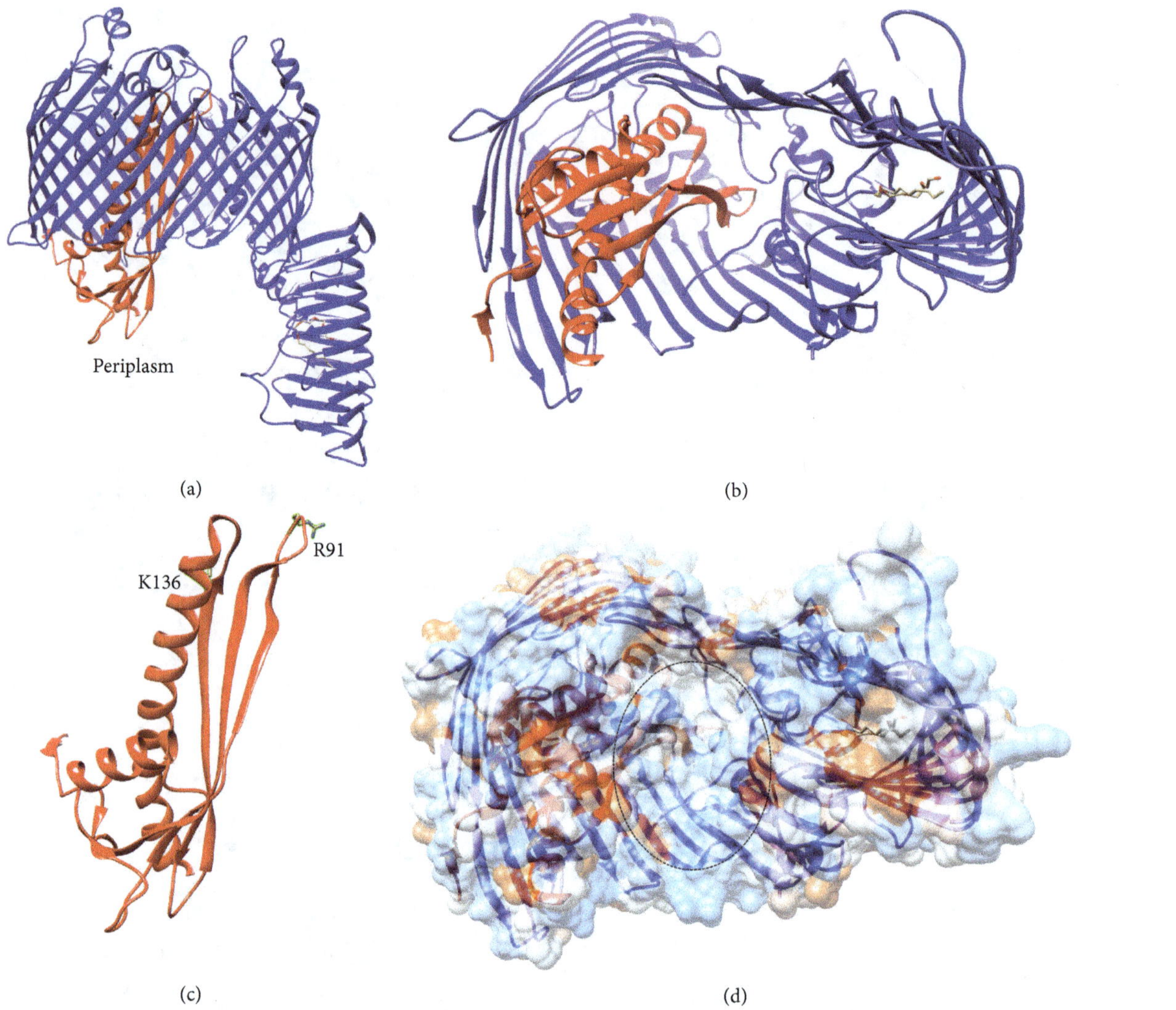

FIGURE 19: LptDE. (a) Ribbon structure of LptDE complex (PDB: 4Q35) [131]. LptD is colored blue, and LptE is colored red. A C8E detergent molecule is shown bound to the N-terminal β-jellyroll domain of LptD. (b) Periplasmic view of (a). (c) LptE from the LptDE complex in (a). Conserved basic residues shown to be important for LptE binding to LPS are highlighted [135]. (d) Hydrophobicity surface of (b). Hydrophobicity is shown in an orange-blue scale with orange as the hydrophobic end. The ellipse marks the periplasmic entrance to the LptDE lumen.

directional transport between proteins in the Lpt system, though these results are also consistent with a soluble carrier mechanism of LptA [130]. However, the stable protein bridge model was supported by the ability of the cytosolic protein LptB and inner membrane proteins LptC and LptF to pull down LptA and the outer membrane proteins LptD and LptE and by the ability of LptA to cause the association of liposomes containing the LptBCFG complex with liposomes containing the LptDE complex [122, 124].

Interactions between the Lpt proteins were further characterized by site-specific crosslinking with the nonnatural amino acid p-benzoyl-L-phenylalanine (Bph) [127]. Crosslinking suggested that the C-terminal end of the LptC β-jellyroll interacted with the N-terminal end of the LptA β-jellyroll, and the C-terminal end of the LptA β-jellyroll interacted with the N-terminal end of the LptD β-jellyroll [127]. In addition, C-terminal mutants of the LptC β-jellyroll were found to be nonfunctional *in vivo* and unable to pull down LptA, LptD, or LptE [125, 126]. On the contrary, the N-terminal transmembrane domain of LptC (a predicted transmembrane helix) seemed to be of limited functional importance as it could be functionally replaced by unrelated transmembrane helices [125].

Crosslinking with Bph was also utilized to provide evidence for binding and directional transfer of LPS [123, 124]. LPS was crosslinked to LptA and LptC when Bph was incorporated in the hydrophobic cleft of their β-jellyrolls but not when Bph was incorporated on the outer surface of their β-jellyrolls thus supporting transport along a continuous hydrophobic slide [123, 124]. In addition, ATP-dependent transfer of LPS from LptBFG to LptC and LptA was observed *in vivo* and in liposomes, and transfer to LptA was enhanced by the inclusion of LptC [123, 124]. Furthermore, LPS was transferred from liposomes containing LptBCFG to liposomes containing LptDE in an ATP- and LptA-dependent manner [124]. Moreover, increase in ATP concentration resulted in LPS crosslinking to LptD at earlier time points and decreased crosslinking to LptD at later time points, which strongly indicated that transfer through the Lpt system was an active process that could be accelerated by an increased rate of ATP hydrolysis by LptB [124]. Therefore, these data support a model, wherein each new LPS molecule pushes the previous LPS molecules along the hydrophobic slide of the Lpt complex until they reach the outer leaflet of the outer membrane [124].

Crystal structures of the $LptB_2FG$ ABC transporter and of the LptB ATPase dimer bound to ATP have been solved and provide some insights into how ATP hydrolysis drives LPS through the Lpt system [128, 132, 134].

The structure of the $LptB_2FG$ complex was solved in the nucleotide-free state and shows that the transmembrane helices of LptF and LptG form a V-shaped pocket at the periplasmic face of the inner membrane (Figure 18) [128, 132]. There are minimal contacts at the interface between TM1 of LptF and TM5 of LptG suggesting that these helices may separate to allow LPS to enter by lateral diffusion (Figure 18) [128, 132]. The V-shaped pocket is lined by highly conserved hydrophobic residues that were found to be important for function *in vivo* [128]. In addition, LptF and LptG have periplasmic β-jellyroll domains similar to those of LptA, LptC, and LptD [128–132]. Mutation of the C-terminal strands of the LptF or LptG β-jellyroll domains caused growth defects in *E. coli*, which suggested that these domains may interact with the *N*-terminal end of the LptC β-jellyroll to form the first unit in the hydrophobic slide [132].

The LptB-E163Q dimer was crystallized with ATP bound showing that ATP binds between the ATPase subunits (Figure 18) [134]. E163 is the predicted catalytic base, and mutation to Gln eliminated ATPase activity [134]. Comparison of the ATP-bound LptB dimer and the nucleotide-free LptB dimer of the $LptB_2FG$ ABC transporter suggests that ATP binding to the transporter will cause the LptB subunits to rotate ~15° [128, 134]. Interactions of LptB with TM1 and the short α-helix between TM2 and TM3 of LptF and LptG likely serve to transmit this motion into the further opening of the space between LptF TM1 and LptG TM5 to allow LPS binding [128]. Other conformational changes associated with the catalytic cycle presumably act to push LPS into the β-jellyroll domain of LptF or LptG, but more work is required to elucidate this mechanism [128].

Finally, crystal structures of the LptDE outer membrane complex increase our understanding of how LPS is transported across the outer membrane and inserted into the extracellular leaflet [131, 136]. The structures reveal that LptD forms a 26-stranded β-barrel, and LptE forms a 4-stranded β-sheet and 2 α-helices (Figure 19) [131, 135, 136]. LptE acts as a plug for this β-barrel, and the extracellular side of the β-barrel is occluded by LptE and the extracellular loops of LptD [131, 136]. On the contrary, the periplasmic side of the β-barrel is open to reveal a large, hydrophilic lumen that may accommodate the polysaccharide of LPS (Figure 19) [131, 136]. LptE has been shown to bind LPS, and conserved basic residues of LptE that were found to be important for LPS binding are located at the extracellular side of the LptDE complex [128, 135, 136]. Therefore, movement of LptE upon LPS binding may allow these residues to interact with the polysaccharide, and this interaction with LptE may help feed the polysaccharide through the β-barrel [131, 135, 136]. The LptDE structures also revealed that the interactions of the first and final strands in the β-barrel are a weak point in the barrel structure thus suggesting that these strands may separate to allow the lipid A moiety of LPS to enter the membrane [131, 136]. As mentioned above, LptD also has an *N*-terminal, periplasmic β-jellyroll domain that likely forms the final subunit in the hydrophobic slide, and this domain was crystallized with a detergent molecule bound in its hydrophobic cleft (Figure 19) [127, 131]. This domain is located near the weak point in the β-barrel [131].

3. Conclusions

This review has summarized what is known about how LPS is synthesized and transported to the outer leaflet of the outer membrane with an emphasis on the structural and biochemical characterization of conserved proteins involved in these processes. While biochemical and structural research have provided a relatively detailed understanding of some aspects of LPS assembly, such as the synthesis of lipid A and the transport of completed LPS from the periplasmic leaflet of the inner membrane to the outer leaflet of the outer membrane, other aspects remain obscure, such as the transport and polymerization of O-antigen repeat units in the Wzy dependent pathway. Further biochemical and structural studies, particularly to obtain structures of Wzx and WaaL, will be required to complete our understanding of LPS assembly.

As discussed in Introduction, understanding the assembly of LPS has important implications for human health. The importance for LPS in bacterial viability and virulence makes the enzymes involved in this process promising targets for antibiotics, and the structural characterization of highly conserved enzymes may facilitate the rational design of broad-spectrum antibiotics targeting this pathway. Moreover, specific modifications in LPS alter the immunological response to this glycolipid, which often plays an important role in infection; in addition to modulating the inflammatory response mediated by the MD2/TLR4 complex, specific modifications to LPS can alter bacterial susceptibility to immune offences such as cationic antimicrobial peptides and the complement system [5, 8, 16, 19, 137–148]. Targeting the enzymes that make these specific modifications could lead to the development of antibiotics that selectively target specific pathogens without damaging normal microbiota.

Conflicts of Interest

The authors declare that there are no conflicts of interest regarding the publication of this review.

Acknowledgments

We would like to thank Dr. Doug Ohlendorf for his help in editing the manuscript. UCSF Chimera was utilized to generate protein structure figures and for protein structure analysis. Chimera is developed by the Resource for Biocomputing, Visualization, and Informatics at the University of California, San Francisco (supported by NIGMS P41-GM103311). The present work was supported by NIH grant GM118047 to HA.

References

[1] C. R. H. Raetz and C. Whitfield, "Lipopolysaccharide endotoxins," *Annual Review of Biochemistry*, vol. 71, no. 1, pp. 635–700, 2002.

[2] J. C. Malinverni and T. J. Silhavy, "An ABC transport system that maintains lipid asymmetry in the gram-negative outer membrane," *Proceedings of the National Academy of Sciences*, vol. 106, no. 19, pp. 8009–8014, 2009.

[3] G. Klein, B. Lindner, H. Brade, and S. Raina, "Molecular basis of lipopolysaccharide heterogeneity in *Escherichia coli*: envelope stress-responsive regulators control the incorporation of glycoforms with a third 3-deoxy-α-D-manno-oct-2-ulosonic acid and rhamnose," *Journal of Biological Chemistry*, vol. 286, no. 50, pp. 42787–42807, 2011.

[4] G. Klein, S. Müller-Loennies, B. Lindner, N. Kobylak, H. Brade, and S. Raina, "Molecular and structural basis of inner core lipopolysaccharide alterations in *Escherichia coli*: incorporation of glucuronic acid and phosphoethanolamine in the heptose region," *Journal of Biological Chemistry*, vol. 288, no. 12, pp. 8111–8127, 2013.

[5] R. F. Maldonado, I. Sá-Correia, and M. A. Valvano, "Lipopolysaccharide modification in Gram-negative bacteria during chronic infection," *FEMS Microbiology Reviews*, vol. 40, no. 4, pp. 480–493, 2016.

[6] J. Bojkovic, D. L. Richie, D. A. Six et al., "Characterization of an Acinetobacter baumannii lptD deletion strain: permeability defects and response to inhibition of lipopolysaccharide and fatty acid biosynthesis," *Journal of Bacteriology*, vol. 198, no. 4, pp. 731–741, 2015.

[7] C. M. Reynolds and C. R. H. Raetz, "Replacement of lipopolysaccharide with free lipid A molecules in *Escherichia coli* mutants lacking all core sugars," *Biochemistry*, vol. 48, no. 40, pp. 9627–9640, 2009.

[8] A. Beceiro, A. Moreno, N. Fernández et al., "Biological cost of different mechanisms of colistin resistance and their impact on virulence in Acinetobacter baumannii," *Antimicrobial Agents and Chemotherapy*, vol. 58, no. 1, pp. 518–526, 2014.

[9] K. Miyake, "Roles for accessory molecules in microbial recognition by Toll-like receptors," *Journal of Endotoxin Research*, vol. 12, no. 4, pp. 195–204, 2006.

[10] J.-I. Kim, C. J. Lee, M. S. Jin et al., "Crystal structure of CD14 and its implications for lipopolysaccharide signaling," *Journal of Biological Chemistry*, vol. 280, no. 12, pp. 11347–11351, 2005.

[11] U. Ohto, K. Fukase, K. Miyake, and Y. Satow, "Crystal structures of human MD-2 and its complex with antiendotoxic lipid IVa," *Science*, vol. 316, no. 5831, pp. 1632–1634, 2007.

[12] B. S. Park, D. H. Song, H. M. Kim, B.-S. Choi, H. Lee, and J.-O. Lee, "The structural basis of lipopolysaccharide recognition by the TLR4-MD-2 complex," *Nature*, vol. 458, no. 7242, pp. 1191–1195, 2009.

[13] R. Shimazu, S. Akashi, H. Ogata et al., "MD-2, a molecule that confers lipopolysaccharide responsiveness on Toll-like receptor 4," *Journal of Experimental Medicine*, vol. 189, no. 11, pp. 1777–1782, 1999.

[14] A. Poltorak, X. He, I. Smirnova et al., "Defective LPS signaling in C3H/HeJ and C57BL/10ScCr mice: mutations in Tlr4 gene," *Science*, vol. 282, no. 5396, pp. 2085–2088, 1998.

[15] M. J. Hall, S. N. Williams, C. J. DeFrances, and A. Golosinskiy, "Inpatient care for septicemia or sepsis: a challenge for patients and hospitals," *NCHS Data Brief*, vol. 62, pp. 1–8, 2011.

[16] C. M. John, N. J. Phillips, D. C. Stein, and G. A. Jarvis, "Innate immune response to lipooligosaccharide: pivotal regulator of the pathobiology of invasive *Neisseria meningitidis* infections," *Pathogens and Disease*, vol. 75, no. 3, 2017.

[17] S. O. Opiyo, R. L. Pardy, H. Moriyama, and E. N. Moriyama, "Evolution of the Kdo2-lipid A biosynthesis in bacteria," *BMC Evolutionary Biology*, vol. 10, no. 1, 2010.

[18] C. L. Ventola, "The antibiotic resistance crisis: part 2: management strategies and new agents," *P & T: A Peer-Reviewed Journal for Formulary Management*, vol. 40, no. 5, pp. 344–352, 2015.

[19] C. R. H. Raetz, C. M. Reynolds, M. S. Trent, and R. E. Bishop, "Lipid A modification systems in gram-negative bacteria," *Annual Review of Biochemistry*, vol. 76, no. 1, pp. 295–329, 2007.

[20] M. S. Anderson, C. E. Bulawa, and C. R. Raetz, "The biosynthesis of gram-negative endotoxin. Formation of lipid A precursors from UDP-GlcNAc in extracts of *Escherichia coli*," *Journal of Biological Chemistry*, vol. 260, pp. 15536–15541, 1985.

[21] M. S. Anderson and C. R. Raetz, "Biosynthesis of lipid A precursors in *Escherichia coli*. A cytoplasmic acyltransferase that converts UDP-N-acetylglucosamine to UDP-3-O-(R-3-hydroxymyristoyl)-N-acetylglucosamine," *Journal of Biological Chemistry*, vol. 262, no. 11, pp. 5159–5169, 1987.

[22] M. S. Anderson, H. G. Bull, S. M. Galloway et al., "UDP-N-acetylglucosamine acyltransferase of *Escherichia coli*. The first step of endotoxin biosynthesis is thermodynamically unfavorable," *Journal of Biological Chemistry*, vol. 268, no. 26, pp. 19858–19865, 1993.

[23] A. H. Williams and C. R. H. Raetz, "Structural basis for the acyl chain selectivity and mechanism of UDP-N-acetylglucosamine acyltransferase," *Proceedings of the National Academy of Sciences*, vol. 104, no. 34, pp. 13543–13550, 2007.

[24] E. F. Pettersen, T. D. Goddard, C. C. Huang et al., "UCSF Chimera—a visualization system for exploratory research and analysis," *Journal of Computational Chemistry*, vol. 25, no. 13, pp. 1605–1612, 2004.

[25] C. R. H. Raetz and S. L. Roderick, "A left-handed parallel β Helix in the structure of UDP-N-acetylglucosamine acyltransferase," *Science*, vol. 270, no. 5238, pp. 997–1000, 1995.

[26] V. Ulaganathan, L. Buetow, and W. N. Hunter, "Nucleotide substrate recognition by UDP-N-acetylglucosamine acyltransferase (LpxA) in the first step of lipid A biosynthesis," *Journal of Molecular Biology*, vol. 369, no. 2, pp. 305–312, 2007.

[27] T. J. O. Wyckoff and C. R. H. Raetz, "The active site of *Escherichia coli* UDP-N-acetylglucosamine acyltransferase. Chemical modification and site-directed mutagenesis," *Journal of Biological Chemistry*, vol. 274, no. 38, pp. 27047–27055, 1999.

[28] T. J. O. Wyckoff, S. Lin, R. J. Cotter, G. D. Dotson, and C. R. H. Raetz, "Hydrocarbon rulers in UDP-N-acetylglucosamine acyltransferases," *Journal of Biological Chemistry*, vol. 273, no. 49, pp. 32369–32372, 1998.

[29] M. S. Anderson, A. D. Robertson, I. Macher, and C. R. Raetz, "Biosynthesis of lipid A in *Escherichia coli*: identification of UDP-3-O-[(R)-3-hydroxymyristoyl]-alpha-D-glucosamine as a precursor of UDP-N2,O3-bis[(R)-3-hydroxymyristoyl]-alpha-D-glucosamine," *Biochemistry*, vol. 27, no. 6, pp. 1908–1917, 1988.

[30] K. Young, L. L. Silver, D. Bramhill et al., "The envA permeability/cell division gene of *Escherichia coli* encodes the second enzyme of lipid A biosynthesis. UDP-3-O-(R-3-hydroxymyristoyl)-N-acetylglucosamine deacetylase," *Journal of Biological Chemistry*, vol. 270, no. 51, pp. 30384–30391, 1995.

[31] J. E. Jackman, C. R. H. Raetz, and C. A. Fierke, "UDP-3-O-(R-3-hydroxymyristoyl)-N-acetylglucosamine deacetylase of *Escherichia coli* is a zinc metalloenzyme," *Biochemistry*, vol. 38, no. 6, pp. 1902–1911, 1999.

[32] T. Ogura, K. Inoue, T. Tatsuta et al., "Balanced biosynthesis of major membrane components through regulated degradation of the committed enzyme of lipid A biosynthesis by the AAA protease FtsH (HflB) in *Escherichia coli*," *Molecular Microbiology*, vol. 31, no. 3, pp. 833–844, 1999.

[33] A. Emiola, S. S. Andrews, C. Heller, and J. George, "Crosstalk between the lipopolysaccharide and phospholipid pathways during outer membrane biogenesis in *Escherichia coli*," *Proceedings of the National Academy of Sciences*, vol. 113, no. 11, pp. 3108–3113, 2016.

[34] G. Klein, N. Kobylak, B. Lindner, A. Stupak, and S. Raina, "Assembly of lipopolysaccharide in *Escherichia coli* requires the essential LapB heat shock protein," *Journal of Biological Chemistry*, vol. 289, no. 21, pp. 14829–14853, 2014.

[35] D. A. Whittington, K. M. Rusche, H. Shin, C. A. Fierke, and D. W. Christianson, "Crystal structure of LpxC, a zinc-dependent deacetylase essential for endotoxin biosynthesis," *Proceedings of the National Academy of Sciences*, vol. 100, no. 14, pp. 8146–8150, 2003.

[36] G. M. Clayton, D. J. Klein, K. W. Rickert et al., "Structure of the bacterial deacetylase LpxC bound to the nucleotide reaction product reveals mechanisms of oxyanion stabilization and proton transfer," *Journal of Biological Chemistry*, vol. 288, no. 47, pp. 34073–34080, 2013.

[37] B. E. Coggins, X. Li, A. L. McClerren, O. Hindsgaul, C. R. H. Raetz, and P. Zhou, "Structure of the LpxC deacetylase with a bound substrate-analog inhibitor," *Nature Structural & Molecular Biology*, vol. 10, no. 8, pp. 645–651, 2003.

[38] J. E. Jackman, C. R. H. Raetz, and C. A. Fierke, "Site-directed mutagenesis of the bacterial metalloamidase UDP-(3-O-acyl)-N-acetylglucosamine deacetylase (LpxC). Identification of the zinc binding site," *Biochemistry*, vol. 40, no. 2, pp. 514–523, 2001.

[39] M. Hernick and C. A. Fierke, "Catalytic mechanism and molecular recognition of *E. coli* UDP-3-O-(R-3-hydroxymyristoyl)-N-acetylglucosamine deacetylase probed by mutagenesis," *Biochemistry*, vol. 45, no. 51, pp. 15240–15248, 2006.

[40] T. M. Kelly, S. A. Stachula, C. R. Raetz, and M. S. Anderson, "The firA gene of *Escherichia coli* encodes UDP-3-O-(R-3-hydroxymyristoyl)-glucosamine N-acyltransferase. The third step of endotoxin biosynthesis," *Journal of Biological Chemistry*, vol. 268, no. 26, pp. 19866–19874, 1993.

[41] C. M. Bartling and C. R. H. Raetz, "Steady-state kinetics and mechanism of LpxD, the N-acyltransferase of lipid A biosynthesis," *Biochemistry*, vol. 47, no. 19, pp. 5290–5302, 2008.

[42] L. Buetow, T. K. Smith, A. Dawson, S. Fyffe, and W. N. Hunter, "Structure and reactivity of LpxD, the N-acyltransferase of lipid A biosynthesis," *Proceedings of the National Academy of Sciences*, vol. 104, no. 11, pp. 4321–4326, 2007.

[43] C. M. Bartling and C. R. H. Raetz, "Crystal structure and acyl chain selectivity of *Escherichia coli* LpxD, the N-acyltransferase of lipid A biosynthesis," *Biochemistry*, vol. 48, no. 36, pp. 8672–8683, 2009.

[44] A. Masoudi, C. R. H. Raetz, P. Zhou, and C. W. Pemble 4th, "Chasing acyl carrier protein through a catalytic cycle of lipid A production," *Nature*, vol. 505, no. 7483, pp. 422–426, 2014.

[45] R. A. Laskowski and M. B. Swindells, "LigPlot+: multiple ligand-protein interaction diagrams for drug discovery," *Journal of Chemical Information and Modeling*, vol. 51, no. 10, pp. 2778–2786, 2011.

[46] C. E. Bulawa and C. R. Raetz, "The biosynthesis of gram-negative endotoxin. Identification and function of UDP-2,3-diacylglucosamine in *Escherichia coli*," *Journal of Biological Chemistry*, vol. 259, no. 8, pp. 4846–4851, 1984.

[47] K. J. Babinski, S. J. Kanjilal, and C. R. H. Raetz, "Accumulation of the lipid A precursor UDP-2,3-diacylglucosamine in an *Escherichia coli* mutant lacking the lpxH gene," *Journal of Biological Chemistry*, vol. 277, no. 29, pp. 25947–25956, 2002.

[48] K. J. Babinski, A. A. Ribeiro, and C. R. H. Raetz, "The *Escherichia coli* gene encoding the UDP-2,3-diacylglucosamine pyrophosphatase of lipid A biosynthesis," *Journal of Biological Chemistry*, vol. 277, no. 29, pp. 25937–25946, 2002.

[49] H. E. Young, M. P. Donohue, T. I. Smirnova, A. I. Smirnov, and P. Zhou, "The UDP-diacylglucosamine pyrophosphohydrolase LpxH in lipid A biosynthesis utilizes Mn2+ cluster for catalysis," *Journal of Biological Chemistry*, vol. 288, no. 38, pp. 26987–27001, 2013.

[50] C. Okada, H. Wakabayashi, M. Kobayashi, A. Shinoda, I. Tanaka, and M. Yao, "Crystal structures of the UDP-diacylglucosamine pyrophosphohydrase LpxH from *Pseudomonas aeruginosa*," *Scientific Reports*, vol. 6, no. 1, 2016.

[51] J. Cho, C.-J. Lee, J. Zhao, H. E. Young, and P. Zhou, "Structure of the essential Haemophilus influenzae UDP-diacylglucosamine pyrophosphohydrolase LpxH in lipid A biosynthesis," *Nature Microbiology*, vol. 1, no. 11, 2016.

[52] T. E. Bohl, P. Ieong, J. K. Lee et al., "The substrate-binding cap of the UDP-diacylglucosamine pyrophosphatase LpxH is highly flexible, enabling facile substrate binding and product release," *Journal of Biological Chemistry*, vol. 293, no. 21, pp. 7969–7981, 2018.

[53] H. E. Young, J. Zhao, J. R. Barker, Z. Guan, R. H. Valdivia, and P. Zhou, "Discovery of the elusive UDP-diacylglucosamine hydrolase in the lipid A biosynthetic pathway in *Chlamydia trachomatis*," *mBio*, vol. 7, no. 2, 2016.

[54] L. E. Metzger 4th and C. R. H. Raetz, "An alternative route for UDP-diacylglucosamine hydrolysis in bacterial lipid A biosynthesis," *Biochemistry*, vol. 49, no. 31, pp. 6715–6726, 2010.

[55] L. E. Metzger 4th, J. K. Lee, J. S. Finer-Moore, C. R. H. Raetz, and R. M. Stroud, "LpxI structures reveal how a lipid A precursor is synthesized," *Nature Structural & Molecular Biology*, vol. 19, no. 11, pp. 1132–1138, 2012.

[56] B. L. Ray, G. Painter, and C. R. Raetz, "The biosynthesis of gram-negative endotoxin. Formation of lipid A disaccharides from monosaccharide precursors in extracts of *Escherichia coli*,"

Journal of Biological Chemistry, vol. 259, no. 8, pp. 4852–4859, 1984.

[57] M. Nishijima and C. R. Raetz, "Membrane lipid biogenesis in *Escherichia coli*: identification of genetic loci for phosphatidylglycerophosphate synthetase and construction of mutants lacking phosphatidylglycerol," *Journal of Biological Chemistry*, vol. 254, no. 16, pp. 7837–7844, 1979.

[58] M. Nishijima and C. R. Raetz, "Characterization of two membrane-associated glycolipids from an *Escherichia coli* mutant deficient in phosphatidylglycerol," *Journal of Biological Chemistry*, vol. 256, no. 20, pp. 10690–10696, 1981.

[59] K. Takayama, N. Qureshi, P. Mascagni, M. A. Nashed, L. Anderson, and C. R. Raetz, "Fatty acyl derivatives of glucosamine 1-phosphate in *Escherichia coli* and their relation to lipid A. Complete structure of A diacyl GlcN-1-P found in a phosphatidylglycerol-deficient mutant," *Journal of Biological Chemistry*, vol. 258, no. 12, pp. 7379–7385, 1983.

[60] K. Takayama, N. Qureshi, P. Mascagni, L. Anderson, and C. R. Raetz, "Glucosamine-derived phospholipids in *Escherichia coli*. Structure and chemical modification of a triacyl glucosamine 1-phosphate found in a phosphatidylglycerol-deficient mutant," *Journal of Biological Chemistry*, vol. 258, no. 23, pp. 14245–14252, 1983.

[61] L. E. Metzger 4th and C. R. H. Raetz, "Purification and characterization of the lipid A disaccharide synthase (LpxB) from *Escherichia coli*, a peripheral membrane protein," *Biochemistry*, vol. 48, no. 48, pp. 11559–11571, 2009.

[62] T. E. Bohl, K. Shi, J. K. Lee, and H. Aihara, "Crystal structure of lipid A disaccharide synthase LpxB from *Escherichia coli*," *Nature Communications*, vol. 9, no. 1, p. 377, 2018.

[63] D. Albesa-Jové, D. Giganti, M. Jackson, P. M. Alzari, and M. E. Guerin, "Structure–function relationships of membrane-associated GT-B glycosyltransferases," *Glycobiology*, vol. 24, no. 2, pp. 108–124, 2014.

[64] L. L. Lairson, B. Henrissat, G. J. Davies, and S. G. Withers, "Glycosyltransferases: structures, functions, and mechanisms," *Annual Review of Biochemistry*, vol. 77, no. 1, pp. 521–555, 2008.

[65] M. E. Guerin, J. Kordulakova, F. Schaeffer et al., "Molecular recognition and interfacial catalysis by the essential phosphatidylinositol mannosyltransferase PimA from mycobacteria," *Journal of Biological Chemistry*, vol. 282, no. 28, pp. 20705–20714, 2007.

[66] A. Rodrigo-Unzueta, M. A. Martínez, N. Comino, P. M. Alzari, A. Chenal, and M. E. Guerin, "Molecular basis of membrane association by the phosphatidylinositol mannosyltransferase PimA enzyme from mycobacteria," *Journal of Biological Chemistry*, vol. 291, no. 27, pp. 13955–13963, 2016.

[67] B. L. Ray and C. R. Raetz, "The biosynthesis of gram-negative endotoxin. A novel kinase in *Escherichia coli* membranes that incorporates the 4'-phosphate of lipid A," *Journal of Biological Chemistry*, vol. 262, no. 3, pp. 1122–1128, 1987.

[68] T. A. Garrett, J. L. Kadrmas, and C. R. H. Raetz, "Identification of the gene encoding the *Escherichia coli* lipid A 4'-kinase. Facile phosphorylation of endotoxin analogs with recombinant LpxK," *Journal of Biological Chemistry*, vol. 272, no. 35, pp. 21855–21864, 1997.

[69] G. Klein, B. Lindner, W. Brabetz, H. Brade, and S. Raina, "*Escherichia coli* K-12 suppressor-free mutants lacking early glycosyltransferases and late acyltransferases: minimal lipopolysaccharide structure and induction of envelope stress response," *Journal of Biological Chemistry*, vol. 284, no. 23, pp. 15369–15389, 2009.

[70] R. P. Emptage, C. W. Pemble, J. D. York, C. R. H. Raetz, and P. Zhou, "Mechanistic characterization of the tetraacyldisaccharide-1-phosphate 4'-kinase LpxK involved in lipid A biosynthesis," *Biochemistry*, vol. 52, no. 13, pp. 2280–2290, 2013.

[71] R. P. Emptage, K. D. Daughtry, C. W. Pemble, and C. R. H. Raetz, "Crystal structure of LpxK, the 4'-kinase of lipid A biosynthesis and atypical P-loop kinase functioning at the membrane interface," *Proceedings of the National Academy of Sciences*, vol. 109, no. 32, pp. 12956–12961, 2012.

[72] R. P. Emptage, N. K. Tonthat, J. D. York, M. A. Schumacher, and P. Zhou, "Structural basis of lipid binding for the membrane-embedded tetraacyldisaccharide-1-phosphate 4'-kinase LpxK," *Journal of Biological Chemistry*, vol. 289, no. 35, pp. 24059–24068, 2014.

[73] K. A. Brozek, K. Hosaka, A. D. Robertson, and C. R. Raetz, "Biosynthesis of lipopolysaccharide in *Escherichia coli*. Cytoplasmic enzymes that attach 3-deoxy-D-manno-octulosonic acid to lipid A," *Journal of Biological Chemistry*, vol. 264, no. 12, pp. 6956–6966, 1989.

[74] C. J. Belunis and C. R. Raetz, "Biosynthesis of endotoxins. Purification and catalytic properties of 3-deoxy-D-manno-octulosonic acid transferase from *Escherichia coli*," *Journal of Biological Chemistry*, vol. 267, no. 14, pp. 9988–9997, 1992.

[75] W. Brabetz, B. Lindner, and H. Brade, "Comparative analyses of secondary gene products of 3-deoxy-D-manno-oct-2-ulosonic acid transferases from Chlamydiaceae in *Escherichia coli* K-12," *European Journal of Biochemistry*, vol. 267, no. 17, pp. 5458–5465, 2000.

[76] U. Mamat, H. Schmidt, E. Munoz et al., "WaaA of the hyperthermophilic bacterium Aquifex aeolicus is a monofunctional 3-deoxy-D-manno-oct-2-ulosonic acid transferase involved in lipopolysaccharide biosynthesis," *Journal of Biological Chemistry*, vol. 284, no. 33, pp. 22248–22262, 2009.

[77] C. Katz and E. Z. Ron, "Dual role of FtsH in regulating lipopolysaccharide biosynthesis in *Escherichia coli*," *Journal of Bacteriology*, vol. 190, no. 21, pp. 7117–7122, 2008.

[78] H. Schmidt, G. Hansen, S. Singh et al., "Structural and mechanistic analysis of the membrane-embedded glycosyltransferase WaaA required for lipopolysaccharide synthesis," *Proceedings of the National Academy of Sciences*, vol. 109, no. 16, pp. 6253–6258, 2012.

[79] S. Ha, D. Walker, Y. Shi, and S. Walker, "The 1.9 A crystal structure of *Escherichia coli* MurG, a membrane-associated glycosyltransferase involved in peptidoglycan biosynthesis," *Protein Science*, vol. 9, no. 6, pp. 1045–1052, 2000.

[80] Y. Hu, L. Chen, S. Ha et al., "Crystal structure of the MurG: UDP-GlcNAc complex reveals common structural principles of a superfamily of glycosyltransferases," *Proceedings of the National Academy of Sciences*, vol. 100, no. 3, pp. 845–849, 2003.

[81] D. Giganti, D. Albesa-Jové, S. Urresti et al., "Secondary structure reshuffling modulates glycosyltransferase function at the membrane," *Nature Chemical Biology*, vol. 11, no. 1, pp. 16–18, 2015.

[82] T. Clementz, J. J. Bednarski, and C. R. H. Raetz, "Function of the htrB high temperature requirement gene of *Escherichia coli* in the acylation of lipid A: HtrB catalyzed incorporation of laurate," *Journal of Biological Chemistry*, vol. 271, no. 20, pp. 12095–12102, 1996.

[83] D. A. Six, S. M. Carty, Z. Guan, and C. R. H. Raetz, "Purification and mutagenesis of LpxL, the lauroyltransferase of

Escherichia coli lipid A biosynthesis," *Biochemistry*, vol. 47, no. 33, pp. 8623–8637, 2008.

[84] S. M. Carty, K. R. Sreekumar, and C. R. H. Raetz, "Effect of cold shock on lipid A biosynthesis in *Escherichia coli*. Induction at 12°C of an acyltransferase specific for palmitoleoyl-acyl carrier protein," *Journal of Biological Chemistry*, vol. 274, no. 14, pp. 9677–9685, 1999.

[85] M. K. Vorachek-Warren, S. M. Carty, S. Lin, R. J. Cotter, and C. R. H. Raetz, "An *Escherichia coli* mutant lacking the cold shock-induced palmitoleoyltransferase of lipid A biosynthesis: absence of unsaturated acyl chains and antibiotic hypersensitivity at 12°C," *Journal of Biological Chemistry*, vol. 277, no. 16, pp. 14186–14193, 2002.

[86] D. Dovala, C. M. Rath, Q. Hu et al., "Structure-guided enzymology of the lipid A acyltransferase LpxM reveals a dual activity mechanism," *Proceedings of the National Academy of Sciences*, vol. 113, no. 41, pp. E6064–E6071, 2016.

[87] T. Clementz, Z. Zhou, and C. R. H. Raetz, "Function of the *Escherichia coli* msbB gene, a multicopy suppressor of htrB knockouts, in the acylation of lipid A. Acylation by MsbB follows laurate incorporation by HtrB," *Journal of Biological Chemistry*, vol. 272, no. 16, pp. 10353–10360, 1997.

[88] A. Röttig and A. Steinbüchel, "Acyltransferases in bacteria," *Microbiology and Molecular Biology Reviews*, vol. 77, no. 2, pp. 277–321, 2013.

[89] T. Tamada, M. D. Feese, S. R. Ferri et al., "Substrate recognition and selectivity of plant glycerol-3-phosphate acyltransferases (GPATs) from Cucurbita moscata and Spinacea oleracea," *Acta Crystallographica Section D Biological Crystallography*, vol. 60, no. 1, pp. 13–21, 2004.

[90] E. J. Rubin, J. P. O'Brien, P. L. Ivanov, J. S. Brodbelt, and M. S. Trent, "Identification of a broad family of lipid A late acyltransferases with non-canonical substrate specificity," *Molecular Microbiology*, vol. 91, no. 5, pp. 887–899, 2014.

[91] J. V. Hankins, J. A. Madsen, D. K. Giles et al., "Elucidation of a novel *Vibrio cholerae* lipid A secondary hydroxy-acyltransferase and its role in innate immune recognition," *Molecular Microbiology*, vol. 81, no. 5, pp. 1313–1329, 2011.

[92] C. Whitfield and M. S. Trent, "Biosynthesis and export of bacterial lipopolysaccharides," *Annual Review of Biochemistry*, vol. 83, no. 1, pp. 99–128, 2014.

[93] E. Frirdich and C. Whitfield, "Lipopolysaccharide inner core oligosaccharide structure and outer membrane stability in human pathogens belonging to the Enterobacteriaceae," *Journal of Endotoxin Research*, vol. 11, no. 3, pp. 133–144, 2005.

[94] C. Whitfield, N. Kaniuk, and E. Frirdich, "Molecular insights into the assembly and diversity of the outer core oligosaccharide in lipopolysaccharides from *Escherichia coli* and Salmonella," *Journal of Endotoxin Research*, vol. 9, no. 4, pp. 244–249, 2003.

[95] J. Mudapaka and E. A. Taylor, "Cloning and characterization of the *Escherichia coli* Heptosyltransferase III: exploring substrate specificity in lipopolysaccharide core biosynthesis," *FEBS Letters*, vol. 589, no. 13, pp. 1423–1429, 2015.

[96] J. A. Yethon, E. Vinogradov, M. B. Perry, and C. Whitfield, "Mutation of the lipopolysaccharide core glycosyltransferase encoded by waaG destabilizes the outer membrane of *Escherichia coli* by interfering with core phosphorylation," *Journal of Bacteriology*, vol. 182, no. 19, pp. 5620–5623, 2000.

[97] J. Qian, T. A. Garrett, and C. R. H. Raetz, "In vitro assembly of the outer core of the lipopolysaccharide from *Escherichia coli* K-12 and *Salmonella typhimurium*," *Biochemistry*, vol. 53, no. 8, pp. 1250–1262, 2014.

[98] C. M. Reynolds, S. R. Kalb, R. J. Cotter, and C. R. H. Raetz, "A phosphoethanolamine transferase specific for the outer 3-deoxy-D-manno-octulosonic acid residue of *Escherichia coli* lipopolysaccharide. Identification of the eptB gene and Ca2+ hypersensitivity of an eptB deletion mutant," *Journal of Biological Chemistry*, vol. 280, no. 22, pp. 21202–21211, 2005.

[99] D. E. Heinrichs, J. A. Yethon, and C. Whitfield, "Molecular basis for structural diversity in the core regions of the lipopolysaccharides of *Escherichia coli* and *Salmonella enterica*," *Molecular Microbiology*, vol. 30, no. 2, pp. 221–232, 1998.

[100] T. C. Meredith, U. Mamat, Z. Kaczynski, B. Lindner, O. Holst, and R. W. Woodard, "Modification of lipopolysaccharide with colanic acid (M-antigen) repeats in *Escherichia coli*," *Journal of Biological Chemistry*, vol. 282, no. 11, pp. 7790–7798, 2007.

[101] G. Klein, A. Stupak, D. Biernacka, P. Wojtkiewicz, B. Lindner, and S. Raina, "Multiple transcriptional factors regulate transcription of the rpoE gene in *Escherichia coli* under different growth conditions and when the lipopolysaccharide biosynthesis Is defective," *Journal of Biological Chemistry*, vol. 291, no. 44, pp. 22999–23019, 2016.

[102] W. T. Doerrler, H. S. Gibbons, and C. R. H. Raetz, "MsbA-dependent translocation of lipids across the inner membrane of *Escherichia coli*," *Journal of Biological Chemistry*, vol. 279, no. 43, pp. 45102–45109, 2004.

[103] A. Ward, C. L. Reyes, J. Yu, C. B. Roth, and G. Chang, "Flexibility in the ABC transporter MsbA: alternating access with a twist," *Proceedings of the National Academy of Sciences*, vol. 104, no. 48, pp. 19005–19010, 2007.

[104] W. Mi, Y. Li, S. H. Yoon, R. K. Ernst, T. Walz, and M. Liao, "Structural basis of MsbA-mediated lipopolysaccharide transport," *Nature*, vol. 549, no. 7671, pp. 233–237, 2017.

[105] H. Ho, A. Miu, M. K. Alexander et al., "Structural basis for dual-mode inhibition of the ABC transporter MsbA," *Nature*, vol. 557, no. 7704, pp. 196–201, 2018.

[106] C. Whitfield, "Biosynthesis and assembly of capsular polysaccharides in *Escherichia coli*," *Annual Review of Biochemistry*, vol. 75, no. 1, pp. 39–68, 2006.

[107] P. D. Rick, K. Barr, K. Sankaran, J. Kajimura, J. S. Rush, and C. J. Waechter, "Evidence that the wzxE gene of *Escherichia coli* K-12 encodes a protein involved in the transbilayer movement of a trisaccharide-lipid intermediate in the assembly of enterobacterial common antigen," *Journal of Biological Chemistry*, vol. 278, no. 19, pp. 16534–16542, 2003.

[108] S. T. Islam, V. L. Taylor, M. Qi, and J. S. Lam, "Membrane topology mapping of the O-antigen flippase (Wzx), polymerase (Wzy), and ligase (WaaL) from *Pseudomonas aeruginosa* PAO1 reveals novel domain architectures," *mBio*, vol. 1, no. 3, 2010.

[109] X. Ruan, D. E. Loyola, C. L. Marolda, J. M. Perez-Donoso, and M. A. Valvano, "The WaaL O-antigen lipopolysaccharide ligase has features in common with metal ion-independent inverting glycosyltransferases," *Glycobiology*, vol. 22, no. 2, pp. 288–299, 2012.

[110] Y. Hong, M. A. Liu, and P. R. Reeves, "Progress in our understanding of Wzx flippase for translocation of bacterial membrane lipid-linked oligosaccharide," *Journal of Bacteriology*, vol. 200, no. 1, 2018.

[111] S. T. Islam, A. C. Gold, V. L. Taylor, E. M. Anderson, R. C. Ford, and J. S. Lam, "Dual conserved periplasmic loops possess essential charge characteristics that support a catch-

and-release mechanism of O-antigen polymerization by Wzy in *Pseudomonas aeruginosa* PAO1," *Journal of Biological Chemistry*, vol. 286, no. 23, pp. 20600–20605, 2011.

[112] S. T. Islam, S. M. Huszczynski, T. Nugent, A. C. Gold, and J. S. Lam, "Conserved-residue mutations in Wzy affect O-antigen polymerization and Wzz-mediated chain-length regulation in *Pseudomonas aeruginosa* PAO1," *Scientific Reports*, vol. 3, no. 1, 2013.

[113] S. Kalynych, D. Yao, J. Magee, and M. Cygler, "Structural characterization of closely related O-antigen lipopolysaccharide (LPS) chain length regulators," *Journal of Biological Chemistry*, vol. 287, no. 19, pp. 15696–15705, 2012.

[114] G. Hagelueken, B. R. Clarke, H. Huang et al., "A coiled-coil domain acts as a molecular ruler to regulate O-antigen chain length in lipopolysaccharide," *Nature Structural & Molecular Biology*, vol. 22, no. 1, pp. 50–56, 2015.

[115] B. R. Clarke, L. K. Greenfield, C. Bouwman, and C. Whitfield, "Coordination of polymerization, chain termination, and export in assembly of the *Escherichia coli* lipopolysaccharide O9a antigen in an ATP-binding cassette transporter-dependent pathway," *Journal of Biological Chemistry*, vol. 284, no. 44, pp. 30662–30672, 2009.

[116] A. Tocilj, C. Munger, A. Proteau et al., "Bacterial polysaccharide co-polymerases share a common framework for control of polymer length," *Nature Structural & Molecular Biology*, vol. 15, no. 2, pp. 130–138, 2008.

[117] R. F. Collins, V. Kargas, B. R. Clarke et al., "Full-length, oligomeric structure of Wzz determined by cryoelectron microscopy reveals insights into membrane-bound states," *Structure*, vol. 25, no. 5, pp. 806–815, 2017.

[118] L. K. Greenfield and C. Whitfield, "Synthesis of lipopolysaccharide O-antigens by ABC transporter-dependent pathways," *Carbohydrate Research*, vol. 356, pp. 12–24, 2012.

[119] L. Cuthbertson, M. S. Kimber, and C. Whitfield, "Substrate binding by a bacterial ABC transporter involved in polysaccharide export," *Proceedings of the National Academy of Sciences*, vol. 104, no. 49, pp. 19529–19534, 2007.

[120] Y. Bi, E. Mann, C. Whitfield, and J. Zimmer, "Architecture of a channel-forming O-antigen polysaccharide ABC transporter," *Nature*, vol. 553, no. 7688, pp. 361–365, 2018.

[121] B. Ma, C. M. Reynolds, and C. R. H. Raetz, "Periplasmic orientation of nascent lipid A in the inner membrane of an *Escherichia coli* LptA mutant," *Proceedings of the National Academy of Sciences*, vol. 105, no. 37, pp. 13823–13828, 2008.

[122] S.-S. Chng, L. S. Gronenberg, and D. Kahne, "Proteins required for lipopolysaccharide assembly in *Escherichia coli* form a transenvelope complex," *Biochemistry*, vol. 49, no. 22, pp. 4565–4567, 2010.

[123] S. Okuda, E. Freinkman, and D. Kahne, "Cytoplasmic ATP hydrolysis powers transport of lipopolysaccharide across the periplasm in *E. coli*," *Science*, vol. 338, no. 6111, pp. 1214–1217, 2012.

[124] D. J. Sherman, R. Xie, R. J. Taylor et al., "Lipopolysaccharide is transported to the cell surface by a membrane-to-membrane protein bridge," *Science*, vol. 359, no. 6377, pp. 798–801, 2018.

[125] R. Villa, A. M. Martorana, S. Okuda et al., "The *Escherichia coli* Lpt transenvelope protein complex for lipopolysaccharide export is assembled via conserved structurally homologous domains," *Journal of Bacteriology*, vol. 195, no. 5, pp. 1100–1108, 2013.

[126] P. Sperandeo, R. Villa, A. M. Martorana et al., "New insights into the Lpt machinery for lipopolysaccharide transport to the cell surface: LptA-LptC interaction and LptA stability as sensors of a properly assembled transenvelope complex," *Journal of Bacteriology*, vol. 193, no. 5, pp. 1042–1053, 2011.

[127] E. Freinkman, S. Okuda, N. Ruiz, and D. Kahne, "Regulated assembly of the transenvelope protein complex required for lipopolysaccharide export," *Biochemistry*, vol. 51, no. 24, pp. 4800–4806, 2012.

[128] Q. Luo, X. Yang, S. Yu et al., "Structural basis for lipopolysaccharide extraction by ABC transporter LptB2FG," *Nature Structural & Molecular Biology*, vol. 24, no. 5, pp. 469–474, 2017.

[129] M. D. L. Suits, P. Sperandeo, G. Dehò, A. Polissi, and Z. Jia, "Novel structure of the conserved gram-negative lipopolysaccharide transport protein A and mutagenesis analysis," *Journal of Molecular Biology*, vol. 380, no. 3, pp. 476–488, 2008.

[130] A. X. Tran, C. Dong, and C. Whitfield, "Structure and functional analysis of LptC, a conserved membrane protein involved in the lipopolysaccharide export pathway in *Escherichia coli*," *Journal of Biological Chemistry*, vol. 285, no. 43, pp. 33529–33539, 2010.

[131] S. Qiao, Q. Luo, Y. Zhao, X. C. Zhang, and Y. Huang, "Structural basis for lipopolysaccharide insertion in the bacterial outer membrane," *Nature*, vol. 511, no. 7507, pp. 108–111, 2014.

[132] H. Dong, Z. Zhang, X. Tang, N. G. Paterson, and C. Dong, "Structural and functional insights into the lipopolysaccharide ABC transporter LptB2FG," *Nature Communications*, vol. 8, no. 1, 2017.

[133] B. Tefsen, J. Geurtsen, F. Beckers, J. Tommassen, and H. de Cock, "Lipopolysaccharide transport to the bacterial outer membrane in spheroplasts," *Journal of Biological Chemistry*, vol. 280, no. 6, pp. 4504–4509, 2005.

[134] D. J. Sherman, M. B. Lazarus, L. Murphy et al., "Decoupling catalytic activity from biological function of the ATPase that powers lipopolysaccharide transport," *Proceedings of the National Academy of Sciences*, vol. 111, no. 13, pp. 4982–4987, 2014.

[135] G. Malojčić, D. Andres, M. Grabowicz et al., "LptE binds to and alters the physical state of LPS to catalyze its assembly at the cell surface," *Proceedings of the National Academy of Sciences*, vol. 111, no. 26, pp. 9467–9472, 2014.

[136] H. Dong, Q. Xiang, Y. Gu et al., "Structural basis for outer membrane lipopolysaccharide insertion," *Nature*, vol. 511, no. 7507, pp. 52–56, 2014.

[137] E. T. Rietschel, T. Kirikae, F. U. Schade et al., "Bacterial endotoxin: molecular relationships of structure to activity and function," *FASEB Journal*, vol. 8, no. 2, pp. 217–225, 1994.

[138] A. J. Ulmer, H. Heine, W. Feist et al., "Biological activity of synthetic phosphonooxyethyl analogs of lipid A and lipid A partial structures," *Infection and Immunity*, vol. 60, no. 8, pp. 3309–3314, 1992.

[139] D. T. Golenbock, R. Y. Hampton, N. Qureshi, K. Takayama, and C. R. Raetz, "Lipid A-like molecules that antagonize the effects of endotoxins on human monocytes," *Journal of Biological Chemistry*, vol. 266, no. 29, pp. 19490–19498, 1991.

[140] R. E. Bishop, H. S. Gibbons, T. Guina, M. S. Trent, S. I. Miller, and C. R. Raetz, "Transfer of palmitate from phospholipids to lipid A in outer membranes of gram-negative bacteria," *EMBO Journal*, vol. 19, no. 19, pp. 5071–5080, 2000.

[141] L. E. Hittle, J. W. Jones, A. M. Hajjar, R. K. Ernst, and A. Preston, "Bordetella parapertussis PagP mediates the addition of two palmitates to the lipopolysaccharide lipid A," *Journal of Bacteriology*, vol. 197, no. 3, pp. 572–580, 2015.

[142] M.-L. Han, T. Velkov, Y. Zhu et al., "Polymyxin-induced lipid A deacylation in *Pseudomonas aeruginosa* perturbs polymyxin penetration and confers high-level resistance," *ACS Chemical Biology*, vol. 13, no. 1, pp. 121–130, 2018.
[143] J. C. Henderson, C. M. Herrera, and M. S. Trent, "AlmG, responsible for polymyxin resistance in pandemicVibrio cholerae, is a glycyltransferase distantly related to lipid A late acyltransferases," *Journal of Biological Chemistry*, vol. 292, no. 51, pp. 21205–21215, 2017.
[144] L.-A. Segev-Zarko, G. Kapach, M. Josten, Y. A. Klug, H.-G. Sahl, and Y. Shai, "Deficient lipid A remodeling by the arnB gene promotes biofilm formation in antimicrobial peptide susceptible *Pseudomonas aeruginosa*," *Biochemistry*, vol. 57, no. 13, pp. 2024–2034, 2018.
[145] M. Helander Ilkka, I. Kilpeläinen, and V. Martti, "Increased substitution of phosphate groups in lipopolysaccharides and lipid A of the polymyxin-resistant pmrA mutants of *Salmonella typhimurium*: a 31P-NMR study," *Molecular Microbiology*, vol. 11, no. 3, pp. 481–487, 2006.
[146] H. Lee, F.-F. Hsu, J. Turk, and E. A. Groisman, "The PmrA-regulated pmrC gene mediates phosphoethanolamine modification of lipid A and polymyxin resistance in *Salmonella enterica*," *Journal of Bacteriology*, vol. 186, no. 13, pp. 4124–4133, 2004.
[147] L. A. Lewis and S. Ram, "Meningococcal disease and the complement system," *Virulence*, vol. 5, no. 1, pp. 98–126, 2014.
[148] H. Loppnow, H. Brade, I. Dürrbaum et al., "IL-1 induction-capacity of defined lipopolysaccharide partial structures," *Journal of immunology (Baltimore, Md.: 1950)*, vol. 142, no. 9, pp. 3229–3238, 1989.

7

Characterization of Erythromycin and Tetracycline Resistance in *Lactobacillus fermentum* Strains

Elizaveta Anisimova and Dina Yarullina

Department of Microbiology, Kazan Federal University, Kremlevskaya Str. 18, Kazan 420008, Russia

Correspondence should be addressed to Dina Yarullina; kasfes@gmail.com

Academic Editor: Clemencia Chaves-López

Lactobacillus fermentum colonizing gastrointestinal and urogenital tracts of humans and animals is widely used in manufacturing of fermented products and as probiotics. These bacteria may function as vehicles of antibiotic resistance genes, which can be transferred to pathogenic bacteria. Therefore, monitoring and control of transmissible antibiotic resistance determinants in these microorganisms is necessary to approve their safety status. The aim of this study was to characterize erythromycin and tetracycline resistance of *L. fermentum* isolates and to estimate the potential transfer of resistance genes from lactobacilli to the other Gram-positive and Gram-negative bacteria. Among six *L. fermentum* strains isolated from human feces and commercial dairy products, five strains demonstrated phenotypic resistance to tetracycline. PCR screening for antibiotic resistance determinants revealed plasmid-located tetracycline resistance genes *tet*(K) and *tet*(M) in all strains and erythromycin resistance genes *erm*(B) in the chromosome of *L. fermentum* 5-1 and *erm*(C) in the plasmid of *L. fermentum* 3-4. All tested lactobacilli lacked conjugative transposon Tn*916* and were not able to transfer tetracycline resistance genes to *Staphylococcus aureus*, *Staphylococcus epidermidis*, *Listeria monocytogenes*, *Acinetobacter baumannii*, *Citrobacter freundii*, and *Escherichia coli* by filter mating. *Staphylococcus haemolyticus* did not accept erythromycin resistance genes from corresponding *Lactobacillus* strains. Thus, in the present study, *L. fermentum* was not implicated in the spread of erythromycin and tetracycline resistance, but still these strains pose the threat to the environment and human health because they harbored erythromycin and tetracycline resistance genes in their plasmids and therefore should not be used in foods and probiotics.

1. Introduction

Lactobacillus fermentum is a common inhabitant of human gastrointestinal and urogenital tracts where these bacteria are known to exert health-promoting effects [1, 2]. It is naturally present in raw milk, dairy products, and many traditional fermented foods and beverages [3, 4]. Some probiotic isolates, such as *L. fermentum* RC-14 and *L. fermentum* 90 TC-4, are already widely used and produced on an industrial scale [5, 6]. The species is listed in the qualified presumption of safety (QPS) published by the European Food Safety Authority (EFSA) [7], and along with other lactic acid bacteria (LAB), it has been generally recognized as safe (GRAS status) by the U.S. Food and Drug Administration. However, over the last decade, bacteria used as probiotics or in starter cultures have received growing attention with regard to their potential involvement in acquisition and transfer of antibiotic resistance genes to food or gut pathogens [8, 9].

In line with the growing awareness of LAB involvement in the spread of antibiotic resistance, it is imperative that the antibiotic resistance profile of probiotic and starter cultures should be studied both at the physiological level and molecular level. A phenotypically resistant strain may be genotypically "susceptible," and in contrast, a susceptible phenotype may carry silent genes, which are observed with genotyping [10]. Determination of antibiotic resistance among lactobacilli is confounded by the absence of standards for susceptibility testing. Microbiological breakpoints have been determined only for some antibiotics and for particular most commonly used probiotic species [11]. Besides, minimum inhibitory concentration (MIC) breakpoint values can vary depending on the method and media applied, as well as between species of the same genus [12–14].

Bacterial resistance to antibiotics can be of two types: intrinsic resistance, which is a natural property of an organism, having a minimal potential for horizontal spread and posing no risk in nonpathogenic bacteria, and acquired resistance, which results from mutation or, what is more often, via the acquisition of generic materials, such as a plasmid or a transposon [10]. Lactobacilli have been reported to have a high natural resistance to vancomycin, aminoglycosides, and the majority of nucleic acid inhibitors [15]. Resistance to other antibiotics varies greatly among species. Erythromycin resistance and tetracycline resistance have been found to be acquired, and therefore, corresponding genetic determinants are often considered potentially transferable [16, 17]. Acquired resistance to erythromycin encoded by the *erm*(B) gene has been detected in two *L. fermentum* strains NWL24 and NWL26 isolated from Chinese fermented foods (yogurt) [18]. *L. fermentum* CB101 also isolated from Chinese fermented food (cucumbers) showed resistance to both erythromycin and tetracycline with very high MICs (512 μg/ml for erythromycin and more than 128 μg/ml for tetracycline) [19]. From four erythromycin- and tetracycline-resistant *L. fermentum* strains isolated from Indian fermented foods, three strains have been found to harbor the *erm*(B) gene and one harbors the tetracycline efflux genes *tet*(K) and *tet* (L) [20]. Several plasmids potentially implemented in transfer of antibiotic resistance have been identified in *L. fermentum*, e.g., resistance plasmid pLME300 of *L. fermentum* ROT1 [21], tetracycline and erythromycin resistance plasmids of *L. fermentum* LF601 [22], and plasmid pLEM3 conferring erythromycin resistance of *L. fermentum* [23]. The works demonstrating conjugative transfer of antibiotic resistance determinants from lactobacilli to other bacteria especially those belonging to the gut microbiota are very limited [24–26]. So far, there is only one evidence of successful transfer of the *erm*(B) gene from *L. fermentum* NWL24 to *Enterococcus faecalis* 181 in filter mating experiments [18].

The aim of this work was to comprehensively characterize erythromycin and tetracycline resistance profiles of *L. fermentum* strains and evaluate the transferability of corresponding resistance genes to other bacteria. In this study, six *L. fermentum* strains isolated from human feces and commercial dairy products were assayed for susceptibility to erythromycin and tetracycline by the agar disc diffusion method, followed by determination of MICs for these antibiotics by the broth microdilution method. Phenotypic assays of bacterial strains were complemented by direct screening for the presence of erythromycin and tetracycline resistance determinants by PCR amplification. The transferability of these antibiotic resistance genes from *L. fermentum* strains to a number of Gram-positive and Gram-negative bacteria was studied in filter mating experiments. This study provides a reference for the safety assessment and contributes to the evaluation system of probiotics.

2. Materials and Methods

2.1. Isolation of Bacteria and Growth Conditions. Six *Lactobacillus* strains were isolated from the feces of healthy individuals (HF-A1, HF-A4, and HF-B1) and commercial dairy products (3-4, 5-1, and 5-2) by serial dilution of the sample in sterile phosphate-buffered saline (PBS), subsequent plating onto the de Man–Rogosa–Sharpe (MRS) agar (HiMedia, India), and incubation under anaerobic conditions (Anaerogas pack; NIKI MLT, Russia) at 37°C. Single colonies were selected, purified, and maintained on the MRS agar for immediate use and in 20% glycerol for storage at −80°C. Prior to all experiments, bacterial isolates were subcultured at least twice.

2.2. Identification of Bacteria. The bacterial colonies were assigned to the genus *Lactobacillus* by MALDI-TOF mass spectrometry (Bruker Biotyper system, Bruker Daltonics, Germany), as previously described [27]. In brief, a single fresh colony from the MRS agar was smeared onto a ground steel target (Bruker Daltonik), overlaid with 1 μl of a saturated solution of alpha-cyano-4-hydroxycinnamic acid matrix in 50% acetonitrile and 2.5% trifluoroacetic acid, and air-dried at room temperature. For each strain, two preparations of the colony material were analyzed. Mass spectra were recorded according to the manufacturer's instructions. The obtained spectra were compared with reference spectra in the integrated database (version 3.2.1.1). Standard Bruker interpretative criteria were applied. Scores ≥2.3 were accepted for reliable species assignment and scores ≥2.0 but <2.3 for reliable identification to the genus level. Scores below 2.0 were considered unreliable.

For most accurate species identification, the 16S rRNA gene was amplified by the PCR method using universal 16S rRNA bacterial primers 27F and 1392R (Table 1) and PCR program described below. The PCR-amplified 1.4 kbp DNA fragments were resolved by agarose gel electrophoresis, purified, and sequenced on an ABI Prism 3730 sequencer (Applied Biosystems). The bacterial isolates were identified to the species level by comparing their 16S rRNA gene sequences with those in the NCBI databases by BLAST.

2.3. Antibiotic Susceptibility Testing and MIC Determination. Susceptibility to erythromycin and tetracycline was determined by the disc diffusion method, as described earlier [27]. In brief, bacteria were grown in the MRS broth overnight at 37°C in anaerobic conditions (Anaerogas pack; NIKI MLT, Russia) and pour-plated on MRS agar plates (0.5 McFarland after inoculation). Antibiotic discs (Scientific Research Centre of Pharmacotherapy, Russia) were placed on the surface of inoculated plates. After 48 h incubation in anaerobic conditions at 37°C, the diameter of the inhibition zone was measured and interpreted as susceptible (S), moderately susceptible (MS), or resistant (R) according to the method in [27]. The following standards for interpreting the zones of inhibition for antibiotics were used: bacteria were considered resistant (R) to erythromycin (15 μg/disc) when the zone of inhibition (mm) was ≤10, moderately susceptible (MS) at

Table 1: Primers used in PCR amplifications.

Target gene	Primer sequence (5′-3′)	T_a (°C)	Amplicon size (bp)	Reference
erm(A)	F: AAGCGGTAAACCCCTCTGAG R: TCAAAGCCTGTCGGAATTGG	52	441	[28]
erm(B)	ermB1-F: CATTTAACGACGAAACTGGC ermB1-R: GGAACATCTGTGGTATGGCG	60	425	[28]
	ermB2-F: GAAAAGGTACTCAACCAAATA ermB2-R: AGTAACGGTACTTAAATTGTTTAC	59	639	[29]
erm(C)	F: ATCTTTGAAATCGGCTCAGG R: CAAACCCGTATTCCACGATT	49	295	[28]
erm(T)	F: TATTATTGAGATTGGTTCAGGG R: GGATGAAAGTATTCTCTAGGGATTT	55	395	[28]
mef(A)	F: CTATGACAGCCTCAATGCG R: ACCGATTCTATCAGCAAAG	52	1400	[30]
Int	F: GCGTGATTGTATCTCACT R: GACGCTCCTGTTGCTTCT	50	1028	[31]
Tet-int	F: CGGATAGATAAAGTACGATA R: TCACGTCTTTTTTCTGACAT	52	2659	[32]
tet(M)	tetM1-F: GAACTCGAACAAGAGGAAAGC tetM1-R: ATG GAAGCCCAGAAAGGAT	60	740	[32]
	tetM2-F: GGTGAACATCATAGACACGC tetM2-R: CTTGTTCGAGTTCCAATGC	58	401	[33]
tet(L)	tetL1-F: GTMGTTGCGCGCTATATTCC tetL1-R: GTGAAMGRWAGCCCACCTAA	55	696	[31]
	tetL2-F: GTTTCGGGTCGGTAATTGGG tetL2-R: GCTATCATTCCACCAATCGC	45	220	[28]
tet(K)	tetK1-F: TTATGGTGGTTGTAGCTAGAAA tetK1-R: AAAGGGTTAGAAACTCTTGAAA	55	348	[31]
	tetK2-F: GTAGCGACAATAGGTAATAG tetK2-R: GCAACTTCTTCTTCAGAAAG	46	278	[28]
tet(S)	F: GGAGTACAGTCACAAACTCG R: GGATATAAGGAGCAACTTTG	55	335	[28]
tet(W)	F: GAGAGCCTGCTATATGCCAGC R: GGGCGTATCCACAATGTTAAC	55	168	[28]
16S rRNA	27F: GAGTTTGATCCTGGCTCAG 1392R: ACGGTTACCTTGTTACGACTT	51	1400	[34]

11–20, and susceptible (S) at ≥21; bacteria were considered resistant (R) to tetracycline (30 μg/disc) when the zone of inhibition (mm) was ≤16, moderately susceptible (MS) at 17–21, and susceptible (S) at ≥22.

The MIC for antibiotics was determined by the broth microdilution method in the MRS broth in 96-well non-treated cell culture plates (Eppendorf). Tetracycline (Sigma-Aldrich) and erythromycin (Sigma-Aldrich) were tested in the concentration range of 0.12–256 μg/ml obtained after a series of twofold dilutions in the MRS broth. Wells were inoculated with 200 μl of the bacterial culture (3×10^7 CFU/ml) and incubated at 37°C. The MIC was read after 48 h of incubation as the lowest concentration of an antibiotic at which visible growth was inhibited. According to the microbiological breakpoints suggested for *L. fermentum* by EFSA [11], strains with MICs higher than 1 μg/ml for erythromycin and 8 μg/ml for tetracycline were considered resistant.

2.4. DNA Preparation and Manipulations. Bacteria were cultivated overnight at 37°C in 30 ml of the MRS broth. The cells were harvested by centrifugation (5 min, 13 000 RPM) and then resuspended in 3 ml of the lysis buffer (50 mM Tris-HCl and 3 mg/ml lysozyme, pH 8.0). The resuspended pellet was incubated 3 h at 37°C with shaking. After this step, 100 μl of sodium dodecyl sulfate (SDS; 10%, w/v) wad added, and the sample was gently inverted for cell disruption. 500 μl of phenol : chloroform (1 : 1, v/v) was added. After a centrifugation step (5 min, 13 000 RPM, room temperature), the clear supernatant was transferred to a new tube. The DNA was precipitated with 500 μl of cold isopropanol, and the DNA was taken out with a plastic tip. The DNA was washed with ethanol 96% (w/v), and after it has been dried, it was diluted in 100 μl of distilled water. All the DNA samples were stored at −20°C.

Plasmid DNA was obtained using GeneJET Plasmid Miniprep Kit (Thermo Scientific), according to the manufacturer's instructions.

2.5. PCR Detection of Genes. The presence of erythromycin and tetracycline resistance genes was determined by PCR amplification using the primers listed in Table 1. The presence of the transposon Tn*916* was detected by amplification of the integrase-encoding gene (*Int*) and examining the interregion between *tet*(M) and *Int* genes with the primer *Tet-int* described in Table 1.

The PCR reaction was carried out in a total volume of 25 μl containing DNA template, 10 pmol of each primer (Table 1), 1 U Taq DNA polymerase, each of four dNTPs at a concentration of 200 μM, and PCR buffer containing Tris-HCl, KCl, $(NH_4)_2SO_4$, $MgSO_4$, and Triton X-100. The amplification program was as follows: initial denaturation at 95°C for 4 min, 35 cycles at 94°C for 30 s, 46–60°C (according to the annealing temperature for the individual primers; Table 1) for 30 s, and 72°C for 1.5 min, and a final extension step at 72°C for 7 min. Positive and negative controls from our lab were used for all PCR reactions. PCR products (5 μl) were separated by electrophoresis on a 1% agarose gel, which was stained with Midori Green DNA Stain (Nippon Genetics Europe, Germany).

2.6. Filter Mating Experiments. Transferability of antibiotic resistance from lactobacilli was examined by filter mating, as described in [24], on a number of Gram-positive (*Staphylococcus aureus* subsp. *aureus* ATCC® 29213™, *Staphylococcus haemolyticus, Staphylococcus epidermidis* (clinical isolates), and *Listeria monocytogenes* 88K (a gift from Dr. Alexey Vasilchenko, Orenburg State University)) and Gram-negative (*Acinetobacter baumannii, Citrobacter freundii*, and *Escherichia coli* (clinical isolates)) bacteria. Clinical isolates of *S. haemolyticus, S. epidermidis*, and *E. coli* were obtained from the Kazan Institute of Epidemiology and Microbiology (Kazan, Russia). Clinical isolates of *A. baumannii* and *C. freundii* were obtained from the Institute of Medical Microbiology (Giessen, Germany). Cultures of pathogens were incubated in the Luria-Bertani (LB) medium. Antibiotic resistance profiles of recipient strains were characterized by determination of MICs, as described above. The lack of erythromycin and tetracycline resistance genes was proved by PCR, as indicated earlier. *S. haemolyticus* was sensitive to both erythromycin and tetracycline but carried the *tet*(K) gene and therefore was used as the recipient strain for erythromycin resistance genes. Other tested strains were either resistant to erythromycin (all Gram-negative strains) or possessed a silent *erm*(B) gene (*S. aureus, S. epidermidis*, and *L. monocytogenes*). As they were sensitive to tetracycline and lacked *tet* genes, they were used as recipient strains for tetracycline resistance genes. Chloramphenicol (40 μg/ml) was used as a selective antibiotic marker for transconjugants of *S. epidermidis, S. haemolyticus*, and *A. baumannii*, rifampicin (2.5 μg/ml) for *S. aureus, L. monocytogenes*, and *E. coli*, and ampicillin (40 μg/ml) for *C. freundii*. These concentrations were sufficient to completely inhibit growth of *Lactobacillus*, whereas recipient bacteria were unaffected.

For filter mating experiments, the MRS broth and LB broth were inoculated (1 : 100) with overnight grown donor *L. fermentum* cultures and recipient pathogen cultures, correspondingly, and incubated at 37°C for approx. 4 h to the mid-exponential phase of growth. One ml of each culture was mixed and filtered through a sterile 0.45 μm pore-size nitrocellulose membrane filter (Millipore, USA). After that, sterile peptone physiological saline (PPS) solution (8.5 g/l NaCl and 1 g/l bacteriological peptone) was passed through the filter to trap the cells more tightly into the membrane. The filters were incubated overnight on the LB agar at 37°C. The bacteria were washed from the filters with 2 ml PPS. Dilutions of the mating mixtures were spread onto LB agar plates containing 10 μg/ml tetracycline or erythromycin (Sigma-Aldrich) and selective antibiotic (chloramphenicol, rifampicin, or ampicillin, all from Sigma-Aldrich) (double selective medium), agar plates containing 10 μg/ml tetracycline or erythromycin (Sigma-Aldrich) or selective antibiotic (single selective medium), and nonselective medium without antibiotics. Control cultures of donor and recipient strains were also individually plated on three types of agar plates. Plates were incubated at 37°C for 24–48 h.

3. Results and Discussion

3.1. Isolation and Identification of L. fermentum Strains. In our previous work, among 15 *Lactobacillus* strains isolated from feces of healthy volunteers (unpublished data) and 19 *Lactobacillus* strains isolated from commercial dairy products and probiotics [27], six strains were shown to be resistant to either erythromycin or tetracycline by the disc diffusion method. Resistance to these two antibiotics is considered the most threatening one in probiotics because it is often acquired and potentially transferable [16, 17]. Therefore, these six *Lactobacillus* strains were used in this study to comprehensively characterize their erythromycin or tetracycline resistance profiles and evaluate the transferability of corresponding resistance genes to other bacteria. All tested strains were putatively assigned to *L. fermentum* species by MALDI Biotyper as sample mass spectra shared the maximum similarity with the reference mass spectrum of *L. fermentum* from MALDI Biotyper software (score values ranged between 2.019 and 2.083). This finding was verified by 16S rRNA gene analysis after we determined that sample 16S rDNA sequences had a similarity score of ≥99% with that of reference sequences in the NCBI database classified as *L. fermentum*.

3.2. Phenotypic Profile of Antimicrobial Resistance. Phenotypic profile of *L. fermentum* resistance to erythromycin and tetracycline is presented in Table 2. All strains tested in this study were susceptible or moderately susceptible to erythromycin by disc diffusion, except for the strain 5-1, which was resistant. With MICs between 0.25 and 1 μg/ml, all strains were considered susceptible to erythromycin based on the MIC breakpoints established by EFSA [11]. Five tested *L. fermentum* isolates showed resistance to tetracycline, whereas the strain 5-1 was moderately susceptible. In full concordance with the results of the disc diffusion method, these five isolates of *L. fermentum* showed high MICs of 16–64 μg/ml for tetracycline and thus were assigned as resistant to tetracycline according to [11]. Tetracycline resistance common among the human fecal isolates may be explained by the intensive use of tetracyclines, alone or in combination, in medicine for prophylaxis or therapy and in consumed food, which could provide the selective pressure needed for antibiotic-resistant bacteria to develop and spread. The phenotypic profile of antimicrobial

Table 2: Characterization of *L. fermentum* strains with antibiotic resistance for erythromycin and tetracycline.

L. fermentum strains	Source	Erythromycin			Tetracycline			Phenotype[d]	Genotype[e]
		Zone of inhibition (mm)[a]	Characterization[b]	MIC[c] (μg/ml)	Zone of inhibition (mm)[a]	Characterization[b]	MIC[c] (μg/ml)		
HF-A1	Feces of healthy individuals	21.2 ± 2.9	S	0.25	15.7 ± 0.6	R	64	TET	*tet*(M)[P], *tet*(K)[P]
HF-A4	Feces of healthy individuals	21.0 ± 1.4	S	0.25	15.5 ± 0.7	R	64	TET	*tet*(M)[P], *tet*(K)[P]
HF-B1	Feces of healthy individuals	20.2 ± 3.4	MS	0.25	14.7 ± 0.6	R	64	TET	*tet*(M)[P], *tet*(K)[P]
3-4	Sour-milk drink, "Tan," "Chistaya Liniya," Russia	21.0 ± 2.5	S	0.25	6.5 ± 2.1	R	16	TET	*erm*(C)[P], *tet*(M)[P], *tet*(K)[P]
5-1	Fermented baked milk, ryazhenka, "Chistoe pole," Russia	6.5 ± 2.1	R	1	20.3 ± 1.4	MS	8	—	*erm*(B)[c], *tet*(M)[P], *tet*(K)[P]
5-2	Fermented baked milk, ryazhenka, "Nash product," Russia	24.3 ± 2.5	S	0.25	14.5 ± 0.7	R	32	TET	*tet*(M)[P], *tet*(K)[P]

[a]Diameters of the zones of inhibition determined by the disc diffusion method (mean value ± standard deviation) (Figure S1A). [b]Based on standards mentioned in Materials and Methods, *L. fermentum* strains were characterized as either susceptible (S), moderately susceptible (MS), or resistant (R) to each antibiotic tested. [c]The minimum inhibitory concentration (MIC) of antibiotics determined by the broth microdilution method as the lowest concentration of an antibiotic at which visible growth was inhibited. [d]Based on MIC values, *L. fermentum* strains were characterized as erythromycin resistant (ERM) or tetracycline resistant (TET). [e]Antibiotic resistance genes present in isolates of *L. fermentum* as detected using PCR: [P]plasmid-located gene; [c]chromosome-located gene (for details, see Figures S1B–S1D).

resistance characterized by disc diffusion mainly coincided with observed MICs, except for the strain *L. fermentum* 5-1, which was resistant to erythromycin in the disc diffusion method, but ultimately was considered nonresistant according to the MIC value.

Thus, phenotypic assays showed that *L. fermentum* strains HF-A1, HF-A4, HF-B1, 3-4, and 5-2 possess tetracycline resistance which is potentially transferable.

3.3. Antibiotic Resistance Genes. Antibiotic resistance genes were detected by PCR, and the results are presented in Table 2. All the strains were tested for the presence of erythromycin resistance genes *erm*(A), *erm*(B), *erm*(C), *erm* (T), and *mef*(A). Only *L. fermentum* 5-1 was positive for the *erm*(B) gene, giving a 425 bp band, corresponding to its resistance phenotype in the disc diffusion assay (Figure S1B, lane 2). The *erm*(B) gene, which encodes an rRNA methylase acting on the 23S ribosomal subunit, is the most frequently found one of all erythromycin resistance genes in *Lactobacillus* species [15]. According to PCR results, the *erm*(B) gene was attributed to chromosomal DNA in *L. fermentum* 5-1 and thus is not likely to be transferable. However, acquired resistance to erythromycin because of *erm*(B) has often been reported in lactobacilli from different sources [9, 18, 20, 21, 23, 35]. The erythromycin resistance gene *erm* (C) was found in plasmid DNA from *L. fermentum* 3-4 (Figure S1B, lane 3), and to our knowledge, this gene has not been described previously from that particular species. Its lower prevalence among lactobacilli is consistent with previous reports [20, 35]. The *erm*(C) gene has only been detected in few *L. plantarum* strains from fermented dry sausages [36] and several chicken isolates of the species *L. salivarius*, *L. agilis*, *L. crispatus*, *L. reuteri*, and *L. saerimneri* [37]. Other erythromycin resistance determinants (*erm*(T) and *mef*(A)) were not detected in any strain.

In this study, we also tested the presence of tetracycline resistance genes *tet*(K), *tet*(M), *tet*(W), *tet*(S), and *tet*(L). The *tet*(K) and *tet*(M) genes were found in plasmid DNA of all the tested strains (Figures S1C and S1D). These determinants have different modes of action: *tet*(K) encodes efflux pumps, whereas *tet*(M) offers ribosomal protection [16]. The simultaneous occurrence of *tet*(K) and *tet*(M) genes is in agreement with reports of other lactobacilli containing multiple *tet* genes [18, 20, 37–39]. Genetic location of the *tet* (K) gene on small multicopy plasmids and *tet*(M) on conjugative transposons (Tn*916*-Tn*1545* family) promotes the spread of these determinants [16]. Yet in this study, we have shown that all the tested *L. fermentum* strains lack the conjugative transposon Tn*916*, carrying the *tet*(M) gene and capable of horizontal, interspecies transfer. No amplicons were yielded when primer pairs specific for the *int* gene, encoding the integrase protein that allows mobilization of Tn*916*, and for the intervening region between the *tet*(M) and *int* sequences, were used (data not shown). We did not find any of *tet*(W), *tet*(S), and *tet*(L) genes in this study.

Results of PCR screening for antibiotic resistance determinants in the *L. fermentum* strains were only partly consistent with those of phenotypic assays. Detection of *tet* genes in the strains HF-A1, HF-A4, HF-B1, 3-4, and 5-2 was in agreement with their resistance phenotypes. *L. fermentum* 5-1 sensitive to tetracycline was discovered to carry silent genes *tet*(K) and *tet*(M), and *L. fermentum* 3-4 sensitive to erythromycin was positive for *erm*(C). Detection of *erm*(B) in *L. fermentum* 5-1 was in agreement with its erythromycin resistant phenotype shown in the disc diffusion assay but was in contradiction with the results of MIC determination. These controversial results can be explained by dependence of susceptibility results on the method for antibiotic susceptibility assessment. It was shown that an increased inoculum size and an extended incubation time resulted in elevated antibiotic MICs for some species [14].

3.4. Transfer of Antibiotic Resistance Genes. We studied the transfer of tetracycline resistance genes by filter mating from six *L. fermentum* strains positive for *tet*(K) and *tet*(M) to a number of Gram-positive and Gram-negative bacteria—*Staphylococcus aureus*, *Staphylococcus epidermidis*, *Listeria monocytogenes*, *Acinetobacter baumannii*, *Citrobacter freundii*, and *Escherichia coli*. Also, *L. fermentum* 5-1 carrying the *erm*(B) gene and *L. fermentum* 3-4 carrying the *erm*(C) gene were used as donors, while *Staphylococcus haemolyticus* served as a recipient for erythromycin resistance genes. The recipient strains were chosen experimentally and met two criteria necessary for counterselection of transconjugants: (i) lack of resistance to erythromycin or tetracycline both in phenotypes and genotypes and (ii) presence of resistance to antibiotic active against tested *Lactobacillus* strains. Besides, these bacteria are members of the human gastrointestinal microbiota [40]. Coagulase-negative staphylococci *S. epidermidis* and *S. haemolyticus* usually colonize normal human skin but also may be isolated from the human gastrointestinal tract. Isolates from the intestinal tracts have been found to express virulence factors indicating their pathogenicity potentials [41].

The results showed that erythromycin and tetracycline resistance genes from tested *L. fermentum* strains cannot be transferred to the other bacteria used in this study via conjugation. No bacterial growth was observed in plates with the double selective medium, thus indicating that none of the recipient cells harbored erythromycin or tetracycline resistance (data not shown).

Lactobacilli and Gram-negative bacteria is an uncommon pair in studying the transferability of antibiotic resistance determinants. To date, there is only one report on conjugative transfer of antibiotic resistance genes from *L. fermentum* to other bacteria. The *erm*(B) gene from *L. fermentum* NWL24 was successfully transferred to *Enterococcus faecalis* 181 by filter mating [18]. However, it was demonstrated that gene transfer via conjugation can take place from Gram-positive to Gram-negative bacteria, and vice versa [42, 43]. The streptococcal conjugative transposon Tn*916* was shown to be able to cross the barrier between a variety of Gram-positive and Gram-negative bacteria, with subsequent expression in the new host [42]. Thus, absence of transconjugants in performed filter mating experiments coincides with the lack of Tn*916* in tested *L. fermentum* strains.

The disability to transfer genes for antibiotic resistance is considered an important parameter for the selection of the LAB strains for the food industry and probiotics [15]. This property is equally important for lactobacilli of normal human microbiota in order to avoid spread of resistance determinants among different species, including potential and obligate pathogens, in habitats with close association of densely packed microorganisms such as the intestine of humans. In all tested *L. fermentum* strains, *tet*(K) and *tet*(M) genes were shown to be located in plasmids and thus are likely to be acquired. Three *L. fermentum* strains (3-4, 5-1, and 5-2) isolated from dairy products represent a serious safety issue because of their tetracycline-resistant phenotype (strains 3-4 and 5-2) and simultaneous presence of tetracycline resistance genes *tet*(K) and *tet*(M) (all strains) along with erythromycin resistance genes *erm*(B) (strain 5-1) and *erm*(C) (strain 3-4). In this work, we have shown that all tested lactobacilli lack conjugative transposon Tn*916* and are not able to transfer tetracycline resistance genes to *S. aureus*, *S. epidermidis*, *L. monocytogenes*, *A. baumannii*, *C. freundii*, and *E. coli* by filter mating. Mating of *L. fermentum* 5-1 carrying the *erm*(B) gene and *L. fermentum* 3-4 carrying the *erm*(C) gene did not yield transconjugants with *S. haemolyticus* as well. These findings reduce the likelihood of the impact of tested lactobacilli to the spread of erythromycin and tetracycline resistance in the human microbiome and favor their safe status but still do not establish their complete safety. Our results highlight the need to include screening of antibiotic resistance into the current practices of safety evaluation mandatory before application of *Lactobacillus* strains as starter cultures or as probiotics.

Conflicts of Interest

The authors declare that there are no conflicts of interest regarding the publication of this paper.

Authors' Contributions

Elizaveta Anisimova carried out the study and drafted the manuscript. Dina Yarullina designed the experiments, reviewed the manuscript, and supervised the work. All authors read and approved the final manuscript.

Acknowledgments

This work was supported by the Program of Competitive Growth of Kazan Federal University, RFBR grant no. 18-34-00268 and partially by RFBR grant no. 17-00-00456 (strain isolation). The research was performed by using the equipment of the Interdisciplinary Center for Collective Use of Kazan Federal University for cellular, genomic, and postgenomic research in Volga region.

Supplementary Materials

Figure S1: phenotypic (A) and genotypic (B, C, and D) erythromycin and tetracycline resistance testing of *L. fermentum*. (A) Inhibition halos obtained by the agar disc diffusion method for *L. fermentum* 5-1 (1), *L. fermentum* HF-A1 (2), *L. fermentum* HF-A4 (3), *L. fermentum* HF-B1 (4), *L. fermentum* 3-4 (5), and *L. fermentum* 5-2 (6). PCR products of erythromycin (B) and tetracycline (C and D) resistance genes. (B) The PCR products amplified with *erm* (B)1 primer pair in plasmid DNA (lane 1) and total DNA (lane 2) of *L. fermentum* 5-1 and with *erm*(C) primer pair in plasmid DNA (lane 3) and total DNA (lane 4) of *L. fermentum* 3-4. The PCR products amplified with *tet*(K)2 (C) and *tet*(M)1 (D) primer pairs in plasmid DNA of *L. fermentum* HF-A1 (lane 1), *L. fermentum* HF-B1 (lane 2), *L. fermentum* HF-A4 (lane 3), *L. fermentum* 5-1 (lane 4), *L. fermentum* 3-4 (lane 5), and *L. fermentum* 5-2 (lane 6). Lane M, 1 kb DNA ladder. (*Supplementary Materials*)

References

[1] S. Kaewnopparat, N. Dangmanee, N. Kaewnopparat, T. Srichana, M. Chulasiri, and S. Settharaksa, "*In vitro* probiotic properties of *Lactobacillus fermentum* SK5 isolated from vagina of a healthy woman," *Anaerobe*, vol. 22, pp. 6–13, 2013.

[2] V. Liévin-Le Moal and A. L. Servin, "Anti-infective activities of *Lactobacillus* strains in the human intestinal microbiota: from probiotics to gastrointestinal anti-infectious biotherapeutic agents," *Clinical Microbiology Reviews*, vol. 27, no. 2, pp. 167–199, 2014.

[3] C. L. Ramos, L. Thorsen, R. F. Schwan, and L. Jespersen, "Strain-specific probiotics properties of *Lactobacillus fermentum, Lactobacillus plantarum* and *Lactobacillus brevis* isolates from Brazilian food products," *Food Microbiology*, vol. 36, no. 1, pp. 22–29, 2013.

[4] T. A. Melo, T. F. Dos Santos, L. R. Pereira, H. M. Passos, R. P. Rezende, and C. C. Romano, "Functional profile evaluation of *Lactobacillus fermentum* TCUESC01: a new potential probiotic strain isolated during cocoa fermentation," *BioMed Research International*, vol. 2017, Article ID 5165916, 7 pages, 2017.

[5] G. Giraffa, N. Chanishvili, and Y. Widyastuti, "Importance of lactobacilli in food and feed biotechnology," *Research in Microbiology*, vol. 161, no. 6, pp. 480–487, 2010.

[6] I. V. Belova, A. G. Tochilina, I. V. Solovyeva et al., "*Lactobacillus fermentum* 90 TC-4 taxonomic status confirmation using whole genome sequencing and MALDI TOF mass spectrum," *Russian Journal of Genetics*, vol. 52, no. 9, pp. 907–913, 2016.

[7] EFSA Panel on Biological Hazards (BIOHAZ), A. Ricci, A. Allende et al., "Update of the list of QPS-recommended biological agents intentionally added to food or feed as notified to EFSA 7: suitability of taxonomic units notified to EFSA until September 2017," *EFSA Journal*, vol. 16, no. 1, pp. 5131–5143, 2018.

[8] M. Zheng, R. Zhang, X. Tian, X. Zhou, X. Pan, and A. Wong, "Assessing the risk of probiotic dietary supplements in the context of antibiotic resistance," *Frontiers in Microbiology*, vol. 8, p. 908, 2017.

[9] S. Mathur and R. Singh, "Antibiotic resistance in food lactic acid bacteria—a review," *International Journal of Food Microbiology*, vol. 105, no. 3, pp. 281–295, 2005.

[10] P. Sharma, S. K. Tomar, P. Goswami, V. Sangwan, and R. Singh, "Antibiotic resistance among commercially available probiotics," *Food Research International*, vol. 57, pp. 176–195, 2014.

[11] EFSA, "Technical guidance prepared by the panel on additives and Products or substances used in animal feed (FEEDAP) on the update of the criteria used in the assessment of bacterial resistance to antibiotics of human or veterinary importance (question no. EFSA-Q-2008-004)," *EFSA Journal*, vol. 732, pp. 1–15, 2008.

[12] G. Huys, K. D'Haene, and J. Swings, "Influence of the culture medium on antibiotic susceptibility testing of food-associated lactic acid bacteria with the agar overlay disc diffusion method," *Letters in Applied Microbiology*, vol. 34, no. 6, pp. 402–406, 2002.

[13] M. Danielsen and A. Wind, "Susceptibility of *Lactobacillus* spp. to antimicrobial agents," *International Journal of Food Microbiology*, vol. 82, no. 1, pp. 1–11, 2003.

[14] M. Egervärn, H. Lindmark, S. Roos, G. Huys, and S. Lindgren, "Effects of inoculum size and incubation time on broth microdilution susceptibility testing of lactic acid bacteria," *Antimicrobial Agents and Chemotherapy*, vol. 51, no. 1, pp. 394–396, 2007.

[15] M. Gueimonde, B. Sánchez, C. G. de los Reyes-Gavilán, and A. Margolles, "Antibiotic resistance in probiotic bacteria," *Frontiers in Microbiology*, vol. 4, p. 202, 2013.

[16] M. C. Roberts, "Update on acquired tetracycline resistance genes," *FEMS Microbiology Letters*, vol. 245, no. 2, pp. 195–203, 2005.

[17] M. C. Roberts, "Update on macrolide-lincosamide-streptogramin, ketolide, and oxazolidinone resistance genes," *FEMS Microbiology Letters*, vol. 282, no. 2, pp. 147–159, 2008.

[18] M. Nawaz, J. Wang, A. Zhou et al., "Characterization and transfer of antibiotic resistance in lactic acid bacteria from fermented food products," *Current Microbiology*, vol. 62, no. 3, pp. 1081–1089, 2011.

[19] L. Pan, X. Hu, and X. Wang, "Assessment of antibiotic resistance of lactic acid bacteria in Chinese fermented foods," *Food Control*, vol. 22, no. 8, pp. 1316–1321, 2011.

[20] S. C. R. Thumu and P. M. Halami, "Presence of erythromycin and tetracycline resistance genes in lactic acid bacteria from fermented foods of Indian origin," *Antonie van Leeuwenhoek*, vol. 102, no. 4, pp. 541–551, 2012.

[21] K. Y. Gfeller, M. Roth, L. Meile, and M. Teuber, "Sequence and genetic organization of the 19.3-kb erythromycin and dalfopristin-resistance plasmid pLME300 from *Lactobacillus fermentum* ROT1," *Plasmid*, vol. 50, no. 3, pp. 190–201, 2003.

[22] H. Ishiwa and S. Iwata, "Drug resistance plasmids in *Lactobacillus fermentum*," *Journal of General and Applied Microbiology*, vol. 26, no. 1, pp. 71–74, 1979.

[23] M. Fons, T. Hégé, M. Ladiré, P. Raibaud, R. Ducluzeau, and E. Maguin, "Isolation and characterization of a plasmid from *Lactobacillus fermentum* conferring erythromycin resistance," *Plasmid*, vol. 37, no. 3, pp. 199–203, 1997.

[24] D. Gevers, G. Huys, and J. Swings, "*In vitro* conjugal transfer of tetracycline resistance from *Lactobacillus* isolates to other Gram-positive bacteria," *FEMS Microbiology Letters*, vol. 225, no. 1, pp. 125–130, 2003.

[25] L. Jacobsen, A. Wilcks, K. Hammer, G. Huys, D. Gevers, and S. R. Andersen, "Horizontal transfer of *tet*(M) and *erm*(B) resistance plasmids from food strains of *Lactobacillus plantarum* to *Enterococcus faecalis* JH2-2 in the gastrointestinal tract of gnotobiotic rats," *FEMS Microbiology Ecology*, vol. 59, no. 1, pp. 158–166, 2007.

[26] L. I. I. Ouoba, V. Lei, and L. B. Jensen, "Resistance of potential probiotic lactic acid bacteria and bifidobacteria of African and European origin to antimicrobials: determination and transferability of the resistance genes to other bacteria," *International Journal of Food Microbiology*, vol. 121, no. 2, pp. 217–224, 2008.

[27] N. L. Bruslik, D. R. Akhatova, A. A. Toimentseva, S. R. Abdulkhakov, O. N. Ilyinskaya, and D. R. Yarullina, "Estimation of probiotic lactobacilli drug resistance," *Antibiotics and chemoterapy*, vol. 60, no. 3-4, pp. 6–13, 2015.

[28] M. Egervärn, S. Roos, and H. Lindmark, "Identification and characterization of antibiotic resistance genes in *Lactobacillus reuteri* and *Lactobacillus plantarum*," *Journal of Applied Microbiology*, vol. 107, no. 5, pp. 1658–1668, 2009.

[29] J. Sutcliffe, T. Grebe, A. Tait-Kamradt, and L. Wondrack, "Detection of erythromycin-resistant determinants by PCR," *Antimicrobial Agents and Chemotherapy*, vol. 40, no. 11, pp. 2562–2566, 1996.

[30] C. Liu, Z. Y. Zhang, K. Dong, J. P. Yuan, and X. K. Guop, "Antibiotic resistance of probiotic strains of lactic acid bacteria isolated from marketed foods and drugs," *Biomedical and Environmental Sciences*, vol. 22, no. 5, pp. 401–412, 2009.

[31] D. Gevers, M. Danielsen, G. Huys, and J. Swings, "Molecular characterization of *tet*(M) genes in *Lactobacillus* isolates from different types of fermented dry sausage," *Applied and Environmental Microbiology*, vol. 69, no. 2, pp. 1270–1275, 2003.

[32] C. Devirgiliis, D. Coppola, S. Barile, B. Colonna, and G. Perozzi, "Characterization of the Tn916 conjugative transposon in a food-borne strain of *Lactobacillus paracasei*," *Applied and Environmental Microbiology*, vol. 75, no. 12, pp. 3866–3871, 2009.

[33] G. Werner, R. J. L. Willems, B. Hildebrandt, I. Klare, and W. Witte, "Influence of transferable genetic determinants on the outcome of typing methods commonly used for *Enterococcus faecium*," *Journal of Clinical Microbiology*, vol. 41, no. 4, pp. 1499–1506, 2003.

[34] D. J. Lane, "16S/23S rRNA sequencing," in *Nucleic Acid Techniques in Bacterial Systematics*, E. Stackebrandt and M. Goodfellow, Eds., pp. 115–175, John Wiley and Sons, New York, NY, USA, 1991.

[35] A. S. Hummel, C. Hertel, W. H. Holzapfel, and C. M. A. P Franz, "Antibiotic resistances of starter and probiotic strains of lactic acid bacteria," *Applied and Environmental Microbiology*, vol. 73, no. 3, pp. 730–739, 2007.

[36] D. Zonenschain, A. Rebecchi, and L. Morelli, "Erythromycin- and tetracycline-resistant lactobacilli in Italian fermented dry sausages," *Journal of Applied Microbiology*, vol. 107, no. 5, pp. 1559–1568, 2009.

[37] M. Dec, R. Urban-Chmiel, D. Stępień-Pyśniak, and A. Wernicki, "Assessment of antibiotic susceptibility in *Lactobacillus* isolates from chickens," *Gut Pathogens*, vol. 9, no. 1, p. 54, 2017.

[38] E. Songisepp, P. Hütt, M. Rätsep et al., "Safety of a probiotic cheese containing *Lactobacillus plantarum* Tensia according to a variety of health indices in different age groups," *Journal of Dairy Science*, vol. 95, no. 10, pp. 5495–5509, 2012.

[39] M. S. Ammor, M. Gueimonde, M. Danielsen et al., "Two different tetracycline resistance mechanisms, plasmid-carried *tet*(L) and chromosomally located transposon-associated *tet* (M), coexist in *Lactobacillus sakei* Rits 9," *Applied and Environmental Microbiology*, vol. 74, no. 5, pp. 1394–1401, 2008.

[40] M. Rajilić-Stojanović and W. M. de Vos, "The first 1000 cultured species of the human gastrointestinal microbiota,"

FEMS Microbiology Reviews, vol. 38, no. 5, pp. 996–1047, 2014.

[41] E. O. Akinkunmi and A. Lamikanra, "Phenotypic determination of some virulence factors in staphylococci isolated from faecal samples of Nigerian children," *African Journal of Biomedical Research*, vol. 15, no. 2, pp. 123–128, 2012.

[42] J. Bertram, M. Stratz, and P. Durre, "Natural transfer of conjugative transposon Tn*916* between gram-positive and gram-negative bacteria," *Journal of Bacteriology*, vol. 173, no. 2, pp. 443–448, 1991.

[43] P. Courvalin, "Transfer of antibiotic resistance genes between gram-positive and gram-negative bacteria," *Antimicrobial Agents and Chemotherapy*, vol. 38, no. 7, pp. 1447–1451, 1994.

8

Genomic Variability among Field Isolates and Laboratory-Adapted Strains of *Leptospira borgpetersenii* Serovar Hardjo

Alejandro Llanes,[1] **Carlos Mario Restrepo**,[1] **Pablo Riesgo-Ferreiro**,[2] **and Sreekumari Rajeev**[3]

[1]*Centro de Biología Celular y Molecular de Enfermedades, Instituto de Investigaciones Científicas y Servicios de Alta Tecnología (INDICASAT AIP), Ciudad del Saber, Panama, Panama*
[2]*Guanábana Biodata Analytics Inc., Panama, Panama*
[3]*School of Veterinary Medicine, Ross University, Basseterre, Saint Kitts and Nevis*

Correspondence should be addressed to Sreekumari Rajeev; srajeev@rossvet.edu.kn

Academic Editor: David M. Iovannisci

Leptospira borgpetersenii serovar Hardjo colonizes cattle kidneys and may occasionally infect humans and other mammals. Strains belonging to two clonal subtypes (types A and B) with marked differences in their pathogenicity in the hamster experimental model have been described for this serovar. Such differences have been attributed to point mutations in individual genes, although those genes have not yet been characterized. In order to better understand genetic variability among *L. borgpetersenii* serovar Hardjo isolates, we sequenced and compared the genomes of two laboratory-adapted strains and three abattoir-derived field isolates of *L. borgpetersenii* serovar Hardjo. Relatively low genetic variability was observed within isolates of the same subtype, with most of the mutations of moderate or high impact found in the laboratory-adapted isolates. In contrast, several differences regarding gene content and genetic variants were observed between the two subtypes. Putative type-specific genes appear to encode proteins associated with functions that are critical for infection. Some of these genes seem to be involved in transcriptional regulation, possibly leading to a distinct regulatory pattern in each type. These results show that changes in regulatory mechanisms, previously suggested to be critical during *Leptospira* speciation, may occur in *L. borgpetersenii.* In addition, the bioinformatics methodology used in this study for variant calling can be useful to other groups working with nonmodel prokaryotic organisms such as *Leptospira* species.

1. Introduction

Leptospirosis is a zoonotic disease of global importance caused by the spirochete bacteria from the genus *Leptospira*. Pathogenic *Leptospira* inhabit the renal tubules of a wide variety of mammalian species. The infected host can be a maintenance host that do not manifest clinical disease or an incidental host, which develops mild to severe clinical disease. Based on the antigenic structure of the lipopolysaccharide, these bacteria are conveniently grouped into serogroups and within these serogroups, members are further grouped into serovars [1]. Serovar-specific host adaptation is well known among members of the genus *Leptospira* and the serovar Hardjo belonging to *Leptospira interrogans* and *Leptospira borgpetersenii* species is known to infect cattle kidneys and genital tracts [2].

Early studies classified North American and New Zealand strains of serovar Hardjo as Hardjobovis strains based on identical restriction endonuclease analysis patterns [3, 4]. Later, it was found that two distinct strains, namely, Hardjobovis and Hardjoprajitno, were circulating in cattle in Northern Ireland and Scotland [5]. The virulent forms of the isolates obtained from cow's milk of agalactic cases and aborted fetuses belonged to Hardjoprajitno, whereas the majority of the abattoir isolates were Hardjobovis [6], hence indicating a possible difference in virulence and pathogenicity between these two distinct types. Later, using DNA hybridization techniques, Hardjobovis was placed in *L. borgpetersenii* and Hardjoprajitno in *L. interrogans* species [7, 8]. *L. borgpetersenii* serovar Hardjo is seen worldwide in cattle, whereas the ecological niche for *L. interrogans* serovar Hardjo is unknown [9].

Three clonal subtypes (types A, B, and C) of *L. borgpetersenii* serovar Hardjo have been described [10]; however, the subtype C designation is rarely used and it has been later considered to be erroneous [11]. Other studies have attempted to characterize *L. borgpetersenii* strains by various techniques and suggested the existence of heterogenic clonal populations [12, 13]. In a study conducted in the United States in cattle, it was found that 83% of the isolates belonged to serovar Hardjo and the majority (85%) of these isolates were of type A [14, 15]. It has also been shown that both types have differences in pathogenicity in experimental infection in hamsters [16]. Hamsters infected with a subtype A strain established renal colonization and remained asymptomatic with chronic renal infection, while infection with a subtype B strain resulted in a rapidly debilitating disease similar to the acute form typically observed in incidental hosts.

Bulach et al. [17] sequenced and compared the genomes of two *L. borgpetersenii* serovar Hardjo strains, L550 (subtype A) and JB197 (subtype B). The study revealed that these *L. borgpetersenii* genomes are relatively smaller when compared to those of other *Leptospira* species. This genome reduction is associated with an expansion of mobile elements and the loss of several genes involved in signal transduction, exopolysaccharide synthesis, and nitrogen metabolism [18]. It is suggested that this genome erosion in *L. borgpetersenii* has impaired its viability in aquatic environments, resulting in an increased dependence on mammalian hosts for survival. Although several structural differences were observed between both genomes, differences in disease phenotype were attributed to a combination of point mutations in individual genes [17]. Such genes, however, have not yet been characterized.

In recent years, with the advent of next-generation sequencing techniques, several pan-genomic studies have been conducted for the *Leptospira* genus. Fouts et al. [19] sequenced and analyzed the genomes of 20 *Leptospira* species from all major taxonomic groups. This study made a large contribution to our understanding on pathogenesis of *Leptospira*, revealing several virulence-associated features exclusively present in pathogenic species such as CRISPR/Cas systems and specific regulatory and secretory mechanisms. In another comprehensive study, Xu et al. [20] sequenced the genomes of 102 *Leptospira* isolates from different regions of the world, 10 of them belonging to *L. borgpetersenii*. Relatively high genomic variability was observed among different species, with several events of apparent gene gain and loss during the evolution of pathogenic species from those that are saprophytic. Among the genes that were found to be differentially present in pathogenic and nonpathogenic species there are several members of two-component systems (TCSs), prokaryotic systems involved in regulation of gene expression in response to changes in their environment [21]. TCSs are composed of a histidine kinase capable of sensing particular stimuli and subsequently activate a highly specific intracellular response regulator by phosphorylation. Changes in these and several other genes were suggested to be crucial for adaptation to a variety of environmental and host-related conditions. The loss of several genes associated with carbohydrate and nitrogen metabolism further supports the hypothesis that pathogenic species of *Leptospira*, including *L. borgpetersenii*, are evolving to become increasingly host-dependent and almost parasitic.

In this study, we sequenced and compared the genomes of five *L. borgpetersenii* serovar Hardjo specimens, two laboratory-adapted strains belonging to types A and B, respectively, and three field isolates. The field isolates, belonging to subtype A, were obtained from an abattoir located in Georgia, United States [22]. We characterized and compared the genomic variability within and between the isolates of distinct subtype and biological origin, emphasizing on features that are potentially related to their pathogenicity.

2. Materials and Methods

2.1. Bacterial Strains, Culture, and DNA Extraction. The laboratory-adapted strains sequenced in this study derived from strains NVSL S 1343 (here labelled LBH-A) and NVSL S 818 (here labelled LBH-B) from the National Veterinary Services Laboratory (Ames, Iowa, United States). The three field isolates (here labelled BK-6, BK-9, and BK-30) were cultured from abattoir-derived cow kidney samples from Georgia, United States [22]. The isolates were grown and maintained in P80-BA liquid oleic media (National Veterinary Services Laboratory, Ames, Iowa, United States). Up to eight passages were performed and cultures from multiple passages were combined for DNA extraction, in order to obtain the optimum concentration of DNA for whole genome sequencing. Genomic DNA from each specimen was isolated using the MasterPure™ Complete DNA and RNA Purification Kit (Epicentre, Wisconsin, United States) following the manufacturer's instructions.

2.2. Genome Sequencing, Assembly, and Annotation. Genomic DNA from the five specimens was sequenced at the Georgia Genomics Facility of the University of Georgia by using the Illumina MiSeq technology, following standard protocols. Reads were trimmed to 250 bp by using trimmomatic [23] in order to remove low-quality regions towards the 3′ end. Genome sequences for each specimen were generated by first assembling the reads de novo with SPAdes [24] and then contiguating the de novo contigs into scaffolds with ABACAS [25]. In the contiguation step, the genome of *L. borgpetersenii* serovar Hardjo strain L550 [17] was used as a reference for type A specimens (BK-6, BK-9, BK-30, and LBH-A), while that of *L. borgpetersenii* serovar Hardjo strain JB-197 [17] was used for the type B specimen (LBH-B). Gene models annotated in the corresponding reference genomes were transferred to the newly assembled scaffolds by using RATT [26]. Annotation was manually revised to correct transferred gene models by using Artemis and the Artemis Comparison Tool (ACT) [27].

Table 1: Basic features of the five genomes sequenced in this study, compared to those of the reference genomes for *L. borgpetersenii* serovar Hardjo subtypes A and B.

Feature	Reference genomes		Specimens from this study				
			Type A				Type B
	L550 (type A)	JB197 (type B)	BK-6	BK-9	BK-30	LBH-A	LBH-B
Total size (Mbp)	3.93	3.88	3.97	3.95	3.95	3.93	3.88
Chr. I	3.61	3.58	3.65	3.63	3.63	3.61	3.58
Chr. II	0.32	0.30	0.32	0.32	0.32	0.32	0.30
GC-content (%)	40.23	40.24	40.22	40.21	40.21	40.21	40.23
Protein-coding genes	3,623	3,553	3,546	3,531	3,519	3,524	3,522
In mobile elements	154	172	101	100	100	112	140
Transfer RNA genes	37	37	37	37	37	37	37
Ribosomal RNA	5	5	3	3	3	3	3
Pseudogenes	223	232	202	200	191	197	224

2.3. Variant Calling. Illumina reads from the five specimens were mapped to the corresponding reference genomes by using BWA [28]. Read alignments were preprocessed with Picard v. 2.2.2 (http://broadinstitute.github.io/picard) to remove duplicate reads and resolve format conflicts. Reads were further realigned around indels by using the Genome Analysis Toolkit (GATK) v. 3.5 [29]. Single-nucleotide variants (SNVs) and short indels (under 50 bp) were detected by using a pipeline developed in house. To increase sensitivity in variant detection, we use the consensus of three variant calling algorithms, namely, UnifiedGenotyper and HaplotypeCaller from GATK and the Samtools pileup v. 1.3.1 [30]. The experimental methodology for variant calling followed GATK's best practices, adapted to work with bacterial genomes in general and with *Leptospira* genomes in particular. Considering that such genomes are haploid, we set the ploidy to 1 and used the multiallelic and rare variant calling model as opposed to the consensus model, which assumes biallelic sites. To deal with the high abundance of transposons in *Leptospira* genomes, reads mapping in multiple positions were not excluded from the variant calling analysis. The per-Base Alignment Quality (BAQ) algorithm was used to reduce the rate of false positive SNV calls [31]. Due to the lack of a curated set of known variants for the species, variant filtering was performed by using arbitrary cut-off values for the technical parameters relevant for each variant calling algorithm, set by evaluating their distribution on the aggregated variants (see Results). After filtering, variants from the three algorithms were merged and annotated. Only variants predicted by the three algorithms and passing the filtering step were selected for annotation. Functional annotation was done with SnpEff v. 4.2 [32]. CNVnator [33] was further used to detect relatively larger structural variants.

2.4. Functional and Comparative Analysis of Protein-Coding Genes. Genes from the newly annotated genomes and those from the reference strains L550 and JB197 were clustered into ortholog groups by using OrthoMCL [34]. Each ortholog group was tested for evidence of selection by comparing models 1 and 2 of the codeml program from the PAML 4 package [35]. Functional domain prediction in protein-coding genes was done with Interproscan [36]. Gene ontology enrichment analysis was performed with Blast2GO [37]. Suspected virulence of gene models was inferred by comparison against the virulence factors database (VFDB) [38], considering only the core component containing experimentally characterized virulence factors. Regulatory proteins were predicted with P2RP [39].

3. Results

3.1. Genome Sequencing, Annotation, and Variant Calling. Genomic DNA from five *L. borgpetersenii* specimens belonging to the Hardjo serovar was sequenced by using the MiSeq platform from Illumina. De novo assembly with SPAdes resulted in assemblies of ~260 contigs with an N50 size of 45 kb on average. Reference genomes for the corresponding types were then used to contiguate the de novo contigs into pseudochromosomes, with an average size of 3.6 Mb for chromosome I and 0.31 Mb for chromosome II. Chromosome size and GC-content of our assembled genomes are very similar to those of the corresponding reference genomes (Table 1). Genomes assembled as part of this study were submitted to NCBI Genbank under BioProjects PRJNA296689 (LBH-A strain), PRJNA296694 (LBH-B strain), PRJNA296675 (BK-6 isolate), PRJNA296677 (BK-9 isolate), and PRJNA296679 (BK-30 isolate).

Approximately 99.8% of the MiSeq reads from each specimen could be unambiguously mapped to the corresponding reference genomes by using BWA, thus indicating high sample purity. Since the main goal of this work was to compare genomic variability among field isolates and laboratory-adapted strains, we paid special attention to the accuracy of the variant calling step. We used the consensus of three independent algorithms, with an ad hoc variant filtering configuration for each of them. As a starting point for filtering, we used the cut-off values for critical parameters recommended by Jia et al. [40] and then refined those values for each algorithm by evaluating their distribution aggregated across all samples (Figures S1 and S2). This approach allowed us to optimize cut-off values for technical parameters relevant to variant filtering, including the variant calling quality and the depth of coverage per variant position (Table S1). Counts of filtered variants per specimen classified according to their predicted biological impact are shown in Table S2. A principal component analysis (PCA) of SNV data clearly separated the specimens into three categories, corresponding to the subtype A field isolates and the laboratory strains from subtypes A and B, respectively, with 80% of the variability among these categories explained by the first two components (Figure S3).

3.2. Sequence Variants among Members of the Same Type. We found nearly 60 SNVs per specimen on average, 58–60% of which are located within protein-coding genes. Although we did not observe a notable predominance of either synonymous or nonsynonymous SNVs in any specimen, laboratory-adapted strains seem to have a slightly larger number of nonsynonymous SNVs than field isolates. SNVs and small indels located within protein-coding genes are widespread along the chromosomes and several of them are shared by all specimens of the same subtype or origin (Figures 1(b) and 1(d)). Many of these variants affect genes encoding transposases or other proteins associated with mobile elements, which are expected to have a higher mutation rate due to their unstable nature. In summary, we found 30 protein-coding genes, excluding transposases that are affected by variants with moderate or high predicted impact (Figure 2; Table S3). Twenty of these genes are within the *Leptospira* core genome, as recently defined by Xu et al. [20], and ten of them share sequence similarity with previously reported virulence factors in *Leptospira* and other pathogenic bacteria.

We noticed three identical missense SNVs shared by all of our subtype A specimens, which suggests that such variations may be due to mutations specific to the reference strain L550 rather than to those specimens. In addition to these shared variants, results show relatively few missense mutations exclusive to each specimen. The number of such exclusive mutations is relatively higher in laboratory-adapted strains than in field isolates. In fact, we found only two cases in which a single gene appears to have been affected by more than one missense mutation, both in LBH-A. The corresponding genes, LBL_01665 and LBL_12070, respectively, code for an ATP-dependent Clp protease similar to stress protein ClpC from *Listeria monocytogenes* (VFDB accession code VFG0079) and for a serine/threonine protein kinase. Two additional genes from strain JB197 were found to have orthologs with missense mutations in LBH-B, namely, genes LBJ_RS08350 and LBJ_RS17855. Notably, LBJ_RS17855 codes for a histidine kinase similar to the LetS kinase from *Legionella pneumophila* (VFDB accession code VFG1888), a member of a TCS involved in the induction of the transmission phenotype when nutrients become scarce, especially under laboratory conditions [41].

Variants related to mutations with a higher predicted impact, such as nonsense or frameshift mutations, were also mostly observed in the laboratory-adapted strains. One of such variants, a frameshift mutation affecting a gene encoding a chemotaxis protein of the CheD family, is also apparently specific to strain L550 and causes the corresponding gene (LBL_RS08205) to be slightly shorter than those from our four type A specimens. In addition to this shared variant, we found a frameshift mutation in LBH-A which disrupted and possibly inactivated the ortholog of gene LBL_RS195, whose function is unknown. Conversely, in LBH-B, there are three frameshift mutations in the orthologs of genes LBJ_RS00335, LBJ_RS15300, and LBJ_RS11945 from strain JB197, which encode a methyl-accepting serine phosphatase, a DNA polymerase sigma-70 factor and a glucose kinase, respectively. Mutations associated with the last two genes appear to have originated two corresponding pseudogenes in LBH-B. However, the one affecting the serine phosphatase is not likely to have inactivated the corresponding LBH-B ortholog, since it does not affect segments encoding the HAMP and PPM-like phosphatase domains typical of such proteins [42].

Excluding those affecting transposase genes, we only found nonsense mutations prematurely introducing stop codons in two genes from LBH-B, whose orthologs in strain JB197, respectively, code for a transcriptional regulator (LBJ_RS05345) and an integral membrane protein of the TerC family involved in heavy metal ion resistance (LBJ_RS17295). The later gene, LBJ_RS17295, belongs to the *Leptospira* core genome as defined by Xu et al. [20] and its putative ortholog in *E. coli* is considered a virulence factor [43].

Some of the genes mentioned before have been previously found to exhibit positive selection in studies considering orthologs from several *Leptospira* species [20, 44]. However, we did not find evidence of positive selection in any of these genes when considering only *L. borgpetersenii* orthologs from the corresponding subtypes (data not shown).

3.3. Sequence Variants among Members of Different Types. To comparatively describe the differences between type A and type B genomes, we first looked for genes that are putatively specific to each type, whose orthologs were apparently lost or converted into pseudogenes in all the genomes of the opposite type (Table S4). We found 44 genes exclusively present in the type B genomes, half of them coding for transposases or proteins related to mobile elements. Of those genes not associated with mobile elements, 8 are within the *Leptospira* core genome. Three of them, in turn, have sequence similarity to previously reported virulence factors in bacteria. Gene LBJ_RS11430 is similar to the FbpC subunit of an iron (III)-specific ABC transporter from *Neisseria meningitidis* (VFDB accession code VFG1206), while genes LBJ_RS00510 and LBJ_RS13855 are similar to the histidine kinase component of two previously described TCSs. Gene LBJ_RS00510 is similar to a flagellar sensor histidine kinase FleS from *Legionella pneumophila* (VFDB accession code VFG043346) and gene LBJ_RS13855 is similar to kinase YjcC from *Salmonella enterica* (VFDB accession code VFG0584).

In addition to these suspected virulence factors, a gene putatively encoding a glycerate kinase in the subtype B genome (LBJ_RS09170) is apparently inactivated in subtype A, thus suggesting that glycerol metabolism may be affected in this type. Other subtype B genes that have become pseudogenes in subtype A specimens and are associated with metabolic pathways are those encoding a haloacid dehalogenase (LBJ_RS17075) and a methylmalonyl-CoA mutase (LBJ_RS11290), although these genes appear to have putative paralogs with similar function in the genome.

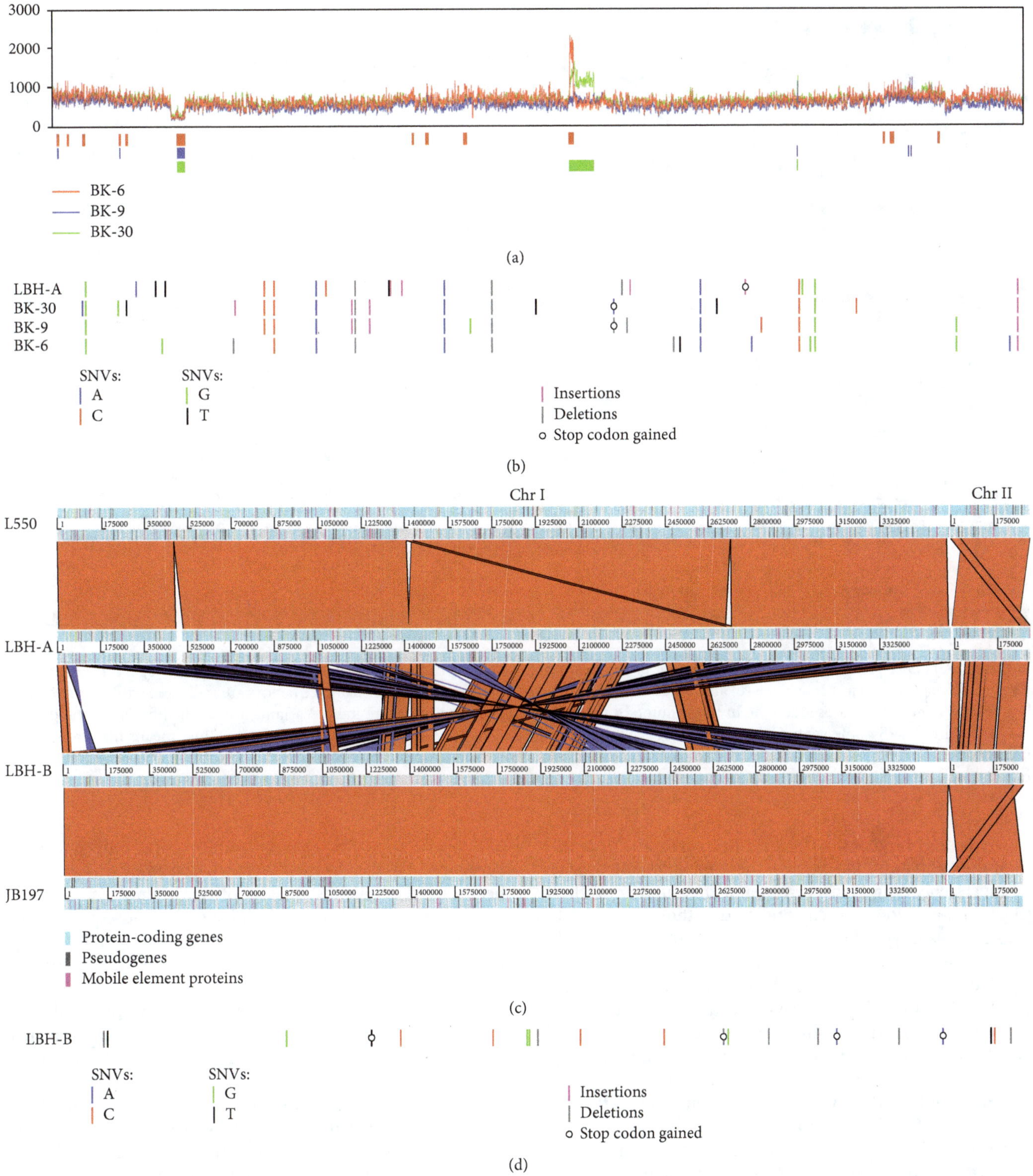

FIGURE 1: *Overview of genomic variability among the specimens included in this study*. Panel (a) shows variations in read depth for the alignment of reads from the three field isolates against the chromosomes of *L. borgpetersenii* strain L550 (subtype A). CNVs (insertions and deletions) larger than 2 kb predicted by CNVnator are shown below the plot. Panel (b) shows nonsynonymous SNVs and short indels located within protein-coding genes, identified for the four subtype A specimens by using strain L550 as a reference. Panel (c) shows chromosome-to-chromosome comparisons among the two reference genomes for subtype A and B and the genomes of our corresponding laboratory strains. In these comparisons, red bands indicate similar regions and blue bands indicate inversions. Panel (d) is similar to panel (b) but for variants identified for the subtype B laboratory strain, using the genome of strain JB197 as reference.

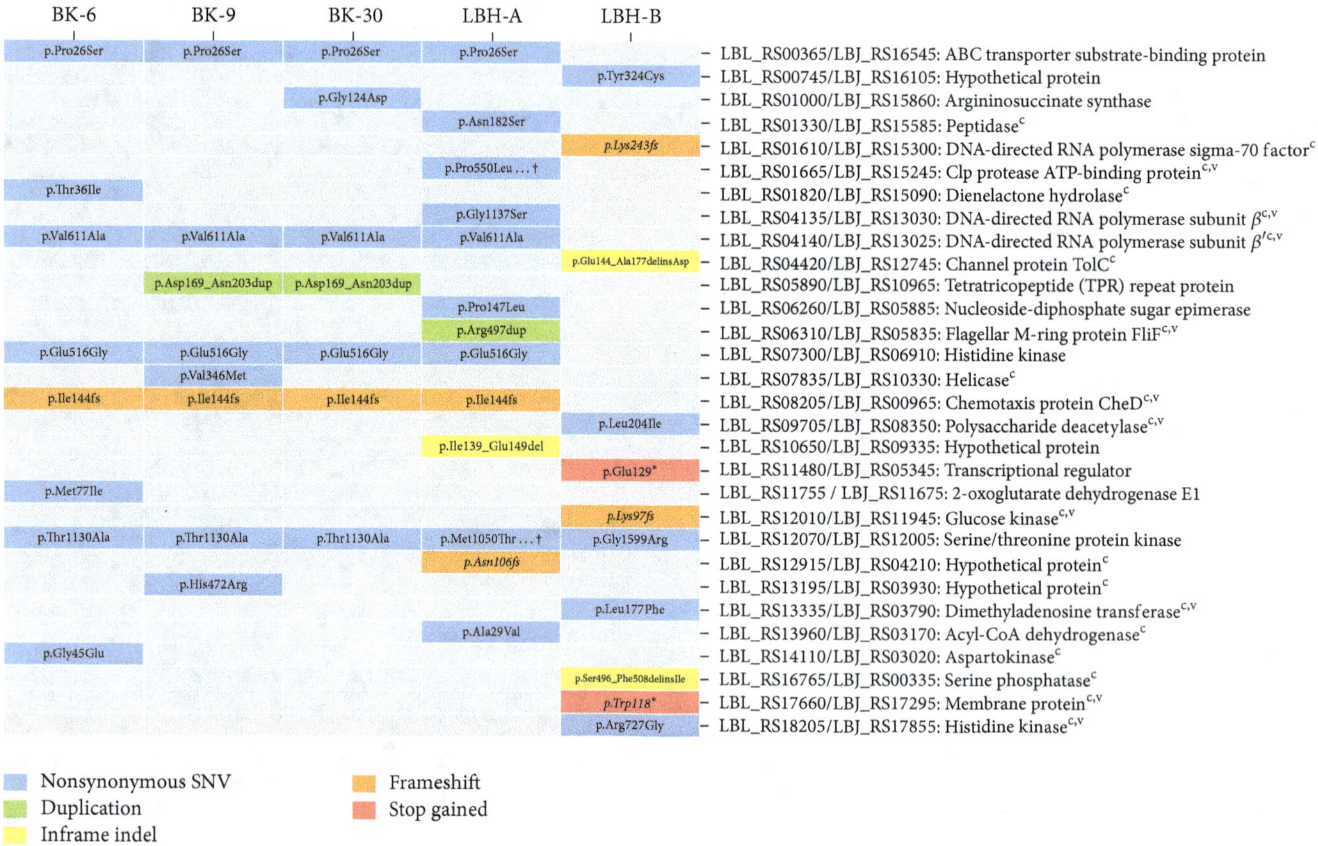

FIGURE 2: *Reference genes with at least one variant of moderate to high predicted impact in our sequenced specimens of the corresponding subtypes.* Core genes are indicated by a (c) and genes with similarity to previously reported virulence factors are indicated with a (v). Variants that putatively generated a pseudogene in the genome of the corresponding specimen are italic. Variants are labelled by using the protein (p.) part of the HGVS nomenclature (http://www.hgvs.org/mutnomen/). In this notation, "del" means deletion, "ins" means insertion, "dup" means duplication, "fs" means frameshift, and "∗" indicates a stop codon gained. † indicates that the corresponding gene has more than one mutation; see Table S3 for the full list of mutations.

We found 48 genes exclusively present in the genomes of all the subtype A specimens. These include only 8 genes putatively encoding transposases and the number of these genes is notably lower when compared to those found in the genomes of subtype B specimens. As in subtype A, 7 of the genes not associated with mobile elements are within the *Leptospira* core genome and two of them, genes LBL_RS13565 and LBL_RS11610, were found to share sequence similarity with previously reported virulence factors. Gene LBL_RS11610 encodes protein similar to PilH from *Pseudomonas aeruginosa* (VFDB accession code VFG1226). PilH acts as the response regulator of a TCS involved in regulation of flagella-independent twitching motility. Conversely, LBL_RS13565 putatively codes for a single-stranded DNA-binding protein (SSB) reported as a virulence factor in *Salmonella enterica* (VFDB accession code VFG000576).

Inactivation of the ortholog of gene LBL_RS13565 in strain JB197 has been recently reported by Martins-Pinheiro et al. [45] in a study conducted to characterize DNA repair genes in *Leptospira*. SSB is involved in recombinational repair of double-strand break damage in DNA and is thought to be associated with natural competence. Naturally competent bacteria appear to express two copies of the gene encoding SSB, while noncompetent bacteria express only one [46]. Presence of the corresponding two copies of this gene has been observed in most species of *Leptospira*, including the saprophytic nonpathogenic *L. biflexa* [18]. Consequently, loss of one of the two copies in strains JB197 and LBH-B suggests that the natural competence of subtype B strains may be affected.

Other subtype A genes that have been apparently lost in subtype B and may have an implication in virulence and survival are those encoding three additional TCSs members (LBL_RS02170, LBL_RS03580, and LBL_RS10150), two transcriptional regulators (LBL_RS04960 and LBL_RS15495), a cation transporter (LBL_RS09835), an adhesin (LBL_RS14550), an endoflagella protein (LBL_RS13610), and two hydrolases (LBL_RS01270 and LBL_RS01810).

In addition to those genes that appear to be exclusively present in subtypes A and B, we also evaluated the differences between both subtypes at the level of point mutations and small indels. In order to find variants potentially specific to subtype A specimens, we used the same methodology described before to align the reads from subtype A specimens against the subtype B reference genome (strain JB197) and to subsequently call for variants from read alignments. As

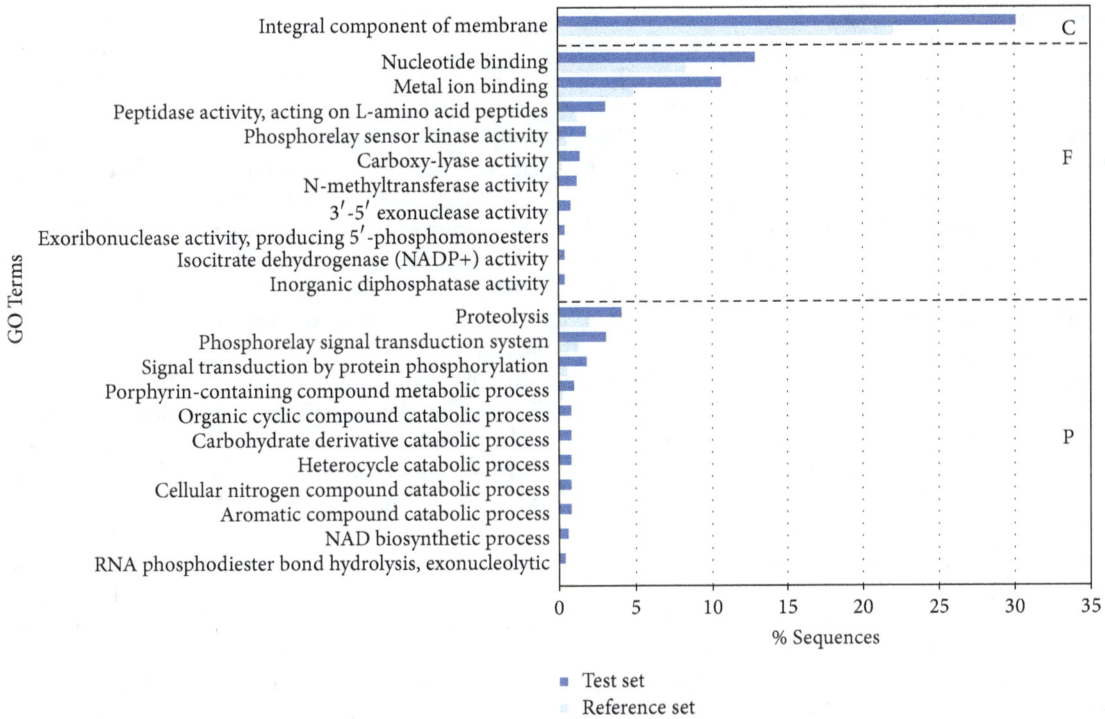

Figure 3: *GO enrichment analysis for the subtype A genes accumulating high-impact mutations.* The analysis was conducted with the genes from the type A specimens having at least one mutation of predicted high impact (test set) and their corresponding orthologs from strain JB197 (reference set). Plot shows GO terms found to be overrepresented with a p value below 0.05, ordered by percent sequences in descending order and grouped by GO class (C: cellular components, F: molecular functions, and P: biological processes).

expected, this resulted in a larger number of variants, including both SNVs and indels (Table S5). Of all the 3,553 genes annotated in the genome of strain JB197, 788 (22%) appear to have an ortholog in at least one of the subtype A specimens with a minimum of one mutation of impact predicted to be high or moderate. When considering only missense mutations altering the chemical nature of the underlying residue, the number of genes with variants was reduced to 483 (Table S6), excluding previously discussed pseudogenes. We found no evidence of positive selection in genes exhibiting variants when considering only *L. borgpetersenii* orthologs.

We also looked for gene ontology (GO) terms enriched in the selected genes when compared to all the genes in the reference genome. The refined enrichment analysis resulted in 22 enriched terms, 10 representing molecular functions, 11 representing biological processes, and one representing a cellular component (Figure 3; Table S7). The only cellular component found to be enriched was that associated with integral membrane proteins. This was also the term encompassing the largest number of proteins in the set, including 147 genes encoding a wide variety of membrane transporters, chemotaxis proteins, and peptidases. Several terms in the categories of molecular functions and biological processes, such as phosphorelay sensor kinase activity, phosphorelay signal transduction systems, and signal transduction by protein phosphorylation, can also be associated with TCSs and other regulatory proteins.

Due to the relatively high variability observed in genes encoding proteins involved in transcriptional regulation, such as members of TCSs, we used the P2RP server to assess the number of genes in the reference genomes of both types putatively encoding transcriptional regulators. Results showed that for TCSs, on average, there are at least 36 loci encoding the histidine kinase component and 26 loci encoding response regulators per genome (Table S8). Of the 62 genes predicted for the subtype A reference genome, 14 are within those found to have variants of predicted high impact and selected for the enrichment analysis described before, most of them encoding sensor histidine kinases. The server predicted nearly 80 loci encoding additional transcription factors per genome (Table S9), including sigma factors of RNA polymerase and other types of proteins with DNA-binding domains. Likewise, 10 of these genes are within those previously found to have at least one high-impact variant.

3.4. Structural Rearrangements and Copy Number Variations (CNVs). In addition to SNVs and small indels, we found relatively few structural variants between subtypes. Genomic regions that could be assembled de novo for our five specimens show high synteny with the corresponding reference genomes, thus maintaining the pattern of large-scale rearrangements previously described for the two reference genomes [17]. Two notable exceptions are the transposition of a segment of 24 kb in chromosome I of LBH-A and another one of ~40 kb from chromosome II of both LBH-A and LBH-B (Figure 1(c)). However, gene content and synteny within these segments are not globally altered by their putative rearrangement.

We additionally looked for duplications or deletions in the genomes of subtype A field isolates by using CNVnator. The program predicted relatively few variations larger than 2 kb (Table S10), many of which appear to be flanked by or at least associated with mobile elements. The only deletion we found in the three isolates correspond to one of the two copies of the cluster encoding ribosomal protein genes, which was previously reported to be duplicated in strain L550 [17]. This finding suggests that our field isolates, and LBH-A, do not have that duplication. Since assembly was performed by using the genome of strain L550 as a reference for subtype A specimens, the region for the first copy in strain L550 was replaced by an assembly gap by ABACAS (Figure 1(c)).

Two notable amplifications were found beginning in approximately the same position in the genomes of isolates BK-6 and BK-30, but with different estimated length and copy number. The presence of such variations was confirmed by the observation of a sudden increase in read depth shown in Figure 1(a). In isolate BK-6, the amplified region spans 15 kb and includes 14 protein-coding genes. According to the estimated read depth, this amplification possibly exists in two copies in addition to the region encoded in the chromosome. A much larger region was found to be amplified in isolate BK-30, beginning in approximately the same position of that reported for BK-6, but spanning around 97 kb and encompassing 84 protein-coding genes. Read depth estimates suggest that this amplification exists in only one additional copy. No amplifications of comparable length and read depth were found in the genomes of the laboratory-adapted strains included in this study.

4. Discussion

Cattle are known to be the maintenance host of *Leptospira borgpetersenii* serovar Hardjo worldwide. Previous studies have found large differences when comparing the genome of *L. borgpetersenii* with those of other pathogenic *Leptospira* species [17, 20]. Two subtypes of *Leptospira borgpetersenii* serovar Hardjo have been described for this serovar, namely subtypes A and B, with marked differences in disease phenotype in the hamster experimental model. Bulach et al. [17] suggested that phenotypic differences between both subtypes are due to a particular combination of point mutations in several genes, but such genes have not yet been characterized. In addition, little is known about the genetic variability within strains of *Leptospira borgpetersenii* belonging to the Hardjo serovar.

In this study, the genomes of three field isolates and two laboratory-adapted strains of *Leptospira borgpetersenii* serovar Hardjo were sequenced and analyzed. Our goal was to study the genetic variation among field isolates in comparison to laboratory-adapted strains, to improve our understanding of the evolutionary dynamics of this pathogenic species. The field isolates used in this study have undergone multiple passages, but at least one of them (BK-6) has been previously shown to induce colonization in hamsters [22]. Multiple passages may increase the chances of mutations in these isolates, but are very unlikely to affect gene content.

To ensure accuracy in exploring genetic variation, we used a pipeline combining three variant calling algorithms, with customized cut-off values for relevant technical parameters used to filter the detected variants before merging the results. This allowed us to increase sensitivity and to discard potentially false positive calls. In terms of SNVs and small indels, we found relatively few variants associated with mutations of predicted moderate or high impact within the subtype A field isolates and between laboratory strains and their corresponding reference genomes.

The number of variants associated with protein-coding genes (excluding those related to mobile elements) was slightly higher in the two laboratory strains than in field isolates. In the laboratory strains, some of these variants are related to high-impact mutations, including frameshifts and nonsense mutations putatively inactivating the corresponding genes. This higher number of variants in laboratory-adapted strains can be attributed to the lack of selective pressure when growing in laboratory conditions, compared to those strains surviving in the environment and hosts, which need to overcome potentially unfavorable environmental conditions and destructive host defense mechanisms. This distinct evolutionary process developed in laboratory and nature has also been reported for *L. interrogans* [47].

Nevertheless, the number of variants found among the subtype A field isolates and laboratory strains in this study are lower than those we previously found between a field isolate and a laboratory-adapted strain of *L. interrogans* serovar Hardjo [44]. Although it seems that there is relatively low genetic variation in terms of SNVs or small indels within specimens of the same subtype or origin, differences observed in this study allowed separation of specimens into well-defined clusters in a principal component analysis. Unfortunately, when describing genetic variation among field isolates, we were limited to subtype A specimens, since we did not obtain any subtype B isolates from our prevalence studies.

It is worth mentioning that we found a region in chromosome I of subtype A field isolates that appears to be amplified with varying length and estimated copy number. Regardless of the specimen, amplified regions appear to be flanked by mobile elements, suggesting their role in amplification. We have no evidence for the nature of such suspected amplifications, either intrachromosomal or episomic (extrachromosomal), but excision of a region of a chromosome in the form of a plasmid-like amplicon has been previously reported in *Leptospira* [48]. We found no gene putatively related to plasmid replication within the reported amplified regions, but the regions contain several genes that are involved in survival and colonization of host tissues, thus suggesting that their amplification may be associated with a higher dosage for those genes and may represent an adaptive advantage.

To better understand the differences in the genetic background of subtypes A and B, we conducted a careful revision of potentially type-specific genes, complementary to that performed by Bulach et al. [17]. Recent progress in microbial molecular biology, especially on the field of bacterial transcriptional regulation, allowed us to assign a function to some of those genes that were apparently lost in each subtype. In fact, some of those genes appear to encode proteins involved in the regulation of gene expression, including canonical transcriptional regulators or members of two-component systems (TCSs). TCSs are prokaryotic systems composed of a histidine kinase able to sense external stimuli, subsequently activating a highly specific response regulator by phosphorylation [21].

For a better characterization of the genetic differences between subtypes, we mapped the reads from all of the subtype A specimens and used the same protocol described above for variant calling. As expected, this resulted in a reasonably larger number of variants, with counts relatively similar for all specimens and variant types. An enrichment analysis conducted with subtype B genes whose orthologs in subtype A have at least one moderate- to high-impact mutation showed enrichment of GO terms also associated with TCSs, including signal transduction by protein phosphorylation and phosphorelay signal transduction systems. The term phosphorelay refers to the ability many hybrid histidine kinases have to autophosphorylate in an internal receiver domain, before transferring the phosphate group to their corresponding response regulator [49]. In addition to these terms associated with transcriptional regulation, other terms that were found to be enriched are those related to proteolytic activity and membrane processes. Such membrane processes include membrane transport, motility and chemotaxis, all of which are critical for bacterial colonization, invasion and survival within the animal host.

Globally, our findings support that the differences in virulence between subtypes A and B are due to several mutations in individual genes, some of which encode regulatory proteins, possibly leading to a distinct regulatory pattern in each subtype. TCSs play a key role in allowing bacteria to sense, respond and adapt to external stimuli, while they are growing in the environment or colonizing a host [21]. It has been recently shown that loss of certain TCSs and gain of new ones have occurred during the evolution of the *Leptospira* genus [20]. It is believed that such variation in TCSs have been a critical step in the evolution of *Leptospira* species from the saprophytic and nonpathogenic phenotype to those with the ability to colonize animal hosts. We also found that variations in these and other genes involved in transcriptional regulation also appear to occur within strains of the same species and serovar in *L. borgpetersenii*. These findings suggest that future research regarding pathogenomics of *L. borgpetersenii* and perhaps other pathogenic *Leptospira* species should be oriented towards unravelling the mechanisms for regulation of gene expression, some of which have been just recently described in other microbial organisms.

5. Conclusions

Our results show relatively low variability at the sequence level within specimens of the same subtype in *L. borgpetersenii* serovar Hardjo, with the majority of the differences observed in laboratory-adapted strains. Several differences were found between subtypes, including putative specific genes and several high- or moderate-impact mutations affecting orthologs. Many of the genes exhibiting differences between subtypes are associated with survival and virulence, including genes encoding proteins involved in transcriptional regulation of gene expression such as canonical regulators or TCSs. These results suggest that changes in regulatory mechanisms also occur in *L. borgpetersenii* at the levels of strains. Genome-wide analysis data from this study will promote our understanding of the evolutionary dynamics and host adaptation factors of this important zoonotic pathogen. In addition, the bioinformatics methodology used in this study for variant calling can be useful to other groups working with nonmodel prokaryotic organisms such as *Leptospira* species.

Disclosure

The present address of Pablo Riesgo-Ferreiro is Genomics England Limited, Dawson Hall 6BQ, Charterhouse Square, London EC1M, UK.

Conflicts of Interest

The authors declare that there are no conflicts of interest regarding the publication of this article.

Authors' Contributions

Sreekumari Rajeev and Alejandro Llanes conceived and designed the study. Alejandro Llanes and Carlos Mario Restrepo assembled and annotated the genomes and performed genomic analyses. Pablo Riesgo-Ferreiro implemented the variant calling pipeline and analyzed the corresponding results. Alejandro Llanes, Pablo Riesgo-Ferreiro, and Sreekumari Rajeev wrote the paper. All authors read and approved the final manuscript.

Acknowledgments

The authors thank UGA Georgia Genomics Facility for the sequencing and Dr. Walt Lorenz from the UGA Quantitative Biology Consulting Group for initial assembly and transfer of data. This paper is funded by University of Georgia, Provost's Summer Research Support (http://research.uga.edu/), and Ross University School of Veterinary Medicine Intramural Grant (http://veterinary.rossu.edu/).

Supplementary Materials

Supplementary 1. Figure S1: distribution of values of relevant technical parameters for predicted SNV aggregated across all samples. Optimized cut-off values used for variant filtering are shown with a blue dashed line. Predicted variants within the regions shaded in pink were discarded after filtering (PDF 131 KB).

Supplementary 2. Figure S2: distribution of values of relevant technical parameters for predicted indels aggregated across all samples. Optimized cut-off values used for variant filtering are shown with a blue dashed line. Predicted variants within the regions shaded in pink were discarded after filtering (PDF 79 KB).

Supplementary 3. Figure S3: principal component analysis (PCA) conducted with moderate- and high-impact variants predicted within samples of each type. The analysis clearly separated field isolates and laboratory strains of the same type (a), with ~80% of the variability from each class explained by the first two principal components (b) (PDF 49 KB).

Supplementary 4. Table S1: cut-off values used for technical parameters in variant filtering (XLSX 10 KB).

Supplementary 5. Table S2: summary of small variants detected between the sequenced specimens and their reference genomes (XLSX 11 KB).

Supplementary 6. Table S3: reference genes with at least one variant of moderate to high impact in the specimens sequenced in this study (XLSX 14 KB).

Supplementary 7. Table S4: putative subtype-specific genes (XLSX 15 KB).

Supplementary 8. Table S5: summary of small variants detected for subtype A genes when using the subtype B genome as reference (XLSX 10 KB).

Supplementary 9. Table S6: genes from *L. borgpetersenii* strain JB197 whose putative orthologs in subtype A specimens have at least one mutation of moderate to high impact (XLSX 38 KB).

Supplementary 10. Table S7: Gene Ontology (GO) terms found to be enriched in the genes from strain JB197 whose orthologs in our subtype A specimens have high-impact mutations (XLSX 15 KB).

Supplementary 11. Table S8: P2PR predictions of members of two-component systems (TCS) in the genomes of *L. borgpetersenii* strains L550 and JB197 (XLSX 15 KB).

Supplementary 12. Table S9: P2PR predictions of transcriptional regulators in the genomes of *L. borgpetersenii* strains L550 and JB197 (XLSX 15 KB).

Supplementary 13. Table S10: structural variants (duplications/deletions) larger than 2 kb predicted by CNVnator in the genomes of the three subtype A field isolates (XLSX 12 KB).

References

[1] P. N. Levett, "Systematics of Leptospiraceae," in *Leptospira and Leptospirosis*, vol. 387 of *Current Topics in Microbiology and Immunology*, pp. 11–20, Springer, Berlin, Germany, 2015.

[2] W. A. Ellis and A. B. Thiermann, "Isolation of leptospires from the genital tracts of Iowa cows.," *American Journal of Veterinary Research*, vol. 47, no. 8, pp. 1694–1696, 1986.

[3] R. B. Marshall, B. E. Wilton, and A. J. Robinson, "Identification of Leptospira serovars by restriction-endonuclease analysis," *Journal of Medical Microbiology*, vol. 14, no. 1, pp. 163–166, 1981.

[4] A. B. Thiermann, A. L. Handsaker, J. W. Foley, F. H. White, and B. F. Kingscote, "Reclassification of North American leptospiral isolates belonging to serogroups Mini and Sejroe by restriction endonuclease analysis.," *American Journal of Veterinary Research*, vol. 47, no. 1, pp. 61–66, 1986.

[5] R. Marshall, P. Winter, A. Thiermann, and W. Ellis, "Genotypes of Leptospira interrogans serovar hardjo in cattle in the UK," *Veterinary Record*, vol. 117, no. 25-26, pp. 669-670, 1985.

[6] W. A. Ellis, A. B. Thiermann, J. Montgomery, A. Handsaker, P. J. Winter, and R. B. Marshall, "Restriction endonuclease analysis of Leptospira interrogans serovar hardjo isolates from cattle.," *Research in Veterinary Science*, vol. 44, no. 3, pp. 375–379, 1988.

[7] P. H. Yasuda, A. G. Steigerwalt, K. R. Sulzer, A. F. Kaufmann, F. Rogers, and D. J. Brenner, "Deoxyribonucleic Acid Relatedness between Serogroups and Serovars in the Family Leptospiraceae with Proposals for Seven New Leptospira Species," *International Journal of Systematic Bacteriology*, vol. 37, no. 4, pp. 407–415, 1987.

[8] P. Ramadass, R. B. Marshall, and B. D. Jarvis, "Species differentiation of Leptospira interrogans serovar hardjo strain Hardjobovis from strain Hardjoprajitno by DNA slot blot hybridisation.," *Research in Veterinary Science*, vol. 49, no. 2, pp. 194–197, 1990.

[9] W. A. Ellis, "Animal Leptospirosis," in *Leptospira and Leptospirosis*, vol. 387 of *Current Topics in Microbiology and Immunology*, pp. 99–137, Springer, Berlin, Germany, 2015.

[10] R. B. LeFebvre, A. B. Thiermann, and J. Foley, "Genetic and antigenic differences of serologically indistinguishable leptospires of serovar hardjo," *Journal of Clinical Microbiology*, vol. 25, no. 11, pp. 2094–2097, 1987.

[11] M. L. Hotka, M. A. Wilson, T. M. Anderson, C. L. Tichenor, and D. A. Miller, "Comparative serological study of Leptospira serovar hardjo genotypes for use in the microscopic agglutination test," *Journal of Veterinary Diagnostic Investigation*, vol. 19, no. 1, pp. 84–87, 2007.

[12] R. L. Zuerner, W. A. Ellis, C. A. Bolin, and J. M. Montgomery, "Restriction fragment length polymorphisms distinguish Leptospira borgpetersenii serovar hardjo type hardjo-bovis isolates from different geographical locations," *Journal of Clinical Microbiology*, vol. 31, no. 3, pp. 578–583, 1993.

[13] P. Perolat, F. Merien, W. A. Ellis, and G. Baranton, "Characterization of Leptospira isolates from serovar hardjo by ribotyping, arbitrarily primed PCR, and mapped restriction site polymorphisms," *Journal of Clinical Microbiology*, vol. 32, no. 8, pp. 1949–1957, 1994.

[14] D. A. Miller, M. A. Wilson, and G. W. Beran, "Relationships between prevalence of Leptospira interrogans in cattle, and regional, climatic, and seasonal factors," *American Journal of Veterinary Research*, vol. 52, no. 11, pp. 1766–1768, 1991.

[15] D. A. Miller, M. A. Wilson, and G. W. Beran, "Survey to estimate prevalence of Leptospira interrogans infection in mature cattle

in the United States," *American Journal of Veterinary Research*, vol. 52, no. 11, pp. 1761–1765, 1991.

[16] R. L. Zuerner, D. P. Alt, and M. V. Palmer, "Development of Chronic and Acute Golden Syrian Hamster Infection Models With Leptospira borgpetersenii Serovar Hardjo," *Veterinary Pathology*, vol. 49, no. 2, pp. 403–411, 2012.

[17] D. M. Bulach, R. L. Zuerner, P. Wilson et al., "Genome reduction in Leptospira borgpetersenii reflects limited transmission potential," *Proceedings of the National Acadamy of Sciences of the United States of America*, vol. 103, no. 39, pp. 14560–14565, 2006.

[18] M. Picardeau, D. M. Bulach, C. Bouchier et al., "Genome sequence of the saprophyte Leptospira biflexa provides insights into the evolution of *Leptospira* and the pathogenesis of leptospirosis," *PLoS ONE*, vol. 3, no. 2, Article ID e1607, 2008.

[19] D. E. Fouts, M. A. Matthias, H. Adhikarla et al., "What Makes a Bacterial Species Pathogenic?:Comparative Genomic Analysis of the Genus Leptospira," *PLOS Neglected Tropical Diseases*, vol. 10, no. 2, Article ID e0004403, 2016.

[20] Y. Xu, Y. Zhu, Y. Wang et al., "Whole genome sequencing revealed host adaptation-focused genomic plasticity of pathogenic Leptospira," *Scientific Reports*, vol. 6, Article ID 20020, 2016.

[21] J. M. Skerker, B. S. Perchuk, A. Siryaporn et al., "Rewiring the Specificity of Two-Component Signal Transduction Systems," *Cell*, vol. 133, no. 6, pp. 1043–1054, 2008.

[22] S. Rajeev, M. Ilha, M. Woldemeskel, R. D. Berghaus, and M. E. Pence, "Detection of asymptomatic renal Leptospira infection in abattoir slaughtered cattle in southeastern Georgia, United States," *SAGE Open Medicine*, vol. 2, 2014.

[23] A. M. Bolger, M. Lohse, and B. Usadel, "Trimmomatic: a flexible trimmer for Illumina sequence data," *Bioinformatics*, vol. 30, no. 15, pp. 2114–2120, 2014.

[24] A. Bankevich, S. Nurk, D. Antipov et al., "SPAdes: a new genome assembly algorithm and its applications to single-cell sequencing," *Journal of Computational Biology*, vol. 19, no. 5, pp. 455–477, 2012.

[25] S. Assefa, T. M. Keane, T. D. Otto, C. Newbold, and M. Berriman, "ABACAS: algorithm-based automatic contiguation of assembled sequences," *Bioinformatics*, vol. 25, no. 15, pp. 1968-1969, 2009.

[26] T. D. Otto, G. P. Dillon, W. S. Degrave, and M. Berriman, "RATT: rapid annotation transfer tool," *Nucleic Acids Research*, vol. 39, no. 9, article e57, 2011.

[27] T. Carver, M. Berriman, A. Tivey et al., "Artemis and ACT: viewing, annotating and comparing sequences stored in a relational database," *Bioinformatics*, vol. 24, no. 23, pp. 2672–2676, 2008.

[28] H. Li and R. Durbin, "Fast and accurate long-read alignment with Burrows-Wheeler transform," *Bioinformatics*, vol. 26, no. 5, pp. 589–595, 2010.

[29] A. McKenna, M. Hanna, E. Banks et al., "The genome analysis toolkit: a MapReduce framework for analyzing next-generation DNA sequencing data," *Genome Research*, vol. 20, no. 9, pp. 1297–1303, 2010.

[30] H. Li, "A statistical framework for SNP calling, mutation discovery, association mapping and population genetical parameter estimation from sequencing data," *Bioinformatics*, vol. 27, no. 21, pp. 2987–2993, 2011.

[31] H. Li, "Improving SNP discovery by base alignment quality," *Bioinformatics*, vol. 27, no. 8, pp. 1157-1158, 2011.

[32] P. Cingolani, A. Platts, L. L. Wang et al., "A program for annotating and predicting the effects of single nucleotide polymorphisms, SnpEff: SNPs in the genome of *Drosophila melanogaster* strain w ^{1118}; iso-2; iso-3," *Flight Journal*, vol. 6, no. 2, pp. 80–92, 2012.

[33] A. Abyzov, A. E. Urban, M. Snyder, and M. Gerstein, "CNVnator: An approach to discover, genotype, and characterize typical and atypical CNVs from family and population genome sequencing," *Genome Research*, vol. 21, no. 6, pp. 974–984, 2011.

[34] L. Li, C. J. Stoeckert Jr., and D. S. Roos, "OrthoMCL: identification of ortholog groups for eukaryotic genomes," *Genome Research*, vol. 13, no. 9, pp. 2178–2189, 2003.

[35] Z. Yang, "PAML 4: phylogenetic analysis by maximum likelihood," *Molecular Biology and Evolution*, vol. 24, no. 8, pp. 1586–1591, 2007.

[36] P. Jones, D. Binns, H.-Y. Chang et al., "InterProScan 5: genome-scale protein function classification," *Bioinformatics*, vol. 30, no. 9, pp. 1236–1240, 2014.

[37] S. Götz, J. M. García-Gómez, J. Terol et al., "High-throughput functional annotation and data mining with the Blast2GO suite," *Nucleic Acids Research*, vol. 36, no. 10, pp. 3420–3435, 2008.

[38] L. Chen, D. Zheng, B. Liu, J. Yang, and Q. Jin, "VFDB 2016: Hierarchical and refined dataset for big data analysis - 10 years on," *Nucleic Acids Research*, vol. 44, no. 1, pp. D694–D697, 2016.

[39] M. Barakat, P. Ortet, and D. E. Whitworth, "P2RP: A web-based framework for the identification and analysis of regulatory proteins in prokaryotic genomes," *BMC Genomics*, vol. 14, no. 1, article no. 269, 2013.

[40] P. Jia, F. Li, J. Xia et al., "Consensus rules in variant detection from next-generation sequencing data," *PLoS ONE*, vol. 7, no. 6, Article ID e38470, 2012.

[41] B. K. Hammer, E. S. Tateda, and M. S. Swanson, "A two-component regulator induces the transmission phenotype of stationary-phase Legionella pneumophila," *Molecular Microbiology*, vol. 44, no. 1, pp. 107–118, 2002.

[42] L. Aravind and C. P. Ponting, "The cytoplasmic helical linker domain of receptor histidine kinase and methyl-accepting proteins is common to many prokaryotic signalling proteins," *FEMS Microbiology Letters*, vol. 176, no. 1, pp. 111–116, 1999.

[43] H. A. Reiland, M. A. Omolo, T. J. Johnson, and D. J. Baumler, "A Survey of *Escherichia coli* O157:H7 Virulence Factors: The First 25 Years and 13 Genomes," *Advances in Microbiology*, vol. 4, no. 7, pp. 390–423, 2014.

[44] A. Llanes, C. M. Restrepo, and S. Rajeev, "Whole genome sequencing allows better understanding of the evolutionary history of leptospira interrogans serovar hardjo," *PLoS ONE*, vol. 11, no. 7, Article ID e0159387, 2016.

[45] M. Martins-Pinheiro, L. Schons-Fonseca, J. B. da Silva et al., "Genomic survey and expression analysis of DNA repair genes in the genus Leptospira," *Molecular genetics and genomics : MGG*, vol. 291, no. 2, pp. 703–722, 2016.

[46] C. Lindner, R. Nijland, M. Van Hartskamp, S. Bron, L. W. Hamoen, and O. P. Kuipers, "Differential Expression of Two Paralogous Genes of Bacillus subtilis Encoding Single-Stranded DNA Binding Protein," *Journal of Bacteriology*, vol. 186, no. 4, pp. 1097–1105, 2004.

[47] Y. Zhong, X. Chang, X.-J. Cao et al., "Comparative proteogenomic analysis of the Leptospira interrogans virulence-attenuated strain IPAV against the pathogenic strain 56601," *Cell Research*, vol. 21, no. 8, pp. 1210–1229, 2011.

[48] P. Bourhy, L. Salaün, A. Lajus, C. Médigue, C. Boursaux-Eude, and M. Picardeau, "A genomic island of the pathogen Leptospira interrogans serovar Lai can excise from its chromosome," *Infection and Immunity*, vol. 75, no. 2, pp. 677–683, 2007.

[49] E. J. Capra and M. T. Laub, "Evolution of two-component signal transduction systems," *Annual Review of Microbiology*, vol. 66, pp. 325–347, 2012.

9

Preinoculation of Soybean Seeds Treated with Agrichemicals up to 30 Days before Sowing: Technological Innovation for Large-Scale Agriculture

Ricardo Silva Araujo,[1] Sonia Purin da Cruz,[2] Edson Luiz Souchie,[3] Thomas Newton Martin,[4] André Shigueyoshi Nakatani,[1] Marco Antonio Nogueira,[5] and Mariangela Hungria[5]

[1]*Total Biotecnologia Indústria e Comércio S/A, Rua Emilio Romani, No. 1190, CIC, Curitiba, PR, Brazil*
[2]*Universidade Federal de Santa Catarina, Campus Curitibanos, Rodovia Ulisses Gaboardi, Km 3, Curitibanos, SC, Brazil*
[3]*Instituto Federal Goiano, Campus Rio Verde, Rodovia Sul Goiana, Km 01, Caixa Postal 66, Rio Verde, GO, Brazil*
[4]*Universidade Federal de Santa Maria, Avenida Roraima, No. 1000, Santa Maria, RS, Brazil*
[5]*Embrapa Soja, CP 231, 86001-970 Londrina, PR, Brazil*

Correspondence should be addressed to Ricardo Silva Araujo; rsaraujo@totalbiotecnologia.com.br

Academic Editor: Dulal Borthakur

The cultivation of soybean in Brazil experienced an expressive growth in the last decades. Soybean is highly demanding on nitrogen (N) that must come from fertilizers or from biological fixation. The N supply to the soybean crop in Brazil relies on the inoculation with elite strains of *Bradyrhizobium japonicum*, *B. elkanii*, and *B. diazoefficiens*, which are able to fulfill the crop's N requirements and enrich the soil for the following crop. The effectiveness of the association between N_2-fixing bacteria and soybean plants depends on the efficacy of the inoculation process. Seed treatment with pesticides, especially fungicides or micronutrients, may rapidly kill the inoculated bacteria, affecting the establishment and outcome of the symbiosis. The development of technologies that allow inoculation to become a successful component of industrial seed treatment represents a valuable tool for the seed industry, as well as for the soybean crop worldwide. In this article, we report the results of new technologies, developed by the company Total Biotecnologia Indústria e Comércio S/A of Brazil, for preinoculation of soybean seeds with bradyrhizobia, in the presence of agrichemicals. Our results demonstrate improved bacterial survival for up to 30 days after inoculation, without compromising nodulation, N_2-fixation, and yield in the field.

1. Introduction

The cultivation of soybean [*Glycine max* (L.) Merr.] in Brazil is one of the economic activities that experienced the most expressive continuous growth in the last decades [1]. Brazil is currently one of the major producers of soybean in the world, along with the United States of America [2]. The crop occupied around 33 million hectares in the 2015/2016 cropping season, with a total grain production of over 95 million tons, and the grain is the leading commodity of the Brazilian agribusiness, accounting for approximately 13% of all Brazilian exportations [3, 4]. The success of the soybean crop in Brazil, however, could have never been achieved without the full utilization of the natural process of symbiotic nitrogen (N_2) fixation [5].

Soybean is highly demanding on nitrogen (N), requiring around 80 kg of N to produce 1,000 kg of grains [6]. Considering an average grain yield of 3,000 kg ha^{-1}, in Brazil, and 33 million hectares grown with soybean, an astounding 7.9×10^6 tons of N (equivalent to 16.5×10^6 tons of urea) would be necessary to fulfill the crop's demand for N. If all that N was to be supplied by chemical fertilizers, whose efficiency of utilization reaches, at most, 50%, an estimate of 33 million tons of urea would be necessary, and the costs of N fertilization alone would render the soybean crop completely unattractive under Brazilian conditions. However, the cultivation of soybean in Brazil relies on the inoculation with elite strains of *Bradyrhizobium japonicum, B. elkanii,* and *B. diazoefficiens,* which are able to supply all the soybean N requirements and enrich the soil for the following crop [5, 6]. This technology generates an economy of over 15 billion US dollars that would have to be spent with chemical fertilizers each season to supply the crop's demands for N [5].

The effectiveness of the association between N_2-fixing bacteria and soybean plants depends on the efficacy of the inoculation process, which consists of coating the seed surface with either solid (peat-based) or liquid bacterial formulations. Many factors may negatively affect the survival of the bacteria after seed inoculation [7, 8], resulting in failures in the nodulation process and, consequently, in N_2 fixation and N supply to the crop. Seed treatment with pesticides, especially fungicides [9–13] or micronutrients [14], may rapidly kill the inoculated bacteria, affecting the establishment and outcome of the symbiosis [15].

In-furrow inoculation has long been postulated as an alternative to traditional seed inoculation in order to avoid damage to the bacteria [17, 18]. After the liquid inoculant formulations have become a consolidated technology, in-furrow inoculation has been successfully employed for soybean [8, 19]. In addition, spray inoculation has been shown to be effective as remedial inoculation, in cases of failures of soybean nodulation [8]. Nevertheless, the most commonly adopted and practical form of inoculation is still on the seeds.

The increasing technological level of agricultural activities, as well as the high cost of labor, along with issues concerning human safety in the manipulation of pesticides, however, has made on-farm seed manipulation unattractive. Industrial seed treatment that combines the application of pesticides, micronutrients, stimulants, and other additives, as well as inoculants, to the seeds, prior to delivery to the farmer is becoming popular [20]. However, such practices may be very deleterious to the inoculated bacteria and, in most cases, bacterial mortality is so high that no viable bacteria are present on the seeds at sowing [21]. The development of technologies that allow inoculation to become a successful component of industrial seed treatment represents a valuable tool for the seed industry, as well as for the soybean crop worldwide. In this article, we report the results of new technologies, developed by the company Total Biotecnologia Indústria e Comércio S/A of Brazil, for preinoculation of soybean seeds with bradyrhizobia, in the presence of agrichemicals. Our results demonstrate improved bacterial survival for up to 30 days after inoculation, without compromising nodulation, N_2-fixation, and yield in the field.

2. Materials and Methods

2.1. Inoculants and Technologies Employed in This Study. All inoculants employed in this study were commercial products manufactured and commercialized by Total Biotecnologia Indústria e Comércio S/A, in Curitiba, State of Paraná, Brazil. They contain the strains SEMIA5079 *(Bradyrhizobium japonicum)* and SEMIA5080 *(B. diazoefficiens),* which are officially authorized for the composition of commercial inoculants for soybean in Brazil. TotalNitro™ is a standard liquid inoculant [5×10^9 colony forming units (CFU) mL^{-1}] and was employed without any additives or seed treatment chemicals, as a control in some experiments. Standard peat-based TotalNitro inoculant (5×10^9 CFU g^{-1}) was employed either with or without chemicals as a second control in some experiments. Cronos™ technology [liquid inoculant (5×10^9 CFU mL^{-1}), along with two additives (A and B)] was employed in a 2-day preinoculation scheme in the presence of the chemical (insecticide/fungicide) Standak Top™ (Fipronil + Pyraclostrobin + Methyl thiophanate), and in a 4-day preinoculation scheme in the absence of the chemical. CronosNod™ Technology [liquid inoculant (7×10^9 CFU mL^{-1}), along with CronosNod protective additive], was employed in a 7-day preinoculation scheme in the presence of the chemical Standak Top. Finally, CronosTSI™ Technology [standard peat-based TotalNitro inoculant (5×10^9 CFU g^{-1}) along with two TSI™ additives, A and B] was employed in a 30-day preinoculation scheme, also in the presence of the chemical Standak Top. All additives employed are bacterial protectors composed of sugars, polymers, and inert substances that exert osmoprotection of the bacteria and potentiate inoculation.

2.2. Laboratory Experiments. Laboratory experiments performed at Embrapa Soja, Londrina, State of Paraná, and at the Instituto Federal Goiano, Rio Verde, State of Goiás, compared the survival of bradyrhizobia on the surface of seeds that had been inoculated with the preinoculation technologies evaluated in this study, in the presence of chemicals, with standard inoculants on untreated seeds. Bacterial survival was estimated as the number of CFU recovered from the surface of inoculated seeds at different times after inoculation. Treatments depended on the institution that performed the tests and were (1) standard liquid (TotalNitro) inoculant, applied to untreated seeds without additives 2 h before evaluation (time zero); (2) Cronos Technology, applied to treated seeds 2 h and 2 and 4 days before evaluation; (3) CronosNod Technology, applied to treated seeds 2 h and 2, 4, 7, 14, and 21 days before evaluation; and (4) CronosTSI Technology, applied to seeds 2 h and 2, 4, 7, 14, 21, 28, and 35 days before evaluation.

For seed treatment and inoculation, all products were applied to the seeds according to their manufacturers' instructions regarding doses and methods of application. In case chemicals were used, they were applied to the seeds prior to inoculation with bradyrhizobia. Sample lots of 1 kg of certified seeds were inoculated for each treatment. Cultivar BMX Potência RR and NS 7209 Pro were employed at Embrapa Soja and at the Instituto Federal Goiano, respectively. After treatment and inoculation, seeds were stored under controlled laboratory conditions, at 60% relative humidity and 25°C ± 2°C.

The recovery of bacterial cells from inoculated seeds was quantified with the methodology established by Brazilian regulations [22]. At the appropriate sampling times, four subsamples of 100 seeds each were withdrawn from each treatment. Each 100-seed subsample was aseptically transferred to a 250-mL Erlenmeyer flask containing 100 mL sterile 0.85% (w/v) NaCl + 0.01% (w/v) Tween 80 solution, in order to obtain the 10^0 dilution. Flasks were then shaken in an orbital shaker at 150 rpm for 15 min and, from each flask, a 10-fold dilution series in sterile 0.85% (w/v) NaCl solution was prepared. 100-μL aliquots of each dilution from each series were spread onto Petri dishes containing solid (1.5% agar) YMA medium [23] with Congo red (25 mg L^{-1}), vancomycin (1 μg mL^{-1}), and cycloheximide (55 mg L^{-1}). Inoculated plates were incubated at 28°C ± 2°C for seven days. Bacterial colonies from all plates presenting between 30 and 300 colonies were counted, and the number of CFU $seed^{-1}$ was estimated based on the average of the colony counts of the two dilutions closest to the counting range.

All experiments were completely randomized. Subsamples A and B were combined in one single replicate, so that each treatment had three replicates in the end. Results were subjected to an analysis of variance at $p < 0{,}05$. Further on, data were analyzed by the Dunnett (bilateral) test in order to detect significant differences between each treatment and the time zero control without chemicals, considering a 95% confidence interval. All statistical analyses were performed with the Statistica Version 7 software.

2.3. Field Experiments. Field experiments were conducted at sites located in representative regions of soybean growth in Brazil. The team of Embrapa Soja planted experiments in Rio Verde (17°47′S, 50°55′W; 748 m altitude; Köpen-Geiger's Aw climate) and Cachoeira Dourada (18°29′S, 49°28′W; 450 m altitude; Köpen-Geiger's Aw climate), both in the State of Goiás. Investigators at the Instituto Federal Goiano also set up experiments in Rio Verde. The investigators at Universidade Federal de Santa Catarina conducted an experiment in Curitibanos (27°16′S, 50°35′W; 995 m altitude; Köpen-Geiger's Cfa climate), in the 2013/2014 cropping season, and another in Ponte Alta do Norte (27°09′S, 50°28′W; 959 m altitude, Köpen-Geiger's Cfa climate), in the 2014/2015 cropping season, both in the State of Santa Catarina.

At all locations, samples from the soils of the experimental areas were collected 40 to 60 days prior to sowing for chemical and physical analyses performed as described before [8] and estimation of the naturalized population of soybean-nodulating bradyrhizobia by the most-probable-number method [24, 25]. The chemical, physical, and microbiological characteristics of the soils at the experimental sites are presented in Table 1. Treatments and other relevant information about the experiments are shown in Table 2.

Field plots measured at least 24 m^2 and, in general, had eight 6 m long planting rows. Soybean seeds were sown to obtain plant populations of about 300,000 plants ha^{-1}. The cultivar planted at each site is also described in Table 2. All plots were spaced by 0.5 m-wide lines and 1.5 m terraces to avoid cross contamination caused by runoff of irrigation and rainwater. Weeds, insects, and foliar diseases at all sites were controlled according to recommended agronomic practices for the soybean crop.

Plant samples (five plants) were collected from each plot during the vegetative stage, from 30 to 50 days after emergence (DAE), according to the location and weather conditions, for determination of nodule number, nodule and plant biomass, and total N in the shoots. Samples were processed in the laboratory, where roots were separated from shoots, carefully rinsed, and allowed to dry in a 50°C oven until constant weight. Nodules were then removed from the roots and allowed to dry further until they were counted and weighed. Dry shoots were weighed and ground for determination of the N content as described before [26]. Grain yield was determined at the end of the crop cycle, from the central portion (6 m^2 to 15 m^2) of each plot. Seeds were cleaned and weighed and grain yield was estimated after correction of seed weights to 13% moisture.

All data obtained were tested for the normality of variables and variance homogeneity, followed by an analysis of variance (ANOVA) at $p < 0.05$. In case of significance of the ANOVA, means were compared by the Duncan test, at $p < 0.1$ and $p < 0.05$, or the Tukey test at $p < 0.05$, depending on the institution that performed the experiments. Statistical software such as Sisvar, Assistat, and Statistica 7.0 was employed.

3. Results

3.1. Laboratory Experiments. The Cronos, CronosNod, and CronosTSI Technologies significantly improved the survival of the inoculated bacteria on the seeds in the presence of chemicals, when compared to the standard inoculation of untreated seeds (Table 3). On average, the Cronos and CronosNod Technologies promoted a 36-fold and a 26-fold increase, respectively, while CronosTSI promoted an astounding 366-fold increase in the number of bacteria that adhered to the treated seeds and could be recovered two hours after inoculation, compared to the standard inoculation of untreated seeds (Table 3). All products also significantly improved bacterial survival for several days after inoculation of treated seeds, and the CronosNod and CronosTSI Technologies promoted the survival of a satisfactory population of bacteria on seeds that had been treated with chemicals for up to seven and 35 days, respectively (Table 3). All technologies were, therefore, very effective in increasing the adherence of the inoculants to the seeds, promoting bacterial survival even in the presence of chemicals.

3.2. Field Experiments. The experiments conducted by Embrapa Soja in Rio Verde and Cachoeira Dourada suffered severe drought from sowing to early blooming, affecting early plant development and inducing great variability in the results of the parameters evaluated between 40 and 50 days after plant emergence. Nodulation was very poor at both locations and slightly superior when treated seeds received peat-based standard inoculant at sowing (Table 4). Plant biomass responded better when standard liquid inoculant was applied to untreated seeds, as did N accumulation in the shoots (Table 5). Similar results with nonsignificant differences were observed when the preinoculation technologies were employed, suggesting a positive protective effect even under adverse conditions (Table 5). The same trend could be

Table 1: Chemical properties, granulometry analysis, and bradyrhizobial populations of the soils at the experimental areas.

Location	pH in $CaCl_2$	Chemical properties							Granulometry			Bradyrhizobial population size■
		Al	H + Al	K	Ca	Mg	P	C	Clay	Silt	Sand	MPN cells g^{-1}
				$cmol_c\ dm^{-3}$			$mg\ dm^{-3}$	$g\ dm^{-3}$		%		
Rio Verde●	5.1	0.00	3.64	0.80	1.65	1,78	9.6	22.5	36.3	9.5	54.1	zero
Rio Verde♠	6.1	0.03	5.10	0.65	4.40	1.30	15.2	20.2	46.4	17.4	36.2	2.50×10^4
Rio Verde♣	5.9	0.02	4.80	0.62	4.10	1.20	14.6	18.3	46.4	17.4	36.2	3.10×10^4
Cachoeira Dourada	5.4	0.00	3.07	0.37	3.65	1.73	1.7	18.5	57.7	18.2	24.0	3.57
Curitibanos	5.1	1.58	13.70	0.53	3.16	1.39	0.9	33.1	39.00	ND♥	ND	1.47×10^1
Ponte Alta do Norte	4.7	1.42	7.20	0.59	0.39	0.24	0.9	20.35	ND	ND	ND	3.00×10^3

■Most-probable-number of cells g $soil^{-1}$. ●Experiment conducted by Embrapa Soja. ♠Experiment conducted by Instituto Federal Goiano in 2013/2014. ♣Experiment conducted by Instituto Federal Goiano in 2014/2015. ♥Not determined.

TABLE 2: Treatments, cultivars, numbers of replicates, and other pertinent information about field experiments.

	Rio Verde●	Rio Verde♠	Rio Verde♣	Cachoeira Dourada	Curitibanos	P. Alta do Norte
Treatments■ implanted at each location						
Uninoculated control, treated seeds	Yes	Yes	Yes	Yes	Yes	Yes
N fertilizer control, treated seeds	Yes	Yes	Yes	Yes	Yes	Yes
Standard liquid inoculant, untreated seeds	No	No	Yes	Yes	No	No
Standard peat-based inoculant, treated seeds	Yes	Yes	Yes	Yes	Yes	Yes
Standard peat-based inoculant, untreated seeds	No	No	Yes	Yes	No	No
Cronos Inoculant, two days preinoculation, treated seeds	Yes	Yes	Yes	Yes	No	Yes
Cronos Inoculant, two days preinoculation, untreated seeds	No	No	No	No	Yes	No
Cronos Inoculant, four days preinoculation, treated seeds	No	No	Yes	Yes	No	No
CronosNod Technology, seven days preinoculation, treated seeds	Yes	No	Yes	Yes	Yes	No
Cronos TSI Technology, 30 days preinoculation, treated seeds	Yes	Yes	No	No	Yes	Yes
Other relevant information						
Cultivar	NS 7209 PRO	Anta	BMX-Potência RR	BRS-GO-8360	NA 5909 (Nidera)	NA 5909 (Nidera)
Number of replicates	Four	Four	Six	Six	Five	Five
Basal fertilization	NPK 02-20-10 300 kg ha^{-1} + KCl 100 kg ha^{-1}	NPK 02-20-10 300 kg ha^{-1} + KCl 100 kg ha^{-1}	NPK 0-28-20 300 kg ha^{-1}	NPK 0-28-20 300 kg ha^{-1}	NPK 0-18-18 350 kg ha^{-1}	NPK 0-18-18 350 kg ha^{-1}
Micronutrients	No	No	20 g Mo ha^{-1} + 2 g Co ha^{-1}, foliar spray @ V4♥	20 g Mo ha^{-1} + 2 g Co ha^{-1}, foliar spray @ V4	No	No
Sampling for growth evaluation (DAE♦)	35	35	40	50	32	32

■Unless otherwise stated, seeds were treated with STANDAK TOP (see text), according to manufacturer's instructions. ●Experiment conducted by Instituto Federal Goiano in 2013/2014. ♠Experiment conducted by Instituto Federal Goiano in 2014/2015. ♣Experiment conducted by Embrapa Soja. ♥V4 stage: four nodes on the main stem with fully developed leaves [16]. ♦DAE: days after emergence.

TABLE 3: Recovery of *Bradyrhizobium* spp. cells (number of colony forming units, # CFU, per seed) from soybean seeds inoculated with different products, in the absence or presence of seed-treatment chemicals.

Treatment	Time after treatment/inoculation (duration of preinoculation) # CFU seed$^{-1♠}$							
	2 hours	2 days	4 days	7 days	14 days	21 days	28 days	35 days
	Embrapa Soja							
Standard inoculation	1.52×10^5							
Cronos/on treated seeds	$1.17 \times 10^{6*}$	$7.60 \times 10^{5*}$	$4.61 \times 10^{5*}$					
CronosNod on treated seeds	$1.08 \times 10^{6*}$	$8.17 \times 10^{5*}$	4.02×10^{5ns}	1.96×10^{5ns}				
	Instituto Federal Goiano							
Standard inoculation	1.83×10^5							
Cronos/on treated seeds	$1.24 \times 10^{7*}$	$2.68 \times 10^{5*}$	1.28×10^{5ns}					
CronosNod on treated seeds	$8.66 \times 10^{6*}$	$9.67 \times 10^{5*}$	$5.20 \times 10^{5*}$	3.13×10^{5ns}	7.70×10^4	1.70×10^4	6.00×10^3	
CronosTSI on treated seeds	$6.70 \times 10^{7*}$	$1.40 \times 10^{7*}$	$7.30 \times 10^{6*}$	$2.30 \times 10^{6*}$	$5.50 \times 10^{5*}$	2.40×10^3	2.31×10^3	1.13×10^3

♠∗ denotes a significant and ns a nonsignificant difference relative to the control, standard inoculation, evaluated two hours after inoculation, according to the bilateral Dunnett test ($p < 0.05$). All values followed by ∗ were significantly superior to the control; ns means not significantly different from control. Values not followed by ∗ or ns were not compared.

TABLE 4: Nodule number and biomass of soybean plants from the field experiments.

Treatment■	Nodule number (# plant^{-1})●						Nodule biomass (mg plant^{-1})●					
	RV1♠	RV2	RV3	CD	CB	PA	RV1	RV2	RV3	CD	CB	PA
(1)	16.7^{b}	19.8^{a}	1.07^{d}	1.37^{ab}	0.64^{bc}	0.00^{b}	116.5^{b}	145.5^{a}	25.47^{bcd}	14.8^{a}	3.56^{ab}	0.00^{b}
(2)	15.1^{b}	16.2^{a}	0.23^{d}	0.17^{b}	0.40^{c}	0.04^{b}	105.5^{b}	115.5^{ab}	4.11^{d}	0.4^{b}	0.58^{b}	0.13^{b}
(3)	29.7^{a}	17.1^{a}	21.53^{a}	2.67^{a}	1.24^{ab}	1.76^{a}	207.5^{a}	87.5^{b}	108.69^{a}	18.5^{a}	5.85^{ab}	2.53^{a}
(4)	NA♣	NA	0.73^{d}	2.20^{a}	NA	NA	NA	NA	17.05^{cd}	20.0^{a}	NA	NA
(5)	NA	NA	NA	NA	1.48^{a}	NA	NA	NA	NA	NA	5.40^{ab}	NA
(6)	28.2^{a}	17.9^{a}	3.57^{bc}	2.33^{a}	NA	0.32^{b}	198.2^{a}	96.0^{b}	48.39^{bc}	19.9^{a}	NA	0.12^{b}
(7)	NA	NA	5.23^{b}	1.97^{a}	NA	NA	NA	NA	63.45^{b}	18.7^{a}	NA	NA
(8)	29.3^{a}	NA	2.10^{cd}	1.83^{a}	1.56^{a}	NA	205.6^{a}	NA	28.4^{bcd}	15.6^{a}	8.00^{a}	NA
(9)	28.4^{a}	14.4^{a}	NA	NA	3.76^{a}	0.20^{b}	199.6^{a}	103.0^{ab}	NA	NA	13.06^{a}	1.18^{a}
p value	<0.05	<0.05	<0.1	<0.1	<0.05	<0.05	<0.05	<0.05	<0.1	<0.1	<0.05	<0.05

■Treatment descriptions: (1) Uninoculated, not N-fertilized control, treated seeds; (2) N-fertilized control, treated seeds; (3) Standard peat-based inoculant applied on planting day, treated seeds; (4) Standard liquid inoculant applied on planting day, untreated seeds; (5) Cronos inoculant, two days pre-inoculation, untreated seeds; (6) Cronos inoculant, two days pre-inoculation, treated seeds; (7) Cronos inoculant, four days pre-inoculation, treated seeds; (8) CronosNod technology, seven days pre-inoculation, treated seeds; (9) CronosTSI technology, 30 days pre-inoculation, treated seeds. ●Means followed by the same letter, on the same column, are not significantly different, according to the comparison tests described in the Materials and Methods section. ♠RV1 – experiment conducted in Rio Verde by Instituto Federal Goiano in 2013/2014; RV2 - experiment conducted in Rio Verde by Instituto Federal Goiano in 2014/2015, RV3 – experiment conducted in Rio Verde by Embrapa Soja, CD – experiment conducted in Cachoeira Dourada by Embrapa Soja, CB – experiment conducted in Curitibanos by Universidade Federal de Santa Catarina; PA – experiment conducted in Ponte Alta do Norte by Universidade Federal de Santa Catarina. ♣NA = not applicable; the treatment was not included at the specific experimental site.

TABLE 5: Plant biomass, N accumulation in the shoots, and grain yield of soybean plants from the field experiments.

Treatment■	Plant biomass (g plant^{-1})●						N accumulation (mg N plant^{-1})●						Grain yield (kg ha^{-1})●					
	RV1♠	RV2	RV3	CD	CB	PA	RV1	RV2	RV3	CD	CB	PA	RV1	RV2	RV3	CD	CB	PA
(1)	5.0^{b}	4.7^{b}	6.92^{a}	9.45^{a}	2.69^{ab}	5.34^{a}	108.0^{b}	107.5^{a}	153.9^{abc}	228.5^{ab}	87.8^{ab}	33.5^{a}	2207^{b}	1916^{c}	2552^{b}	2717^{abc}	3195^{b}	2731^{c}
(2)	6.5^{ab}	7.1^{a}	6.53^{ab}	8.41^{ab}	2.69^{ab}	5.39^{a}	115.0^{a}	109.0^{a}	200.1^{a}	270.3^{a}	85.6^{ab}	34.4^{a}	2181^{b}	2337^{bc}	2855^{a}	2531^{bc}	3873^{ab}	4445^{a}
(3)	6.5^{ab}	7.4^{a}	6.46^{ab}	7.97^{ab}	3.34^{a}	5.19^{a}	111.0^{ab}	108.5^{a}	174.2^{abc}	229.1^{ab}	89.2^{a}	36.3^{a}	3833^{a}	3134^{a}	3050^{a}	2658^{abc}	3808^{ab}	2905^{b}
(4)	NA♣	NA	7.86^{a}	9.51^{a}	NA	NA	NA	NA	196.1^{ab}	276.2^{a}	NA	NA	NA	NA	2586^{b}	2477^{c}	NA	NA
(5)	NA	NA	NA	NA	2.24^{b}	NA	NA	NA	NA	NA	60.7^{c}	NA	NA	NA	NA	NA	3956^{a}	NA
(6)	7.0^{a}	7.2^{a}	5.34^{b}	8.93^{ab}	NA	5.43^{a}	113.0^{a}	109.0^{a}	135.4^{c}	248.7^{ab}	NA	35.3^{a}	3745^{a}	3189^{a}	2893^{a}	2880^{ab}	NA	3276^{b}
(7)	NA	NA	7.55^{a}	7.47^{ab}	NA	NA	NA	NA	200.5^{a}	179.7^{b}	NA	NA	NA	NA	2807^{a}	2701^{abc}	NA	NA
(8)	7.0^{a}	NA	6.37^{ab}	7.11^{b}	2.65^{ab}	NA	114.0^{a}	NA	148.8^{bc}	172.8^{b}	68.1^{bc}	NA	3696^{a}	NA	2895^{a}	2950^{a}	3479^{ab}	NA
(9)	6.3^{a}	8.1^{a}	NA	NA	3.02^{a}	5.91^{a}	114.0^{a}	104.5^{a}	NA	NA	66.4^{bc}	36.0^{a}	3594^{a}	2977^{a}	NA	NA	3783^{a}	3023^{b}
p value	<0.05	<0.05	<0.1	<0.1	<0.05	<0.05	<0.05	<0.05	<0.1	<0.1	<0.05	<0.05	<0.05	<0.05	<0.1	<0.1	<0.05	<0.05

■Treatment descriptions: (1) uninoculated, not N-fertilized control, treated seeds; (2) N-fertilized control, treated seeds; (3) standard peat-based inoculant applied on planting day, treated seeds; (4) standard liquid inoculant applied on planting day, untreated seeds; (5) Cronos inoculant, two days preinoculation, untreated seeds; (6) Cronos inoculant, two days preinoculation, treated seeds; (7) Cronos inoculant, four days preinoculation, treated seeds; (8) CronosNod technology, seven days preinoculation, treated seeds; (9) CronosTSI technology, 30 days preinoculation, treated seeds. ●Means followed by the same letter, on the same column, are not significantly different, according to the comparison tests described in the Materials and Methods. ♠RV1: experiment conducted in Rio Verde by Instituto Federal Goiano in 2013/2014; RV2: experiment conducted in Rio Verde by Instituto Federal Goiano in 2014/2015, RV3: experiment conducted in Rio Verde by Embrapa Soja, CD: experiment conducted in Cachoeira Dourada by Embrapa Soja, CB: experiment conducted in Curitibanos by Universidade Federal de Santa Catarina; PA: experiment conducted in Ponte Alta do Norte by Universidade Federal de Santa Catarina. ♣NA = not applicable; the treatment was not included at the specific experimental site.

observed for grain yield, where the preinoculation technologies promoted yields that were not significantly different from the N control or the control standard inoculants (Table 5).

In the case of the experiment conducted by the Instituto Federal Goiano, also in Rio Verde, in 2013/2014, nodulation parameters were slightly better. Preinoculation of treated seeds with all three Cronos Technologies promoted nodulation (number and biomass) that was not significantly different from the peat-based inoculant control applied at planting, but was significantly higher than either the uninoculated or the N controls, even when inoculation was performed as early as seven (CronosNod) or 35 (CronosTSI) days before planting (Table 4). Plant biomass and N accumulation in the shoots obtained with all preinoculation treatments were not significantly different from the standard inoculant and N controls, and yet significantly different from the uninoculated control (Table 5), whereas all inoculant treatments promoted significantly higher grain yields than the N and uninoculated controls (Table 5).

No treatment effects were observed on any nodulation parameters in the 2014/2015 experiment conducted by the Instituto Federal Goiano in Rio Verde (Table 4). All inoculation treatments evaluated in 2014/2015, as well as N fertilizer, however, promoted significant increases in plant biomass relative to the uninoculated control, but no significant effects were observed on N accumulation in the shoots (Table 5). All inoculation treatments produced significantly higher yields than the uninoculated control or the N fertilizer treatment (Table 5).

In Curitibanos and Ponte Alta do Norte, experiments were performed in first-year areas of soybean cropping that had been previously occupied by native pasture. Nodulation and plant parameters were also very poor, even though both nodule number and biomass tended to respond to all methods of inoculation, especially when the CronosTSI Technology was employed in Curitibanos (Table 4). No conclusive effects could be observed for plant biomass, but N accumulation in the shoots of the plants from the uninoculated and N controls and from the standard peat-based inoculant treatments was higher than that of the preinoculation treatments in Curitibanos, but not significantly different in Ponte Alta do Norte (Table 5). All treatments promoted similar grain yields, with a remarkable significant response to N fertilization in Ponte Alta do Norte (Table 4).

4. Discussion

Even though legumes can rely solely on N_2 fixation to obtain all the N necessary to attain high grain yields, the effectiveness of the process depends on a fully functional set of root nodules. Nodulation is highly susceptible to abiotic and biotic factors, including the population of live bacteria on the seeds at planting [18, 27, 28]. Any factor that can cause mortality of the inoculated bacteria on the seed surface will therefore negatively affect the formation of nodules and, in turn, N_2 fixation. Factors such as desiccation, temperature, and seed coat toxicity, for instance, have been implicated in poor inoculant survival on seeds and failure of nodulation and N_2 fixation [7].

The utilization of industrially treated and preinoculated seeds is becoming more popular and desired among farmers, but care must be taken because many of the products employed to treat seeds are toxic, resulting in poor survival of the inoculated bacteria [21]. Agrichemicals such as fungicides applied to seeds have been shown to be very toxic to rhizobia, resulting in bacterial death just a few hours after inoculation [11, 29]. Even the conditions under which preinoculated seeds are stored can be harmful to the bacteria [30]. Therefore, if preinoculated seeds are to become widely deployed in agriculture, it is necessary to develop technologies that ensure the survival of the bacteria on the seeds for longer periods after inoculation, especially in the presence of harmful chemicals.

Many studies have addressed the issues of compatibility with agrichemicals and survival of bacteria on the surface of inoculated seeds (e.g., [21, 31]). However, in most cases, compatibility studies were performed *in vitro*, and when survival on the seeds was the issue of interest, chemicals were not always included, or the conditions under which inoculated treated seeds were stored did not mimic the real situation. Compounds such as polymers and alginate, among others, may improve the survival of bacteria [32], but little is known about their effect in the presence of chemicals. When seeds were stored under refrigeration, bacterial survival in the presence of chemicals was shown to be satisfactory [33], but there is little information about survival at room temperature.

In this study we have tested new technologies (CronosNod and CronosTSI) developed by Total Biotecnologia Indústria e Comércio S/A, which claim to favor the survival of satisfactory populations of inoculant bacteria on the surface of seeds that have been treated with a chemical (Standak Top) that combines fungicide and insecticide and were stored at room temperature. The technologies consist of a combination of a highly concentrated liquid (CronosNod) or peat-based (CronosTSI) inoculant and a protectant to be applied to seeds. The tests included the estimation of bacterial survival on treated seeds in the laboratory, and the evaluation of nodulation, plant growth, and yield under different field conditions, in comparison to standard inoculation and to another commercial preinoculation product (Cronos) of the same company that guarantees bacterial survival for up to four days on the surface of untreated seeds.

When compared to the standard peat-based inoculant, all three technologies improved the adherence of bacteria to the seed surfaces. On average, 36-, 26-, and 366-fold more bacteria could be recovered from seeds two hours after inoculation when Cronos, CronosNod, and CronosTSI were employed, respectively, in the presence of the agrichemical, when compared to the standard peat-based inoculation in the absence of agrichemicals. These results clearly demonstrate the improved adherence of the inoculants and the efficacy of all preinoculation technologies to protect the bacteria from being killed by seed treatment. The protective effect, in the presence of agrichemicals, could still be noticed two days after inoculation, when an average three (Cronos) to five (CronosNod) or even 80-fold (CronosTSI) larger population of bacteria could be recovered from treated seeds when compared to the counts obtained two hours after peat-based inoculation of untreated seeds. Even seven days after inoculation, CronosNod and CronosTSI continued to exert their protective effect, and 1.5- and 12.5-fold more bacteria, respectively, could be recovered from treated seeds than from untreated seeds that received peat-based inoculant only two hours before evaluation. CronosTSI maintained live and functional bacteria on the surface of treated seeds for as long as 35 days.

Another important aspect to be studied is the performance of the bacteria present on preinoculated seeds after planting, regarding the development of a functional set of root nodules, N_2 fixation, and grain yield by soybean. Our study demonstrated that if seeds are to be planted right after inoculation, peat-based inoculant is most often the preferable form of inoculation. However, if preinoculation is persuaded, some protective inoculant technology must be used. Cronos, CronosNod, and CronosTSI have been shown to promote good nodulation, plant development, and grain yield, comparable to what is obtained with peat-based inoculation right before planting, when agrichemical-treated seeds were inoculated up to seven (CronosNod) or 30 days (CronosTSI) ahead of planting time.

5. Conclusions

CronosNod and CronosTSI are technological innovations that promote adherence to the seeds, and survival of a satisfactory population of inoculant bacteria on the surface of seeds that have received treatment with agrichemicals.

Plants grown from seeds treated with both technologies present field performances comparable to those from standard peat-based inoculants applied right before planting.

CronosNod and CronosTSI represent viable alternatives for the preinoculation of soybean seeds up to 30 days ahead of planting time.

Conflicts of Interest

The authors declare that there are no conflicts of interest regarding the publication of this paper.

References

[1] M. H. Hirakuri and J. J. Lazzaroto, "O agronegócio da soja nos contextos mundial e brasileiro. Documentos/Embrapa Soja," *ISSN*, vol. 70, pp. 2176–2937, 2014.

[2] USDA, "United States Deparment of Agriculture," http://www.usdabrazil.org.br/home.

[3] CONAB (Companhia Nacional de Abastecimento), Séries históricas: soja, 2016, http://www.conab.gov.br/conteudos.php?a=1252&&Pagina_objcmsconteudos=3#A_objcms-conteudos.

[4] S. Embrapa, Soja em números. Safra 2014/2015. https://www.embrapa.br/soja/cultivos/soja1/dados-economicos.

[5] M. Hungria and I. C. Mendes, "Nitrogen fixation with soybean: the perfect symbiosis?" in *Biological Nitrogen Fixation*, F. J. de Bruijn, Ed., pp. 1005–1019, Wiley & Sons, Inc, NJ, USA, 2015.

[6] M. Hungria, J. C. Franchini, R. J. Campo, and P. H. Graham, "The importance of nitrogen fixation to soybean cropping in South America," in *Nitrogen fixation: origins, applications and research progress, v.4*, W. E. Newton, Ed., pp. 25–42, Springer, Dordrecht, Amsterdam, 2005.

[7] R. Deaker, R. J. Roughley, and I. R. Kennedy, "Legume seed inoculation technology - A review," *Soil Biology and Biochemistry*, vol. 36, no. 8, pp. 1275–1288, 2004.

[8] H. Mariangela, A. N. Marco, and S. A. Ricardo, "Alternative methods of soybean inoculation to overcome adverse conditions at sowing," *African Journal of Agricultural Research*, vol. 10, no. 23, pp. 2329–2338, 2015.

[9] K. Annapurna, "*Bradyrhizobium japonicum*: Survival and nodulation of soybean as influenced by fungicide treatment," *Indian Journal of Microbiology*, vol. 45, no. 4, pp. 305–307, 2005.

[10] T. J. Schulz and K. D. Thelen, "Soybean seed inoculant and fungicidal seed treatment effects on soybean," *Crop Science*, vol. 48, no. 5, pp. 1975–1983, 2008.

[11] R. J. Campo, R. S. Araujo, and M. Hungria, "Nitrogen fixation with the soybean crop in Brazil: Compatibility between seed treatment with fungicides and bradyrhizobial inoculants," *Symbiosis*, vol. 48, no. 1-3, pp. 154–163, 2009.

[12] J. É. Zilli, K. G. Ribeiro, R. J. Campo, and M. Hungria, "Influence of fungicide seed treatment on soybean nodulation and grain yield," *Revista Brasileira de Ciencia do Solo*, vol. 33, no. 4, pp. 917–923, 2009.

[13] Kunal and P. Sharma, "Influence of pesticide-treated seeds on survival of *Mesorhizobium* sp. *Cicer*, symbiotic efficiency and yield in chickpea," *Plant Protection Science*, vol. 47, no. 1, pp. 37–43, 2011.

[14] R. J. Campo, R. S. Araujo, and M. Hungria, "Molybdenum-enriched soybean seeds enhance N accumulation, seed yield, and seed protein content in Brazil," *Field Crops Research*, vol. 110, no. 3, pp. 219–224, 2009.

[15] A. Bikrol, N. Saxena, and K. Singh, "Response of Glycine max in relation to nitrogen fixation as influenced by fungicide seed treatment," *African Journal of Biotechnology*, vol. 4, no. 7, pp. 667–671, 2005.

[16] W. R. Fehr and C. E. Caviness, "Stages of soybean development," Tech. Rep. 80, Iowa State University, Ames,Iowa , US, 1977.

[17] E. S. Jensen, "Inoculation of pea by application of Rhizobium in the planting furrow," *Plant and Soil*, vol. 97, no. 1, pp. 63–70, 1987.

[18] J. Brockwell, R. R. Gault, D. F. Herridge, L. J. Morthorpe, and R. J. Roughley, "Studies on alternative means of legume inoculation: Microbiological and agronomic appraisals of commercial procedures for inoculating soybeans with Bradyrhizobium japonicum," *Australian Journal of Agricultural Research*, vol. 39, no. 6, pp. 965–972, 1988.

[19] R. J. Campo, R. S. Araujo, F. L. Mostasso, and M. Hungria, "In-furrow inoculation of soybean as alternative to fungicide and micronutrient seed treatment," *Revista Brasileira de Ciencia do Solo*, vol. 34, no. 4, pp. 1103–1112, 2010.

[20] J. B. França-Neto, A. A. Henning, F. C. Krzyzanowski, F. A. Henning, and I. Lorini, "Adoτπo do tratamento industrial de sementes de soja no Brasil, safra 2014/15," *Informativo Abrates*, vol. 1, pp. 26–29, 2015.

[21] E. J. Hartley, L. G. Gemell, and R. Deaker, "Some factors that contribute to poor survival of rhizobia on preinoculated legume seed," *Crop and Pasture Science*, vol. 63, no. 9, pp. 858–865, 2012.

[22] MAPA, Ministério da Agricultura, Pecuária e Abastecimento, Instrução Normativa nº 4, November 12, 2010. http://sistemasweb.agricultura.gov.br/sislegis/action/detalhaAto.do?

method=gravarAtoPDF&tipo=INM&numeroAto=00000030&seqAto=000&valorAno=2010&orgao=SDA/MAPA&codTipo=&desItem=&desItemFim.

[23] M. Hungria, G. W. O'Hara, and J. E. Zilli, "Isolation and growth of rhizobia," in *Working with rhizobia, ACIAR: Canberra*, J. G. Howieson and M. J. Dilworth, Eds., pp. 39–60, ACIAR, Canberra, Australia, 2016.

[24] J. M. Vincent, *A Manual for The Practical Study of Root-Nodule Bacteria*, Oxford: Blackwell Scientific Publications, 1970.

[25] G. W. O'Hara, M. Hungria, P. Woomer, and J. G. Howieson, "Working with rhizobia," in *Working with rhizobia, ACIAR:*, J. G. Howieson and M. J. Dilworth, Eds., pp. 109–124, ACIAR, Canberra, Australia, 2016.

[26] M. Hungria, J. C. Franchini, R. J. Campo et al., "Nitrogen nutrition of soybean in Brazil: Contributions of biological N 2 fixation and N fertilizer to grain yield," *Canadian Journal of Plant Science*, vol. 86, no. 4, pp. 927–939, 2006.

[27] H. G. Thornton, "The influence of the number of nodule bacteria applied to the seed upon nodule formation in legumes," *The Journal of Agricultural Science*, vol. 19, no. 2, pp. 373–381, 1929.

[28] H. N. Patrick and W. L. Lowther, "Influence of the number of rhizobia on the nodulation and establishment of Trifolium ambiguum," *Soil Biology and Biochemistry*, vol. 27, no. 4-5, pp. 717–720, 1995.

[29] S. Martyniuk, J. Orón, and M. Martyniuk, "Interaction between chemical seed dressing and Bradyrhizobium inoculant on lupine seeds," *Botanica. Lithuanica Supplement*, vol. 3, pp. 95–98, 2013.

[30] R. Deaker, E. Hartley, and G. Gemell, "Conditions affecting shelf-life of inoculated legume seed," *Agriculture*, vol. 2, no. 4, pp. 38–51, 2012.

[31] F. J. Temprano, M. Albareda, M. Camacho, A. Daza, C. Santamaría, and D. N. Rodríguez-Navarro, "Survival of several Rhizobium/Bradyrhizobium strains on different inoculant formulations and inoculated seeds," *International Microbiology*, vol. 5, no. 2, pp. 81–86, 2002.

[32] M. Hungria, M. F. Loureiro, I. C. Mendes, R. J. Campo, and P. H. Graham, "Inoculant preparation, production and application," in *Nitrogen Fixation in Agriculture, Forestry, Ecology and the Environment*, W. E. Newton, Ed., pp. 223–254, Springer, Dordrecht, Amsterdam, 2005.

[33] E. Ferreira, M. A. A. Nogueira, J. Fukami, R. B. Conceição, and M. Hungria, "Nova legislação, recomendação de doses de inoculantes e pré-inoculação: riscos ao sucesso da contribuição da fixação biológica do nitrogênio para a cultura da soja," in *Reunião de Pesquisa da Soja na Região Central do Brasil*, vol. 23, São Pedro, Brazil, 2011.

10

Staphylococcus saprophyticus Recovered from Humans, Food, and Recreational Waters in Rio de Janeiro, Brazil

Viviane Santos de Sousa,[1] Ana Paula de Souza da-Silva,[1] Leif Sorenson,[2] Raphael Paiva Paschoal,[1] Renata Fernandes Rabello,[3] Eloiza Helena Campana,[1] Márcia Soares Pinheiro,[3] Lyssa Oliveira Ferreira dos Santos,[1] Natacha Martins,[1] Ana Carolina Nunes Botelho,[1] Renata Cristina Picão,[1] Sérgio Eduardo Longo Fracalanzza,[1] Lee Woodland Riley,[2] George Sensabaugh,[2] and Beatriz Meurer Moreira[1]

[1]*Laboratório de Investigação em Microbiologia Médica, Universidade Federal do Rio de Janeiro, Rio de Janeiro, RJ, Brazil*
[2]*School of Public Health, University of California, Berkeley, Berkeley, CA, USA*
[3]*Instituto Biomédico, Universidade Federal Fluminense, Niterói, RJ, Brazil*

Correspondence should be addressed to Beatriz Meurer Moreira; bmeurer@micro.ufrj.br

Academic Editor: Todd R. Callaway

Staphylococcus saprophyticus is an important agent of urinary tract infection (UTI) in young women, but information about this pathogen in human microbiota and in common environment is lacking. The aim of this study was to characterize *S. saprophyticus* isolates from genitoanal microbiota of 621 pregnant women, 10 minas cheese packs, and five beaches in Rio de Janeiro city and compare PFGE profiles of these isolates with five UTI PFGE clusters described in this city. We investigated 65 *S. saprophyticus* isolates from microbiota, 13 from minas cheese, and 30 from beaches and 32 UTI isolates. Antimicrobial resistance was determined by disk diffusion, MIC by agar dilution, and PCR. Erythromycin-resistance genes *erm*(C), *msr*(A), *msr*(B), *mph*(C), and *lin*(A) were found in 93% of isolates. Trimethoprim-sulfamethoxazole resistance correlated with *dfrG* or *dfrA* genes. Three cefoxitin-resistant isolates carried the *mecA* gene. All isolates obtained from cheese were susceptible to all antimicrobial agents. Six of 10 pregnant women with >1 isolate had monoclonal colonization. Isolates from pregnant women shared 100% similarity with UTI PFGE cluster types A and E obtained almost 10 years previously, suggesting temporal persistence of *S. saprophyticus*. Antimicrobial resistance of beach isolates reflected the profiles of human isolates. Taken together, results indicate a shared source for human and environmental isolates.

1. Introduction

The source or reservoir of *Staphylococcus saprophyticus* for humans is not fully known. This coagulase-negative microorganism is recognized to cause urinary tract infection (UTI) in sexually active young women [1]. Despite the numerous reports of this microorganism in food [2–4], the relationship between these findings and the occurrence of UTI in humans has not been demonstrated [5]. In Brazil, *S. saprophyticus* was described in minas cheese [6], one of the most popular kinds of fresh cheese in the country, and in the water of a polluted river [7]. In addition, reports of *S. saprophyticus* in the marine environment [8] and food derived from fish [9, 10] draw attention to the spread of this microorganism. These may indicate that nonhuman sources of *S. saprophyticus* colonization may include, in addition to food, contact with the marine environment, an unexplored phenomenon.

In a previous report, we described the distribution of pulsed-field gel electrophoresis (PFGE) types in a population of 32 *S. saprophyticus* isolates from community-acquired UTI in Rio de Janeiro from 2005 to 2006 [11]. Clusters of isolates with indistinguishable PFGE patterns were observed among unrelated individuals, indicating possible point sources of this uropathogen. Such putative sources were not investigated in the previous study, and much of the epidemiology of *S. saprophyticus* UTI is yet to be described.

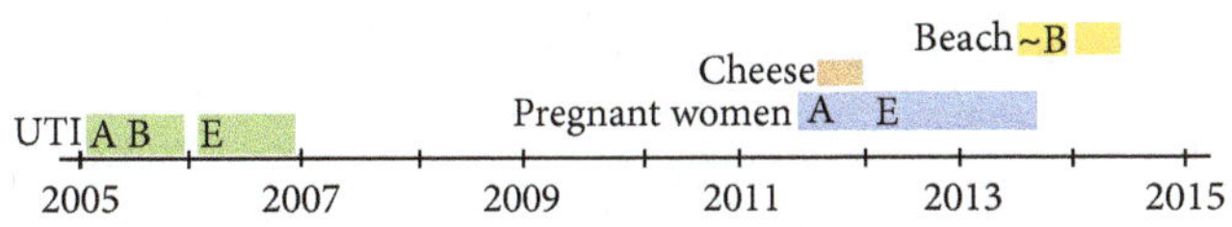

FIGURE 1: Distribution of isolates obtained from different sources in Rio de Janeiro according to time. Capital letters represent the urinary tract infection (UTI) PFGE types. "~B": beach isolate with 82% similarity with UTI type B.

The emergence of antimicrobial resistance in uropathogens is a worldwide concern. The characterization of resistance determinants in UTI pathogens is important for the clinical management of UTI. Although *S. saprophyticus* may carry many resistance genes, most studies include this organism in a coagulase-negative group or deal with low numbers of samples [12–17], and the resistance determinants in clinical isolates are rarely described. Reports of the presence in the urogenital microbiota associated with UTI by this microorganism date from the 1980s [18]; therefore, recent evidence of this colonization in the human microbiota is lacking.

The aim of the present study was to characterize antimicrobial resistance genes in *S. saprophyticus* isolates from UTI cases, the microbiota of pregnant women, and two potential colonization sources: artisanal minas cheese and the waters of five beaches in the city of Rio de Janeiro. In addition, we aimed to compare the PFGE band profiles of these isolates with five uropathogenic *S. saprophyticus* clusters previously described in this city [11].

2. Materials and Methods

2.1. Bacterial Isolation and Identification. We obtained isolates from four sources: UTI [11], pregnant women's microbiota, minas cheese, and beach waters. We considered as suspect of *S. saprophyticus* the colonies from cultures of cheese and the microbiota samples that grew on Mannitol Salt Agar (BD, New Jersey, USA) with 100 μg/mL of novobiocin (Inlab, São Paulo, Brazil) and the colonies from beach water samples that grew on the same selective medium with addition of 0.005% sodium azide [28]. The distribution of isolates according to time (years) and source of detection is shown in Figure 1.

Isolates from UTI were obtained from 32 patients of two walk-in clinics in Rio de Janeiro city during March–November 2005 and March–November 2006 [11]. PFGE typing revealed the presence of five time-based clusters (types A–E) in isolates of patients coming from places spread in the community. One isolate of each cluster type was used in comparisons with PFGE types obtained in the present study. Isolates from pregnant women's microbiota were obtained from vaginal and anal swab specimens stored during surveillance for *Streptococcus agalactiae* colonization. Samples were collected from 621 healthy pregnant women in the 35–37th weeks of gestation, from August/2011 to August/2013; five colonies from vaginal and anal sites of each patient were selected for *S. saprophyticus* screening. Ten minas cheese packs from three different brands were obtained in November 2011 from different stores in eight neighborhoods in Rio de Janeiro city, taken to the lab in original packaging, and coded with roman numbers. Analysis was performed in 25 g samples from each cheese pack as described [29]. After dilutions, the material was seeded in Baird-Parker Agar (BD, New Jersey, USA); up to five colonies of each morphology were selected, isolated, and preserved. In this first step, we obtained 75 colonies from the ten cheese packs. In a second step, colonies were seeded onto *S. saprophyticus* selective medium.

Water samples were obtained in two time points, December 2013 and March 2014 (single-day collection each), from beaches of Botafogo, Copacabana, Flamengo, Ipanema, and Leblon, in Rio de Janeiro city. Water was collected in collaboration with previous study [30] with modifications in the process of obtaining isolates. Briefly, in the first sampling section, water volumes inoculated in each Petri dish were 200 μL without filtering and 10 mL, 20 mL, 50 mL, and 100 mL filtered in 0.22 μm Millipore membranes (Merck, Darmstadt, Germany). In the second sampling section, after observing the high numbers of *S. saprophyticus* isolates obtained only from Leblon beach in the first section, only waters from this beach were sown in the five Petri dishes with different inoculum volumes. Waters from the other four beaches required just the 50 mL membrane filtration. Up to 16 colonies in each Petri dish were chosen and named according to the source and morphology.

All colonies were identified by biochemical tests [31] and matrix-assisted laser desorption ionization time of flight mass spectrometry (MALDI-TOF), using MALDI Biotyper 3.1 software (Bruker Daltonics) [32]. *S. saprophyticus* American Type Culture Collection (ATCC) 15305 was the control test. Isolates identified as *S. saprophyticus* were stored in suspensions containing 10% (w/v) skimmed milk and 10% glycerol (v/v) in a freezer at −20°C.

2.2. Antimicrobial Susceptibility Test. The following antimicrobial agents (OXOID, Hampshire, England) were tested by disk diffusion method according to the Clinical and Laboratory Standards Institute (CLSI) [33, 34]: ciprofloxacin (5 μg), clindamycin (2 μg), erythromycin (15 μg), gentamicin (10 μg), levofloxacin (5 μg), linezolid (30 μg), nitrofurantoin (300 μg), norfloxacin (10 μg), penicillin (10 UI), and trimethoprim-sulfamethoxazole (1,25/23,75 μg). Cefoxitin (30 μg) disks predicted oxacillin resistance. Clindamycin inducible resistance was determined by the disk diffusion *D*-test. For antimicrobials with any isolate showing resistance in disk diffusion tests, minimum inhibitory concentrations (MIC) [33, 34] were determined by agar dilution.

2.3. Detection of Antimicrobial Resistance Genes. Resistance determinants investigated and primers used in this study are described in Table 1. We screened for the presence of genes related to antimicrobial agents showing resistance in disk diffusion tests, including genes described in *S. saprophyticus* or other *Staphylococcus* species. We confirmed one PCR product of each gene by sequencing and comparisons in GenBank with BLAST. Sequencing was additionally performed for some of the UTI isolates to verify the presence

Table 1: Primers used for screening of resistance determinants in *Staphylococcus saprophyticus*.

Antimicrobial	Gene	Sequence 5′-3′	Size (bp)	Ref.
Gentamicin	*aac(6′)-Ie+aph(2″)*	CCAAGAGCAATAAGGGCATACC CACACTATCATAACCATCACCG	347	[19]
	ant(4′)-Ia	CTGCTAAATCGGTAGAAGC CAGACCAATCAACATGGCACC	172	
	aph(3′)-IIIa	CTGATCGAAAAATACCGCTGC TCATACTCTTCCGAGCAAAGG	268	
Erythromycin	*erm*(A)	TATCTTATCGTTGAGAAGGGATT CTACACTTGGCTTAGGATGAAA	533	[20]
	erm(B)	CTATCTGATTGTTGAAGAAGGATT GTTTACTCTTGGTTTAGGATGAAA	359	[21]
	erm(C)	CTTGTTGATCACGATAATTTCC ATCTTTTAGCAAACCCGTATTC	190	[22]
	msr(A)	GGCACAATAAGAGTGTTTAAAGG AAGTTATATCATGAATAGATTGTCCTGTT	940	[21]
	msr(B)	TATGATATCCATAATAATTATCCAATC AAGTTATATCATGAATAGATTGTCCTGTT	595	
	lin(A)	GGTGGCTGGGGGGTAGATGTATTAACTGG GCTTCTTTTGAAATACATGGTATTTTTCGATC	323	
	mph(C)	GAGACTACCAAGAAGACCTGACG CATACGCCGATTCTCCTGAT	722	[23]
	mef(A)	AGTATCATTAATCACTAGTGC TTCTTCTGGTACAAAAGTGG	348	[24]
Oxacillin	*mecA*	GTGAAGATATACCAAGTGATT ATGCGCTATAGATTGAAAGGAT	147	[25]
Penicillin[a]	*blaZ*	ATGTAATTCAAACAGTTCACATGCC ATAGGTTCAGATTGGCCCTTAGG	701	[26]
	pbp1A			This study
	Fragment 1	GGCGAACATTCCACAGTGTTGACT GGGTGTCAGTAAGGCGTTCAAACCA	627	
	Fragment 2	TCTCACCGTCACCAATAACGATTGG TACAGGCGGATCGCTTGGG	680	
	Fragment 3	TCAGCTACTTGTGCGGTACCTGT TTTGCAGTAGTGATGGACGCACA	758	
	Fragment 4	GCCGCCTGTAGCTTCACCGT TGAAAGGTGCTATGGGCGTAGAGAA	650	
	Fragment 5	TGTGCGTCCATCACTACTGCAAA GGGGCAGTCCTACTCATCCTTGGA	788	
	Fragment 6	GCCGACATATCAACAACGGTAGCCA AGACCAAAACCGTAAAACGCAAGGT	683	
	pbp2A			This study
	Fragment 1	ACGTGCTGCACTTTGTTGGTTACT TGCCGGAAACGATGCACCAA	775	
	Fragment 2	TTGGTGCATCGTTTCCGGCA GCGGAATCTGCTTATCTTGCTGGT	763	
	Fragment 3	CGCGATCTTCAGATGAACGTTGGA GCGAGGACCGCGTATGACGG	869	

Table 1: Continued.

Antimicrobial	Gene	Sequence 5′-3′	Size (bp)	Ref.
	pbp3A			This study
	Fragment 1	GACGATTATGACGGCCTTTT TGGATCAAGTGCAGAACCAG	702	
	Fragment 2	CCTGACAGTCAACGAAGCTG AAGATCATCGCCACGTGAAC	769	
	Fragment 3	GCATTTTCGGTGACGTTTCT CGCTCGTAATATCGGTTGGT	714	
	Fragment 4	TGATTGATGAGCCGCTTAAA ATGGCTCAGGTACTGGTTGG	744	
	Fragment 5	GGCATTTAACGAAAAAGATGGA TACGTTTACGCGCATGCTAA	410	
	pbp4A			This study
	Fragment 1	AATCCAGCGACAAACATCCCATTCA AAGCGCAGCAGCATTACTAGAGTT	835	
	Fragment 2	GGCGCCATTAATGTTTCGCAAACA CGCCGGCGCCCATGATAACT	718	
	Fragment 3	GGTGCCGATATGAGCCTAGAAGGT TGGGCCATGATTGGGAAGGCG	859	
Trimethoprim	*dfrD*	CCCTGCTATTAAAGCACC CATGACCAGATAACTC	606	[27]
	dfrG	TGCTGCGATGGATAAGAA TGGGCAAATACCTCATTCC	405	[26]
	dfrK	CAAGAGATAAGGGGTTCAGC ACAGATACTTCGTTCCACTC	229	
	dfrS1(dfrA)	CACTTGTAATGGCACGGAAA CGAATGTGTATGGTGGAAAG	270	
	dfr[b]	AATGGACATCGGTTGGGTTGCCT CGCACCGGATTCAAATGTCTCGC	484	This study
Norfloxacin	*gyrA*	CGAGTGAGATGCGCGAGTCATTCTT ACGTTGACGACCGCCACCAC	731	This study
	parC	ACGTTCGTGATGGGCTCAAACCT ACGTAATCCAGTACGGTCTGTCTCA	797	

Ref.: reference. [a]PCR primers were designed to target smaller *pbp* 1-4 gene fragments and allow for the respective whole gene sequence amplification; [b]chromosomal *dihydrofolate reductase*.

of mutations in *gyrA*, *par*C, and *pbp* (1 to 4) and the chromosomal dihydrofolate reductase. The *gyrA* and *par*C genes were characterized in the quinolone resistant isolate and four susceptible isolates randomly selected in order to observe specific amino acid substitutions [35, 36]. For the *pbp* genes, overlapping primer sequences were designed as to cover smaller fragments, using the *S. saprophyticus* ATCC 15305 as a template, to allow the determination of the entire gene sequences of the 32 UTI isolates.

2.4. Pulsed-Field Gel Electrophoresis. PFGE was performed with *Sma*I (20U, New England, Ipswich, USA), as described [37] with the following modifications: (i) bacterial suspensions were prepared in 500 μL PIV buffer [NaCl 1 M and Tris HCl 10 mM (pH 7,6)] so that an aliquot of the 1 : 10 diluted suspension in distilled water exhibited an OD of 1.5 in a spectrophotometer with 600 nm wavelength; (ii) after solidification, blocks were transferred to lysing solution (Tris HCl 6 mM [pH 7,6], NaCl 1 M, EDTA 100 mM [pH 7,5], Brij 58 at 0.5%, sodium lauroyl sarcosinate at 0.5%, 0.5 mg/ml of lysozyme [Sigma], and 0.05 mg/ml lysostaphin [Sigma]) for 18–24 h at 37°C, under gentle agitation. Analysis of band profiles was performed with BioNumerics version 7.1 (Applied Maths, Kortrijk, Belgium) and dendrograms were built with Dice coefficient, 1.0% tolerance, using UPGMA. Clusters were named with capital letters (A–E) when belonging to the ITU genotypes. Clusters included isolates with ≥90% similarity or with one band difference.

2.5. Data Analysis. Data were analyzed with the free-access program OpenEpi version 3.03 [38]. Differences between categorical variables were analyzed by chi-square or Fisher's

TABLE 2: Distribution of *Staphylococcus saprophyticus* isolates obtained from pregnant women's microbiota, beaches, and minas cheese in Rio de Janeiro city.

Sample source	Number and % of *S. saprophyticus* isolates
Microbiota (23 positive women)	
Vaginal (3)	11, 17% (1–5 per woman)
Anal (16)	27, 41.5% (1–4 per woman)
Both (4)	27, 41.5% (3–9 per woman)
Total	65
Beach water (10 samples)	
Botafogo (2)	6, 20% (1–5 per sample)
Copacabana (2)	1, 3% (0-1 per sample)
Flamengo (2)	1, 3%
Ipanema (2)	1, 3%
Leblon (2)	21*, 70%
Total	30
Minas cheese (10 cheeses)	
Brand A (4)	9, 69.2% (0–9 per cheese)
Brand B (3)	4, 30.7% (0–4 per cheese)
Brand C (3)	0
Total	13, 100%
Total of *S. saprophyticus* isolates	108

*The comparison among *S. saprophyticus* obtained from each beach by Fisher's exact test showed that Leblon had significantly higher numbers of isolates than Botafogo, Copacabana, Flamengo, and Ipanema ($p \leq 0.01$).

exact test. A p value ≤ 0.05 was defined as statistically significant.

3. Results and Discussion

3.1. Identification of S. saprophyticus from UTI, Pregnant Women's Microbiota, Minas Cheese, and Beach Water. The *S. saprophyticus* study collection included 32 isolates from UTI [11], 65 isolates from 23 (4% of 621) pregnant women, 30 isolates from waters of five beaches, and 13 isolates from two minas cheese packs (Table 2).

The finding of 4% pregnant women colonized with *S. saprophyticus* was below the colonization occurrence described in the three papers previously published, which reported (i) 11% of 123 women from a Kidney Clinic [18], (ii) 14% of 44 UTI cases [39], and (iii) 7% of 276 women in routine gynecological care [40]. However, our study population is higher (621 women), and the 95% confidence interval (CI) is the lowest among the others: CI of 0.02–0.05 from our study, (i) CI of 0.009–0.09, (ii) CI of 0.06–0.19, and (iii) CI of 0.03–0.08. Our finding is robust and suggests a reliable result. The study is refined by genotypic characterization of those isolates and reflects the presence of genes encoding antimicrobial resistance circulating in the human microbiota.

The 30 isolates identified as *S. saprophyticus* on the five beaches were obtained in both periods: 11 (37%) in December 2013 and 19 (63%) in March 2014. In recent years, *S. saprophyticus* has been described in aquatic environments taken as polluted, such as the marine environment in Lebanon [8] and river flood in Porto Alegre (Brazil) [7]. Here, we observed a high number ($p \leq 0.01$) of *S. saprophyticus* isolates in Leblon beach, known to receive waters periodically from two fluvial channels carrying untreated domestic sewage. In periods of rain, the channels' floodgates are opened several times to decrease the water level, leading sewage into the sea. There were records of heavy rains during the days prior to sampling, indicating that this practice had recently happened; contamination of seawaters with untreated domestic sewage could explain the high numbers of *S. saprophyticus* in Leblon. In addition, pregnant women and beach water isolates shared similar resistance gene profiles, indicating a link between these sources.

MALDI-TOF MS confirmed identification of all 140 isolates with scores >2 in 84% and <2 in 16%. The scores were consistent with the biochemical tests used to identify *S. saprophyticus*, including values lower than the threshold of 2.0 for species identification (≥1.844), as described by others [32]. For coagulase-negatives, in general, this technique is capable of identifying species and subspecies with good performance [41].

The use of selective culture medium was instrumental. Mannitol Salt agar inhibited Gram-negative bacteria, and novobiocin improved the chances to obtain *S. saprophyticus* from cheese and the genital-anal microbiota; moreover, the addition of sodium azide [28] in this medium was essential to isolate *S. saprophyticus* from beach waters. Sodium azide shows bacteriostatic effect to most Gram-negative bacteria; Gram-positive bacteria are usually resistant to this compound [42].

3.2. Antimicrobial Resistance Phenotypes and Genes. Disk diffusion and MIC tests were congruent for susceptibility category. Table 3 shows the analytical findings regarding phenotypic resistance and detection of antimicrobial resistance genes. Comparison of resistance genes present in isolates from the four collections showed a higher concentration of these genes in isolates from pregnant women and beach water (Table 4). This study is the first to report the resistance determinants *dfrA* and *dfrG* and *msr*(B) and *mph*(C) in *S. saprophyticus* obtained from humans.

Trimethoprim-sulfamethoxazole resistance, in 17% ($n = 17$) of isolates, correlated with the *dfrG* or *dfrA* genes, present in all resistant isolates, and only 10% ($n = 10$) of susceptible isolates ($p < 0.0001$). The sequence of the chromosomal *dfr* gene in all isolates was identical to that of trimethoprim susceptible ATCC 15305 strain, and the plasmid genes *dfrG* and *dfrA* shared 100% nucleotide identity with the respective variants described in other staphylococcal species including *S. aureus* (*dfrA*: ACSN01000071.1; *dfrG*: ABFA0100044.1), *S. epidermidis* (*dfrA*: NC_002976.3), and *S. pseudintermedius* (*dfrG*: NC_014925.1).

TABLE 3: Antimicrobial susceptibility and resistance determinants in 98 *Staphylococcus saprophyticus* isolates.

Antimicrobial agent	Disk diffusion halo (mm)/interpretation	MIC (μg/mL)/interpretation	Resistance determinant: Gene	Resistance determinant: Number (%) of isolates
Sulfamethoxazole-trimethoprim	6–12/R	4–64/R	*dfrG*	15 (15.3)
	6/R	32/R	*dfrA*	1 (1)
	11/I	8/R	*dfrG*	1 (1)
	18–35/S	0.125–4/S	*dfrG*	10 (10.2)
	29–37/S	0.125–2/S	None	71 (72.4)
MLS_B				
Erythromycin	6/R	4–≥64/R	*erm*(C)	7 (7.1)
	6/R	32–>64/R	*erm*(C), *msr*(B)	2 (2)
	6/R	>64/R	*erm*(C), *mph*(C)	2 (2)
	6/R	≥64/R	*msr*(A), *msr*(B)	1 (1)
	6/R	≥64/R	*erm*(C), *msr*(A), *msr*(B)	9 (9.1)
	6–10/R	32–≥64/R	*erm*(C), *msr*(A), *msr*(B), *mph*(C)	10 (10.2)
	19/I	0.125/S	*erm*(C)	1 (1)
	22–33/S	0.125–1/S	*erm*(C)	44 (44.8)
	30/S	1/S	*msr*(A)	1 (1)
	28–32/S	0.25–1/S	*erm*(C), *msr*(A)	7 (7.1)
	28–34/S	1/S	*erm*(C), *msr*(B)	3 (3)
	27–30/S	1/S	*erm*(C), *mph*(C)	3 (3)
	28–30/S	1/S	*erm*(C), *msr*(A), *msr*(B)	1 (1)
	26–32/S	0.125–1/S	None	7 (7.1)
Clindamycin	30–33/IR[a]	0.125–2/S	*erm*(C)	6 (6.1)
	34/IR[a]	0.25/S	*erm*(C), *msr*(B)	1 (1)
	28–30/IR[a]	0.25/S	*erm*(C), *msr*(A), *msr*(B)	2 (2)
	28–32/IR[a]	0.25/S	*erm*(C), *msr*(A), *msr*(B), *mph*(C)	4 (4)
	25/S	0.25/S	*lin*(A), *erm*(C), *msr*(A), *msr*(B), *mph*(C)	1 (1)
	25/S	0.25/S	*lin*(A), *erm*(C)	1 (1)
	25–35/S	0.125–0.25/S	None	83 (84.6)
Oxacillin	6–9/R	32–>64/R	*mecA*	3 (3)
	12/R	1/R	*mecA*	1 (1)
	20–22/R	0.5–1/R	None	5 (5.1)
	23–35/S	0.5–2/R	None	84 (85.7)
	29–44/S	0.125–0.25/S	None	5 (5.1)
Penicillin	6/R	8–16/R	*mecA*[b]	3 (3)
	15/R	2/R	None	1 (1)
	26–28/R	0.25–1/R	None	2 (2)
	22/R	0.125/S	*mecA*[b]	1 (1)
	17–27/R	0.125/S	None	9 (9.1)
	30–45/S	≤0.125/S	None	82 (83.6)
Gentamicin	14/I	8/I	None	1 (1)
	15/S	8/I	None	1 (1)
	15–37/S	0.125/S	None	96 (97.9)
Norfloxacin	15/I	0.125/S	None	1 (1)
	17–33/S	0.125/S	None	97 (98.9)

The 98 isolates in this table were selected from the 140 obtained to include only one isolate per patient. MLS_B: *macrolide-lincosamide-streptogramin* B; R: resistant; I: intermediate resistance; S: susceptible; IR: inducible resistance; MIC: minimum inhibitory concentration. [a]All inducible resistance detected in *D*-test. [b]Same isolates resistant to oxacillin.

TABLE 4: Distribution of the antimicrobial resistance genes by sample source among the 98 *Staphylococcus saprophyticus* isolates.

Gene	Number and % in each collection				
	UTI/2005-2006 $N = 32$	Pregnant women/2011–2013 $N = 23$	Beach water/2013-2014 $N = 30$	Cheese/2011 $N = 13$	Total $N = 98$
MLS_B					
ermC	24 (75)	6 (26)	10 (33.3)	11 (84.6)	78 (79.5)
msrA	0 (0)	0 (0)	1 (3.3)	0	34 (34.6)
msrB	0 (0)	0 (0)	0 (0)	0	26 (26.5)
mphC	0 (0)	0 (0)	0 (0)	0	18 (18.3)
ermC, msrA	0	4 (17.3)	3 (10)	0	7 (7.1)
ermC, msrB	0	2 (8.6)	1 (3.3)	0	3 (3)
ermC, linA	0	0	1 (3.3)	0	1 (1)
ermC, mphC	0	2 (8.6)	3 (10)	0	5 (5.1)
msrA, msrB	0	0	1 (3.3)	0	1 (1)
ermC, msrA, msrB	5 (16)	2 (8.6)	3 (10)	0	10 (10.2)
ermC, msrA, mphC	0	2 (8.6)	0	0	2 (2)
ermC, msrA, msrB, mphC	0	4 (17.3)	5 (16.6)	0	10 (10.2)
ermC, msrA, msrB, mphC, linA	0	0	1 (3.3)	0	1 (1)
>1 gene[a]	5 (15.6)	16 (69.5)	20 (20.4)	0	41 (41.8)
Other					
dfrG	16 (50)	7 (30.4)	3 (10)	0	26 (26.5)
dfrA	1 (3.1)	0 (0)	0 (0)	0	1 (1)
mecA	0 (0)	2 (8.6)	2 (6.6)	0	4 (4)

MLS_B: macrolide-lincosamide-streptogramin B. [a]$p < 0.001$ for comparison of >1 gene present among isolates from UTI or cheese and pregnant women and from UTI or cheese and beach water.

Although MLS_B antimicrobial agents are not recommended for UTI treatment [43], several genes that encode resistance to these agents were reported in *S. saprophyticus*. The most frequently described in this species is *erm*(C) [13, 14, 23, 44], almost universally present among our UTI isolates (n = 29/91%; Table 4). MLS_B resistance genes were present in resistant and susceptible isolates alike (Table 3). The *erm*(C) gene shared 100% nucleotide identity with those described in *S. aureus* (NC_007792.1) and *S. haemolyticus* (NC_007170.1). Both *msr*(A) and *msr*(B) genes shared 99% of nucleotide identity with *S. aureus* (CP002141.1) and *S. xylosus* (M81802.1), respectively. The *mph*(C) gene had 99% identity with those described in *S. aureus* (CP007177.1) and *S. epidermidis* (GQ900458.1).

In the more recent collections, a combination of genes replaced the *erm*(C) gene as a sole MLS_B resistance determinant, a difference certainly due to the time lag between strains' isolation periods (2005-2006 for UTI and 2011–2013 for other isolates) (Table 4). Concerning clindamycin resistance, only inducible phenotype was observed in 13% of isolates (n = 13). Resistant isolates had different combinations of *erm*(C), *msr*(A), *msr*(B), *mph*(C), and *lin*(A). The *lin*(A) gene shared 100% and 99% identity with the genes already described in *S. aureus* (EU350088.1) and *S. haemolyticus* (M14039.1), respectively.

β-Lactam resistance was rarely supported by molecular mechanism (Table 3). Oxacillin resistance was confirmed in only four isolates (4%) with the *mecA* gene, though 9% of isolates were resistant by disk diffusion and 93% by MIC determination. We observed 89 non-*mecA* isolates with oxacillin-resistant breakpoints, though with low MIC values (0.5 μg/mL–2 μg/mL). It has been suggested that the most appropriate breakpoints to cefoxitin disk diffusion and oxacillin MIC should be ≤19 mm and ≥16 μg/mL, respectively, to indicate *mecA*-positive *S. saprophyticus* [45]. With these criteria, all 89 *S. saprophyticus* isolates would be classified as susceptible. However, this MIC breakpoint [45] would be inappropriate for one of the *mecA*-positive isolates from beach waters, with MIC 1 μg/ml, confirmed in triplicate. On the other hand, the suggested cefoxitin inhibition zone <19 mm would fit, since this *mecA*-positive isolate had a 12 mm halo.

Penicillin resistance by MIC correlated with the presence of *mecA* gene in only three isolates (Table 3). Other three penicillin-resistant isolates (MIC ≤ 1 μg/mL) had no resistance determinant, including *blaZ* or PBP nucleotide sequence polymorphisms that distinguished resistant from susceptible isolates. By comparing PBP sequences of the 32 UTI isolates with those of *S. saprophyticus* ATCC 15305, we found only substitutions unrelated to phenotypic penicillin resistance in the PBP1 deduced amino acid sequence: V625M in 31 isolates and E731K in all 32 isolates. The contribution of the two substitutions in the deduced amino acid sequence of PBP1 to antimicrobial resistance is yet to be unraveled.

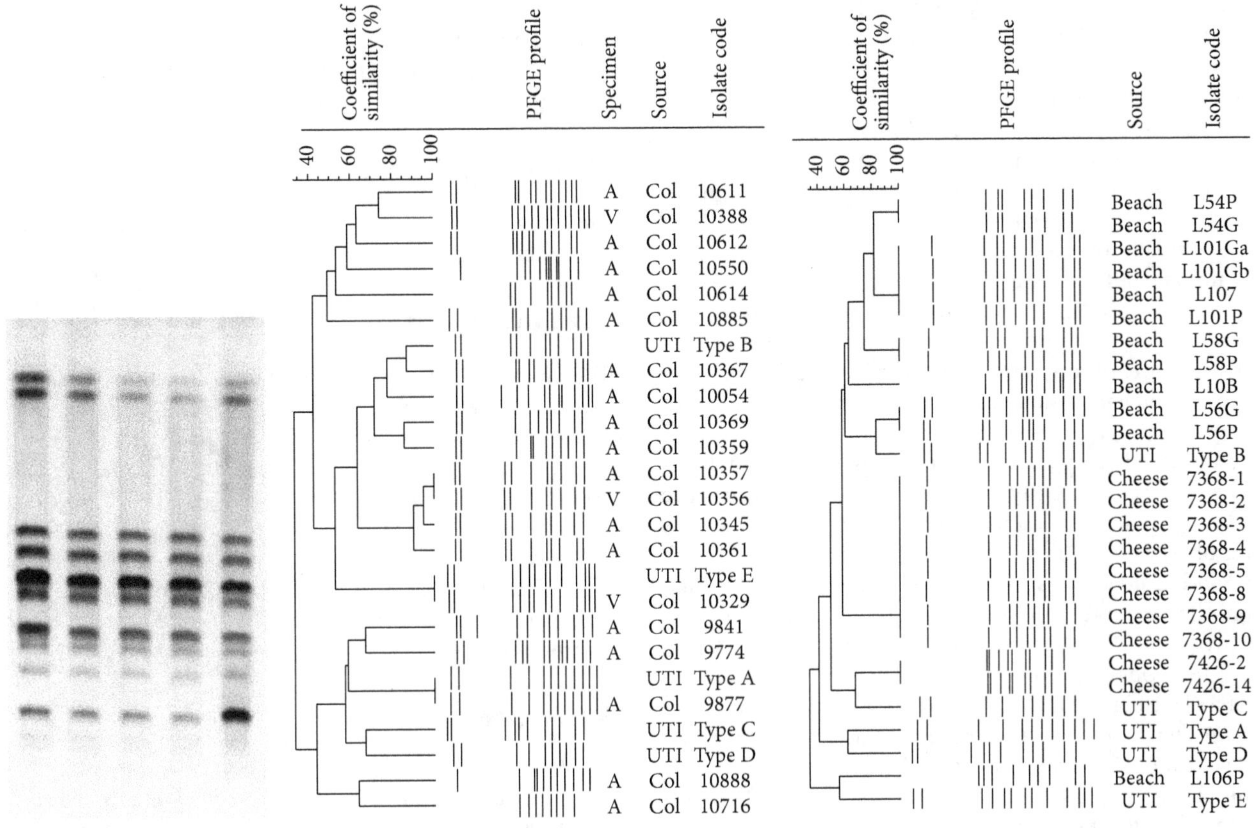

FIGURE 2: Pulsed-field gel electrophoresis profiles of the *Staphylococcus saprophyticus* isolates obtained from pregnant women, minas cheese, and beach waters of Rio de Janeiro. PFGE band patterns of (a) five isolates obtained from the microbiota of a single pregnant women; (b) comparison of the five uropathogenic PFGE clusters with one isolate per pregnant woman (n = 21) and (c) 10 isolates from minas cheese and 12 from beach waters. "A": anal; "V": vaginal.

Two of the *mecA*-positive isolates were obtained from the microbiota of pregnant women (Table 4). The presence of this isolate may indicate risk of UTI by oxacillin-resistant *S. saprophyticus* to this group. Although *S. saprophyticus* is uncommon UTI agent in pregnant women [46], treatment of such infections in this population is performed with first-generation cephalosporins, such as cephalexin [47], a β-lactam without activity for *mecA*-carrying isolates. In addition, colonized pregnant women may represent a risk for transmission of resistant microorganisms to the newborn [48], as already shown to occur with resistant enterobacteria [49, 50].

Regarding other drugs, even though one isolate was intermediately resistant to norfloxacin, no *gyrA* and *parC* gene mutations were found (Table 3). It is possible that the result of the disk diffusion test was false-positive, since the MIC was susceptible and the corresponding mutations were not found with the resistance phenotype. In the case of the two isolates with intermediate resistance to gentamicin (Table 3), it was not possible to elucidate mechanisms of resistance through the aminoglycoside-modifying enzyme-encoding genes studied. This suggests that additional mechanisms of resistance may be involved in the expression of this phenotype for this microorganism.

3.3. PFGE Typing. We performed typing to investigate if multiple isolates from colonization in microbiota would be clonal and to compare our isolates from the multiple sources with the five PFGE clusters we detected previously.

From the total 23 pregnant women with *S. saprophyticus*, isolates from 21 were typeable. This collection included 43 isolates from 10 women with more than one isolate and one from each of 11 women with a single isolate (54 isolates in total). PFGE band profiles from a single individual were mostly clonal: six women had profiles 90%–100% similar (Figure 2(a)), and the other four were colonized with isolates 76%–100% similar (three band differences at most). Our strategy to type more than one colony per women to study the microbiota may be helpful for future epidemiological studies, as we demonstrated that *S. saprophyticus* isolates from the same person are usually clonal.

Comparison of PFGE profiles from different women (Figure 2(b)) showed that four of them shared profiles >90% similar, and comparison with the previously detected clones revealed that each of two women had isolates 100% similar with UTI type A and others with UTI type E. This result indicates that *S. saprophyticus* clones can stay successfully established in the environment and among people; UTI and pregnant women's microbiota PFGE types were obtained almost a decade apart. Similar results were previously observed in Germany but only among UTI isolates [51].

Of the 10 typeable isolates from cheese, only those from the same cheese pack were clonal (100% similar, Figure 2(c)). Our results indicate single original contamination in the cheese sample, but this may not be the rule, as another study in France showed polyclonal contamination in this same food [3]. It is possible that a large number of cheese packs from more brands would show a different clonal composition per cheese as the study in France. More studies regarding this issue are needed to support this food as a possible source of UTI. Despite the difference between the UTI and the cheese types in the present study, this food could be a brief point source of *S. saprophyticus* to humans.

We obtained 12 typeable isolates from beach water (all from Leblon beach). Only isolates from the same culture plate were clonal (100% similar, Figure 2(c)). Although we did not observe clusters with >90% similarity among beach waters and UTI isolates, one *S. saprophyticus* from beach and another from UTI (type B pattern) had similarity of 82% (two band differences). Further studies are needed to investigate a possible connection among beach water and human UTI isolates. Indeed, the association between ITU by *S. saprophyticus* and swimming activities was described long ago [52]. However, only a technique such as whole genome sequencing could propose the transmission direction of strains from food to humans.

PFGE performance was unsuccessful for 18 *S. saprophyticus* isolates from beach waters. The reason for such failure may relate to bacterial survival strategy in saturated environment such as beach waters. Possible changes in *S. saprophyticus* cell wall could differentiate isolates in beach waters. Such differences were demonstrated in marine *Pseudomonas aeruginosa* [53]. Indeed, PFGE protocols are validated for clinical isolates and may be less suitable for environmental organisms.

Comparison among PFGE profiles of all isolates showed that the similarity detected in smaller groups of isolates remained (data not shown).

4. Conclusions

The presence of similar genotypic characteristics (high presence of resistance genes) in isolates from microbiota and beach waters indicates that *S. saprophyticus* may transit between both sites. Likewise, the two isolates with *mecA* gene and full resistance to oxacillin in beach waters alert to the possibility of transmission of this highly relevant resistance determinant to other humans. On the other hand, the finding of *S. saprophyticus* fully susceptible to antimicrobial agents in minas cheese suggests that these isolates represent microorganisms in their wild state in the environment.

Resistance to oxacillin and gentamicin was little related to genes screened in this study. Additional investigations are needed to assess the discrepancies observed here. The unconfirmed resistance to norfloxacin by MIC or mutations in QRDR regions indicated that the disk diffusion test was inaccurate.

The PFGE association of isolates from microbiota with 100% similarity, compared to two PFGE UTI cluster types obtained one decade apart, shows long-term persistence of uropathogenic types.

Conflicts of Interest

The authors declare that there are no conflicts of interest regarding the publication of this paper.

Acknowledgments

The authors thank Ana Maria Nunes Botelho, Maria Cícera Carvalho, and Agnes Marie de Sá Figueiredo for the genotypic screening of the cefoxitin-resistant isolates. This research was supported by Conselho Nacional de Desenvolvimento Científico e Tecnológico (CNPq), Coordenação de Aperfeiçoamento de Pessoal de Nível Superior (CAPES), Fundação de Amparo à Pesquisa do Estado do Rio de Janeiro (FAPERJ) of Brazil, and the Fogarty International Program in Global Infectious Diseases (TW006563) of the National Institutes of Health.

References

[1] A. Eriksson, C. G. Giske, and A. Ternhag, "The relative importance of *Staphylococcus saprophyticus* as a urinary tract pathogen: distribution of bacteria among urinary samples analysed during 1 year at a major Swedish laboratory," *Acta Pathologica, Microbiologica, et Immunologica Scandinavica*, vol. 121, no. 1, pp. 72–78, 2013.

[2] J. C. Soares, M. R. Marques, F. K. Tavaria, J. O. Pereira, F. X. Malcata, and M. M. Pintado, "Biodiversity and characterization of *Staphylococcus* species isolated from a small manufacturing dairy plant in Portugal," *International Journal of Food Microbiology*, vol. 146, no. 2, pp. 123–129, 2011.

[3] E. Coton, M.-H. Desmonts, S. Leroy et al., "Biodiversity of coagulase-negative *Staphylococci* in french cheeses, dry fermented sausages, processing environments and clinical samples," *International Journal of Food Microbiology*, vol. 137, no. 2-3, pp. 221–229, 2010.

[4] A. M. M. Abd-Elall, M. E. M. Mohamed, and M. A. I. Awadallah, "Potential airborne microbial hazards for workers on dairy and beef cattle farms in Egypt," *Veterinaria Italiana*, vol. 45, no. 2, pp. 275–285, 2009.

[5] S. Leroy, P. Giammarinaro, J.-P. Chacornac, I. Lebert, and R. Talon, "Biodiversity of indigenous staphylococci of naturally fermented dry sausages and manufacturing environments of small-scale processing units," *Food Microbiology*, vol. 27, no. 2, pp. 294–301, 2010.

[6] V. L. M. Rall, J. M. Sforcin, M. F. R. de Deus et al., "Polymerase chain reaction detection of enterotoxins genes in coagulase-negative staphylococci isolated from Brazilian minas cheese," *Foodborne Pathogens and Disease*, vol. 7, no. 9, pp. 1121–1123, 2010.

[7] A. P. Basso, P. D. Martins, G. Nachtigall, S. van der Sand, T. M. de Moura, and A. P. G. Frazzon, "Antibiotic resistance and enterotoxin genes in *Staphylococcus* sp. Isolates from polluted water in southern Brazil," *Anais da Academia Brasileira de Ciencias*, vol. 86, no. 4, pp. 1813–1820, 2014.

[8] S. Harakeh, H. Yassine, S. Hajjar, and M. El-Fadel, "Isolates of *Staphylococcus aureus* and *saprophyticus* resistant to antimicrobials isolated from the Lebanese aquatic environment," *Marine Pollution Bulletin*, vol. 52, no. 8, pp. 912–919, 2006.

[9] D. Sergelidis, A. Abrahim, T. Papadopoulos et al., "Isolation of methicillin-resistant *Staphylococcus* spp. from ready-to-eat fish products," *Letters in Applied Microbiology*, vol. 59, no. 5, pp. 500–506, 2014.

[10] B.-S. Kim, C.-T. Kim, B. H. Park et al., "Draft genome sequence of *Staphylococcus saprophyticus* subsp. *saprophyticus* M1-1, isolated from the gills of a Korean rockfish, sebastes schlegeli hilgendorf, after high hydrostatic pressure processing," *Journal of Bacteriology*, vol. 194, no. 16, pp. 4441-4442, 2012.

[11] V. S. de Sousa, R. F. Rabello, R. C. D. S. Dias et al., "Time-based distribution of *Staphylococcus saprophyticus* pulsed field gel-electrophoresis clusters in community-acquired urinary tract infections," *Memorias do Instituto Oswaldo Cruz*, vol. 108, no. 1, pp. 73–76, 2013.

[12] M. Sousa, N. Silva, G. Igrejas et al., "Antimicrobial resistance determinants in *Staphylococcus* spp. recovered from birds of prey in Portugal," *Veterinary Microbiology*, vol. 171, no. 3-4, pp. 436–440, 2014.

[13] M. A. Fernández-Fuentes, H. Abriouel, E. Ortega Morente, R. Pérez Pulido, and A. Gálvez, "Genetic determinants of antimicrobial resistance in Gram positive bacteria from organic foods," *International Journal of Food Microbiology*, vol. 172, pp. 49–56, 2014.

[14] A. Le Bouter, R. Leclercq, and V. Cattoir, "Molecular basis of resistance to macrolides, lincosamides and streptogramins in *Staphylococcus saprophyticus* clinical isolates," *International Journal of Antimicrobial Agents*, vol. 37, no. 2, pp. 118–123, 2011.

[15] O. C. Sampimon, T. J. G. M. Lam, D. J. Mevius, Y. H. Schukken, and R. N. Zadoks, "Antimicrobial susceptibility of coagulase-negative staphylococci isolated from bovine milk samples," *Veterinary Microbiology*, vol. 150, no. 1-2, pp. 173–179, 2011.

[16] B. Söderquist and C. Berglund, "Methicillin-resistant *Staphylococcus saprophyticus* in Sweden carries various types of staphylococcal cassette chromosome *mec* (SCC*mec*)," *Clinical Microbiology and Infection*, vol. 15, no. 12, pp. 1176–1178, 2009.

[17] D. T. Dubin, J. E. Fitzgibbon, M. D. Nahvi, and J. F. John, "Topoisomerase sequences of coagulase-negative staphylococcal isolates resistant to ciprofloxacin or trovafloxacin," *Antimicrobial Agents and Chemotherapy*, vol. 43, no. 7, pp. 1631–1637, 1999.

[18] P. A. Jordan, A. Iravani, G. A. Richard, and H. Baer, "Urinary tract infection caused by *Staphylococcus saprophyticus*," *Journal of Infectious Diseases*, vol. 142, no. 4, pp. 510–515, 1980.

[19] F.-J. Schmitz, A. C. Fluit, M. Gondolf et al., "The prevalence of aminoglycoside resistance and corresponding resistance genes in clinical isolates of staphylococci from 19 European hospitals," *Journal of Antimicrobial Chemotherapy*, vol. 43, no. 2, pp. 253–259, 1999.

[20] M. C. P. Nguyen, P.-L. Woerther, M. Bouvet, A. Andremont, R. Leclercq, and A. Canu, "*Escherichia coli* as reservoir for macrolide resistance genes," *Emerging Infectious Diseases*, vol. 15, no. 10, pp. 1648–1650, 2009.

[21] G. Lina, A. Quaglia, M.-E. Reverdy, R. Leclercq, F. Vandenesch, and J. Etienne, "Distribution of genes encoding resistance to macrolides, lincosamides, and streptogramins among staphylococci," *Antimicrobial Agents and Chemotherapy*, vol. 43, no. 5, pp. 1062–1066, 1999.

[22] F. Martineau, F. J. Picard, N. Lansac et al., "Correlation between the resistance genotype determined by multiplex PCR assays and the antibiotic susceptibility patterns of *Staphylococcus aureus* and *Staphylococcus epidermidis*," *Antimicrobial Agents and Chemotherapy*, vol. 44, no. 2, pp. 231–238, 2000.

[23] P. Lüthje and S. Schwarz, "Antimicrobial resistance of coagulase-negative staphylococci from bovine subclinical mastitis with particular reference to macrolide-lincosamide resistance phenotypes and genotypes," *Journal of Antimicrobial Chemotherapy*, vol. 57, no. 5, pp. 966–969, 2006.

[24] M. Santagati, F. Iannelli, M. R. Oggioni, S. Stefani, and G. Pozzi, "Characterization of a genetic element carrying the macrolide efflux gene *mef*(A) in *Streptococcus pneumoniae*," *Antimicrobial Agents and Chemotherapy*, vol. 44, no. 9, pp. 2585–2587, 2000.

[25] K. Zhang, J.-A. McClure, S. Elsayed, T. Louie, and J. M. Conly, "Novel multiplex PCR assay for characterization and concomitant subtyping of staphylococcal cassette chromosome *mec* types I to V in methicillin-resistant *Staphylococcus aureus*," *Journal of Clinical Microbiology*, vol. 43, no. 10, pp. 5026–5033, 2005.

[26] M. A. Argudin, B. A. Tenhagen, A. Fetsch et al., "Virulence and resistance determinants of german *Staphylococcus aureus* ST398 isolates from nonhuman sources," *Applied and Environmental Microbiology*, vol. 77, no. 9, pp. 3052–3060, 2011.

[27] G. E. Dale, H. Langen, M. G. P. Page, R. L. Then, and D. Stuber, "Cloning and characterization of a novel, plasmid-encoded trimethoprim- resistant *dihydrofolate reductase* from *Staphylococcus haemolyticus* MUR313," *Antimicrobial Agents and Chemotherapy*, vol. 39, no. 9, pp. 1920–1924, 1995.

[28] S. R. Stengren and M. J. Starzyk, "A modified medium for the recovery of *Staphylococcus* from water," *Microbios*, vol. 41, no. 165-166, pp. 191–203, 1984.

[29] Y. Salfinger and M. L. Tortorello, *Compendium of Methods for the Microbiological Examination of Foods*, American Public Health Association, Washington, DC, USA, 2001.

[30] L. F. Montezzi, E. H. Campana, L. L. Corrêa et al., "Occurrence of carbapenemase-producing bacteria in coastal recreational waters," *International Journal of Antimicrobial Agents*, vol. 45, no. 2, pp. 174–177, 2015.

[31] T. L. Bannerman and S. J. Peacock, "*Staphylococcus Micrococcus*, and other catalase- positive cocci," in *Manual of Clintcal Microbiology*, P. R. Murray, E. L. Baron, J. H. Jorgensen, M. L. Landry, and M. A. Pfaller, Eds., pp. 390–401, Washington, DC, USA, 2007.

[32] T.-F. Lee, H. Lee, C.-M. Chen et al., "Comparison of the accuracy of matrix-assisted laser desorption ionization-time of flight mass spectrometry with that of other commercial identification systems for identifying *Saphylococcus saprophyticus* in urine," *Journal of Clinical Microbiology*, vol. 51, no. 5, pp. 1563–1566, 2013.

[33] CLSI, *Methods for Dilution Antimicrobial Susceptibility Tests for Bacteria That Grow Aerobically: Approved Standard*, Clinical and Laboratory Standards Institute, Pa, USA, 9th, (M07-A9) edition, 2012.

[34] CLSI., *Performance Standards for Antimicrobial Susceptibility Testing; Twenty-Fifth Informational Supplement*, Clinical and Laboratory Standards Institute, Pa, USA, 2015.

[35] M. Yamada, J. Yoshida, S. Hatou, T. Yoshida, and Y. Minagawa, "Mutations in the quinolone resistance determining region in *Staphylococcus epidermidis* recovered from conjunctiva and their association with susceptibility to various fluoroquinolones," *British Journal of Ophthalmology*, vol. 92, no. 6, pp. 848–851, 2008.

[36] L. J. V. Piddock, "Mechanisms of fluoroquinolone resistance: an update 1994–1998," *Drugs*, vol. 58, no. 2, pp. 11–18, 1999.

[37] R. F. Rabello, C. R. V. M. Souza, R. S. Duarte, R. M. M. Lopes, L. M. Teixeira, and A. C. D. Castro, "Characterization of *Staphylococcus aureus* isolates recovered from bovine mastitis in Rio de Janeiro, Brazil," *Journal of Dairy Science*, vol. 88, no. 9, pp. 3211–3219, 2005.

[38] K. M. Sullivan, A. Dean, and M. M. Soe, "OpenEpi: a web-based epidemiologic and statistical calculator for public health," *Public Health Reports*, vol. 124, no. 3, pp. 471–474, 2009.

[39] P. Hedman, O. Ringertz, K. Olsson, and R. Wollin, "Plasmid-identified *Staphylococcus saprophyticus* isolated from the rectum of patients with urinary tract infections," *Scandinavian Journal of Infectious Diseases*, vol. 23, no. 5, pp. 569–572, 1991.

[40] M. E. Rupp, D. E. Soper, and G. L. Archer, "Colonization of the female genital tract with *Staphylococcus saprophyticus*," *Journal of Clinical Microbiology*, vol. 30, no. 11, pp. 2975–2979, 1992.

[41] T. Spanu, E. De Carolis, B. Fiori et al., "Evaluation of matrix-assisted laser desorption ionization-time-of-flight mass spectrometry in comparison to *rpoB* gene sequencing for species identification of bloodstream infection staphylococcal isolates," *Clinical Microbiology and Infection*, vol. 17, no. 1, pp. 44–49, 2011.

[42] H. C. Lichstein and M. H. Soule, "Studies of the effect of sodium azide on microbic growth and respiration," *Journal of Bacteriology*, vol. 47, no. 3, pp. 221–230, 1944.

[43] K. Gupta et al., "International clinical practice guidelines for the treatment of acute uncomplicated cystitis and pyelonephritis in women: a 2010 update by the infectious diseases society of america and the european society for microbiology and infectious diseases," *Clinical Infectious Diseases*, vol. 52, no. 5, pp. e103–e120, 2011.

[44] C. Strauss, Y. Hu, A. Coates, and V. Perreten, "A novel *erm*(44) gene variant from a human *Staphylococcus saprophyticus* confers resistance to macrolides, lincosamides but not streptogramins," *Antimicrobial Agents and Chemotherapy*, vol. 61, no. 1, Article ID e01655-16, 2016.

[45] M. Higashide, M. Kuroda, S. Ohkawa, and T. Ohta, "Evaluation of a cefoxitin disk diffusion test for the detection of *mecA*-positive methicillin-resistant *Staphylococcus saprophyticus*," *International Journal of Antimicrobial Agents*, vol. 27, no. 6, pp. 500–504, 2006.

[46] J. Matuszkiewicz-Rowińska, J. Małyszko, and M. Wieliczko, "Urinary tract infections in pregnancy: old and new unresolved diagnostic and therapeutic problems," *Archives of Medical Science*, vol. 11, no. 1, pp. 67–77, 2015.

[47] M. Lee, P. Bozzo, A. Einarson, and G. Koren, "Urinary tract infections in pregnancy," *Canadian Family Physician*, vol. 54, no. 6, pp. 853-854, 2008.

[48] F. Chereau, P. Herindrainy, B. Garin et al., "ESBL- and NDM-1-producing *Enterobacteriaceae* colonization among pregnant women in the community in a low income country: a potential reservoir for transmission of multi-resistant *Enterobacteriaceae* to neonates," *Antimicrobial Agents and Chemotherapy*, vol. 59, no. 6, pp. 3652–3655, 2015.

[49] L. A. Denkel, F. Schwab, A. Kola et al., "The mother as most important risk factor for colonization of very low birth weight (VLBW) infants with extended-spectrum β-lactamase-producing Enterobacteriaceae (ESBL-E)," *Journal of Antimicrobial Chemotherapy*, vol. 69, no. 8, Article ID dku097, pp. 2230–2237, 2014.

[50] A. Pathak, S. P. Chandran, K. Mahadik, R. Macaden, and C. S. Lundborg, "Frequency and factors associated with carriage of multi-drug resistant commensal *Escherichia coli* among women attending antenatal clinics in Central India," *BMC Infectious Diseases*, vol. 13, no. 1, article 199, 2013.

[51] M. Widerström, J. Wiström, S. Ferry, C. Karlsson, and T. Monsen, "Molecular epidemiology of *Staphylococcus saprophyticus* isolated from women with uncomplicated community-acquired urinary tract infection," *Journal of Clinical Microbiology*, vol. 45, no. 5, pp. 1561–1564, 2007.

[52] P. Hedman and O. Ringertz, "Urinary tract infections caused by *Staphylococcus saprophyticus*. a matched case control study," *Journal of Infection*, vol. 23, no. 2, pp. 145–153, 1991.

[53] N. H. Khan, M. Ahsan, W. D. Taylor, and K. Kogure, "Culturability and survival of marine, freshwater and clinical *Pseudomonas aeruginosa*," *Microbes and Environments*, vol. 25, no. 4, pp. 266–274, 2010.

11

Prevalence of Multidrug Resistance and Extended-Spectrum β-Lactamase Carriage of Clinical Uropathogenic *Escherichia coli* Isolates in Riyadh, Saudi Arabia

Abdulaziz Alqasim,[1] Ahmad Abu Jaffal,[2] and Abdullah A. Alyousef[1]

[1]*Microbiology Research Group, Department of Clinical Laboratory Sciences, College of Applied Medical Sciences, King Saud University, P.O. Box 10219, Riyadh 11433, Saudi Arabia*
[2]*Department of Clinical Laboratory Sciences, King Saud Bin Abdulaziz University for Health Sciences, Riyadh, Saudi Arabia*

Correspondence should be addressed to Abdulaziz Alqasim; aalqasim@ksu.edu.sa

Academic Editor: Todd R. Callaway

The prevalence of extended-spectrum β-lactamase-producing *Escherichia coli* (ESBL-producing *E. coli*) has recently increased worldwide. This study aims at determining the antimicrobial susceptibility patterns of a collection of clinical *E. coli* urine isolates and evaluating the ESBL carriage of these isolates at phenotypic and genotypic levels. A total of 100 *E. coli* urine isolates were collected at a tertiary healthcare centre in Riyadh from January 2018 to March 2018. Antimicrobial susceptibility testing was carried out for all isolates. ESBL production was characterized at phenotypic and genotypic levels using double-disc synergy test and polymerase chain reaction, respectively. Detection of different ESBL variants was performed using DNA sequencing. Of 100 *E. coli* isolates, 67 were associated with multidrug resistance (MDR) phenotype. All isolates showed variable resistance levels to all antibiotics used here expect to imipenem, where they were all imipenem-sensitive. 33 out of 100 *E. coli* isolates were positive for ESBLs by phenotypic and genotypic methods. Among all ESBL-positive *E. coli* isolates, the CTX-M was the most prevalent ESBL type (31/33 isolates; 93.94%). CTX-M-15 variant was detected in all isolates associated with CTX-M carriage. Multiple ESBL gene carriage was detected in 15/33 isolates (45.45%), where 11 (33.33%) isolates produced two different ESBL types while 4 isolates (12.12%) associated with carrying three different ESBL types. Our study documented the high antimicrobial resistance of ESBL-producing *E. coli* to many front-line antibiotics currently used to treat UTI patients, and this implies the need to continuously revise the local guidelines used for optimal empirical therapy for UTI patients. It also showed the high prevalence of ESBL carriage in *E. coli* urine isolates, with CTX-M-15 being the most predominant CTX-M variant.

1. Introduction

Urinary tract infections (UTIs) are among the most common bacterial infections acquired in community and hospital settings [1]. UTIs are a main cause of hospital admissions and are associated with high morbidity, mortality, and economic costs [2, 3]. Worldwide, it is estimated that UTIs affect about 150 million people each year, costing the global economy more than 6 billion US dollars [4]. UTIs occur in all age groups and in both genders [5]. However their incidence increases with age [6], and the annual incidence of UTIs in the elderly population ranges from 10% in the community to as high as 30% of hospitalized patients [2]. UTIs are more common in women than in men [7], with an estimated 50–60% of women suffering from at least one UTI during their lifetime [3].

It is well known that *Escherichia coli* (*E. coli*) is the main causative agent of UTIs [8], and the uropathogenic *E. coli* (UPEC) group have the capability of causing around 90% of community-acquired UTIs and up to 50% of nosocomial UTIs [9, 10]. Over the past two decades, UPEC strains have shown an increased level of antimicrobial resistance to front-line antibiotics such as trimethoprim-sulfamethoxazole and ciprofloxacin. Several surveillance studies during the 2000s across Europe, North America, and South America have demonstrated that resistance to these antibiotics has been observed in as many as 20–45% of UPEC isolates [1, 11].

UPEC have also been associated with a high level of extended-spectrum β-lactamase (ESBL) gene carriage [12]. ESBLs comprise many plasmid-mediated derivatives such as TEM, OXA, and SHV [13]. Since 2000, a new group of ESBLs, called CTX-M (i.e., "active on CefoTaXime, first isolated in Munich"), has emerged [14]. Since then, CTX-M β-lactamases have been the predominant ESBL type worldwide [15]. Within the CTX-M family, CTX-M-15 is currently the most prevalent CTX-M genotype [14, 15]. This group of ESBLs has been associated with an extensive pattern of antimicrobial resistance to many antibiotics, including β-lactam agents such as penicillins, cephalosporins, monobactams, and carbapenems [16–19]. In addition, CTX-M-producing *E. coli* strains are often associated with coresistance to other large antimicrobial families such as aminoglycosides and fluoroquinolones [19]. This increasing level of antimicrobial resistance of UPEC is of great concern since it can limit the therapeutic choices used for treating common bacterial infections such as UTIs and highlights the growing threat of the emergence of pandrug resistance in UPEC.

Since treatment of UTI is frequently started empirically, prior information on the prevalence of causative agents as well as the antimicrobial susceptibility profiles in a particular setting is essential [20]. However, the distribution of UTI causative agents and their antibiotic susceptibility profiles differs regionally [21, 22]. Therefore, the evaluation of the local etiology and antimicrobial susceptibility profiles is essential to achieve the most effective empirical therapy [8].

In Saudi Arabia, previous studies have been conducted to determine the prevalence of ESBL-producing *Enterobacteriaceae* (EPE), with CTX-M enzymes being the most common ESBL type [23–30]. However, there are a limited number of studies on the prevalence of ESBL-producing *E. coli*. Additionally, there are few detailed reports on characterizing the genotypes of ESBL enzymes have been published.

This study aims at determining the antimicrobial susceptibility patterns of a collection of clinical *E. coli* urine isolates, to phenotypically assess the ESBL carriage of these isolates and to characterize the ESBL genotypes of these isolates. This study is important for providing clinicians with information required to facilitate the effective treatment and management of UTI patients.

2. Materials and Methods

2.1. Bacterial Isolates. One hundred nonconsecutive, nonduplicate clinical *E. coli* urine isolates were obtained from urine samples of inpatients hospitalized at a tertiary healthcare centre in Riyadh, Saudi Arabia. Midstream clean catch urine samples showing significant bacterial growth, of $>10^5$ colony-forming units (CFU/mL) with a single type of bacteria, were considered positive for UTIs. Urine isolates were collected over a period of 3 months from January 2018 to March 2018 and were identified as *E. coli* using the Vitek 2 identification system (Vitek2-ID-GNB, BioMerieux). The isolates were stored at −80°C in Luria Bertani broth (HiMedia Labs, India) with 20% glycerol (v/v).

2.2. Antimicrobial Susceptibility Testing. Antimicrobial susceptibility testing (AST) for all isolates was carried out on Mueller–Hinton agar (HiMedia Labs, India) using disk diffusion method according to the guidelines of Clinical and Laboratory Standards Institute (CLSI) [31] using a panel of 10 commercially available antibiotics (LIOFILCHEM, Italy). Information on these antibiotics and their concentrations are shown in Table 1. *E. coli* ATCC 25922 was used as a control strain.

2.3. Phenotypic Detection of ESBL Production. ESBL production was assessed using the CLSI recommendations for ESBL screening and phenotypic confirmation tests [31]. For initial ESBL screening, UPEC isolates showing an inhibition zone size of ≤22 mm with ceftazidime (30 μg) were identified as potential ESBL producers. The double-disc synergy test (DDST) was carried out for the phenotypic confirmation of ESBL production. For this test, a ceftazidime disc (30 μg) was placed 25 mm away from a combination disc containing ceftazidime-clavulanic acid (30/10 μg). When the zone of inhibition between the combination disc and the corresponding single antibiotic disc differed by ≥5 mm, the strain was identified as an ESBL producer. *E. coli* ATCC 25922 and *Klebsiella pneumoniae* ATCC 700603 were used as negative and positive control strains, respectively.

2.4. Polymerase Chain Reaction (PCR) of ESBL-Encoding Genes. Bacterial genomic DNA was extracted using the ONE-4-ALL Genomic DNA Mini-Preps Kit (Bio Basic Inc., Canada) according to manufacturer's instructions. PCR assays were performed to determine the presence of ESBL-encoding genes: bla_{OXA}, bla_{TEM}, bla_{SHV}, and $bla_{CTX\text{-}M}$ Groups 1, 2 and 9 using multiplex PCR primer sets (Table 2) and conditions previously described [32].

2.5. DNA Sequencing for Identification of CTX-M ESBL Gene Variants. Automatic DNA sequencing for both strands of all PCR products was carried out using the 3730*xl* DNA analyzer (ThermoFisher Scientific, United States) to identify the CTX-M ESBL variants. These variants were identified by comparison with the sequences in GenBank (http://blast.ncbi.nlm.nih.gov/Blast.cgi).

3. Results

3.1. Demographic Characteristics of Study Population. A total of 100 urine samples were collected from inpatients during this study period. Among these samples, 76 (76%) belonged to female patients while 24 (24%) belonged to male patients. With regard to the age categories of patients, a total of 60 (60%) samples were collected from adults, 21 (21%) samples belonged to elderly population, and 19 (19%) were obtained from children. Demographic characteristics of patients are shown in Table 3.

Table 1: Details on antibiotics used in this study.

Antibiotic group	Antibiotic name	Concentration (μg/disc)
Penicillins	Ampicillin (AM)	10 μg
β-lactam/β-lactamase inhibitors combination	Amoxicillin-clavulanic acid (AUG)	20/10 μg
Aminoglycosides	Gentamicin (GM)	10 μg
Second-generation cephalosporins	Cefoxitin (FOX)	30 μg
Third-generation cephalosporins	Ceftazidime (CAZ)	30 μg
Tetracyclines	Tetracycline (T)	30 μg
Folate pathway inhibitors	Trimethoprim-sulfamethoxazole (SXT)	1.25/23.75 μg
Carbapenems	Imipenem (IMP)	10 μg
Fluoroquinolones	Ciprofloxacin (CIP)	5 μg
Nitrofurans	Nitrofurantoin (NI)	300 μg

Table 2: Multiplex PCR-specific primers used for ESBL gene detection in *Enterobacteriacaea.*

PCR name	β-lactamase targeted	Sequence (5′– 3′)	Product (bp)	Reference
Multiplex I	TEM variants including TEM-1 and TEM-2	CATTTCCGTGTCGCCCTTATTC CGTTCATCCATAGTTGCCTGAC	800	[32]
	SHV variants including SHV-1	AGCCGCTTGAGCAAATTAAAC ATCCCGCAGATAAATCACCAC	713	
	OXA-1, OXA-4, and OXA-30	GGCACCAGATTCAACTTTCAAG GACCCCAAGTTTCCTGTAAGTG	564	
Multiplex II	Variants of CTX-M group 1 including CTX-M-1, CTX-M-3, and CTX-M-15	TTAGGAARTGTGCCGCTGYA CGATATCGTTGGTGGTRCCAT	688	
	Variants of CTX-M group 2 including CTX-M-2	CGTTAACGGCACGATGAC CGATATCGTTGGTGGTRCCAT	404	
	Variants of CTX-M group 9 including CTX-M-9 and CTX-M-14	TCAAGCCTGCCGATCTGGT TGATTCTCGCCGCTGAAG	561	

Table 3: Demographic data of study population.

Age category	Gender		Total
	Female	Male	
Children*	16	3	19
Adults**	46	14	60
Elderly***	14	7	21
Total	76	24	100

*Children: 0–17 years, **adults: 18–64 years, ***elderly: ≥65 years.

3.2. Antimicrobial Susceptibility Profiles and ESBL Prevalence. Antimicrobial susceptibility testing results are shown in Table 4. Among all antibiotics tested in this study, imipenem was the most active agent as all *E. coli* isolates were imipenem-susceptible. Out of the 100 *E. coli* isolates, 92% were resistant to ampicillin, 55% to amoxicillin-clavulanic acid, and 12% to gentamicin. The resistance rates for ceftazidime, cefoxitin, tetracycline, and trimethoprim-sulfamethoxazole were 29%, 13%, 49%, and 54%, respectively. Additionally, 40% and 15% of the isolates were nonsusceptible to ciprofloxacin and nitrofurantoin, respectively.

With regard to ESBL production, initial ESBL screening demonstrated that 41% of all *E. coli* isolates had an inhibition zone size of ≤22 mm with ceftazidime (30 μg) and therefore initially identified as potential ESBL producers. However, phenotypic confirmation of ESBL production by DDST showed that 33 (33%) of all isolates were confirmed to be ESBL producers, while 67 (67%) isolates were non-ESBL producers.

The susceptibility profiles for the ESBL-producing *E. coli* isolates are shown in Table 3. Of all the 33 *E. coli* isolates identified as ESBL producers, 29 (88%) were resistant to amoxicillin-clavulanic acid, 28 (85%) to ceftazidime, 27 (82%) to trimethoprim-sulfamethoxazole, and 25 (76%) to ciprofloxacin. Additionally, 9 (27%) isolates were resistant to gentamicin, 6 (18%) to cefoxitin, 23 (70%) to tetracycline, and 7 (21%) to nitrofurantoin. All tested isolates were susceptible to imipenem, while they were all ampicillin-resistant.

For non-ESBL-producing *E. coli* isolates, the antimicrobial resistance levels were lower than that of ESBL-producing isolates for all antibiotics tested expect imipenem. Of all the 67 non-ESBL-producing *E. coli* isolates, 59 (88%) isolates were resistant to ampicillin, 27 (40%) to trimethoprim-sulfamethoxazole, 26 (39%) to amoxicillin-clavulanic acid, and 26 (39%) for tetracycline. The antimicrobial resistance rates for other antibiotics ranged from 1.5% for ceftazidime to 22% to ciprofloxacin. All non-ESBL-producing *E. coli* isolates were imipenem-susceptible.

3.3. Multidrug Resistance (MDR) Phenotype of E. coli Isolates. Percentage of the clinical *E. coli* isolates showing MDR phenotype, which is defined as exhibiting resistance to at least 1 agent in ≥3 antimicrobial categories/groups [33], is shown in Figure 1. Our data showed a high level of multidrug resistance among the tested isolates with 67 (67%) of all *E. coli* isolates showing the MDR phenotype. Of these isolates, 18 (26.86%) isolates were resistant to 3 out of 10 antibiotic groups employed, 17 (25.37%) were nonsusceptible to 4 antibiotic groups, and 10 (14.92%) were

Table 4: Antibiotic resistance rates of *E. coli* isolates tested in this study.

Antibiotic	Number (%) of resistant isolates		
	ESBL producing (n = 33)	Non-ESBL producing (n = 67)	Total (n = 100)
Ampicillin	33 (100%)	59 (88%)	92 (92%)
Amoxicillin-clavulanic acid	29 (88%)	26 (39%)	55 (55%)
Gentamicin	9 (27%)	3 (4.5%)	12 (12%)
Ceftazidime	28 (85%)	1 (1.5%)	29 (29%)
Cefoxitin	6 (18%)	7 (10.5%)	13 (13%)
Tetracycline	23 (70%)	26 (39%)	49 (49%)
Trimethoprim-sulfamethoxazole	27 (82%)	27 (40%)	54 (54%)
Imipenem	0 (0%)	0 (0%)	0 (0%)
Ciprofloxacin	25 (76%)	15 (22%)	40 (40%)
Nitrofurantoin	7 (21%)	8 (12%)	15 (15%)

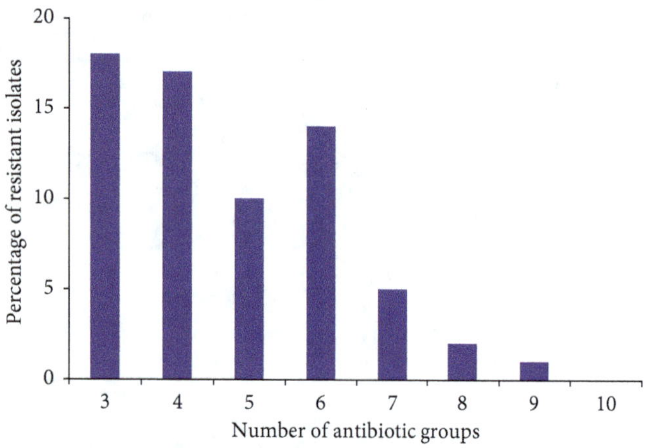

Figure 1: Percentage of clinical *E. coli* urine isolates showing multidrug resistance (MDR) phenotype.

resistant to 5 antibiotic groups. Additionally, 14 (20.89%), 5 (7.46%), and 2 (2.99%) were resistant to 6, 7, and 8 antibiotics, respectively. There was only 1 (1.49%) isolate that showed resistance to 9 antibiotics while none of the isolates had the ability to show resistance to all the 10 antibiotics used in this study. With respect to the association between ESBL production and the MDR phenotype, our data showed that all the 33 ESBL-positive *E. coli* isolates (100%) were MDR; however, only 34 out of the 67 non-ESBL-producing isolates (50.7%) were MDR.

3.4. Distribution of ESBL Gene Carriage in E. coli Isolates. 33 ESBL-producing *E. coli* isolates were tested for the production of the major ESBL genes (bla_{OXA}, bla_{TEM}, bla_{SHV}, and bla_{CTX-M} Groups 1, 2, and 9) by PCR. These isolates were all tested positive for ESBL production. The ESBL types detected in the *E. coli* isolates belonged to the CTX-M-Group 1 family, TEM and OXA ESBL genes, while the SHV-type, CTX-M-Group 2, and CTX-M-Group 9 were not detected in any of the tested isolates. The detected ESBL types were either solely or concomitantly harbored by the 33 ESBL-producing *E. coli* isolates. Our data demonstrated that 31 (93.94%) isolates were found to carry the gene encoding for CTX-M-Group 1 and 4 isolates (12.12%) produced the TEM-type ESBL. The CTX-M-Group 1 ESBL was solely carried by 17 isolates, whereas a sole TEM carriage was detected in only 1 isolate.

However, the gene encoding for OXA-type ESBL was not solely detected in any of the tested isolates but rather it was concomitantly produced by 8 isolates (24.24%) that either carry the genes encoding for CTX-M-Group 1, TEM, or both. Additionally, among all ESBL-positive isolates, 15 (45.45%) isolates were associated with multiple ESBL gene carriage. Of these, 11 (33.33%) were able to produce two different ESBL types, while 4 (12.12%) were able to concomitantly produce all the three ESBL types detected in this study. With respect to the CTX-M-Group 1 gene variants, our data showed that the CTX-M-15 variant was present in all the CTX-M-Group 1 producing isolates (Table 5).

4. Discussion

The current emergence of multidrug-resistant *E. coli* is becoming a global concern, and infections caused by MDR ESBL-producing *E. coli* represent a major challenge to clinicians and public health worldwide [34]. This study focuses on determining the antimicrobial susceptibility patterns of 100 *E. coli* urine isolates and on characterizing the ESBL carriage of these isolates at both phenotypic and genotypic levels.

Among the 100 urine samples collected in this study, 76 (76%) belonged to female population while 24 (24%) belonged to male patients. The gender distribution of patients in our study is consistent with many previous local [35, 36] and international [37, 38] reports showing the high incidence of UTI in women compared to men across all age groups. It has been demonstrated that the high UTI incidence in women can be attributed to many reasons such as anatomical factors that allow quick access of bacteria to the bladder, poor hygiene, sexual activity, and use of contraceptives [39]. With respect to age group of patients, our data showed that 60% of all study population was adults, followed by elderly (21%). This is in agreement with a recent report demonstrating that UTIs were more common in adults than any other age category [36]. Additionally, adult women were among the most dominant patient group in our study, and they accounted for 46% of total patient number. This concurs with previous studies showing that adult women are 30 times more likely to develop a UTI than men, with almost

Table 5: The distribution of β-lactamase genes among ESBL-producing *E. coli* isolates.

Isolate ID	β-lactamase types						CTX-M G1 variants
	CTX-M G1	CTX-M G2	CTX-M G9	TEM	OXA	SHV	
U1	+	−	−	−	−	−	CTX-M-15
U4	+	−	−	−	−	−	CTX-M-15
U7	+	−	−	−	−	−	CTX-M-15
U9	+	−	−	−	−	−	CTX-M-15
U10	+	−	−	−	−	−	CTX-M-15
U11	+	−	−	−	−	−	CTX-M-15
U12	+	−	−	−	−	−	CTX-M-15
U15	+	−	−	−	+	−	CTX-M-15
U16	+	−	−	−	−	−	CTX-M-15
U17	+	−	−	−	−	−	CTX-M-15
U20	+	−	−	−	−	−	CTX-M-15
U24	+	−	−	−	−	−	CTX-M-15
U27	+	−	−	−	−	−	CTX-M-15
U28	+	−	−	−	+	−	CTX-M-15
U31	+	−	−	−	+	−	CTX-M-15
U46	+	−	−	−	+	−	CTX-M-15
U55	+	−	−	+	+	−	CTX-M-15
U57	+	−	−	−	−	−	CTX-M-15
U63	+	−	−	−	+	−	CTX-M-15
U65	+	−	−	+	+	−	CTX-M-15
U66	+	−	−	+	−	−	CTX-M-15
U68	+	−	−	−	−	−	CTX-M-15
U71	+	−	−	−	−	−	CTX-M-15
U73	+	−	−	−	+	−	CTX-M-15
U74	+	−	−	−	−	−	CTX-M-15
U75	−	−	−	+	−	−	NA[1]
U78	+	−	−	−	−	−	CTX-M-15
U82	+	−	−	+	+	−	CTX-M-15
U85	+	−	−	+	+	−	CTX-M-15
U87	+	−	−	−	+	−	CTX-M-15
U93	+	−	−	+	−	−	CTX-M-15
U95	−	−	−	+	+	−	NA
U98	+	−	−	+	−	−	CTX-M-15

[1]Nonapplicable.

half of them experiencing at least one UTI episode during their lifetime [3].

The antimicrobial susceptibility patterns were determined for all the 100 *E. coli* urine isolates. The ESBL-producing *E. coli* isolates showed higher levels of resistance to all antibiotics compared to the non-ESBL-producing isolates expect for imipenem, where all *E. coli* isolates tested in this study were imipenem-sensitive. This is not surprising given that ESBL-producing *E. coli* strains are frequently associated with coresistance to other antimicrobial agents such as aminoglycosides and fluoroquinolones [19]. The ESBL-producing *E. coli* showed the greatest resistance to ampicillin, amoxicillin-clavulanic acid, ceftazidime, trimethoprim-sulphamethoxazole, and ciprofloxacin. This finding is in agreement with many previous reports [30, 40–42] showing that ESBL-producing *E. coli* isolates were more resistant to trimethoprim-sulphamethoxazole, ciprofloxacin, and third-generation cephalosporins compared to non-ESBL-producing *E. coli*, although this resistance did not affect imipenem. This increased resistance levels of ESBL-producing *E. coli* isolates to some front-line antibiotics such as cephalosporins, trimethoprim-sulphamethoxazole, and ciprofloxacin, could be ascribed to the inappropriate and excessive use of these agents in the empirical treatment of UTIs [43]. It has been shown that these antibiotics are commonly used for the treatment of UTIs worldwide [44], and the currently reported increase in resistance trends among uropathogens, primarily *E. coli*, has not only led to high rates of morbidity and mortality [45], but it has also complicated the management of UTIs [46].

In an attempt to achieve the effective treatment of UTIs, several guidelines were reviewed, in the late 2000s, to reposition nitrofurantoin as first-line therapy for community-acquired and nosocomial uncomplicated lower UTI [47, 48]. With respect to *E. coli* resistance to nitrofurantoin, although previous reports showed that it was low in China (1.6%) [49] and in North America (1.1%) [50], it has been reported to be high in Latin American hospitals (13%) [51].

An important worrisome finding in this study is the high rate of resistance to nitrofurantoin (21%) in the ESBL-producing *E. coli*. This is in contrary to many previous local [28] and international studies [49, 50] showing a low prevalence of nitrofurantoin resistance in *E. coli* isolates. The possible explanation of this high nitrofurantoin resistance is the currently reported exponential global increase in nitrofurantoin prescribing for the empirical treatment of hospitalized patients with positive urine cultures [52].

Based on the suggestion that if the resistance to a particular antibiotic is higher than 20%, that antibiotic should not be prescribed in the empirical antimicrobial treatment [53], and since our data show that the resistance levels of ESBL-producing *E. coli* isolates to 9 out of 10 antibiotics were more than 20%, we propose that carbapenems, such as imipenem, could be the drug of choice to treat UTI patients given that all *E. coli* isolates tested in this study were fully sensitive to imipenem, and this is the case with many recently published reports in Saudi Arabia showing *E. coli* urine isolates with full sensitivity to carbapenems [30, 54].

Additionally, we believe that it is crucial to assess the local antimicrobial resistance rates for specific pathogens and to revise the current guidelines used for optimal treatment regimens for ESBL-producing uropathogens, particularly *E. coli*, causing uncomplicated UTI in order to achieve the effective treatment.

Considering the level of multidrug resistance of *E. coli* isolates, this study found that 67 (67%) of all *E. coli* isolates were MDR. This is considered a high level of multidrug resistance when compared to a recently published report from Saudi Arabia showing that the MDR phenotype was determined in only 22.8% of all tested *E. coli* isolates [36]. Additionally, multidrug resistance was higher in ESBL-producing *E. coli* compared to non-ESBL *E. coli*, which is in total agreement with a previous local study showing the same finding [28], although the sample size here was low compared to that study.

Drug resistance of *E. coli* is mediated by ESBLs, mainly of the CTX-M family [19]. According to the Pan European Antimicrobial Resistance using Local Surveillance (PEARLS) study (2001–2002), the percentages of ESBL production among *E. coli* was 5.4% for all the study geographical sites [55]. Worldwide, many previous studies have shown that the prevalence of ESBL-producing *E. coli* varies across different geographical regions with low rates of about 1.5% reported in Denmark [56] and 5% in Canada [57] compared to much higher prevalence rates documented in reports from Turkey (36.7%) [58] and India (69%) [59].

In our study, we found that 33 (33%) isolates were ESBL producers, while 67 (67%) isolates were non-ESBL producers. The ESBL types detected in the *E. coli* isolates belonged to the CTX-M-Group 1 family, TEM and OXA, while the SHV-type, CTX-M-Group 2 and CTX-M-Group 9 were not detected in any of the tested isolates. Of these 33 ESBL-producing isolates, 31 (93.94%) were found to carry the gene encoding for CTX-M-Group 1, and DNA sequencing data showed that the CTX-M-15, a variant of CTX-M-Group 1, was detected in all of these isolates. Additionally, among all ESBL-positive isolates, 15 (45.45%) isolates were associated with multiple ESBL gene carriage. Of these, 11 (33.33%) were associated with a concomitant production of two different ESBL types, while 4 (12.12%) were associated with a concomitant production of all the three ESBL types detected in this study.

In this regard, previous studies in Saudi Arabia showed that the prevalence rates of ESBL-producing *E. coli* were 6.5% and 10.3% in 2002 and 2004, respectively [25]. However, recent reports have demonstrated a higher prevalence of ESBL-producing *E. coli* causing UTI. A 2-year retrospective observational study on the role of ESBL-producing *E. coli* in UTIs found that 27.4% of all *E. coli* strains isolated from urine samples were ESBL producers [60]. Another study found that the prevalence of ESBL-producing *E. coli* was 33.3% [28]. However, these studies did not determine the ESBL types present in *E. coli* isolates. Al-Agamy and colleagues determined the ESBL carriage and types of a collection of *E. coli* isolates and found that 20.4% of the total *E. coli* isolates were ESBL producers. They also found that CTX-M-15 was detected, concomitantly with the narrow-spectrum β-lactamase TEM-1, in all ESBL-producing *E. coli* [30].

Currently, CTX-M enzymes have replaced the traditional ESBL types such as SHV and TEM enzymes as the most prevalent ESBL type in *E. coli* [15]. Our study clearly shows that CTX-M enzymes are the dominant ESBLs in Saudi Arabia and that the CTX-M-15 was the most common variant among all CTX-M-Group 1, which is consistent with the previous finding reported by Al-Agamy and colleagues [30].

It has been shown that the successful dissemination of CTX-M-15 enzyme has been attributed to the spread of genetic elements, through horizontal gene transfer, and the clonal expansion of a pandemic *E. coli* clone, *E. coli* ST131 [19]. *E. coli* ST131 has been reported worldwide as the most common extraintestinal pathogenic *E. coli* (ExPEC) clone, and this clone is often MDR and is commonly associated with carrying CTX-M-15 enzyme [61, 62]. In the future, it would be important to study the molecular epidemiology of MDR *E. coli* strains to determine the population structure and clonal diversity of currently emerging MDR *E. coli* clones.

Our study has limitations: firstly, it used a low number of *E. coli* isolates that were obtained from a single healthcare facility in Saudi Arabia. Additionally, this study was performed on urine samples collected from hospitalized patients, which affects the accurate assessment of epidemiological changes in the community. Also important is that our study was performed in Riyadh City and this does not necessarily reflect the antimicrobial resistance trends in other regions within Saudi Arabia and that the clinical information of patients included in this study are very scarce.

However, this study shows a high prevalence of ESBL-producing *E. coli* compared to what has been previously reported in Saudi Arabia, with CTX-M-type enzymes being the most predominant ESBL type. Furthermore, our results provide evidence for the predominance of CTX-M-15-producing *E. coli* among UPEC isolates in Saudi Arabia, and this agrees with many reports demonstrating the changing epidemiology of ESBL-producing *E. coli* worldwide.

To conclude, this study demonstrates that *E. coli* isolates are associated with high multidrug resistance rates as well as high ESBL carriage. It also shows that ESBL-producing *E. coli* have high resistance levels to antibiotics, particularly to those used for empirical therapy of UTI patients. This highlights the need to evaluate the local antimicrobial resistance rates for specific pathogens and to review the current guidelines used for empirical treatment regimens for ESBL-producing uropathogens in order to ensure effective treatment of infections caused by these pathogens.

Abbreviations

ESBL-producing *E. coli*:	Extended-spectrum β-lactamase-producing *Escherichia coli*
MDR:	Multidrug resistance
UTIs:	Urinary tract infections
UPEC:	Uropathogenic *E. coli*
CTX-M:	"Active on CefoTaXime, first isolated in Munich"
TEM:	ESBL variant isolated from a patient named Temoneira
SHV:	Sulfhydryl variable
OXA:	Oxacillin hydrolyzing capabilities
CLSI:	Clinical and Laboratory Standards Institute
DDST:	Double-disc synergy test
AST:	Antimicrobial susceptibility testing
PCR:	Polymerase chain reaction
μg:	Microgram
ST:	Sequence type.

Conflicts of Interest

All authors declare that they have no competing interests.

Authors' Contributions

Abdulaziz Alqasim and Abdullah A. Alyousef contributed to the study design, Abdulaziz Alqasim collected the samples. Abdulaziz Alqasim and Ahmad Abu Jaffal contributed to the laboratory experiments. All the authors contributed to the data interpretation. Abdulaziz Alqasim and Abdullah A. Alyousef drafted and revised the manuscript. All the authors read and approved the final manuscript.

Acknowledgments

The authors extend their appreciation to the Deanship of Scientific Research at King Saud University for funding this work through the Research Project no. R17-02-42.

References

[1] B. Foxaman, "The epidemiology of urinary tract infection," *Nature Reviews Urology*, vol. 7, pp. 653–660, 2010.

[2] A. Cove Smith and M. Almond, "Management of urinary tract infections in the elderly," *Trends in Urology, Gynaecology & Sexual Health*, vol. 12, no. 4, pp. 31–34, 2007.

[3] B. Foxman, "Epidemiology of urinary tract infections: incidence, morbidity, and economic costs," *The American Journal of Medicine*, vol. 113, no. 1, pp. 5–13, 2002.

[4] C. M. Gonzalez and A. J. Schaeffer, "Treatment of urinary tract infection: what's old, what's new, and what works," *World Journal of Urology*, vol. 17, no. 6, pp. 372–382, 1999.

[5] J.-S. Huh, "The prevalence of urinary tract infections in institutionalized vs. noninstitutionalized elderly persons," *Urogenital Tract Infection*, vol. 11, no. 2, pp. 56–61, 2016.

[6] A. Stapleton, "Prevention of recurrent urinary-tract infections in women," *The Lancet*, vol. 353, no. 9146, pp. 7-8, 1999.

[7] R. D. Harrington and T. M. Hooton, "Urinary tract infection risk factors and gender," *The Journal of Gender-Specific Medicine*, vol. 3, pp. 27–34, 2000.

[8] S. Farajnia, M. Y. Alikhani, R. Ghotaslou, B. Naghili, and A. Nakhlband, "Causative agents and antimicrobial susceptibilities of urinary tract infections in the northwest of Iran," *International Journal of Infectious Diseases*, vol. 13, no. 2, pp. 140–144, 2009.

[9] R. Kucheria, P. Dasgupta, S. Sacks, M. Khan, and N. Sheerin, "Urinary tract infections: new insights into a common problem," *Postgraduate Medical Journal*, vol. 81, no. 952, pp. 83–86, 2005.

[10] L. Zhang and B. Foxman, "Molecular epidemiology of *Escherichia coli* mediated urinary tract infections," *Frontiers in Bioscience*, vol. 8, no. 5, pp. e235–e244, 2003.

[11] G. Croxall, V. Weston, S. Joseph, G. Manning, P. Cheetham, and A. Mcnally, "Increased human pathogenic potential of *Escherichia coli* from polymicrobial urinary tract infections in comparison to isolates from monomicrobial culture samples," *Journal of Medical Microbiology*, vol. 60, no. 1, pp. 102–109, 2011.

[12] J. D. Pitout, P. Nordmann, K. B. Laupland, and L. Poirel, "Emergence of *Enterobacteriaceae* producing extended-spectrum β-lactamases (ESBLs) in the community," *Journal of Antimicrobial Chemotherapy*, vol. 56, pp. 52–59, 2005.

[13] M. H. Nicolas-Chanoine, J. Blanco, V. Leflon-Guibout et al., "Intercontinental emergence of *Escherichia coli* clone O25: H4-ST131 producing CTX-M-15," *Journal of Antimicrobial Chemotherapy*, vol. 61, no. 2, pp. 273–281, 2008.

[14] G. Peirano and J. D. D. Pitout, "Molecular epidemiology of *Escherichia coli* producing CTX-M [beta]-lactamases: the worldwide emergence of clone ST131 O25: H4," *International Journal of Antimicrobial Agents*, vol. 35, no. 4, pp. 316–321, 2010.

[15] R. Cantón and T. M. Coque, "The CTX-M β-lactamase pandemic," *Current Opinion in Microbiology*, vol. 9, no. 5, pp. 466–475, 2006.

[16] M. Accogli, T. Giani, M. Monaco et al., "Emergence of *Escherichia coli* ST131 sub-clone H30 producing VIM-1 and KPC-3 carbapenemases, Italy," *Journal of Antimicrobial Chemotherapy*, vol. 69, no. 8, pp. 2293–2296, 2014.

[17] J. C. Cai, R. Zhang, Y. Y. Hu, H. W. Zhou, and G.-X. Chen, "Emergence of *Escherichia coli* sequence type 131 isolates producing KPC-2 carbapenemase in China," *Antimicrobial Agents and Chemotherapy*, vol. 58, no. 2, pp. 1146–1152, 2014.

[18] T. J. Johnson, M. Hargreaves, K. Shaw et al., "Complete genome sequence of a carbapenem-resistant extraintestinal pathogenic *Escherichia coli* strain belonging to the sequence type 131 H30R subclade," *Genome Announcements*, vol. 3, no. 2, pp. 1-2, 2015.

[19] B. A. Rogers, H. E. Sidjabat, and D. L. Paterson, "*Escherichia coli* O25b-ST131: a pandemic, multiresistant, community-associated strain," *Journal of Antimicrobial Chemotherapy*, vol. 66, no. 1, pp. 1–14, 2011.

[20] J. A. Dias Neto, L. D. M. D. Silva, A. C. P. Martins et al., "Prevalence and bacterial susceptibility of hospital acquired urinary tract infection," *Acta Cirurgica Brasileira*, vol. 18, no. 5, pp. 36–38, 2003.

[21] D. Farrell, I. Morrissey, D. de Rubeis, M. Robbins, and D. Felmingham, "A UK multicentre study of the antimicrobial susceptibility of bacterial pathogens causing urinary tract infection," *Journal of Infection*, vol. 46, no. 2, pp. 94–100, 2003.

[22] D. Mathai, R. Jones, M. Pfaller, and T. S. P. G. N. America, "Epidemiology and frequency of resistance among pathogens causing urinary tract infections in 1,510 hospitalized patients: a report from the SENTRY Antimicrobial Surveillance Program (North America)," *Diagnostic Microbiology and Infectious Disease*, vol. 40, no. 3, pp. 129–136, 2001.

[23] S. Ahmad, N. F. Al-Juaid, F. Q. Alenzi, E. H. Mattar, and O. E.-S. Bakheet, "Prevalence, antibiotic susceptibility pattern and production of extended-spectrum β-lactamases amongst clinical isolates of *Klebsiella pneumoniae* at Armed Forces Hospital in Saudi Arabia," *Journal of the College of Physicians and Surgeons Pakistan*, vol. 19, pp. 264-265, 2009.

[24] M. Al-Agamy, A. Shibl, and A. Tawfik, "Prevalence and molecular characterization of extended-spectrum [beta]-lactamase-producing *Klebsiella pneumoniae* in Riyadh, Saudi Arabia," *Annals of Saudi Medicine*, vol. 29, no. 4, pp. 253–257, 2009.

[25] A. A. Kader and A. Kumar, "Prevalence and antimicrobial susceptibility of extended-spectrum beta-lactamase-producing *Escherichia coli* and *Klebsiella pneumoniae* in a general hospital," *Annals of Saudi Medicine*, vol. 25, no. 3, pp. 239–242, 2005.

[26] A. M. Shibl, M. H. Al-Agamy, H. Khubnani, A. C. Senok, A. F. Tawfik, and D. M. Livermore, "High prevalence of acquired quinolone-resistance genes among *Enterobacteriaceae* from Saudi Arabia with CTX-M-15 β-lactamase," *Diagnostic Microbiology and Infectious Disease*, vol. 73, no. 4, pp. 350–353, 2012.

[27] A. F. Tawfik, A. M. Alswailem, A. M. Shibl, and M. H. Al-Agamy, "Prevalence and genetic characteristics of TEM, SHV, and CTX-M in clinical *Klebsiella pneumoniae* isolates from Saudi Arabia," *Microbial Drug Resistance*, vol. 17, no. 3, pp. 383–388, 2011.

[28] F. E. Al Otaibi and E. E. Bukhari, "Clinical and laboratory profiles of urinary tract infections caused by extended-spectrum beta-lactamase-producing *Escherichia coli* in a tertiary care center in central Saudi Arabia," *Saudi Medical Journal*, vol. 34, pp. 171–176, 2013.

[29] M. H. Al-Agamy, A. M. Shibl, N. A. Elkhizzi, D. Meunier, J. F. Turton, and D. M. Livermore, "Persistence of *Klebsiella pneumoniae* clones with OXA-48 or NDM carbapenemases causing bacteraemias in a Riyadh hospital," *Diagnostic Microbiology and Infectious Disease*, vol. 76, no. 2, pp. 214–216, 2013.

[30] M. H. Al-Agamy, A. M. Shibl, M. M. Hafez, M. N. Al-Ahdal, Z. A. Memish, and H. Khubnani, "Molecular characteristics of extended-spectrum β-lactamase-producing *Escherichia coli* in Riyadh: emergence of CTX-M-15-producing *E. coli* ST131," *Annals of Clinical Microbiology and Antimicrobials*, vol. 13, no. 1, pp. 1–7, 2014.

[31] Clinical and Laboratory Standards Institute, *Performance Standards for Antimicrobial Susceptibility Testing 28th Information Supplement (M100-S28)*, Clinical and Laboratory Standards Institute, Wayne, PA, USA, 2018.

[32] C. Dallenne, A. da Costa, D. Decré, C. Favier, and G. Arlet, "Development of a set of multiplex PCR assays for the detection of genes encoding important β-lactamases in *Enterobacteriaceae*," *Journal of Antimicrobial Chemotherapy*, vol. 65, no. 3, pp. 490–495, 2010.

[33] A. P. Magiorakos, A. Srinivasan, R. Carey et al., "Multidrug-resistant, extensively drug-resistant and pandrug-resistant bacteria: an international expert proposal for interim standard definitions for acquired resistance," *Clinical Microbiology and Infection*, vol. 18, no. 3, pp. 268–281, 2012.

[34] D. Paterson, "Recommendation for treatment of severe infections caused by *Enterobacteriaceae* producing extended-spectrum β-lactamases (ESBLs)," *Clinical Microbiology and Infection*, vol. 6, no. 9, pp. 460–463, 2000.

[35] S. Al-Mijalli, "Bacterial uropathogens in urinary tract infection and antibiotic susceptibility pattern in Riyadh Hospital, Saudi Arabia," *Cellular & Molecular Medicine*, vol. 3, no. 1, pp. 1–6, 2017.

[36] M. Q. Alanazi, F. Y. Alqahtani, and F. S. Aleanizy, "An evaluation of *E. coli* in urinary tract infection in emergency department at KAMC in Riyadh, Saudi Arabia: retrospective study," *Annals of Clinical Microbiology and Antimicrobials*, vol. 17, no. 1, pp. 1–7, 2018.

[37] M. C. Raynor, "Urinary infections in men," *The Medical Clinics of North America*, vol. 95, no. 1, pp. 43–54, 2011.

[38] T. A. Rowe and M. Juthani-Mehta, "Urinary tract infection in older adults," *Aging Health*, vol. 9, no. 5, pp. 519–528, 2013.

[39] A. S. John, C. I. Mboto, and B. Agbo, "A review on the prevalence and predisposing factors responsible for urinary tract infection among adults," *European Journal of Experimental Biology*, vol. 6, pp. 7–11, 2016.

[40] A. A. Ahmed, H. Osman, A. M. Mansour et al., "Antimicrobial agent resistance in bacterial isolates from patients with diarrhea and urinary tract infection in the Sudan," *The American Journal of Tropical Medicine and Hygiene*, vol. 63, no. 5, pp. 259–263, 2000.

[41] Ö. Azap, H. Arslan, K. Şerefhanoğlu et al., "Risk factors for extended-spectrum β-lactamase positivity in uropathogenic *Escherichia coli* isolated from community-acquired urinary tract infections," *Clinical Microbiology and Infection*, vol. 16, no. 2, pp. 147–151, 2010.

[42] S. J. Moyo, S. Aboud, M. Kasubi, E. F. Lyamuya, and S. Y. Maselle, "Antimicrobial resistance among producers and non-producers of extended spectrum beta-lactamases in urinary isolates at a tertiary Hospital in Tanzania," *BMC Research Notes*, vol. 3, no. 1, p. 348, 2010.

[43] A. Abduzaimovic, M. Aljicevic, V. Rebic, S. M. Vranic, K. Abduzaimovic, and S. Sestic, "Antibiotic resistance in urinary isolates of *Escherichia coli*," *Materia Socio-Medica*, vol. 28, no. 6, p. 416, 2016.

[44] Y. Taur and M. A. Smith, "Adherence to the Infectious Diseases Society of America guidelines in the treatment of uncomplicated urinary tract infection," *Clinical Infectious Diseases*, vol. 44, no. 6, pp. 769–774, 2007.

[45] M. Tumbarello, M. Sanguinetti, E. Montuaori et al., "Predictors of mortality in patients with bloodstream infections caused by extended-spectrum-β-lactamase-producing *Enterobacteriaceae*: importance of inadequate initial antimicrobial treatment," *Antimicrobial Agents and Chemotherapy*, vol. 51, no. 6, pp. 1987–1994, 2007.

[46] T. M. Hooton, R. Besser, B. Foxman, T. R. Fritsche, and L. E. Nicolle, "Acute uncomplicated cystitis in an era of increasing antibiotic resistance: a proposed approach to empirical therapy," *Clinical Infectious Diseases*, vol. 39, no. 1, pp. 75–80, 2004.

[47] J. Beckford-Ball, "Management of suspected bacterial uri-

nary tract infection," *Nursing Times*, vol. 102, pp. 25-26, 2006.

[48] K. Gupta, T. M. Hooton, K. G. Naber et al., "International clinical practice guidelines for the treatment of acute uncomplicated cystitis and pyelonephritis in women: a 2010 update by the Infectious Diseases Society of America and the European Society for Microbiology and Infectious Diseases," *Clinical Infectious Diseases*, vol. 52, no. 5, pp. e103–e120, 2011.

[49] P.-L. Ho, Y. P.-S. Chu, W.-U. Lo et al., "High prevalence of *Escherichia coli* sequence type 131 among antimicrobial-resistant *E. coli* isolates from geriatric patients," *Journal of Medical Microbiology*, vol. 64, no. 3, pp. 243–247, 2015.

[50] G. G. Zhanel, T. L. Hisanaga, N. M. Laing et al., "Antibiotic resistance in outpatient urinary isolates: final results from the North American Urinary Tract Infection Collaborative Alliance (NAUTICA)," *International Journal of Antimicrobial Agents*, vol. 26, no. 5, pp. 380–388, 2005.

[51] A. C. Gales, H. S. Sader, R. N. Jones, and S. P. Group, "Urinary tract infection trends in Latin American hospitals: report from the SENTRY antimicrobial surveillance program (1997–2000)," *Diagnostic Microbiology and Infectious Disease*, vol. 44, no. 3, pp. 289–299, 2002.

[52] A. Huttner, E. M. Verhaegh, S. Harbarth, A. E. Muller, U. Theuretzbacher, and J. W. Mouton, "Nitrofurantoin revisited: a systematic review and meta-analysis of controlled trials," *Journal of Antimicrobial Chemotherapy*, vol. 70, no. 9, pp. 2456–2464, 2015.

[53] L. E. Nicolle, "Urinary tract infection: traditional pharmacologic therapies," *The American Journal of Medicine*, vol. 113, no. 1, pp. 35–44, 2002.

[54] E. J. Alyamani, A. M. Khiyami, R. Y. Booq, M. A. Majrashi, F. S. Bahwerth, and E. Rechkina, "The occurrence of ESBL-producing *Escherichia coli* carrying aminoglycoside resistance genes in urinary tract infections in Saudi Arabia," *Annals of Clinical Microbiology and Antimicrobials*, vol. 16, no. 1, pp. 1–13, 2017.

[55] S. Bouchillon, B. Johnson, D. Hoban et al., "Determining incidence of extended spectrum β-lactamase producing *Enterobacteriaceae*, vancomycin-resistant *Enterococcus faecium* and methicillin-resistant *Staphylococcus aureus* in 38 centres from 17 countries: the PEARLS study 2001–2002," *International Journal of Antimicrobial Agents*, vol. 24, no. 2, pp. 119–124, 2004.

[56] D. S. Hansen, H. Schumacher, F. Hansen et al., "Extended-spectrum β-lactamase (ESBL) in Danish clinical isolates of *Escherichia coli* and *Klebsiella pneumoniae*: Prevalence, β-lactamase distribution, phylogroups, and co-resistance," *Scandinavian Journal of Infectious Diseases*, vol. 44, no. 3, pp. 174–181, 2012.

[57] P. J. Simner, G. G. Zhanel, J. Pitout et al., "Prevalence and characterization of extended-spectrum β-lactamase–and AmpC β-lactamase–producing *Escherichia coli*: results of the CANWARD 2007–2009 study," *Diagnostic Microbiology and Infectious Disease*, vol. 69, no. 3, pp. 326–334, 2011.

[58] A. D. Celik, Z. Yulugkural, F. Kuloglu et al., "CTX-M type extended spectrum β-lactamases in *Escherichia coli* isolates from community acquired upper urinary tract infections at a university in the european part of Turkey," *Journal of Microbiology, Immunology and Infection*, vol. 43, no. 2, pp. 163–167, 2010.

[59] A. Pathak, Y. Marothi, V. Kekre, K. Mahadik, R. Macaden, and C. S. Lundborg, "High prevalence of extended-spectrum β-lactamase-producing pathogens: results of a surveillance study in two hospitals in Ujjain, India," *Infection and Drug Resistance*, vol. 5, pp. 65–73, 2012.

[60] N. Al Mously, O. Al Arfaj, L. Al Fadhil, and S. Mukaddam, "Antimicrobial susceptibility patterns of ESBL *Escherichia coli* isolated from community and hospital-acquired urinary tract infections," *Journal of Health Specialties*, vol. 4, no. 2, pp. 133–139, 2016.

[61] H. Dhanji, M. Doumith, O. Clermont et al., "Real-time PCR for detection of the O25b-ST131 clone of *Escherichia coli* and its CTX-M-15-like extended-spectrum β-lactamases," *International Journal of Antimicrobial Agents*, vol. 36, no. 4, pp. 355–358, 2010.

[62] S. Vimont, A. Boyd, A. Bleibtreu et al., "The CTX-M-15-producing *Escherichia coli* clone O25b: H4-ST131 has high intestine colonization and urinary tract infection abilities," *PloS One*, vol. 7, no. 9, pp. 1–10, 2012.

12

Environmentally Friendly Production of D(−) Lactic Acid by *Sporolactobacillus nakayamae*: Investigation of Fermentation Parameters and Fed-Batch Strategies

Susan Michelz Beitel,[1] Luciana Fontes Coelho,[1] Daiane Cristina Sass,[1] and Jonas Contiero[1,2]

[1] *Department of Biochemistry and Microbiology, Institute of Bioscience, São Paulo State University (UNESP), São Paulo, SP, Brazil*
[2] *Associate Laboratory IPBEN, UNESP, Av. 24A 1515 CEP, 13506-900, Rio Claro, SP, Brazil*

Correspondence should be addressed to Jonas Contiero; jconti@rc.unesp.br

Academic Editor: Joseph Falkinham

The interest in the production of lactic acid has increased due to its wide range of applications. In the present study, the variables that affect fermentative D(−) lactic acid production were investigated: neutralizing agents, pH, temperature, inoculum percentage, agitation, and concentration of the medium components. An experimental design was applied to determine the optimal concentrations of the medium components and fermentation was studied using different feeding strategies. High production (122.41 g/L) and productivity (3.65 g/L·h) were efficiently achieved by *Sporolactobacillus nakayamae* in 54 h using a multipulse fed-batch technique with an initial medium containing 35 g/L of yeast extract (byproduct of alcohol production), 60 g/L of crystallized sugar, and 7.5 mL/L of salts. The fermentation process was conducted at 35°C and pH 6.0 controlled by NaOH with a 20% volume of inoculum and agitation at 125 rpm. The production of a high optically pure concentration of D(−) lactic acid combined with an environmentally friendly NaOH-based process demonstrates that *S. nakayamae* is a promising strain for D(−) lactic acid production.

1. Introduction

Lactic acid production has received greater attention due to its considerable potential in biotechnological applications in a wide range of fields as well as the increasing need for new biomaterials, such as biodegradable, biocompatible, and polylactic products [1–3]. This compound is used in the food, chemical, pharmaceutical, and cosmetic industries, is amenable to several chemical conversions, and is used as a precursor for various chemicals and materials [4].

The two optically active isomeric forms of lactic acid are D(−) and L(+). Lactic acid can be obtained by chemical synthesis, which inevitably leads to the production of a racemic mixture, whereas the pure enantiomers L(+) or D(−) lactic acid can be produced through fermentative processes. The industrial production of L(+) lactic acid has been extensively reported using several substrates, but few studies have addressed the commercial production of D(−) lactic acid [5].

Lactic acid is becoming important as an intermediate feedstock used as a monomer for the synthesis of biodegradable polymers [6]. Lactic acid polymer [polylactic acid (PLA)] has three different forms: poly-L-lactic acid, poly-D-lactic acid, and poly-DL-lactic acid [7]. PLA is used in the field of biomedicine in the form of implants, sutures, bone fixation material, and microsphere drug delivery systems due to its excellent biocompatibility and biodegradability [8, 9].

For the production of lactic acid, the most important carbohydrates are the disaccharides lactose, maltose, and sucrose [10]. Sucrose enters cells through a specific permease system and is split into glucose and fructose [11]. Since PLA manufacturers require large amount of lactic acid at a relatively low cost, the determination of inexpensive raw materials is important to the feasibility of the microbial production of lactic acid [12]. However, the raw materials used as the carbon source must undergo hydrolysis in order to make the sugars assimilable to the bacteria [13]. Thus, pure sugars facilitate downstream processing, whereas inexpensive raw materials as alternative substrates result in an increase in the downstream processing cost due to their relatively high amount of impurities [14].

The fermentation rate depends mainly on pH, temperature, agitation, initial substrate concentration, and the concentration of nitrogenous nutrients [15]. Batch fermentation is widely used in lactic acid production. However, the long fermentation times required with this technique results in low productivity as well as low cell concentrations. The inhibitory effects of the initial substrate concentration on production are also considered a major drawback of this fermentation method. To solve such problems, fed-batch fermentation has been investigated to improve the production, productivity, and yield of lactic acid [3].

Lactic acid is produced by several microorganisms, especially lactic acid bacteria, including genetically modified strains, and some fungi [15–18]. The species *Sporolactobacillus nakayamae* belongs to the family Sporolactobacillaceae and is an endospore-forming, microaerophilic, mesophilic, homofermentative, and Gram-positive bacterium that exclusively produces D(−) lactic acid [19]. The aims of the present study were to investigate the influence of the concentration of culture medium components as well the effects of temperature, pH, inoculum percentage, agitation conditions, and pH controlling agents on D(−) lactic acid production by *S. nakayamae*. Fed-batch strategies were also studied to increase production and productivity levels of D(−) lactic acid.

2. Materials and Methods

2.1. Bacterium Strain and Culture Medium. The producer strain was the previously identified *as Sporolactobacillus nakayamae* by Macrogen (Korea). A glucose, yeast extract, and peptone (GYP) medium containing 20% (v/v) glycerol was used for the maintenance of the strains at −80°C. The cells were propagated three times in GYP broth before use and incubated at 35 ± 1°C under stationary conditions until the optical density (OD_{600}) reached 2.5 (about 24 h).

2.2. Optimization of Fermentation Media

2.2.1. Response Surface Methodology. The response surface methodology (RSM) is a set of experimental strategies, mathematical methods, and statistical inferences that enable efficient empirical exploration in a reduced number of experiments [20]. To evaluate the influence of crystallized sugar, yeast extract, and the salts solution (4% $MgSO_4{\cdot}7H_2O$, 0.16% $MnSO_4{\cdot}4H_2O$, 0.2% $FeSO_4{\cdot}7H_2O$, and 0.2% NaCl) on lactic acid production by *S. nakayamae*, the RSM was used with three replicates at the center point, totaling 17 experiments. The medium used for fermentation was formulated based on the previously used GYP medium. Crystallized sugar is a commercial table sugar found at an affordable price in Brazil. The yeast extract was from Zilor (a major ethanol-producing company in Brazil) and is a byproduct of ethanol fermentation, which is processed and available at a very low cost. The levels used for encoding the independent variables are shown in Table 1.

Table 1: Variables and respective levels for D(−) lactic acid production using response surface methodology.

Variable		Levels				
		−1.68	−1	0	1	1.68
Crystallized sugar (g/L)	X_1	52.8	80	120	160	187.2
Yeast extract (g/L)	X_2	1.4	15	35	55	68.6
GYP salt solution (ml/L)	X_3	0.9	3	6	9	11.04

Fermentation runs were carried out in 125 mL Erlenmeyer flasks with an inoculum concentration of 10%, initial pH at 6.0, and 5% $CaCO_3$ used as the pH neutralizer. The flasks were incubated for 48 hours at 35°C under stationary conditions. D(−) lactic acid and residual sugar were quantified at the end of fermentation. To validate the optimization of the medium composition, five repetitions were carried out using the optimized conditions (120 g/L of crystallized sugar, 35 g/L of yeast extract, and 7.5 mL/L of GYP salt solution) to confirm the results of the response surface analysis. The results were expressed as mean values. The experimental design was determined using the Statistica 7 software program (StatSoft, Tulsa, USA), which was also used for the analysis of the results.

2.3. Effect of Culture Conditions on Lactic Acid Production. Variations of different process parameters (pH, pH controlling agent, temperature, agitation, and inoculum volume) were tested to optimize D(−) lactic acid production. Fermentation was conducted in a bioreactor (Infors HT Multifors 2) with a working volume of 300 mL of modified GYP medium (35 g/L of yeast extract, 120 g/L of crystallized sugar, and 7.5 mL/L of GYP salts solution) and 10% (v/v) of inoculum (except in tests on the influence of inoculum size on lactic acid production). The effect of pH on lactic acid production was evaluated from 5.0 to 7.0. The influence of the pH controlling agent was evaluated throughout fermentation by the automatic addition of controlling agents. The influence of temperature was determined from 31 to 39°C and the influence of agitation was evaluated from 0 to 150 rpm. The production medium was inoculated with different inoculum levels (5 to 25% v/v). Fermentation was stopped after 48 hours. Samples were withdrawn periodically from the bioreactor to determine the concentrations of sucrose and D(−) lactic acid. The final fermentation volume was considered for the calculation of lactic acid normalization (production g/L × final volume mL/initial volume mL).

2.4. Fed-Batch Fermentation Strategies. Fed-batch fermentation was performed in a 750 mL bioreactor (Infors HT Multifors 2) with a working volume of 300 ml of optimized medium containing 35 g/L of yeast extract, 7.5 mL/L of salts solution, and various amounts of crystallized sugar, as required. Fermentation was run at 35°C and 125 rpm with an inoculum volume of 20% (v/v). NaOH was added automatically to control the pH at 6.0. The feed solution contained 730 g/L of crystallized sugar and 1% of yeast extract. In fed-batch fermentation, the feed solution was pumped into the fermenter using a peristaltic pump coupled to a computer. Samples were withdrawn at established intervals for the determination of lactic acid production, sugar consumption, and cell growth. Batch fermentation was conducted with an initial crystallized sugar concentration of 120 g/L and NaOH was added throughout the process to control the pH.

Different feeding strategies were tested to enhance lactic acid production. In the single-pulse fed-batch fermentation, the initial concentration of crystallized sugar was 120 g/L. The feed solution was supplied once when the residual sugar concentration decreased to 60 g/L (at 18 hours) to bring the concentration up to approximately the initial value. Multipulse fed-batch fermentation was conducted with an initial crystallized sugar concentration of 60 g/L, followed by several additions of the feed solution when the residual sugar concentration was below 40 g/L. In the constant fed-batch fermentation, the feed solution was added when the residual sugar concentration decreased from the initial 120 g/L to 60 g/L (at 18 hours) and pumped at a rate of 3 ml/h until 30 hours. In exponential fed-batch fermentation using the Irirs 6 software, a control strategy developed by Nor et al. (2001) [21] was used, in which the feed rate (F) was determined by (B), which takes into account a mass balance, assuming a constant cell yield on substrate as well as a constant maintenance coefficient throughout the fermentation.

$$F = \left(\frac{\mu}{Y_{X/S}\,(Si - S)} \right) V_0 X_0 \exp\left(\mu t\right) \qquad \text{(B)}$$

in which S and Si are, respectively, the crystallized sugar concentration in the medium at the beginning of the feed (g/l) and in the feed solution (g/l), t is culture time (h), V_0 is initial culture volume (L), X and X_0 are, respectively, the cell concentration at the start of feeding (g/l) and initial cell concentration (g/l), $Y_{X/S}$ is cell yield on sucrose (g of cells/g of crystallized sugar), and μ is the specific cell growth rate (h^{-1}). Feeding was begun after 18 hours and lasted for 10 hours.

2.5. Analytical Techniques. The fermented broth was used for the determination of D(−) lactic acid and residual sucrose. The samples were centrifuged at 7000 ×g for 15 minutes. The supernatant was filtered through a 0.22 μm membrane and used for analysis. The concentrations were determined using high-performance liquid chromatography (HPLC) equipped with a Rezex ROA column (Phenomenex, USA) and a differential refracting index detector (RID-A, Shimadzu). The mobile phase (0.005 M H_2SO_4) was fed at a flow rate of 0.6 mL/min and temperature was maintained at 65°C. After the medium optimization process, the optical purity of the lactic acid produced was determined by HPLC using a Chirex 3126 phenomenex column with 1 mM of $CuSO_4$ as the mobile phase at 1 mL/min (26°C). Cell growth was determined based on turbidity at 600 nm. The absorbance of the sample measured in a spectrophotometer was correlated to the dry weight mass.

3. Results and Discussion

3.1. Influence of Concentration of Medium Components on Lactic Acid Production Using Response Surface Methodology. To obtain greater lactic acid production, the culture conditions were optimized using a factorial design. The independent variables selected for the study were crystallized sugar, yeast extract, and GYP salt solution distributed on two levels, two axial points, and three central points. The results obtained after fermentation were expressed as g/l for the amount of lactic acid produced and residual sugar, g/g for yield, and g/L·h for productivity (Table 2). After 48 hours of fermentation, the greatest production was 88.24 g/L of D(−) lactic acid (run 10). In this run, residual sucrose was 25.49 g/L, productivity was 1.84 g/L·h, and the yield was 0.93 g/g.

Table 3 displays the results of the analysis of variance (ANOVA) for the selected quadratic model. The coefficient of determination (R^2 = 0.84) suggests that the model explains 84% of the total variation in the response. Fisher's test demonstrated that the model was significant [Fcalc (4.39) > Ft (3.68)] and has a very low probability value (Pmodel = 0.03). Among the model terms, interactions X_1X_1 and X_2X_2 were significant with a 95% probability, demonstrating a negative effect, likely due to catabolite repression by the substrates. The results were submitted to multiple regression analysis methods and yielded the regression equation (C):

$$\begin{aligned} Y = {} & 85.34 + 3.93X_1 - 9.62X_1X_1 + 3.44X_2 \\ & - 13.78X_2X_2 - 0.02X_3 - 2.88X_3X_3 - 4{,}89X_1X_2 \qquad \text{(C)} \\ & + 4{,}89X_1X_3 + 5.01X_2X_3 \end{aligned}$$

in which Y is the predicted response (D(−) lactic acid production) and X_1, X_2, X_3 are, respectively, the coded values of the test variables crystallized sugar, yeast extract, and salt solution. The response for the regression equation is plotted in Figure 1. The graphs shows the interaction of the variables and optimum levels for D(−) lactic acid production.

Based on the response surface graphs, the medium components that induced the greatest D(−) lactic acid production were 120 g/L of crystallized sugar, 35 g/L of yeast extract, and 7.5 mL/L of salt solution. The concentration of crystallized sugar was selected considering residual sucrose, which should not exceed 10 g/L. The validation of the model for lactic acid production optimization was performed using the selected concentrations in five replicates and the results were expressed as mean values. Under these conditions, production was 89.69 g/L of D(−) lactic acid, productivity was 1.87 g/Lh, the yield was 0.96 g/g, and residual sucrose was 26.67 g/L. These results confirm the validity and usefulness of the model equation. The optimized culture medium was analyzed using HPLC with a chiral column to determine the optical purity of the lactic acid produced. This analysis revealed 98.97% D(−) lactic acid and 1.03% L(+) lactic acid, thereby demonstrating the high optical purity characteristic of *S. nakayamae*.

3.2. Effects of Neutralizing Agents on D(−) Lactic Acid Production. A drop in pH occurs during lactic acid fermentation

TABLE 2: RSM design with real values and experimental results for lactic acid production, productivity, yield, and residual sugar concentration.

	Experimental factors			Response variable			
Run	Crystallized sugar (g/L)	Yeast extract (g/L)	Salt solution (mL/L)	Lactic acid	Productivity	$Y_{p/s}$	Residual sugar
	X_1	X_2	X_3	g/L	g/Lh	g/g	g/L
1	80.0	15.0	3.0	65.13	1.36	0.81	0
2	160.0	15.0	3.0	59.02	1.23	0.75	81.41
3	80.0	55.0	3.0	61.32	1.28	0.97	17.00
4	160.0	55.0	3.0	58.36	1.22	0.88	93.34
5	80.0	15.0	9.0	26.64	0.56	0.97	52.44
6	160.0	15.0	9.0	62.84	1.31	0.84	85.00
7	80.0	55.0	9.0	65.59	1.37	0.99	14.03
8	160.0	55.0	9.0	59.51	1.24	0.82	87.35
9	120.0	35.0	0.9	71.01	1.48	0.98	47.20
10	120.0	35.0	11.0	88.24	1.84	0.93	25.49
11	120.0	1.40	6.0	44.10	0.92	0.98	74.86
12	120.0	68.5	6.0	53.53	1.12	0.99	65.84
13	52.8	35.0	6.0	50.86	1.06	0.96	0.00
14	187.2	35.0	6.0	70.31	1.46	0.77	96.34
15	120.0	35.0	6.0	87.98	1.83	0.98	30.64
16	120.0	35.0	6.0	81.85	1.71	0.94	33.34
17	120.0	35.0	6.0	85.33	1.78	0.89	24.35

X_1 = crystallized sugar; X_2 = yeast extract; X_3 = salt solution. Culture conditions: 35°C, initial pH at 6.0 neutralized by 5% $CaCO_3$, stationary fermentation, and 10% inoculum. Fermentation was conducted for 48 hours.

TABLE 3: ANOVA with estimated regression coefficients for D(−) lactic acid production.

Source	Coefficient	Sum of squares	Degrees of freedom	Mean square	F ratio	P
X_1	3.93	211.632	1	211.632	2.351	0.169
X_1X_1	−9.62	1043.505	1	1043.505	11.594	0.011*
X_2	3.44	161.814	1	161.814	1.798	0.221
X_2X_2	−13.78	2141.408	1	2141.408	23.794	0.001*
X_3	−0.02	0.005	1	0.005	0.000	0.994
X_3X_3	−2.88	94.113	1	94.113	1.045	0.340
X_1X_2	−4.89	191.395	1	191.395	2.126	0.188
X_1X_3	4.89	191.982	1	191.982	2.133	0.187
X_2X_3	5.01	200.901	1	200.901	2.232	0.178
Model	—	3557.58	9	395.28	4.39	0.03
Error	—	629.97	7	89.99	—	—
Lack of fit	—	611.07	5	122.21	12.93	0.07
Pure error	—	18.90	2	9.45	—	—
Total	—	4187.55	16	—	—	—

X_1 = crystallized sugar; X_2 = yeast extract; X_3 = salt solution. *Statistically significant at 95% probability level.

and the microorganism is unable to continue the fermentation. Thus, neutralization is essential and several bases can be used for this purpose, which is closely related to downstream processing [22]. The effect of different neutralizing agents (KOH 10 N, $Ca(OH)_2$ 6 N, NaOH 10 N, 5% $CaCO_3$, 6% $CaCO_3$, and 27% NH_4OH) on lactic acid production was investigated. Fermentation was conducted using the optimized culture medium at 35°C and pH 6.0 controlled by neutralizing agents automatically added to the process, except $CaCO_3$, which was added to the fermentation vessel at the beginning of the process. Fermentation was conducted for 48 hours.

Table 4 displays the results of the pH neutralizing tests. Based on these data, NaOH was selected as the neutralizing agent considering the practicality of the process and the maximum volumetric lactic acid production (103.71 g/L), productivity (2.16 g/L·h), and yield (0.94 g/g). No residual sugar was found in this experiment, which is interesting from the standpoint of further lactic acid purification and polymerization. Moreover, NaOH used as a pH controller in

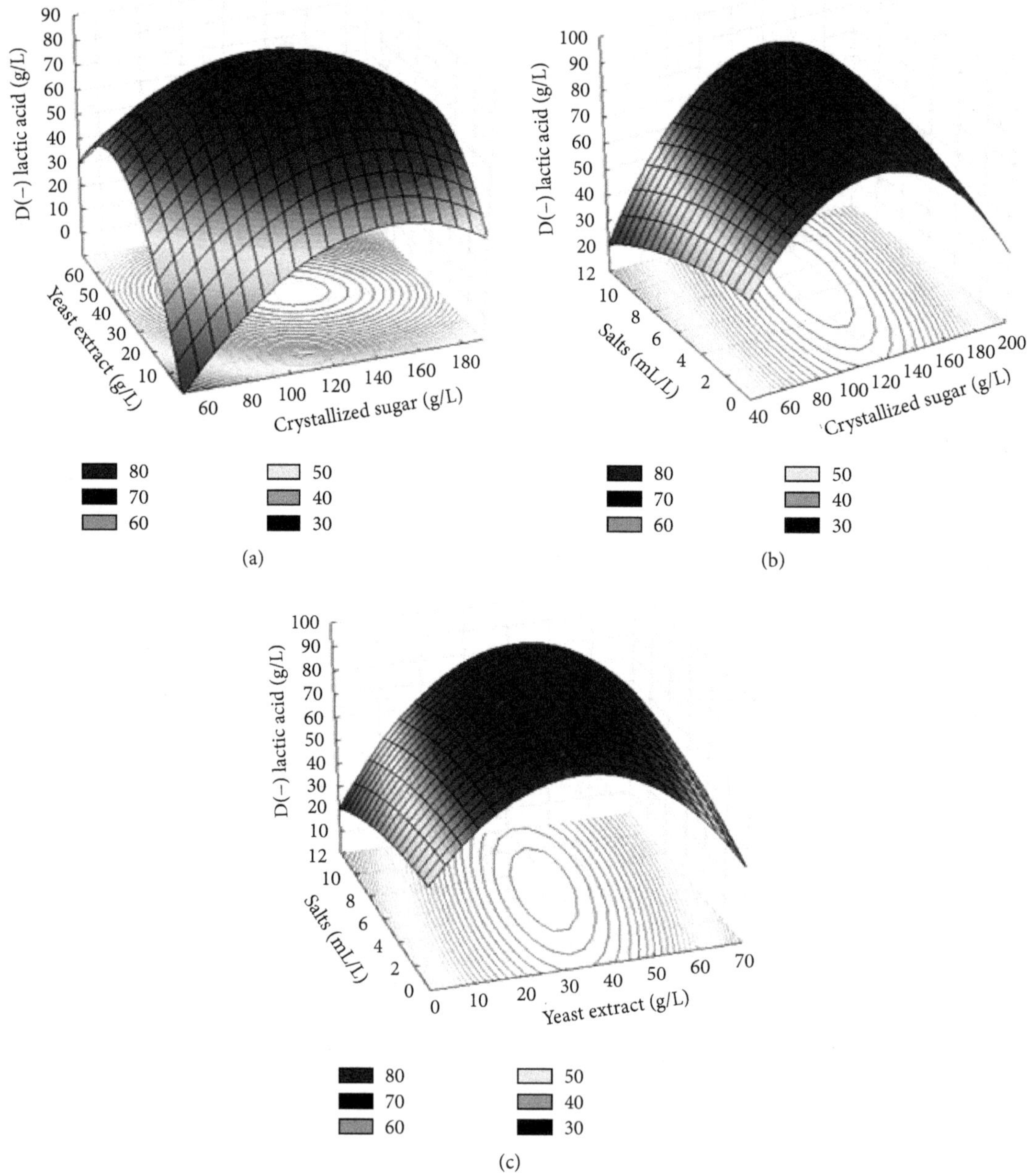

FIGURE 1: Response surface for lactic acid production by *S. nakayamae*. Interactions between (a) yeast extract and crystallized sugar, (b) salts and crystallized sugar, and (c) yeast extract and salts.

TABLE 4: Comparison of D-lactic acid fermentation by *S. nakayamae* using different neutralizing agents.

Neutralizing agent	Price R$/kg	Used in fermentation g/300 mL	Cost in fermentation R$	D(−) lactic acid* g/L	P g/L·h	$Y_{p/s}$ g/g	RS* g/L
$CaCO_3$ 5%	44.00	15.00	0.66	91.17	1.90	0.90	11.76
$Ca(OH)_2$	40.00	11.11	0.44	88.95	1.85	0.81	3.52
NaOH	51.00	12.00	0.61	103.71	2.16	0.94	0.00
KOH	86.00	5.61	0.48	84.45	1.98	0.79	0.00
NH_4OH	17.20	9.17	0.17	16.05	0.33	0.13	88.85
$CaCO_3$ 6%	44.00	18.00	0.79	37.46	0.78	0.85	65.99

P = productivity; $Y_{p/s}$ = Yield; RS = residual sucrose. *Normalized lactic acid and sucrose titer was calculated from measured titer (not shown) from fermentation broth with dilution ratio of neutralizing agent used. Productivity and yield were calculated considering normalized data.

fermentation processes does not generate precipitated waste, making it environmentally friendly [23]. In contrast, the ammoniacal solution was not an appropriate neutralizing agent, as demonstrated by the low lactic acid production (16.05 g/L). This result is in agreement with data reported by other authors studying lactic acid production by *Rhizopus oryzae*, with 24.9 g/L of lactic acid produced when an ammoniacal solution was used as neutralizing agent, as a high concentration of ammonia can be toxic to microbial cells [24].

Calcium carbonate ($CaCO_3$) is often used as such a pH controlling agent [25] and yielded the second highest D(−) lactic acid production (91.17 g/L) when used at a proportion of 5% in the present study. In contrast, the 6% proportion inhibited microorganism growth. The low solubility of $CaCO_3$ in the fermentation broth could cause problems in the subsequent purification process, as the most used recovery method consumes lime as well as sulfuric acid and also produces a large quantity of calcium sulfate sludge as solid waste [26].

When $Ca(OH)_2$ 6 N was used, 88.95 g/L of D(−) lactic acid was achieved. Although this neutralizing agent is affordable, it is necessary to stir the solution constantly using a magnetic plate/stove during fermentation to prevent clogs in the base feeding tubes, resulting in an added cost to the process. A previous study reports D(−) lactic acid production of 127 g/L using $Ca(OH)_2$ as the neutralizing agent on the laboratory scale; the author states that the lower osmotic pressure derived from this neutralizing base may be partially responsible for the facilitation of D-lactic acid fermentation [27].

As shown in Figure 2, the process ended without any residual sugar when NaOH or KOH was used as the neutralizing agent. Besides lactic acid production of 84.45 g/L using KOH neutralization, this agent is unaffordable for large scale production. The figure shows the time course of lactic acid and residual sucrose concentration using different neutralizing agents for 48 hours.

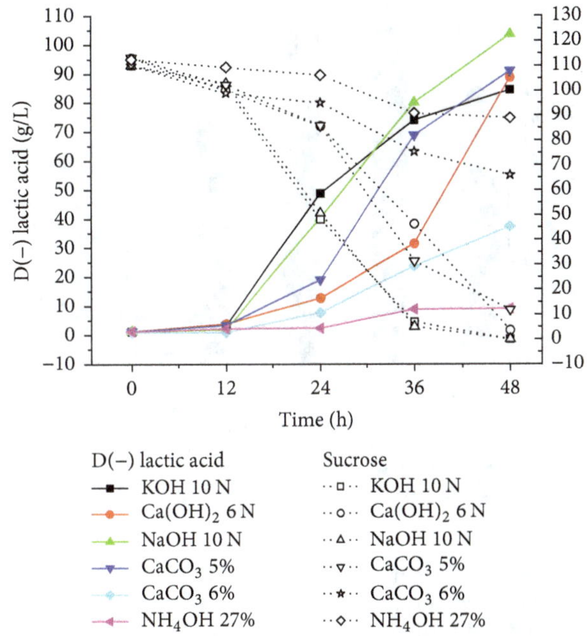

FIGURE 2: Time course of lactic acid and residual sucrose concentration using different neutralizing agents [KOH, $Ca(OH)_2$, NaOH, CaCO, and NH_4OH]. Culture conditions: GYP modified medium, 35°C pH 6.0, 10% (v/v) inoculums, 100 rpm.

3.3. *Effects of Culture Conditions on Lactic Acid Production*

3.3.1. Influence of Temperature and pH on D(−) Lactic Acid Production. The capacity of microorganisms to produce lactic acid is influenced by conditions, such as temperature, pH, agitation, and inoculum percentage, the optimization of which is essential. Temperature and pH are important parameters that affect the fermentation process [28]. D(−) lactic acid produced by the *S. nakayamae* was efficient at pH 6.0, with an optimum temperature at 35°C (Figures 3(a) and 3(b)), reaching a production of 99.43 g/L with a low level of residual sucrose (3.45 g/L). The productivity and yield under these conditions were 2.07 g/L·h and 0.91 g/g, respectively (data not shown). Other authors report achieving a high D-lactate concentration using strains of *Sporolactobacillus* sp. at 42°C [29, 30]. Temperature exerts an influence on the activity of metabolic cells. Most bacteria that convert sugar into lactic acid are classified as either thermophilic or mesophilic, with optimum growth between 20 and 40°C [31].

The optimum pH for lactic acid production by microorganisms ranges from 5.0 to 7.0 and is dependent on the species of microorganism [32]. In the present study, lactic acid production decreased significantly when the pH was lower than 6.0 (Figure 3(b)). Likewise, previous authors report 52.37 g/L of D(−) lactic acid by *Lactobacillus* sp. LMI8 when fermentation was conducted at pH 6.0 [33].

3.3.2. Effect of Agitation and Inoculum Size on D(−) Lactic Acid Fermentation. The influence of agitation and inoculum size on D( ) lactic acid production was investigated, the results of which are shown in Figure 4. High D(−) lactic acid production (107.67 g/L) was found when agitation was 125 rpm. Moreover, high productivity (2.24 g/Lh) and yield (0.98 g/g) (data not shown) were achieved and no residual sucrose was found under this condition. Stationary conditions were inefficient for lactic acid production (65.03 g/L), possibly due to insufficient homogenization of the culture medium. In contrast, a previous study reported no difference in lactic acid production by *Lactobacillus casei* under stationary conditions or agitation at 100 rpm [34].

Maximum lactic acid production (113.73 g/L) was obtained with low residual sucrose (3.37 g/L) when using 20% (v/v) of inoculum in the culture medium. Moreover, productivity was 2.37 g/Lh and yield was 0.98 g/g under this condition (data not shown). The low lactic acid production (82.34 g/L) with 5% (v/v) inoculum could be attributed to the low density of the starter culture. In a previous study addressing the influence of inoculum size on L(+) lactic acid production by *Rhizopus oryzae* ASC081, production decreased when the inoculum size was less than 10% due to an inadequate enzymatic efficiency [35].

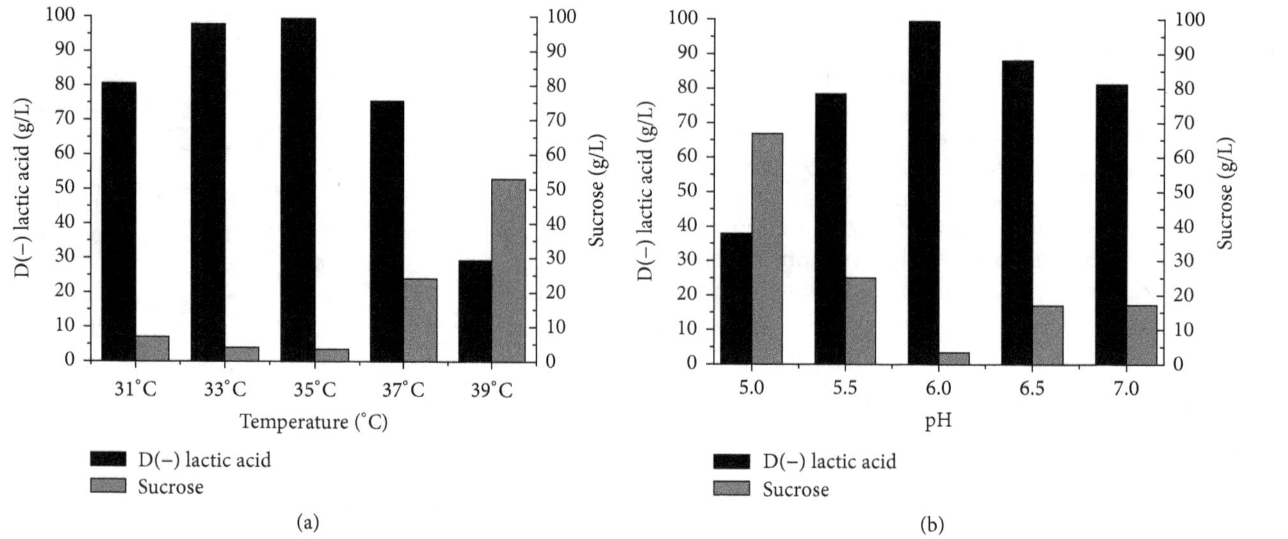

FIGURE 3: Effect of (a) temperature and (b) pH on D(−) lactic acid production and sucrose consumption by *S. nakayamae*. Culture conditions: GYP modified medium (120 g/L of crystallized sugar, 35 g/L of yeast extract, and 7.5 salt solution) with NaOH 10 N as pH neutralizing agent. (a) GYP modified medium at pH 6.0, under 100 rpm, with 10% inoculum; (b) GYP modified medium at 35°C, under 100 rpm, with 10% inoculum.

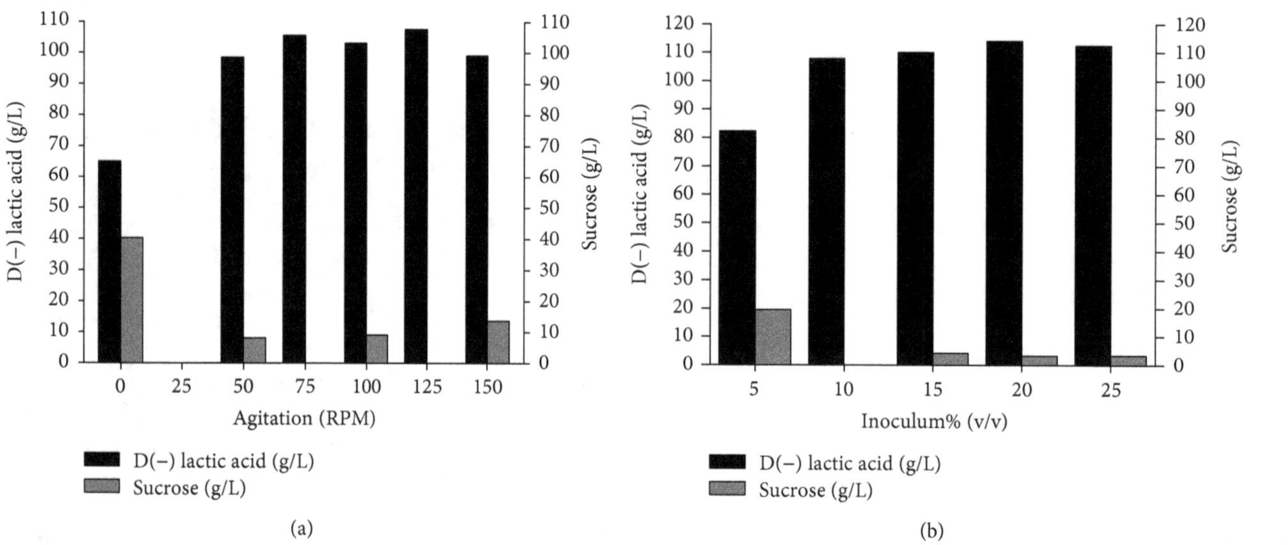

FIGURE 4: Effect of (a) agitation and (b) inoculum size on D(−) lactic acid production and sucrose consumption by *S. nakayamae*. Culture conditions: GYP modified medium (120 g/L crystallized sugar, 35 g/L yeast extract, and 7.5 mL/L salt solution) with NaOH 10 N as pH neutralizing agent. (a) GYP modified medium at 35°C, pH 6.0, with 10% inoculum; (b) GYP modified medium at 35°C and pH 6.0 under 125 rpm.

3.4. Fed-Batch Fermentation Strategies. Fed-batch cultures were developed to maximize the production and productivity of D(−) lactic acid by *S. nakayamae*. The time courses of cell growth, sucrose consumption, and D(−) lactic acid production during the batch and fed-batch cultures (pulse and multipulse) are shown in Figure 5.

Maximum D(−) lactic acid production (126.64 g/L) was obtained at the end of the single-pulse fed-batch (Figure 5(b)). The highest cell density (11.67 g/L) was also found in this culture. However, it took 78 hours to achieve these results, with productivity of 1.62 g/L·h. Using pulse fed-batch to improve D(−) lactic acid productivity by *Sporolactobacillus inulinus* YBS1-5, Bai et al. (2016) [36] achieved 107 g/L of lactic acid and productivity of 1.19 g/L·h at 90 h.

The fed-batch fermentation with two pulses (Figure 5(c)) initiated at low levels of sucrose demonstrated promising results, achieving the second greatest lactic acid production (122.64 g/L) at 54 hours; at this time, the residual sugar concentration was 13.47 g/L. Moreover, this process had the highest productivity at 13 hours (3.65 g/L·h) in comparison to

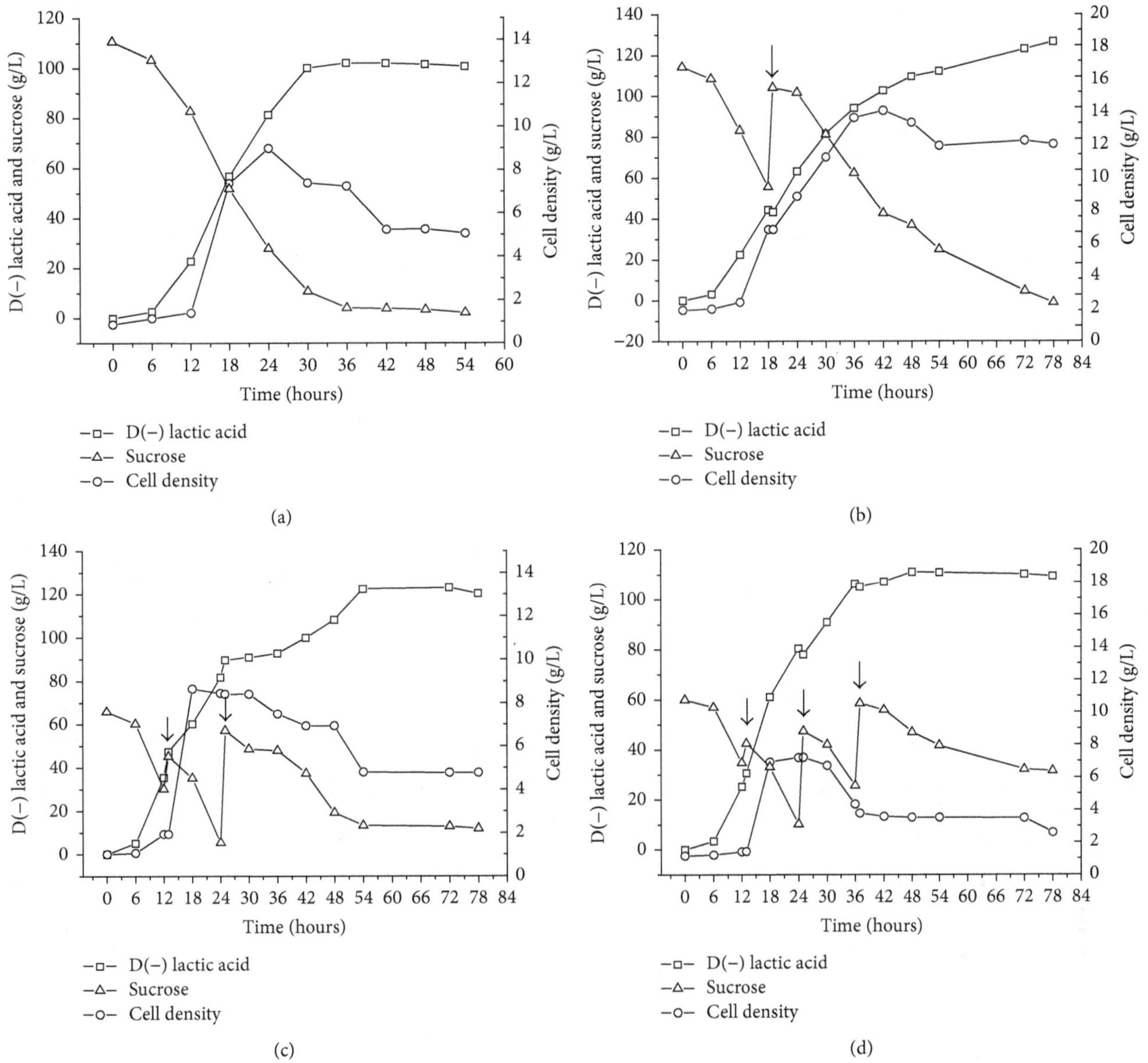

FIGURE 5: Time courses of sucrose consumption, D(−) lactic acid production, and cell growth in batch fermentation (a), pulse fed-batch (b), and multipulse fed-batch fermentation (c and d) by *S. nakayamae*. Experimental conditions: medium containing 35 g/L of yeast extract and 7.5 mL/L of GYP salts and crystallized sugar (as required); pH 6.0, NaOH 10 N as controlling agent, temperature at 35°C, 125 rpm, and 20% of inoculum (v/v).

the other methods. Bai et al. (2003) [37] report similar results studying fed-batch fermentation for L(+) lactic production by *Lactobacillus lactis*.

The multipulse, fed-batch culture with three pulses (Figure 5(d)) was not satisfactory. The amount of sugar added likely caused high osmotic pressure in the cells, resulting in plasmolysis and decreasing both the fermentation rate and sugar utilization. This finding is in agreement with data described by Kotzamanidis et al. (2002) [38]. Other authors report greater production by *Sporolactobacillus* sp. CASD using a multipulse strategy in comparison to a single pulse [30].

The final D(−) lactic acid concentration with the constant feed rate and exponential fed-batch systems was 119.03 g/L and 121.02 g/L, respectively (Figures 6(a) and 6(b)). These strategies were not satisfactory, as the sucrose concentration was greater than 20 g/L at the end of the process. Although productivity values were low at the end of the process, the highest productivity was 3.15 g/L·h with the exponential fed-batch method at 24 hrs and 3.06 g/L·h with the constant feed rate at 30 hrs. Bernardo et al. (2016) [39] report similar results using a constant feed rate for L(+) lactic acid production by *Lactobacillus rhamnosus* B103. The authors report a slight increase in lactic acid production, but a high residual substrate concentration at the end of the process. Other authors report a 56.5% improvement in L(+) lactic acid production using exponential feeding compared to traditional batch culture, achieving 157.5 g/L of L(+) lactic

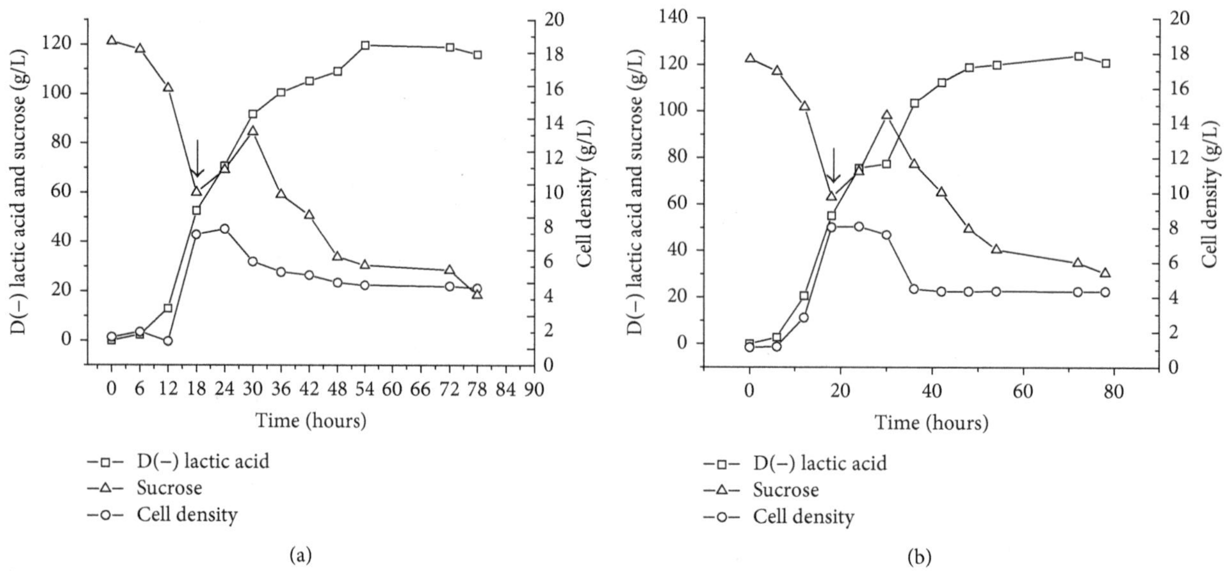

Figure 6: Time courses of sucrose consumption, D(−) lactic acid production, and cell growth in constant feed rate (a) and exponential fed-batch fermentation (b) by *S. nakayamae*. Experimental conditions: medium containing 35 g/L of yeast extract and 7.5 mL/L of GYP salts and crystallized sugar (as required); pH 6.0, NaOH 10 N as controlling agent, temperature at 35°C, 125 rpm, and 20% of inoculum (v/v).

acid and productivity of 1.88 g/L·h by *Lactobacillus casei* after 84 h of fermentation [40].

Except for the three-pulse feeding strategy, all other fed-batch strategies were able to increase the production and productivity of D(−) lactic acid when compared to batch fermentation (Figure 5(a)), which achieved maximum production and productivity of 106.95 g/L and 2.67 g/L·h, respectively.

The fed-batch process could increase the total substrate content in the bioreactor by maintaining a low substrate concentration during fermentation and avoiding the inhibitory effects of sugar on lactic acid production as well as reducing the negative effects of osmotic pressure on bacterial cells. Other advantages of this process are higher cell concentration and productivity as well as a high titer of lactic acid production at the end of the fermentation process. The final product was not an inhibitory factor, as the lactic acid was presented as a salt (sodium lactate) when NaOH was used as the neutralizing agent. However, the osmotic pressure may be high at the end of the process, which can lead to a reduction in lactic acid production. This decline also could be related to decreasing metabolic potential of an aging microbial biocatalyst, as reported before by Ou et al., 2011 [41]. This paper is a part of a doctoral thesis [41].

4. Conclusion

D(−) lactic acid was successfully produced by *S. nakayamae* using fermentation under optimized conditions. The experimental design was very useful for determining the optimal concentrations of constituents that have significant effects on D(−) lactic acid production. This study has shown that NaOH is an effective, environmentally friendly pH controlling agent. Moreover, pH, temperature, agitation, and inoculum size all exerted a significant influence on the production process. The fed-batch process was able to increase both the production and productivity of D(−) lactic acid. These findings demonstrate that *S. nakayamae* is a promising strain for D(−) lactic acid production.

Conflicts of Interest

The authors declare that there are no conflicts of interest regarding the publication of this paper.

Acknowledgments

The authors are grateful to the São Paulo Research Foundation (FAPESP), Braskem, and the National Council for Scientific and Technological Development (CNPq process302935/2015-0) for fellowships and financial support.

References

[1] T. Ghaffar, M. Irshad, Z. Anwar et al., "Recent trends in lactic acid biotechnology: a brief review on production to purification," *Journal of Radiation Research and Applied Sciences*, vol. 7, no. 2, pp. 222–229, 2014.

[2] C. Gao, C. Ma, and P. Xu, "Biotechnological routes based on lactic acid production from biomass," *Biotechnology Advances*, vol. 29, no. 6, pp. 930–939, 2011.

[3] M. A. Abdel-Rahman, Y. Tashiro, and K. Sonomoto, "Recent advances in lactic acid production by microbial fermentation processes," *Biotechnology Advances*, vol. 31, no. 6, pp. 877–902, 2013.

[4] A. Corma Canos, S. Iborra, and A. Velty, "Chemical routes for the transformation of biomass into chemicals," *Chemical Reviews*, vol. 107, no. 6, pp. 2411–2502, 2007.

[5] Y. Tokiwa and B. P. Calabia, "Biological production of functional chemicals from renewable resources," *Canadian Journal of Chemistry*, vol. 86, no. 6, pp. 548–555, 2008.

[6] J. R. M. Almeida, L. C. L. Fávaro, and B. F. Quirino, "Biodiesel biorefinery: Opportunities and challenges for microbial production of fuels and chemicals from glycerol waste," *Biotechnology for Biofuels*, vol. 5, article no. 48, 2012.

[7] Y. Tokiwa and B. P. Calabia, "Biodegradability and biodegradation of poly(lactide)," *Applied Microbiology and Biotechnology*, vol. 72, no. 2, pp. 244–251, 2006.

[8] A. Södergård and M. Stolt, "Properties of lactic acid based polymers and their correlation with composition," *Progress in Polymer Science (Oxford)*, vol. 27, no. 6, pp. 1123–1163, 2002.

[9] K. Madhavan Nampoothiri, N. R. Nair, and R. P. John, "An overview of the recent developments in polylactide (PLA) research," *Bioresource Technology*, vol. 101, no. 22, pp. 8493–8501, 2010.

[10] C. Idler, J. Venus, and B. Kamm, "Microorganisms for the Production of Lactic Acid and Organic Lactates," in *Microorganisms in Biorefineries*, vol. 26 of *Microbiology Monographs*, pp. 225–273, Springer Berlin Heidelberg, Berlin, Heidelberg, 2015.

[11] A. Von Wright, L. Axelsson, S. Lahtinen, A. C. Ouwehand, S. Salminen, and A. von Wright, "LA bacteria —microbiological and functional aspects," in *LA Bacteria Microbiological And Functional Aspects*, A. Von Wright, L. Axelsson, S. Lahtinen, A. C. Ouwehand, S. Salminen, and A. von Wright, Eds., pp. 1–16, CRC Press/Taylor Francis Group, Boca Raton, 2012.

[12] R. Datta and M. Henry, "Lactic acid: recent advances in products, processes and technologies—a review," *Journal of Chemical Technology and Biotechnology*, vol. 81, no. 7, pp. 1119–1129, 2006.

[13] J. Narita, S. Nakahara, H. Fukuda, and A. Kondo, "Efficient production of L-(+)-lactic acid from raw starch by Streptococcus bovis 148," *Journal of Bioscience and Bioengineering*, vol. 97, no. 6, pp. 423–425, 2004.

[14] Y. Li, A. Shahbazi, K. Williams, and C. Wan, "Separate and concentrate lactic acid using combination of nanofiltration and reverse osmosis membranes," *Applied Biochemistry and Biotechnology*, vol. 147, no. 1-3, pp. 1–9, 2008.

[15] S. Zhou, T. B. Causey, A. Hasona, K. T. Shanmugam, and L. O. Ingram, "Production of optically pure D-lactic acid in mineral salts medium by metabolically engineered Escherichia coli W3110," *Applied and Environmental Microbiology*, vol. 69, no. 1, pp. 399–407, 2003.

[16] S. Saitoh, N. Ishida, T. Onishi et al., "Genetically engineered wine yeast produces a high concentration of L-lactic acid of extremely high optical purity," *Applied and Environmental Microbiology*, vol. 71, no. 5, pp. 2789–2792, 2005.

[17] R. P. John, K. M. Nampoothiri, and A. Pandey, "Solid-state fermentation for L-lactic acid production from agro wastes using Lactobacillus delbrueckii," *Process Biochemistry*, vol. 41, no. 4, pp. 759–763, 2006.

[18] R. Ganguly, P. Dwivedi, and R. P. Singh, "Production of lactic acid with loofa sponge immobilized Rhizopus oryzae RBU2-10," *Bioresource Technology*, vol. 98, no. 6, pp. 1246–1251, 2007.

[19] W. H. Holzapfel and S. J. Botha, "Physiology of Sporolactobacillus strains isolated from different habitats and the indication of in vitro antagonism against Bacillus species," *International Journal of Food Microbiology*, vol. 7, no. 2, pp. 161–168, 1988.

[20] D. C. Montgomery, *Design and Analysis of Experiments*, John Wiley & Sons, 5th edition, 2012.

[21] Z. M. Nor, M. I. Tamer, J. M. Scharer, M. Moo-Young, and E. J. Jervis, "Automated fed-batch culture of Kluyveromyces fragilis based on a novel method for on-line estimation of cell specific growth rate," *Biochemical Engineering Journal*, vol. 9, no. 3, pp. 221–231, 2001.

[22] R. Auras, L.-T. Lim, S. E. M. Selke, and H. Tsuji, "Preface," *Poly(Lactic Acid): Synthesis, Structures, Properties, Processing, and Applications*, pp. xvii–xix, 2010.

[23] N. Assavasirijinda, D. Ge, B. Yu, Y. Xue, and Y. Ma, "Efficient fermentative production of polymer-grade d-lactate by an engineered alkaliphilic Bacillus sp. strain under non-sterile conditions," *Microbial Cell Factories*, vol. 15, no. 1, article no. 3, 2016.

[24] H. W. Yien, T. J. Chen, W. C. Pan, and H. J. Wu, "Effects of neutralizing agents on lactic acid production by sweet potato starch," *J. Microbiol. Biotechnol*, vol. 26, pp. 437–441, 2010.

[25] S. Nakano, C. U. Ugwu, and Y. Tokiwa, "Efficient production of d-(-)-lactic acid from broken rice by Lactobacillus delbrueckii using Ca(OH)2 as a neutralizing agent," *Bioresource Technology*, vol. 104, pp. 791–794, 2012.

[26] R. N. Shreve and J. A. Brink, *Chemical Process Industries*, McGraw-Hill, New York, 1st edition, 1977.

[27] Y. Liu, W. Gao, X. Zhao et al., "Pilot scale demonstration of d-lactic acid fermentation facilitated by Ca(OH)2 using a metabolically engineered Escherichia coli," *Bioresource Technology*, vol. 169, pp. 559–565, 2014.

[28] S. D. Yuwono and T. Kokugan, "Study of the effects of temperature and pH on lactic acid production from fresh cassava roots in tofu liquid waste by Streptococcus bovis," *Biochemical Engineering Journal*, vol. 40, no. 1, pp. 175–183, 2008.

[29] B. Zhao, L. Wang, F. Li et al., "Kinetics of D-lactic acid production by sporolactobacillus sp. strain CASD using repeated batch fermentation," *Bioresource Technology*, vol. 101, no. 16, pp. 6499–6505, 2010.

[30] L. Wang, B. Zhao, F. Li et al., "Highly efficient production of d-lactate by Sporolactobacillus sp. CASD with simultaneous enzymatic hydrolysis of peanut meal," *Applied Microbiology and Biotechnology*, vol. 89, no. 4, pp. 1009–1017, 2011.

[31] A. H. Jawad, A. F. M. Alkarkhi, O. C. Jason, A. M. Easa, and N. A. Nik Norulaini, "Production of the lactic acid from mango peel waste—factorial experiment," *Journal of King Saud University—Science*, vol. 25, no. 1, pp. 39–45, 2013.

[32] K. Hofvendahl and B. Hahn-Hägerdal, "Factors affecting the fermentative lactic acid production from renewable resources," *Enzyme and Microbial Technology*, vol. 26, no. 2–4, pp. 87–107, 2000.

[33] C. J. B. de Lima, L. F. Coelho, and J. Contiero, "The use of response surface methodology in optimization of lactic acid production: Focus on medium supplementation, temperature and pH control," *Food Technology and Biotechnology*, vol. 48, no. 2, pp. 175–181, 2010.

[34] P. S. Panesar, J. F. Kennedy, C. J. Knill, and M. Kosseva, "Production of L(+) Lactic Acid using Lactobacillus casei from Whey," *Brazilian Archives of Biology and Technology*, vol. 53, no. 1, pp. 219–226, 2010.

[35] G. Shi, G. Wang, X. Chen, and C. Li, "Optically pure L-lactic acid production directly from leftover bits and pieces of potato starch using an amylolytic pellet-form complex Rhizopus oryzae ASC081," *Journal of Applied Science and Engineering*, vol. 16, no. 2, pp. 205–210, 2013.

[36] Z. Bai, Z. Gao, J. Sun, B. Wu, and B. He, "D-Lactic acid production by Sporolactobacillus inulinus YBS1-5 with simultaneous

utilization of cottonseed meal and corncob residue," *Bioresource Technology*, vol. 207, pp. 346–352, 2016.

[37] D.-M. Bai, Q. Wei, Z.-H. Yan, X.-M. Zhao, X.-G. Li, and S.-M. Xu, "Fed-batch fermentation of Lactobacillus lactis for hyper-production of L-lactic acid," *Biotechnology Letters*, vol. 25, no. 21, pp. 1833–1835, 2003.

[38] C. Kotzamanidis, T. Roukas, and G. Skaracis, "Optimization of lactic acid production from beet molasses by Lactobacillus delbrueckii NCIMB 8130," *World Journal of Microbiology and Biotechnology*, vol. 18, no. 5, pp. 441–448, 2002.

[39] M. P. Bernardo, L. F. Coelho, D. C. Sass, and J. Contiero, "L-(+)-Lactic acid production by Lactobacillus rhamnosus B103 from dairy industry waste," *Brazilian Journal of Microbiology*, vol. 47, no. 3, pp. 640–646, 2016.

[40] S. Ding and T. Tan, "l-lactic acid production by Lactobacillus casei fermentation using different fed-batch feeding strategies," *Process Biochemistry*, vol. 41, no. 6, pp. 1451–1454, 2006.

[41] M. S. Ou, L. O. Ingram, and K. T. Shanmugam, "L(+)-Lactic acid production from non-food carbohydrates by thermotolerant Bacillus coagulans," *Journal of Industrial Microbiology and Biotechnology*, vol. 38, no. 5, pp. 599–605, 2011.

13

Bacterial Profile, Antibacterial Resistance Pattern, and Associated Factors from Women Attending Postnatal Health Service at University of Gondar Teaching Hospital, Northwest Ethiopia

Abebaw Bitew Kifilie,[1] Mulat Dagnew,[2] Birhanemeskel Tegenie,[2] Biruk Yeshitela,[3] Rawleigh Howe,[3] and Ebba Abate[4]

[1]*Department of Medical Laboratory Science, College of Medicine and Health Sciences, Debre Markos University, Debre Markos, Ethiopia*
[2]*Department of Medical Microbiology, School of Biomedical and Laboratory Science, College of Medicine and Health Science, University of Gondar, Gondar, Ethiopia*
[3]*Armauer Hansen Research Institute, Addis Ababa, Ethiopia*
[4]*Department of Immunology and Molecular Biology, School of Biomedical and Laboratory Science, College of Medicine and Health Science, University of Gondar, Gondar, Ethiopia*

Correspondence should be addressed to Abebaw Bitew Kifilie; abebawbit05@gmail.com

Academic Editor: Joseph Falkinham

Introduction. Surgical site infection is a vital cause of maternal mortality and morbidity, especially in resource-limited countries. The rise of antibiotic resistance bacterial infection poses a big threat to this vulnerable population. However, there is lack of studies around the study area. *Objective.* The purpose of this study was to identify bacterial profile, antibacterial resistance pattern, and associated factors among mothers attending postnatal care health service. *Methods.* Institutional based cross-sectional study was conducted on 107 study participants at University of Gondar Teaching Hospital from 1 January 2016 to 30 May 2016. Wound swab, aspirate, and biopsy were collected and performed for culture and drug resistance testing. Data were entered and analyzed by using SPSS version 20. Bivariate and multivariate logistic regression models were fitted to determine the associated factors for bacterial infection. Odds ratio (95% CI) was calculated to determine the strength of statistically significant associated factors. *Result.* Bacterial growth was confirmed in 90 (84.1%) of 107 study participants suspected to have surgical site infection. The predominant bacterial isolates were *S. aureus* (41.6%), *E. coli* (19.8%), *K. pneumoniae* (13.9%), *coagulase negative Staphylococcus* (12.9%), and *Enterobacter* spp. (4%). The majority of isolates were resistant to ampicillin, amoxicillin, and tetracycline but susceptible to ceftriaxone and amikacin. Multidrug-resistant bacteria species were isolated. Using a procedure such as cesarean section and episiotomy for delivery and premature rapture of membrane had strong association with bacterial infection. *Conclusion.* The high prevalence of bacterial profile and isolation of multidrug-resistant bacteria pose a big threat to postnatal mothers and their children. Factors such as cesarean section, episiotomy for delivery, and premature rapture of membrane were predictors for bacterial infection. Therefore, there should be done a continuous surveillance as well as rational use of antibiotics and a longitudinal study using phenotypic and genotypic methods will be done.

1. Introduction

Surgical site infection (SSI) is an infection that occurs on the skin and subcutaneous tissue within 30 days of surgical incision or deep tissue surgical procedure [1]. It is characterized by the basic sign of redness, pain, swelling, raised incision tissue temperature, and systemic fever [2, 3]. It also causes infection to women pelvic organs when normal flora of the women's genital and gastrointestinal tract contaminates the sterile amniotic fluid [4]. Women who give birth by cesarean section delivery have 5-fold to 20-fold risk of bacterial infection than women who give birth vaginally [5]. The predominant pathogenic bacteria isolated from infected wound include *Escherichia coli (E. coli)*, *Streptococcus* species, *Enterococcus faecalis* (*E. faecalis*), *Staphylococcus aureus (S. aureus)*, Coagulase negative *Staphylococcus* (CoNS), *Gardnerella vaginalis*, and genital mycoplasmas [6, 7]. Bacteria that are isolated from SSI are too common to resist multiple classes of antibiotics. The downfall of efficient antibacterials will weaken our capacity to manage contagious disease in vulnerable patients undergoing cesarean section delivery [8]. World health leaders have declared that antibiotic-resistant bacteria are hurtful bacteria that cause a devastating risk to people in the world [9].

Generally bacterial infections during cesarean section delivery are among the principal sources of maternal mortality, which accounts for nearly one-tenth of global maternal death [10]. Annually the global estimate of SSIs is from 0.5% to 15%; this leads to an estimate of 358,000 maternal deaths in the world, of which 99% were in developing countries and half of which were in Sub-Saharan Africa; besides death, women who are exposed to peripartum infections are at risk of severe morbidity and long-term disabilities such as chronic pelvic pain, fallopian tube blockage, secondary infertility, and prematurity of the child [11, 12]. The risk of SSI was associated with premature rupture of membranes, prolonged labour, malnutrition, diabetes mellitus, obesity, lack of preincision antimicrobial care, which increases patients hospital stay, readmission, cost of care, and mortality [13, 14]. Decreasing the number of deliveries by C/S and identifying associated factors for SSI could contribute to decreasing maternal morbidity [7, 15].

Screening and treating bacterial infection during postnatal periods can improve future pregnancy and delivery outcomes. However, as far as our knowledge is concerned, there is a lack of studies conducted on bacterial profile and antibiotic susceptibility pattern among women attending postnatal service in developing countries including Ethiopia; as a result, the magnitude is not well known and is not comprehensive. To confirm proper treatment, recent information about organism, source of maternal infection, and its drug sensitivity test is mandatory. The information compiled here will increase the awareness of significant bacterial infection during postnatal periods, so periodic evaluation of bacterial profile and its drug response is needed to bring updated information. Therefore, this study determines bacterial profile, antibacterial susceptibility pattern, and associated factors among women attending postnatal health services.

2. Materials and Methods

2.1. Study Design, Area, and Period. An institutional based cross-sectional study was conducted at University of Gondar Teaching Hospital from January 2016 to May 2016. Gondar town is located 738 km far away from Addis Ababa, the capital city of Ethiopia.

2.2. Sample Size and Sampling Technique. A total of 107 women who have an infection due to cesarean section (CS) and episiotomy delivery were self-reporting to the hospital and were sampled for laboratory culture confirmation during the study period.

2.3. Sociodemographic Data and Specimen Collection. Sociodemographic variables such as age, residence, marital status, educational level, and occupation and other relevant clinical data such as gravidity, parity, and mode of delivery were collected using structured questionnaire. The specimen was collected by trained midwife nurses after getting ethical clearance from the School of Biomedical and Laboratory Science ethical review committee and informed written consent from study participants. Each biological specimen was collected for bacteriological laboratory analysis such as culture, Gram stain, biochemical test, and antibacterial susceptibility test.

2.4. Wound Specimen Collection and Laboratory Processing. The area around surgical and episiotomy site was cleaned with normal saline in order to reduce contamination. Then 107 wound specimens were collected by wound swabbing, Levine technique, and Z-technique by using sterile cotton tipped applicator stick, aspirating by needle and cutting wound biopsy depending on the wound type. Wound exudates were collected prior to wound cleaning during swabbing, but in Levine technique and Z-technique wound exudates were collected after wound cleaning.

Wound swabbing using Levine technique involves the rotation of swab over 1 cm^2 area with sufficient pressure to fast fluid release from wound, whereas Z-technique requires the rotation of swab between 2 fingers by zigzag motion from margin to margin of wound at 10 points [16, 17].

Cotton tipped applicator stick that contains specimen was socked to a test tube containing Tryptone Soya broth in order to prevent drying and transported to microbiology laboratory.

The sample was inoculated onto blood agar plate (BAP) (Oxoid), MacConkey agar (MaC) (Oxoid), and Mannitol Salt Agar (MSA) (Oxoid) and incubated aerobically for 18–24 hours at 37°C. Identification of bacteria was performed based on colony morphology such as size, pigment, and edge with the naked eye from culture media and Gram stain. Gram stain was used to distinguish Gram-positive and Gram-negative isolates. Biochemical characteristics such as catalase, coagulase, bacitracin, novobiocin, and optochin were used to identify Gram-positive pathogenic bacterial species, whereas triple sugar iron agar, indole test, motility test, urea test, hydrogen sulfide production, citrate test, and lysine decarboxylase test results were used to identify Gram-negative pathogenic bacterial species.

2.5. Antibacterial Susceptibility Test. Antibacterial susceptibility testing was done by using modified Kirby-Bauer disk diffusion method and interpreted according to Clinical and Laboratory Standard Institute (CLSI, 2016) guidelines [18]. Around 3–5 pure colonies of bacteria were taken and transferred to a tube containing 5 ml sterile nutrient broth (Oxoid) and mixed gently until a homogenous suspension was formed and incubated for 3–5 hours until the turbidity of the suspension becomes adjusted to the density of 0.5% McFarland standards. Then they were inoculated to Müller-Hinton agar (MHA, PH = 7.2–7.4) (Oxoid) by using sterile swab evenly with 60°. The inoculums were allowed to dry for 5–15 minutes. Antibacterial discs were distributed 15 mm away from the edge and ≥24 mm apart from each other.

Antibiotic discs penicillin (10 IU), cloxacillin (30 μg), tetracycline (30 μg), clindamycin (2 μg), gentamycin (10 μg), nalidixic acid (30 μg), ciprofloxacin (5 μg), amikacin (30 μg) ampicillin (10 μg), amoxicillin (10 μg), cefoxitin (30 μg), cefixime (5 μg), ceftriaxone (30 μg), trimethoprim sulfamethoxazole (1.25/23.75 μg), and ceftazidime (30 μg) were used for SSI. Then the plates were incubated at 37°C for 24 hours and the results were interpreted according to the most recent version of Clinical Laboratory Standard Institute (CLSI 2016) [18]. The criteria to select the antimicrobial agents were based on availability, CLSI guide line, the organisms' Gram reaction, and frequent prescription of drugs for the management of postnatal infections.

2.6. Data Management, Analysis, and Presentation. The collected data were coded and transferred from a questionnaire to a computer file. Data were entered and statistically analyzed by SPSS version 20 software. Descriptive statistics were used to analyze frequency, mean, range, and standard deviation of sociodemographic characteristics and obstetric and clinical variables of study. Tables and figures were used to present the findings.

Bivariate and multivariate logistic regression models were fit to determine possible associated factors with bacterial infection. Odds ratio was used as a measure of the strength of association and reported with 95% confidence intervals (95% CI) to determine statistically significant association of risk factors with bacterial infection. p value ≤ 0.05 was considered to be statistically significant.

2.7. Data Quality Control. All steps in data collection and recording were monitored. The reagents were checked for expiry date and appropriate storage of temperature and humidity. SOPs were prepared and strictly followed. The quality of culture media and antimicrobial susceptibility testing was checked by using quality control standard strains of *E. coli* ATCC 25922, *S. aureus* ATCC 25923, *E. faecalis* ATCC 29212, and *K. pneumoniae* ATCC@BAA1705. 0.5% McFarland standards were used to standardize the inoculum density of bacterial suspension for susceptibility test. The acceptance range of 0.5% McFarland optical density is 0.08–0.1 [18].

2.8. Ethical Consideration. Ethical clearance was obtained from University of Gondar, School of Biomedical and Laboratory Sciences (SBMLS) ethical review committee for the initiation of the study. There was no additional sample to be taken from the study participants but only for the sake of this study. A written informed consent was obtained from mothers after explaining the purpose and objective of the study to them. Participants would have full right to continue or withdraw from the study. All information was kept confidential by assigning code and assessed only by principal investigator. The laboratory results were communicated timely with physicians and nurses for better patient management.

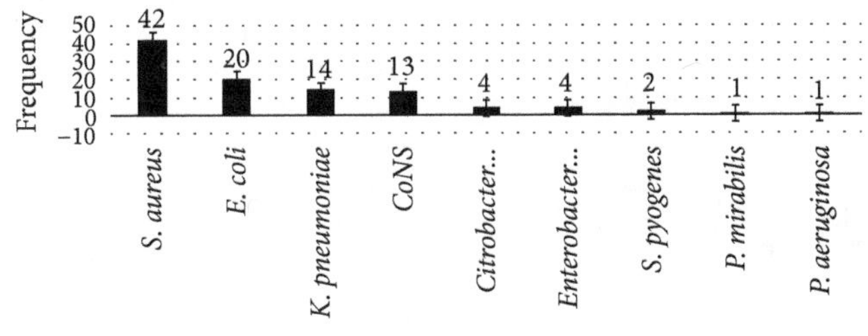

FIGURE 1: Profile and frequency of bacterial isolates from women having SSI during postnatal period at the University of Gondar Teaching Hospital from 1 January 2016 to 30 May 2016.

3. Result

3.1. Sociodemographic, Obstetric, and Clinical Variables. A total of 107 women with symptoms of infections during postnatal periods were investigated in this study. The mean age of the study participants was 26.21 (±5.5) and ranged from 15 to 44. The majority of the participants were urban dwellers (77.6%), orthodox Christian (90.7%), married (98.1%), and housewives (60.7%). A total of 107 postnatal clinic (PNC) participants were delivered from health institution, 52 (48.6%) of them had history of prolonged labour, 43 (40.2%) had history of premature rapture of membrane, and 81 (48.5%) were delivered cesarean section. Most of those participants had a history of 1–3 gravida (88.8%) (Table 1).

3.2. Isolation of Pathogenic Bacteria from SSI. Of all the study participants, 107 wound specimens were collected based on physician request at the University of Gondar Teaching Hospital. From these suspected clinical specimens, the overall pathogenic bacterial infections were 90 (84.1%). Among these, 79 (87.8%) infections were due to single bacterial isolates, whereas 11 (12.2%) infections were due to mixed bacterial isolates. So a total of 101 bacterial isolates were isolated. Majority of them were Gram-positive and the predominant were *S. aureus*, 42/101 (41.6%), and CoNS, 13/101 (12.9%), followed by Gram-negative *E. coli*, 20/101 (19.8%), and *K. pneumoniae*, 14/101 (13.9%) (Figure 1).

3.3. Antibacterial Susceptibility Pattern for Bacterial Pathogen from SSI. Majority of isolates from SSI were resistant to regularly used antibacterials. *S. aureus* was resistant to ampicillin (71.4%), amoxicillin (66.7%), and trimethoprim-sulfamethoxazole (61.9%). Likewise, CoNS was resistant to ampicillin (84.6%), amoxicillin (84.6%), trimethoprim-sulfamethoxazole (61.5%), penicillin (61.5%), cloxacillin (38.5%), and cefoxidime (38.5%). In contrast, clindamycin (84.2%), cefoxitin (82.5%), cefixime (73.7%), and ceftriaxone (68.4%) were susceptible to Gram-positive isolates (Table 2).

TABLE 1: Sociodemographic, obstetric, and clinical variables from women attending antenatal and postnatal services at the University of Gondar Teaching Hospital from 1 January 2016 to 30 May 2016.

Characteristics	Number (%)
Age (years)	
15–24	39 (36.5%)
25–34	56 (52.3%)
35–44	12 (11.2%)
Residence	
Urban	83 (77.6%)
Rural	24 (23.4%)
Religion	
Orthodox	97 (90.7%)
Muslim	7 (6.5%)
Protestant	3 (2.8%)
Marital status	
Married	105 (98.1)
Single	2 (1.9%)
Educational level	
Illiterate	35 (32.7%)
Primary school (1–8)	14 (13.1%)
Secondary school	31 (29.0%)
Higher education	27 (25.2%)
Occupational status	
Housewives	65 (60.7%)
Self-employees	17 (15.9%)
Government employees	20 (18.7%)
Students	3 (2.8%)
Daily laborer	2 (1.9%)
Gravidity	
1–3	95 (88.8%)
4–6	10 (9.3%)
7–9	2 (1.9%)
Prolonged labour	
Yes	52 (48.6)
No	55 (51.4%)
Premature rapture of membrane	
Yes	43 (40.2%)
No	64 (59.8%)
Mode of delivery	
Vaginal delivery	69 (41.3%)
Cesarean delivery	81 (48.5%)
Instrumental delivery	17 (10.2%)
Place of delivery	
Health institution	107 (100%)
Diabetes mellitus	
No	98 (91.6%)
Yes	9 (8.4%)
HIV/AIDS	
No	103 (96.3%)
Yes	4 (3.7%)

HIV: human immunodeficiency virus.

On the other hand, Gram-negative isolates were resistant to ampicillin (88.6%), amoxicillin (79.5%), and trimethoprim-sulfamethoxazole (54.5%). *E. coli* was resistant to ampicillin (80%), amoxicillin (70%), and ceftriaxone (60%). *K. pneumoniae* was shown to be resistant to ampicillin (100%), amoxicillin (100%), trimethoprim-sulfamethoxazole (64.3%), ceftriaxone (57.1%), ceftazidime (28.6%), gentamycin (21.4%), and ciprofloxacin (14.3%). *Citrobacter* spp. were resistant to ampicillin (100%). In contrast, most Gram-negative isolates were susceptible to amikacin (95.5%) and ciprofloxacin (83.7%) (Table 3).

3.4. Multidrug Resistance Pattern of Bacterial Pathogens. Taking all bacterial isolates from SSI sample were shown 75% and 82.5% multidrug resistance (MDR = resistance of ≥2 drugs with different classes) for Gram negative and Gram positive bacteria respectively. Gram-negative bacteria such as *K. pneumoniae* were 12 (85.7%) resistant to ≥2 drugs in different class. Likewise, CoNS were 11 (84.6%) resistant to ≥2 drugs in different class (Table 4).

3.5. Characteristics of Risk Factors Attributed to Bacterial Infections. In bivariate logistic analysis, demographic and clinical factors such as age, occupation, level of education, gravidity, parity, prolonged labour, and history of diabetes mellitus were not significantly associated with bacterial infection in this study; however, mode of delivery and premature rapture of membrane were strongly associated in both bivariate and multivariate analyses for bacterial infection ($p \leq 0.05$). Delivery by cesarean section and episiotomy were 102 and 86 times at risk for bacterial infection than instrumental deliveries (AOR (95% CI) = 102 (5.2, 2038, $p = 0.002$) and 86 (5, 1436, $p = 0.002$), resp.) (Table 5).

4. Discussion

Maternal infection during postnatal period was a vital factor for morbidity and mortality [3, 19]. In the current study, the possible etiologic agents causing morbidity among mothers during postnatal period were isolated. Three different biological wound specimens such as wound swab, aspirate, and biopsy were collected to ascertain possible pathogenic bacteria. From this study, the total prevalence of confirmed bacteria was 84.1%. This showed that women who delivered by cesarean section and episiotomy were vulnerable to various bacterial infections and this has continued even after delivery up to 30 days [1]. Considering the fact that postnatal women are part of vulnerable population and the effect could also influence the health status of the neonates, the current finding has shown high public health consequence. This is similar to studies conducted in Ethiopia and elsewhere [2, 20–22].

In the current study, the rate of pathogenic bacteria which were isolated from postnatal women was highly considerable for SSI. The predominant pathogens that cause SSIs are *S. aureus*, *E. coli*, *K. pneumoniae*, and CoNS. The reason for the predominance of those organisms may be the fact that most of the infected patients in this study had undergone cesarean section and episiotomy procedure for delivery which is favorable for different bacterial colonization commonly reported during SSI [3, 4]. This study was in line with studies

TABLE 2: Antibacterial susceptibility pattern for Gram-positive SSI causing pathogens from women attending postnatal service at University of Gondar Teaching Hospital from 1 January 2016 to 30 May 2016.

Bacterial isolates		Antibacterial susceptibility pattern									
		AMP	AMO	CRO	P	CAZ	SXT	CXM	CXC	DA	CXT
S. aureus ($n = 42$)	S%	12 (28.6)	14 (33.3)	27 (64.3)	21 (50)	24 (57.1)	16 (38.1)	31 (73.8)	20 (47.6)	35 (83.3)	35 (83.3)
	R%	30 (71.4)	28 (66.7)	15 (35.7)	21 (50)	18 (42.9)	26 (61.9)	11 (26.2)	22 (52.4)	7 (16.7)	7 (16.7)
CoNS ($n = 13$)	S%	2 (15.4)	2 (14.3)	8 (61.5)	5 (38.5)	8 (61.5)	5 (38.5)	10 (76.9)	8 (61.5)	12 (92.3)	11 (84.6)
	R%	11 (84.6)	11 (85.7)	5 (38.5)	8 (61.5)	5 (38.5)	8 (61.5)	3 (23.1)	5 (38.5)	1 (7.7)	2 (15.4)
S. pyogenes ($n = 2$)	S%	0 (0)	1 (50)	4 (100)	2 (100)	1 (50)	1 (50)	1 (50)	1 (50)	1 (50)	1 (50)
	R%	2 (100)	1 (50)	0 (0)	0 (0)	1 (50)	1 (50)	1 (50)	1 (50)	1 (50)	1 (50)
Total ($n = 57$)	S%	*14 (24.6)*	*17 (29.8)*	*39 (68.4)*	*28 (49.1)*	*33 (57.9)*	*22 (38.6)*	*42 (73.7)*	*29 (50.9)*	*48 (84.2)*	*47 (82.5)*
	R%	*43 (75.4)*	*40 (70.2)*	*18 (31.6)*	*29 (50.9)*	*24 (42.1)*	*35 (61.4)*	*15 (26.3)*	*28 (49.1)*	*9 (15.8)*	*10 (17.5)*

P: penicillin; CAZ: ceftazidime; CXM: cefixime; CXC: cloxacillin; CD: clindamycin; CXT: cefoxitin; *S. pyogenes: Streptococcus pyogenes.*

conducted in Bahir Dar (83.3%) [9], Addis Ababa (84.1%) [3], and Nizwa Hospital, Oman (77.72%) [23]. But the current findings were higher compared to studies conducted in Gondar (31.5%) [24], Tanzania (61.8%) [22], Nepal (62.4%) [5], and Estonian University (6.2%) [7]. The possible explanation for this discrepancy might be due to poor wound care and aseptic technique during C/S and episiotomy procedure.

It is known that, from wound, there are different microbial contaminations. Wound cleaning without regular handwashing antiseptics such as 70% alcohol and/or normal saline worsens the complications. To prevent further contamination, cleaning wounds with normal saline, covering with clean and dry bandage, giving patient education about personal hygiene, avoiding cross infection by restricting visitors, providing routine education regarding infection control measures, and following necessary SOP during sample collection should be applied [23]. The prevalence of pathogenic bacteria from wound was different, even though the study participants were similar. The difference might be related to accessibility of diagnostic materials, the use of aseptic technique during sample processing, and sample size. A study showed in Jimma [14] that the specimens were collected from women infected due to cesarean section for delivery including ovarian tumor operation and Myoma, but in the present study the specimens were collected from women infected due to cesarean section and episiotomy delivery.

Nowadays, emerging of high rate of bacteria resistant to multiple antibiotics is becoming a global threat by crossing worldwide borders and spreading between regions with significant speed. World health leaders have declared that drug-resistant organisms are frightening organisms that have a disastrous risk to persons in each country in the world [25]. Several factors augment this problem. Understanding bacterial profile and its antibiotic pattern is of paramount importance for proper management and minimizing the circulation of common pathogenic resistance strains in the community [13]. In this study, we tried to assess the sensitivity pattern of the isolated bacteria to different drugs in vitro. In the present finding, *S. aureus, E. coli, CoNS,* and *K. pneumoniae* are the commonest pathogens isolated from wound and were resistant to many antibiotics, indicating the big threat they poses. On the other hand, Gram-negative bacteria were susceptible to strong antibiotics like amikacin, ciprofloxacin, and others. This result is similar to studies in Bahir Dar [9], Addis Ababa [3], Mek'ele [26], and Palestine [19]. However, as long as there is high rate of irrational use of antibiotics as well as poor adherence in the society, these antibiotics may no longer be effective to pathogens. Different antibacterial susceptibility patterns were observed for bacterial isolates collected from wound sites. One of the commonest isolated bacteria from wound site was *S. aureus*, which was resistant to ampicillin (71.4%), amoxicillin (66.7%), and trimethoprim-sulfamethoxazole (61.9%). More or less this is in agreement with a study done in Bahir Dar [9]. Similarly, CoNS was highly resistant to ampicillin (84.6%), amoxicillin (84.6%), and trimethoprim-sulfamethoxazole (63.6%). This is in agreement with a study done in Bahir Dar [9]. Among Gram-negative bacteria, *E. coli* was highly resistant to ampicillin (80%), amoxicillin (70%), and ceftriaxone (60%). This finding is in line with a study done in Addis Ababa [3, 23]. In addition, *K. pneumoniae* was shown to be fully resistant to amoxicillin (100%) and ampicillin (100%) and strongly resistant to trimethoprim-sulfamethoxazole (64.3%) and ceftriaxone (57.1%). This is in agreement with a study done in Addis Ababa and Debre Markos [3, 25].

In the current study, multidrug resistance (MDR) bacterial isolate showed 75% and 82.5% resistance to Gram-negative and Gram-positive bacteria, respectively; this is in line with a study done in Addis Ababa [3]. Several reasons could have supported such level of resistance; this includes mismanagement of drugs by health specialists, unexperienced experts, and untrained individuals. In addition to this, most drugs in the study area can be bought deprived of laboratory confirmation, which leads to misuse of drugs by the community and thus causes the occurrence and spread of antibacterial tolerance. Additional fundamental reason could be reduced hospital sanitation, contributing to the spread of those drug-resistant bacteria in the area [24, 25]. This situation raises serious concerns. Moreover, the development of high resistance gene pool may increase antibacterial resistance. Taken together, these findings clearly show how resistance strains are expanding at an alarming rate in the area. With this trend, an antibiotic that was effective a year ago might no longer be used. This creates great burden, especially

Table 3: Antibacterial susceptibility pattern for Gram-negative SSI causing pathogens from women attending postnatal service at University of Gondar Teaching Hospital from 1 January 2016 to 30 May 2016.

Bacterial isolates		Antibacterial susceptibility pattern											
		AMX	AMP	CRO	CIP	CXT	CAZ	SXT	CXM	GEN	NA	AMK	TTC
E. coli ($n = 20$)	S%	6 (30)	4 (20)	8 (40)	16 (80)	16 (80)	12 (60)	10 (50)	15 (75)	13 (65)	15 (75)	19 (95)	6 (30)
	R%	14 (70)	16 (80)	12 (60)	4 (20)	4 (20)	8 (40)	10 (50)	5 (25)	7 (35)	5 (25)	1 (5)	14 (70)
K. pneumoniae	S%	0 (0)	0 (0)	6 (42.9)	12 (85.7)	13 (92.9)	10 (71.4)	5 (35.7)	11 (78.6)	11 (78.6)	13 (92.9)	14 (100)	2 (14.3)
($n = 14$)	R%	14 (100)	14 (100)	8 (57.1)	2 (14.3)	1 (7.1)	4 (28.6)	9 (64.3)	3 (21.4)	3 (21.4)	1 (7.1)	0 (0)	12 (86)
Citrobacter	S%	2 (50)	0 (0)	4 (100)	4 (100)	2 (50)	2 (50)	3 (75)	2 (50)	2 (50)	3 (75)	3 (75)	2 (50)
spp.($n = 4$)	R%	2 (50)	4 (100)	0 (0)	0 (0)	2 (50)	2 (50)	1 (25)	2 (50)	2 (50)	1 (25)	1 (25)	2 (50)
Enterobacter	S%	1 (25)	1 (25)	3 (25)	3 (75)	4 (100)	3 (75)	2 (50)	4 (100)	3 (75)	3 (75)	4 (100)	2 (50)
spp.($n = 4$)	R%	3 (75)	3 (75)	1 (25)	1 (25)	0 (0)	1 (25)	2 (50)	0 (0)	1 (25)	1 (25)	0 (0)	2 (50)
P. aeruginosa ($n =$	S%	0 (0)	0 (0)	1 (100)	1 (100)	1 (100)	1 (100)	0 (0)	1 (100)	0 (0)	1 (100)	1 (100)	0 (0)
1)	R%	1 (100)	1 (100)	0 (0)	0 (0)	0 (0)	0 (0)	1 (100)	0 (0)	1 (100)	0 (0)	(0)	1 (100)
P. mirabilis ($n = 1$)	S%	0 (0)	0 (0)	1 (100)	1 (100)	0 (0)	0 (0)	0 (0)	0 (0)	0 (0)	1 (100)	1 (100)	0 (0)
	R%	1 (100)	1 (100)	0 (0)	0 (0)	1 (100)	1 (100)	1 (100)	1 (100)	1 (100)	0 (0)	0 (0)	1 (100)
Total = 44	S%	*9 (20.5)*	*5 (11.4)*	*23 (52.3)*	*37 (83.7)*	*36 (80.1)*	*28 (63.6)*	*20 (45.5)*	*33 (75)*	*29 (65.9)*	*36 (81.8)*	*42 (95.5)*	*12 (27.3)*
	R%	*35 (79.5)*	*39 (88.6)*	*21 (47.7)*	*7 (16.3)*	*8 (19.9)*	*16 (36.4)*	*24 (54.5)*	*11 (25)*	*15 (34.1)*	*8 (18.2)*	*2 (4.5)*	*32 (72.7)*

CIP: ciprofloxacin; GEN: gentamycin; AMK: amikacin; NA: nalidixic acid; TTC: tetracycline; SXT: sulfamethoxazole trimethoprim; R: resistance; S: sensitivity.

TABLE 4: Multidrug resistance (MDR) of Gram-negative and Gram-positive bacterial isolates from women attending antenatal and postnatal service.

(a)

Antibiograms		Gram-negative bacterial isolates					
	Total = *44*	*E. coli* = 20	*K. pneumoniae* = 14	*Citrobacter* spp. = (4)	*Enterobacter* spp. = (4)	*P. mirabilis* = 1	*P. aeruginosa* = *1*
AMP, TTC	5 (11.4)	2 (10)	3 (21.4)	—	—	—	—
AMO, SXT	1 (2.3)	—	1 (11.1)	—	—	—	—
AMP, TTC, GEN	1 (2.3)	1 (5)	—	—	—	—	—
AMP, TTC, SXT	3 (6.8)	1 (5)	2 (14.3)	—	—	—	—
AMO, SXT, CRO	1 (2.3)	1 (5)	—	—	—	—	—
AMX, TTC, CRO	1 (2.3)	1 (5)	—	—	—	—	—
AMX, TTC, CRO	1 (2.3)	1 (5)	—	—	—	—	—
AMX, CRO, GEN, TTC	1 (2.3)	1 (5)	—	—	—	—	—
AMX, TTC, SXT, CRO	5 (11.4)	1 (5)	2 (14.3)	—	1 (25)	—	1 (100)
AMP, SXT, CRO, NAL	2 (4.5)	2 (10)	—	—	—	—	—
AMO, SXT, CXM, TTC	1 (2.3)	—	—	1 (25)	—	—	—
AMO, CPR, CRO, NAL	1 (2.3)	1 (5)	—	—	—	—	—
AMO, TTC, SXT, GEN, CRO	4 (9.1)	1 (5)	3 (21.4)	—	—	—	—
AMO, CXM, AMK, TTC, NAL	1 (2.3)	—	—	1 (25)	—	—	—
AMO, SXT, CXM, GEN, TTC	1 (2.3)	—	—	—	—	1 (100)	—
AMO, CRO, CPR, SXT, TTC, NAL	1 (2.3)	1 (5)	—	—	—	—	—
AMO, CPR, SXT, CXM, TTC, NAL	1 (2.3)	—	1 (11.1)	—	—	—	—
AMO, CRO, CPR, SXT, GEN, TTC, NAL	3 (6.8)	2 (10)	—	—	1 (25)	—	—
Total	*33 (75)*	*15 (75)*	*12 (85.7)*	*2 (50)*	*2 (50)*	*1 (100)*	*1 (100)*

(b)

Antibiograms		Gram-positive bacterial isolates		
	Total = 57	*S. aureus* = 42	CoNS = 13	*S. pyogenes* = 2
P, CXM	1 (1.8)	1 (2.4)	—	—
P, CRO	1 (1.8)	—	1 (7.7)	—
AMP, CXC	2 (3.5)	1 (2.4)	1 (7.7)	—
AMP, CAZ	2 (3.5)	1 (4.8)	1 (7.7)	—
AMO, SXT	6 (10.5)	5 (11.9)	1 (7.7)	—
P, CRO, SXT	1 (1.8)	1 (2.4)	—	—
AMP, SXT, CAZ	3 (5.3)	1 (2.4)	2 (15.4)	—
AMO, CAZ, CXC	3 (5.3)	3 (7.1)	—	—
AMO, CRO, CXC	4 (7)	4 (9.5)	—	—
AMO, CRO, SXT	2 (3.5)	2 (4.8)	—	—
AMP, SXT, CXC	2 (3.5)	1 (2.4)	1 (7.7)	—
P, SXT, CXC, CD	2 (3.5)	2 (4.8)	—	—
P, CRO, SXT, CXC	6 (10.5)	4 (9.5)	2 (15.4)	—
AMP, CAZ, SXT, CXC	4 (7)	2 (4.8)	1 (7.7)	1 (50)
AMP, CRO, CXC, SXT	3 (5.3)	3 (7.1)	—	—
P, CRO, SXT, CXC, DA	1 (1.8)	1 (2.4)	—	—
P, SXT, CAZ, CXC, DA	2 (3.5)	1 (2.4)	1 (7.7)	—
AMO, CRO, SXT, CXC, DA	2 (3.5)	2 (4.8)	—	—
Total	*47 (82.5)*	*35 (83.3)*	*11 (84.6)*	*1 (50)*

TTC: tetracycline; GEN: gentamycin; NA: nalidixic acid; CPR: ciprofloxacin; CoNS: coagulase negative *Staphylococcus* species; GAS: group A *Streptococcus-pyogenes*; CD: clindamycin; CXC: cloxacillin.

TABLE 5: Bivariate and multivariate analysis for the assessment of factors associated with bacterial infection from women attending antenatal and postnatal service at the University of Gondar Teaching Hospital from 1 January 2016 to 30 May 2016.

Characteristics	Culture		Bivariate analysis	Multivariate analysis	p value
	Positive number (%)	Negative number (%)	COR (95% CI)	AOR (95% CI)	
Age					
15–24	34 (37.8)	5 (29.4)	1		
25–34	49 (54,4)	7 (41.2)	2.267 (0.454, 11.327)		
35–44	7 (7.8)	5 (29.4)	1.741 (0.393, 7.713)		
Residence					
Urban	70 (77.8)	13 (76.5)	0.923 (0.273, 3.164)		
Rural	20 (22.2)	4 (23.5)	1		
Occupation					
Self-employer	85 (94,4)	17 (100)	1		
Government	5 (5.6)	0 (0)	0.982 (0.585, 1.648)		
Education					
Illiterate	28 (31.1)	7 (41.2)	1		
Primary	12 (13.3)	2 (11.8)	0.696 (0.181, 2.674)		
Secondary	27 (30)	4 (23.5)	1.043 (0.167, 6.59)		
Higher education	23 (25.6)	4 (23.5)	1.174 (0.264, 5.226)		
Marital status					
Married	88 (97.8)	17 (100)	1		
Single	2 (2.2)	0 (0)	0.691 (0.227, 2.101)		
Mode of delivery					
VDE	34 (37.8)	3 (17.6)	15.111 (3.095, 73.77)	102 (5.2, 2028)	0.002
C/S	50 (55.6)	6 (35.3)	11.111 (2.865, 43.099)	86 (5, 1436)	0.002
Instrumental	6 (6.7)	8 (47.1)	1		
Prolonged labour					
No	45 (50)	10 (58.8)	1		
Yes	45 (50)	7 (41.2)	0.700 (0.245, 2.001)		
PRM					
No	50 (55.6)	14 (82.4)	1		
Yes	40 (44.4)	3 (17.6)	0.268 (0.072, 0.997)	0.012 (0.001, 0.305)	0.008
Gravidity					
1–3	80 (88.9)	15 (88.2)	1.746 (0.513, 5.940)		
4–6	9 (10)	1 (5.9)	1.290 (0.343, 4.856)		
7–9	1 (1.1)	1 (5.9)	1		

Note. ∗ indicates statistical significance at $p \leq 0.05$, in both bivariate and multivariate analyses. AOR: adjusted odds ratio; COR: crude odds ratio; 1: reference group; 95% CI: 95% confidence interval; UTI: urinary tract infection; C/S: cesarean section; VDE: vaginal delivery with episiotomy; PRM: premature rapture of membrane.

to people living in resource-poor countries, where they could not ensure their daily bread let alone for medication. The cost of new antibiotics is also high, which in turn poses great burden for poor countries [27].

In addition to the identification of the common bacteria and their resistance pattern, this study tried to evaluate the association of different sociodemographic and clinical parameters with bacterial infection. In the present study, the prevalence of bacteria was associated with obstetric parameters like mode of delivery through cesarean section and vaginal delivery by using episiotomy ($p = 0.002$). This could be attributed to the physiological change, immune shift, high bleeding, and contamination; this is in line with studies conducted in Bahir Dar [9], Jimma [4, 14], and Estonian University [10]. There was also an association of developing bacterial infection with premature rapture of membrane ($p = 0.018$). This is in line with studies conducted in Jimma [4], Tanzania [11], Hong Kong [28], and Brazil [29]. This might be due to the fact that having a premature rapture of membrane could expose women to various injuries, which again enable pathogenic bacteria to have suitable environment to cause infection. The presence of wound infection following cesarean section and episiotomy was aggravated due to improper wound care and improper treatment. In addition, women who have got premature rapture of membrane could get exposed to various injuries and minor tears, which again enable pathogenic bacteria to have convenient environment to cause infection.

The limitation of anaerobic culture facility that is used to cultivate anaerobic bacteria might increase the quantity of bacterial identification. Since it is a cross-sectional study, it has limited capacity to assess risk factors.

5. Conclusion and Recommendations

The prevalence of bacterial infection was too increased among women who attended postnatal care. The major bacteria were *S. aureus*, *E. coli*, CoNS, *K. pneumoniae*, *Citrobacter*, *Enterobacter*, and *S. pyogenes*. Gram-negative and Gram-positive bacteria revealed resistance to the most frequently used drugs tested in vitro. Multidrug-resistant bacteria show big threat posed by antibiotic resistant strains in vulnerable mothers. Drug-resistant bacterial infection leads to increased patient hospital stay, health care costs, and death rate. Factors such as cesarean section, episiotomy for delivery, and premature rapture of membrane were predictors for bacterial infection among postnatal mothers. Reviewing the nature and cause of bacterial infection and its drug resistance is necessary to overcome bacterial infection. To confirm applicable treatment, the existing information about bacteria that cause maternal infections and their drug resistance pattern is essential.

Conflicts of Interest

The authors declare that there are no conflicts of interest regarding the publication of this study.

Acknowledgments

The study was funded by Armauer Hansen Research institute. The authors thank study members for volunteering to give a specimen and Dr. Dawit Kassahun. They also would like to thank Professor Feleke Moges, Department of Medical Microbiology, University of Gondar, and Gondar University Teaching Hospital. Finally, they want to acknowledge Amhara Health Bureau, Armauer Hansen Research Institute, and Federal Ministry of Health.

References

[1] S. M. Pear, "Patient Risk Factors and Best Practices for Surgical Site Infection Prevention," in *Managing Infection Control*, p. 56, 2007.

[2] W. Dessie, G. Mulugeta, S. Fentaw, A. Mihret, M. Hassen, and E. Abebe, "Pattern of bacterial pathogens and their susceptibility isolated from surgical site infections at selected referral hospitals, Addis Ababa, Ethiopia," *International Journal of Microbiology*, vol. 2016, Article ID 2418902, 8 pages, 2016.

[3] D. Amenu, T. Belachew, and F. Araya, "Surgical site infection rate and risk factors among obstetric cases of Jimma University Specialized Hospital, Southwest Ethiopia," *Ethiopian Journal of Health Sciences*, vol. 21, no. 2, pp. 91–100, 2011.

[4] J. Amatya, M. Rijal, and R. Baidya, "Bacteriological study of the postoperative wound samples and antibiotic susceptibility pattern of the isolates in BB hospital," *The Journal of Sexual Medicine*, vol. 3, no. 1, p. 1019, 2015.

[5] K. Conroy, A. F. Koenig, Y. Yan-Hong, A. Courtney, H. J. Lee, and E. R. Norwitz, "Infectious morbidity after cesarean delivery: 10 Strategies to reduce risk," *Reviews in Obstetrics & Gynecology*, vol. 5, pp. 69–77, 2012.

[6] DM. Kabau, S. Wanjala, and F. Kagema, "Incidence and determinants of surgical site infection after Caesarean Delivery at Kenyatta National Hospital," *Incidence and determinants of surgical site infection after Caesarean Delivery at Kenyatta National Hospital*, p. 51, 2014.

[7] P. Mitt, K. Lang, A. Peri, and M. Maimets, "Surgical-site infections following cesarean section in an Estonian University Hospital: Postdischarge surveillance and analysis of risk factors," *Infection Control and Hospital Epidemiology*, vol. 26, no. 5, pp. 449–454, 2005.

[8] F. Kasolo, A. A. Yahaya, J. B. Ndihokubwayo, B. Impouma, C. J. Oxenford, and S. Cognat, "Guide for establishing laboratory-based surveillance for antimicrobial resistance; World Health Organization Regional Office for Africa," *Guide for establishing laboratory-based surveillance for antimicrobial resistance*, pp. 1–25, 2013.

[9] W. Mulu, G. Kibru, G. Beyene, and M. Damtie, "Postoperative nosocomial infections and antimicrobial resistance pattern of bacteria isolates among patients admitted at Felege Hiwot Referral Hospital, Bahirdar, Ethiopia," *EthiopianJournal of Health Sciences*, vol. 22, no. 1, pp. 7–18, 2012.

[10] J. Neu and J. Rushing, "Cesarean versus vaginal delivery: long-term infant outcomes and the hygiene hypothesis," *Clinics in Perinatology*, vol. 38, no. 2, pp. 321–331, 2011.

[11] B. H. Rimawi, *Infectious Comorbidities Encountered in Obstetrics and Neonatology*, OMICS International, 2014, http://www.esciencecentral.org/ebooks.

[12] R. Simões, R. C. Cavalli, W. M. Bernardo, A. J. Salomão, and E. C. Baracat, "Cesarean delivery and prematurity," *Revista da Associação Médica Brasileira*, vol. 61, no. 6, pp. 489–494, 2015.

[13] G. Adwan, N. Abu Hasan, I. Sabra, D. Sabra, S. Al-butmah, and S. Odeh, "Detection of bacterial pathogens in surgical site infections and their antibiotic sensitivity profile," *International Journal of Medical Research and Health Sciences*, vol. 5, no. 5, pp. 75–82, 2016, http://www.ijmrhs.com.

[14] T. Sahile, S. Esseye, G. Beyene, and S. Ali, "Post-surgical Infection and Antibiotic Susceptibility Patterns of Bacteria Isolated from Admitted Patients with Signs of Infection at Jimma University Specialized Hospital, Jimma, Ethiopia," *International Journal of TROPICAL DISEASE & Health*, vol. 17, no. 4, pp. 1–12, 2016.

[15] L. Charrier, P. Serafini, A. Ribatti, A. Castella, G. Rabacchi, and C. M. Zotti, "Post-partum surgical wound infections: Incidence after caesarean section in an Italian hospital," *Journal of Preventive Medicine and Hygiene*, vol. 50, no. 3, pp. 159–163, 2009.

[16] R. Cooper, "Practice development how to take a wound swab," *Wounds International*, vol. 1, no. 3, pp. 19-20, 2010.

[17] S. E. Gardner, R. A. Frantz, C. L. Saltzman, S. L. Hillis, H. Park, and M. Scherubel, "Diagnostic validity of three swab techniques for identifying chronic wound infection," *Wound Repair and Regeneration*, vol. 14, no. 5, pp. 548–557, 2006.

[18] Clinical and Laboratory Standards Institute, *Performance standards for antimicrobial susceptibility testing: twenty-fourth informational supplement M100-S24*, CLSI, Wayne, PA, USA, 2016.

[19] V. Filippi, D. Chou, C. Ronsmans, W. Graham, and L. Say, *Levels and Causes of Maternal Morbidity and Mortality*, chapter 3, 1-31, 2013.

[20] S. Sriskandan, "In infectious diseases, national centre for infection prevention and management, department of infectious diseases immunity, imperial college London, UK," *J R Coll Physicians Edinb*, vol. 41, pp. 339-46, 2011.

[21] O. Issa, EC. Mgongo, and J. W. Gichuhi, "Post caesarean section wound sepsis at mnazi mmoja hospital Zanzibar," *Journal of Obstetrics and Gynaecology*, pp. 1–9, 2010.

[22] F. J. Mpogoro, S. E. Mshana, M. M. Mirambo, B. R. Kidenya, B. Gumodoka, and C. Imirzalioglu, "Incidence and predictors of surgical site infections following caesarean sections at Bugando Medical Centre, Mwanza, Tanzania," *Antimicrobial Resistance and Infection Control*, vol. 3, no. 1, article no. 25, 2014.

[23] H. Dhar, I. Al-Busaidi, B. Rathi, A. Nimre, V. Sachdeva, and I. Hamdi, "A Study of Post-Caesaren Section Wound Infections in a Regional Referral Hospital Oman," *SQU Medical Journal*, vol. 14, no. 2, pp. 212–216, 2014.

[24] B. Amare, Z. Abdurrahman, B. Moges, and J. Ali, "Postoperative surgical site bacterial infections and drug susceptibility patterns at Gondar University Teaching Hospital, Northwest Ethiopia," *Journal of Bacteriology & Parasitology*, vol. 2, no. 8, 2011.

[25] T. Frieden, *Communicable Disease Control (CDC): Antibiotic Resistance threats in the United States*, 2013.

[26] R. E. Mengesha, B. G.-S. Kasa, M. Saravanan, D. F. Berhe, and A. G. Wasihun, "Aerobic bacteria in post surgical wound infections and pattern of their antimicrobial susceptibility in Ayder Teaching and Referral Hospital, Mekelle, Ethiopia," *BMC Research Notes*, vol. 7, no. 1, article 575, 2014.

[27] H. Inoue and R. Minghui, "Antimicrobial resistance: translating political commitment into national action," *Bulletin of the World Health Organization*, vol. 95, no. 4, pp. 242-242, 2017.

[28] W. W. Y. Lam, "Abdominal wound infection after caesarean delivery in a district Hospital at University of Hong Kong," *HKU Scholars Hub*, pp. 3–75, 2006.

[29] T. C. F. Farret, J. Dallé, V. da Silva Monteiro, C. V. W. Riche, and V. S. Antonello, "Risk factors for surgical site infection following cesarean section in a Brazilian Women's Hospital: A case-control study," *The Brazilian Journal of Infectious Diseases*, vol. 19, no. 2, pp. 113–117, 2015.

14

Isolation and Physiomorphological Characterization of *Escherichia coli* O157:H7-Infecting Bacteriophages Recovered from Beef Cattle Operations

Pushpinder Kaur Litt and Divya Jaroni

Department of Animal Science and Food and Agricultural Products Center, Oklahoma State University, 148 FAPC, Monroe St., Stillwater, OK 74078, USA

Correspondence should be addressed to Divya Jaroni; divya.jaroni@okstate.edu

Academic Editor: Karl Drlica

Bacteriophages, recovered from beef cattle environment and specifically targeting *Escherichia coli* O157:H7, were examined for their physiological and morphological characteristics. Degree of bacterial lysis and host range of isolated bacteriophages was determined against 55 isolates of *E. coli* O157:H7. Morphology of phages was examined under transmission electron microscope. Phage growth parameters, particularly rate of adsorption, rise period, latent period, and burst size were also determined. The stability of isolated phages was tested at acidic and alkaline pH, at high temperatures, and in cold storage. A total of 7 phages were isolated which showed lytic activity against 50 out of 55 isolates of *E. coli* O157:H7. Based on the morphology, phages were classified into Myoviridae or Siphoviridae family. Phages had a rise period between 19 and 40 min, a short latent period between 12 and 30 min, and a large burst size (89–631 virions per infected cell), indicating high lytic activity. Phages remained stable for 24 h at a wide pH (1–11) and temperature range (40–60°C) and for 90 d in cold storage. Characterization of bacteriophages, with a diverse host range of *E. coli* O157:H7, could aid in the development of effective biocontrol strategies for this pathogen in the food industry.

1. Introduction

Escherichia coli O157:H7 is an important foodborne pathogen that lives commensally in the rumen of cattle and other food animals such as sheep and goat [1–5]. Additionally, deer, horses, dogs, and birds can also transiently harbor this pathogen [4, 6–8]. Direct or indirect contact with animals or manure of animals carrying *E. coli* O157:H7 could mediate its transfer to water and food products, which could result in human infections upon consumption. In a typical year in the United States (US), *E. coli* O157:H7 causes an estimated 63,000 foodborne illnesses, 2,100 hospitalizations, and 20 deaths, imposing an economic burden of $271 million [9, 10]. Advanced stages of pathogen infection include hemorrhagic colitis, which could progress into hemolytic uremic syndrome. Severe complications are characterized by renal failure, hemolytic anemia, and thrombocytopenia, which could be fatal [11–13]. Some high-risk food commodities associated with these illnesses include beef and meat products, fresh produce, unpasteurized apple juice, and dairy products [13–28].

Reduction of *E. coli* O157:H7 at the preharvest and processing level could play a significant role in preventing the introduction of this pathogen into the food chain. Various approaches, such as diet and probiotic therapy, vaccination, and antibiotics, against *E. coli* O157:H7 have been evaluated to control its prevalence in food animals [3, 10, 29]. However, recolonization of previously infected animals, with the same or different strains of *E. coli* O157:H7, limits the effectiveness of these strategies [3, 30]. Similarly, at the processing level, various control measures such as organic acids, chlorine, or other antimicrobial washes have been used to control *E. coli* O157:H7 persistence in foods [31–34]. However, these interventions remain insufficient, as is evident from the continuing foodborne illness outbreaks and food recalls associated with this pathogen. It is therefore important to develop effective alternatives to control *E. coli* O157:H7 in the food and animal industry.

Targeted use of bacteriophages against *E. coli* O157:H7 could reduce this problem. Bacteriophages are ubiquitous, naturally abundant viruses, which invade and kill specific bacteria. They are recognized as gastrointestinal (GI) commensal microorganisms and have been isolated from various sources [35, 36]. Almost a century ago, d'Herelle [37] demonstrated the efficacy of bacteriophages against human pathogens. However, more recently, due to the emergence of antibiotic-resistant microorganisms, the long-lost phage therapy has regained interest for use against foodborne pathogens. Under *in vitro* conditions, virulent bacteriophages have shown the potential for selective elimination of *E. coli* O157:H7 [38, 39]. Additionally, studies in sheep have shown that gut colonization by *E. coli* O157:H7 can be prevented by oral administration of O157:H7-infecting phages [39]. Efficacy of bacteriophages against *E. coli* O157:H7 has also been documented in various foods such as cantaloupe, spinach, lettuce, tomato, broccoli, and ground beef [40–42]. In a study by O'Flynn et al. [43], a cocktail of *E. coli* O157:H7-specific bacteriophages completely reduced pathogen counts on the surface of beef steaks. Additionally, studies have shown that phages can effectively eliminate *E. coli* O157:H7 in biofilms on various food-processing surfaces such as stainless steel and high-density polyethylene [42, 44]. Even though studies have demonstrated the potential of phage-based technologies to control *E. coli* O157:H7, very little is known about the morphology, physiology, and characteristics of specific phages. At the same time, selection of effective bacteriophages for food and animal industry applications also requires an understanding of their survival under various stress conditions. The present study was aimed at isolating *E. coli* O157:H7-specific bacteriophages from beef cattle operations and determining their physiomorphological characteristics, including survival under various pH, high temperatures, and cold storage conditions.

2. Materials and Methods

2.1. Bacterial Cultures. Isolation of bacteriophages was carried out using the host strain, *E. coli* O157:H7 ATCC 43895 (Table 1). Host range of isolated phages was determined using 55 *E. coli* O157:H7 isolates (Table 1). Two of these isolates were clinical strains (ATCC 43895, ATCC 43888) while the rest were wild-type (WT) isolates, retrieved from our laboratory culture collection, originally isolated from bovine feces or beef cattle farm environment [45]. Prior to an experiment, overnight culture of each *E. coli* O157:H7 isolate was prepared in tryptic soy broth (TSB; Bacto™, BD, Sparks, MD) and incubated statically at 37°C for 18 h.

2.2. Bacteriophage Isolation. Over a period of two years, water ($n = 67$) and bovine fecal ($n = 60$) samples were collected from beef cattle operations in Oklahoma, during the summer months, and used for isolation of bacteriophages specific to *E. coli* O157:H7 [45]. Water (10 mL) and fecal (10 g) samples were enriched in 25 mL double strength NZ-amine casamino acid yeast extract sodium chloride magnesium sulfate (NZCYM; RPI Corp, IL) broth for 18 h at 37°C, along with 10 mL of host bacterial culture (*E. coli* O157:H7 ATCC 43895). After incubation, 2 mL suspension was centrifuged at 12,000 rpm for 10 min to remove cellular debris and fecal material. The supernatant was filtered using 0.45 μ syringe filter (EMD Millipore Millex™, Carrigtwohill, Ireland) and the filtrate plated on NZCYM agar (Fisher Scientific, NJ) via double-layer agar method, as described by Sambrook et al. [46]. The presence of phage was confirmed by plaque formation on the agar plate and the phage further concentrated by the method described by Chandra et al. [47]. Briefly, an overnight culture (100 μL) of *E. coli* O157:H7 was suspended in molten (0.75%) NZCYM agar and poured onto an NZCYM agar plate which was allowed to solidify for 2–5 min. The plate was then streaked with the phage, as horizontal and vertical lines, using a sterile platinum loop and incubated at 37°C for 18–20 h. After incubation, 5 mL SM buffer (10 mM Tris–HCl, pH 7.5; 100 mM NaCl; 10 mM $MgSO_4$; Fisher Scientific, NJ) was poured onto the agar plates to elute any plaques along the lines. The agar containing the plaques was then scraped with a sterile platinum loop to release the phages and the elution was centrifuged at 12,000 rpm for 15 min. The resulting supernatant was filtered using a 0.22 μ syringe filter, to which 0.1% chloroform (Fisher Scientific, NJ) was added to make a phage working-stock solution which was stored at 4°C until further use. Plaque size (mm) of individual phages was determined by performing plaque assay [48] and by measuring their diameter using Vernier Calipers (Fisherbrand™ Traceable™ Digital Calipers, NJ).

Prior to an experiment, phage titers were determined (Table 2), as plaque forming units (PFU) mL^{-1}, by serially diluting the phage working-stock in phosphate buffered saline (PBS; pH 7.4; sodium chloride, Fisher Scientific, NJ; potassium chloride, sodium phosphate monobasic and sodium phosphate dibasic, Sigma-Aldrich, MO) and by performing plaque assay [48].

2.3. Bacteriophage Host Range. Host range of all the phages was tested against 55 *E. coli* O157:H7 isolates (Table 1), using spot-on-lawn assay [48]. Overnight culture (100 μL) of each bacterial isolate was added to 5 mL of molten (0.75%) NZCYM agar and was overlaid onto an NZCYM agar plate. Ten microliters of the respective phage solution (~10^8 PFU mL^{-1}) was spotted onto each bacterial agar plate. The plates were incubated at 37°C for 18 h and examined for clear zones on the bacterial lawn, to determine the host range of the respective phages. Bacterial sensitivity to a bacteriophage was established by bacterial lysis at the spot where the phage was deposited. Based on the degree of clarity on the bacterial lawn, the spots were differentiated into three categories: clear (+++), turbid (++), or no reaction (—) (Table 1 and Figure 1).

2.4. Bacteriophage Morphology. Phage morphology was determined using transmission electron microscopy. Sample was prepared by suspending an overnight culture (100 μL) of *E. coli* O157:H7 in 5 mL molten (0.75%) NZCYM agar and was poured onto a fresh NZCYM agar plate. Phage stock (prepared above) was serially diluted in SM buffer

TABLE 1: Host range* of phages against *E. coli* O157:H7 isolates.

Strains	P-1	P-2	P-3	P-4	P-5	P-6	P-7
ATCC 43895	+++	+++	+++	+++	+++	+++	+++
ATCC 43888	+++	+++	+++	+++	+++	+++	+++
O157:H7 RF1	+++	+++	+++	+++	+++	+++	+++
O157:H7 RF2	+++	+++	+++	+++	+++	+++	+++
O157:H7 RF3	+++	+++	+++	+++	+++	+++	+++
O157:H7 RF4	+++	+++	+++	+++	+++	—	+++
O157:H7 RF5	+++	+++	+++	+++	+++	+++	+++
O157:H7 RF6	—	—	—	—	—	—	—
O157:H7 RF7	—	—	—	—	—	—	—
O157:H7 RF8	+++	+++	+++	+++	+++	+++	+++
O157:H7 RF9	+++	+++	+++	+++	+++	+++	+++
O157:H7 RF10	+++	+++	+++	+++	+++	+++	+++
O157:H7 RF11	+++	+++	+++	+++	+++	+++	+++
O157:H7 RF12	+++	+++	+++	+++	+++	+++	+++
O157:H7 RF13	+++	+++	+++	+++	+++	+++	+++
O157:H7 RF14	+++	+++	+++	+++	+++	+++	+++
O157:H7 RF15	+++	+++	+++	+++	+++	+++	+++
O157:H7 SF1	+++	+++	+++	+++	+++	+++	+++
O157:H7 SF2	+++	+++	+++	+++	+++	+++	+++
O157:H7 SF4	+++	+++	+++	+++	+++	+++	+++
O157:H7 SF5	+++	+++	+++	+++	+++	+++	+++
O157:H7 SF7	+++	+++	+++	+++	+++	+++	+++
O157:H7 SF8	+++	+++	+++	+++	+++	+++	+++
O157:H7 SF10	+++	+++	+++	+++	+++	+++	+++
O157:H7 SF12	+++	+++	+++	+++	+++	+++	+++
O157:H7 SF14	+++	+++	+++	+++	+++	+++	+++
O157:H7 SF15	+++	+++	+++	+++	+++	+++	+++
O157:H7 SW3	—	—	—	—	—	—	—
O157:H7 TW3	—	—	—	—	—	—	—
O157:H7 TE1	++	++	++	++	++	++	++
O157:H7 TE2	+++	+++	+++	+++	+++	+++	+++
O157:H7 TF1	+++	+++	+++	+++	+++	+++	+++
O157:H7 TF2	+++	+++	+++	+++	+++	+++	+++
O157:H7 TF3	+++	+++	+++	+++	+++	+++	+++
O157:H7 TF4	+++	+++	+++	+++	+++	+++	+++
O157:H7 TF6	+++	+++	+++	+++	+++	+++	+++
O157:H7 TF7	+++	+++	+++	+++	+++	+++	+++
O157:H7 TF8	+++	+++	+++	+++	+++	+++	+++
O157:H7 TF9	+++	+++	+++	+++	+++	+++	+++
O157:H7 TF10	+++	+++	+++	+++	+++	+++	+++
O157:H7 TF11	+++	+++	+++	+++	+++	+++	+++
O157:H7 TF12	+++	+++	+++	+++	+++	+++	+++
O157:H7 TF13	+++	+++	+++	+++	+++	+++	+++
O157:H7 TF14	+++	+++	+++	+++	+++	+++	+++
O157:H7 RF15	+++	+++	+++	+++	+++	+++	+++
O157:H7 XW5	+++	+++	+++	+++	+++	+++	+++
O157:H7 JF4	+++	+++	—	—	+++	—	+++
O157:H7 LF4	+++	+++	+++	+++	+++	+++	+++
O157:H7 KF10	+++	+++	+++	+++	+++	+++	+++
O157:H7 JEQ1	+++	+++	+++	+++	+++	+++	+++
O157:H7 JF6	+++	+++	+++	+++	+++	+++	+++
O157:H7 KF7	+++	+++	+++	+++	+++	+++	+++
O157:H7 LF5	+++	+++	+++	+++	+++	+++	+++
O157:H7 EF2	—	—	—	—	—	—	—

*Based on the degree of clarity on the bacterial lawn, the spots were differentiated into three categories: clear (+++), turbid (++), or no reaction (—).

Table 2: Titer (PFU mL^{-1}) and plaque size (mm) of isolated phages.

Phages	Titer (PFU mL^{-1})	Plaque size (mm)
P-1	7.9	0.3
P-2	7.4	0.4
P-3	7.8	0.5
P-4	7.7	0.5
P-5	7.5	0.4
P-6	8.0	0.5
P-7	7.5	0.5

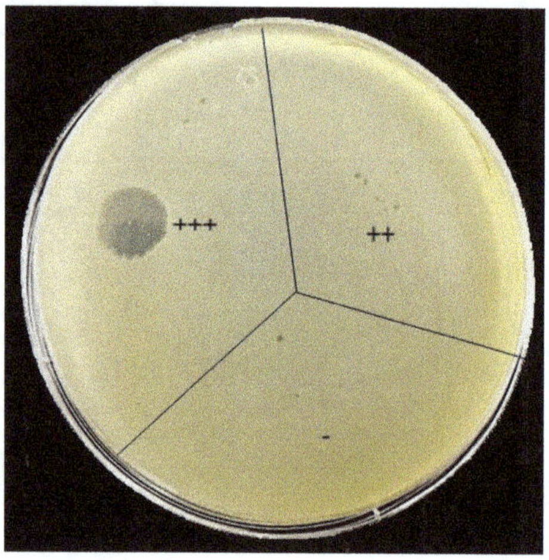

Figure 1: Representative NZCYM agar plate showing the three categories of phage spots: clear (+++), turbid (++), and no reaction (—), on the lawn of host *E. coli* O157:H7.

supplemented with 1 M calcium (SM-Ca; Fisher Scientific, NJ). 10 μL of each dilution was spotted onto the bacterial agar plate (prepared above) and incubated at 37°C for 18 h. Following incubation, phages were eluted in 10 μL SM-Ca by carefully pipetting up and down (7–10 times) onto the webbed spot. Each eluted phage sample was negatively stained with 2% phosphotungstic acid (Emsdiasum, Hatfield, PA) on carbon-coated grids (Emsdiasum, Hatfield, PA) and examined under the transmission electron microscope (TEM; JEM-2100TEM, JEOL). Morphology (shape), capsid diameter (nm), and tail length (nm) of the phages were determined by electron micrographs at a magnification of 50,000x (Oklahoma Technology and Research Park Venture, Oklahoma State University, Stillwater, OK).

2.5. Bacteriophage Stability at Acidic and Alkaline pH, at High Temperatures, and in Cold Storage. Stability of isolated phages was tested against a wide pH (1–11) and temperature (−80 to 90°C) range to determine their survival under acidic, alkaline, high temperature, and cold storage conditions. All phage stability studies were conducted with a phage population of about 10^8 PFU mL^{-1} (Table 2).

Stability studies for acidic and alkaline conditions were conducted according to the methods described by Niu et al. [49]. Briefly, phages were suspended in PBS adjusted with 1 M NaOH or HCl (Fisher Scientific, NJ), to yield a pH range of 1–11, and incubated at 37°C to determine their survival at 1, 2, 4, 6, 12, and 24 h.

For thermal stability [50], working stock of each phage (100 μL) was suspended in 900 μL PBS and the suspension was incubated at 40, 60, 70, and 90°C for 60 min in a dry bath (Fisher Scientific, IA). Phage population at each temperature was determined every 10 min during the 60-minute incubation.

For cold storage stability [50], 1 mL stock solution of each isolated phage was stored at 4, −20, and −80°C and sampled at 0, 1, 30, 60, and 90 d. Surviving phage populations for each study were determined by serially diluting the samples in PBS and plating on NZCYM agar, using double-layer agar method.

2.6. Phage Adsorption and One-Step Growth Kinetics. Phage adsorption assay [51] was performed to determine the time taken by the phages to adsorb to the host bacterial surface. Their rise period and latent period, along with their burst size, were determined through one-step growth kinetics study. Overnight culture (1 mL) of host *E. coli* O157:H7 (10^9 CFU mL^{-1}) was centrifuged at 12,000 rpm for 2 min. After discarding the supernatant, the pellet was resuspended in 900 μL PBS and 100 μL of the phage-stock solution (10^8 PFU mL^{-1}) added to the suspension to achieve a 0.1 multiplicity of infection (MOI). This bacteria-phage suspension was used for the adsorption and one-step growth kinetics study.

For phage adsorption study, the bacteria-phage suspension was incubated at 37°C for 80 min. After taking an initial sample at 0 min, a 100 μL aliquot was sampled every 20 min by adding to 900 μL PBS and centrifuging for 2 min at 12,000 rpm. The supernatant containing unadsorbed phages was filtered through 0.2 μ filter and plated using double-layer agar technique. Percent of unadsorbed phages, at each 20-minute interval, was calculated as the ratio of PFU mL^{-1} in the supernatant to the initial PFU mL^{-1} at 0 min. The percent of adsorbed phages was determined by subtracting the percent of unadsorbed phages from 100.

For one-step growth experiment, the bacteria-phage suspension was incubated at 37°C for 10 min to allow the phages to adsorb to the bacterial cell [52]. After incubation, the suspension was centrifuged at 13,000 rpm for 1 min and the supernatant subjected to plaque assay to determine the titer of the unadsorbed phages. The pellet containing infected bacterial cells was immediately resuspended in 10 mL of prewarmed (37°C) TSB and, after taking an initial sample (100 μL) at 0 min, incubated at 37°C for 60 min. To obtain the one-step growth curve, a 100 μL sample was collected every 5 min, serially diluted, and plated on NZCYM agar using double-layer agar method. The rise period and latent period, as well as the burst size of each phage, were calculated by fitting the one-step growth curve into a 4-parameter sigmoidal model [53] as follows: $P_t = P_o + \alpha/(1 + e^{-((t-t_r)/b)})$, where P_t was the log phage-density (PFU mL^{-1}) at time t (min), P_o was the initial log phage-density at time 0 (min), t_r was the rise period (min), α was the log density (PFU mL^{-1}) of virions per infected cell, and b was the maximum rate of exponential

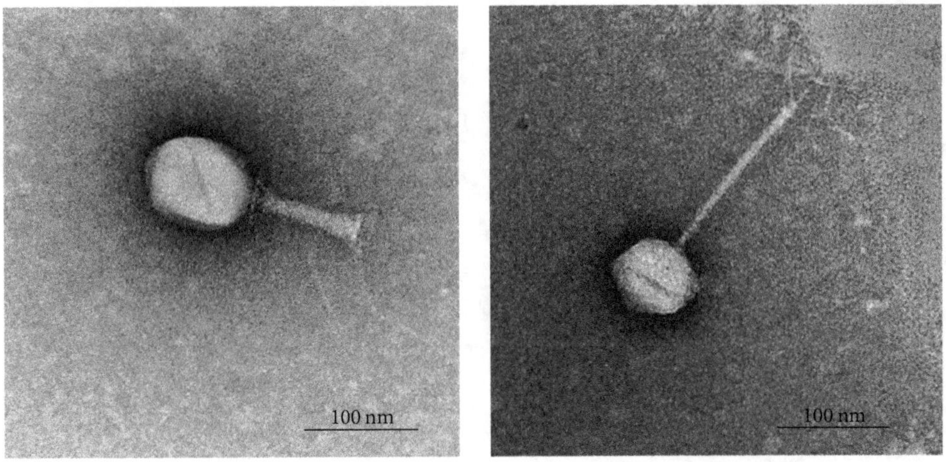

(a) (b)

FIGURE 2: Transmission electron micrograph of phages P-1, P-2, and P-5 belonging to Myoviridae family (a), and P-3, P-4, P-6, and P-7 categorized as Siphoviridae family (b). Scale bar represents 100 nm.

phage growth (min^{-1}). The burst size (PFU $cell^{-1}$) and latent period (min) were calculated as 10^{α} and $t_r - 1/b$, respectively. The latent period was defined as the time-interval between the end of adsorption and the beginning of the first burst (i.e., the constant period), indicated by minimal increase in phage titer [54, 55]. The rise period (after latent period) was identified as the time over which infected bacterial cells lysed, with significant increase in phage titer [54]. The burst size was defined as the maximum yield of phages obtained from the appearance of first phage progeny until complete bacterial cell lysis [48, 56].

2.7. Statistical Analysis. All the experiments were repeated three times and the mean values of the three replicates obtained. Surviving phage populations, obtained in each study, were converted to $\log_{10}$ PFU mL^{-1}. Statistical analysis was performed to determine the effect of pH and high temperature on phage survival. Data were analyzed to determine the analysis of variance (ANOVA) using Duncan's multiple range test (JMP v.12 software; SAS Inst., Cary, NC, USA). Significant differences between results were estimated at $P < 0.05$.

3. Results

3.1. Isolation and Host Range Determination of Bacteriophages. From the 127 environmental samples tested, seven phages, specific to *E. coli* O157:H7, were isolated and designated as P-1 to P-7. The isolated phages formed medium-sized (ca. 0.3–0.5 mm in diameter) clear plaques on the lawn of its host bacteria (Table 2).

Host range of the isolated phages was determined against 55 *E. coli* O157:H7 isolates (ATCC and WT). All the isolated phages formed clear (+++) spots on the lawn of 49 (89%) *E. coli* O157:H7 isolates, indicating strong lytic activity and a wide host range (Table 1). However, phage P-6 did not lyse isolates RF4 and JF4, and phages P-3 and P-4 did not lyse isolate JF4. All seven phages formed turbid (++) spots against the isolate TE1 and showed no reaction (—) against the WT isolates EF2, RF6, RF7, SW3, and TW3 (Table 1).

3.2. Bacteriophage Morphology. The electron micrographs showed that phages P-1, P-2, and P-5 had an icosahedral head (about 89 nm diameter) and a short contractile tail (ca. 115 nm) with a base plate and several tail fibers (Figure 2(a)). The overall morphology indicated that these phages were T4-like phages, belonging to the Myoviridae family of the Caudovirales order. Phages P-3, P-4, P-6, and P-7 had a hexagonal head (approximately 89 nm diameter) with an extremely long flexible tail (ca. 198 nm) and no visible collar or terminal knobs (Figure 2(b)). They were identified as the Siphoviridae family phages of the Caudovirales order.

3.3. Bacteriophage Stability at Acidic and Alkaline pH, at High Temperatures, and in Cold Storage. All the isolated phages were tested against a wide pH range (1–11), over a period of 24 h, to determine their stability under acidic and alkaline conditions (Figures 3(a) and 3(b)). In the acidic pH range (1–5), phages exhibited the most stability at pH 5, with minimal loss in viability (0.1–1.7 logs PFU mL^{-1}) after 2 h of incubation. Of the seven phages tested, three (P-2, P-5, and P-6) survived at pH 1 over a period of 24 h, with a population of 4.4–5.9 logs PFU mL^{-1} (Figure 3(a)), while P-3 survived at pH 1 for 1 h. All the phages survived at pH 2 over a period of 24 h, with limited activity loss (0.2–3.7 logs PFU mL^{-1}) after 1 h of incubation (Figure 3(a)). Among all the phages, P-3 was the most stable at pH 2, with no significant ($P < 0.05$) loss in viability (0.9 logs PFU mL^{-1}) after 24 h of incubation. All the phages survived well at pH 7 and 9 with no significant ($P < 0.05$) loss in viability (0.1–1.9 logs PFU mL^{-1}) after 24 h of incubation (Figure 3(b)). Phages were also stable at pH 11 for up to 12 h and had minimal reduction (0.8–2.3 logs PFU mL^{-1}) in titers after 24 h of incubation (Figure 3(b)). Among

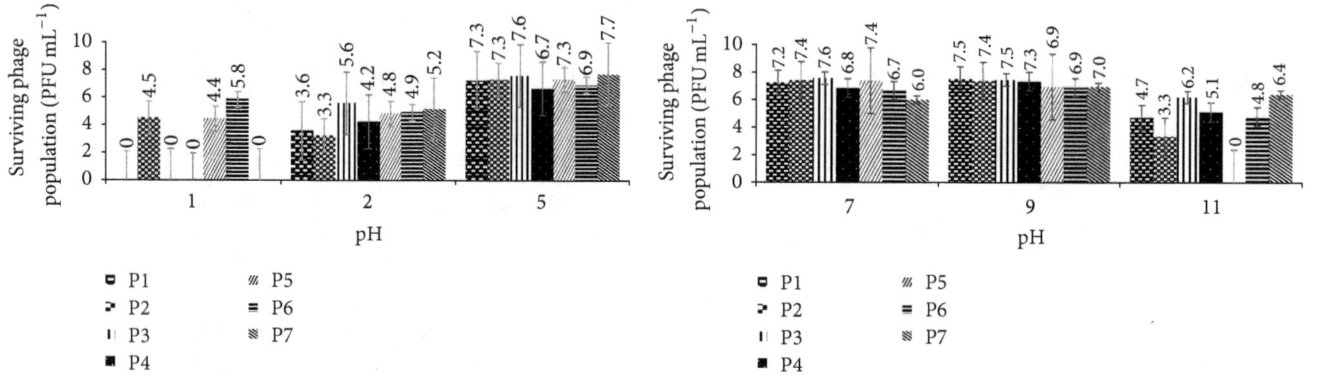

(a) Stability of phages at acidic pH (1, 2, 5) over 24 h. Data expressed as mean ± SD for three independent experiments. Data was analyzed at a significance level of $P < 0.05$

(b) Stability of phages at alkaline pH (7, 9, 11) over 24 h. Data expressed as mean ± SD for three independent experiments. Data was analyzed at a significance level of $P < 0.05$

FIGURE 3

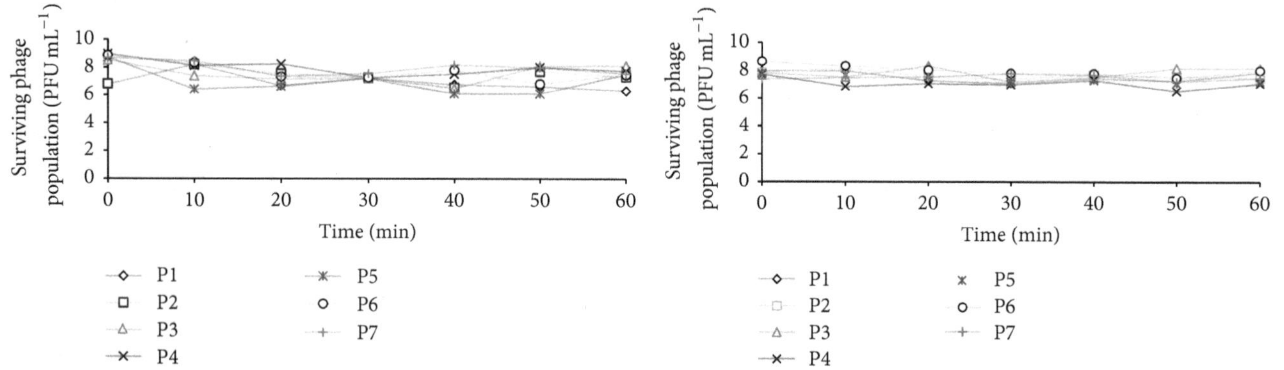

(a) Thermal stability of phages at 40°C for 60 min. Data was expressed as mean ± SD for three independent experiments. Data was analyzed at a significance level of $P < 0.05$

(b) Thermal stability of phages at 60°C for 60 min. Data was expressed as mean ± SD for three independent experiments. Data was analyzed at a significance level of $P < 0.05$

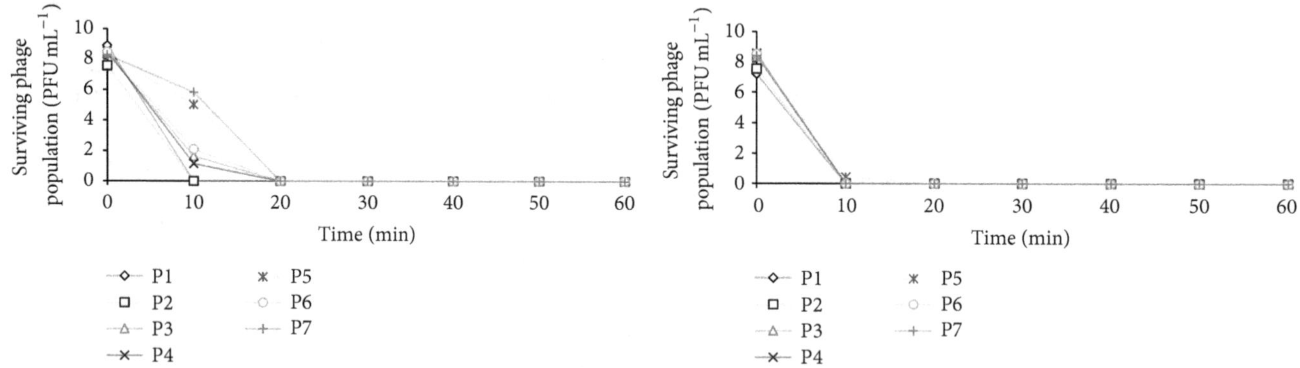

(c) Thermal stability of phages at 70°C for 60 min. Data was expressed as mean ± SD for three independent experiments. Data was analyzed at a significance level of $P < 0.05$

(d) Thermal stability of phages at 90°C for 60 min. Data was expressed as mean ± SD for three independent experiments. Data was analyzed at a significance level of $P < 0.05$

FIGURE 4

all the seven phages, P-7 was the most stable at pH 11 with slight reductions in population (0.4–1.6 PFU mL^{-1}) during the 24 h period.

Results from the thermal stability studies revealed that all the tested phages were stable in the 40–60°C temperature range and did not lose viability after 60 min of incubation at the respective temperatures (Figures 4(a) and 4(b)). Phage viability was significantly ($P < 0.05$) affected at higher temperatures of 70 and 90°C (Figures 4(c) and 4(d)). At 70°C, the populations of phages P-5 and P-7 were reduced by 2.5 and 3.3

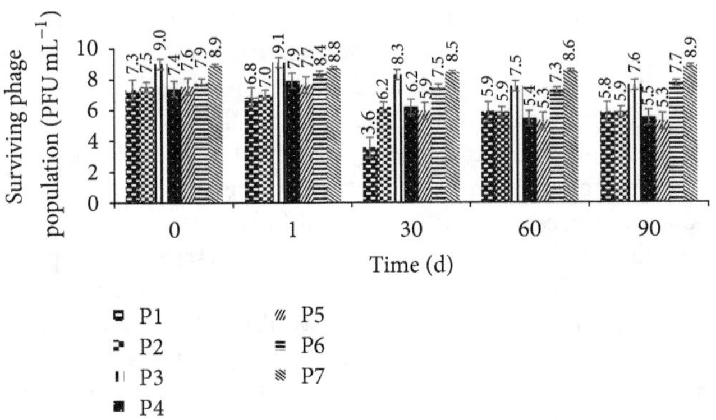

(a) Phage survival at 4°C over 90 d. Data was expressed as mean ± SD for three independent experiments

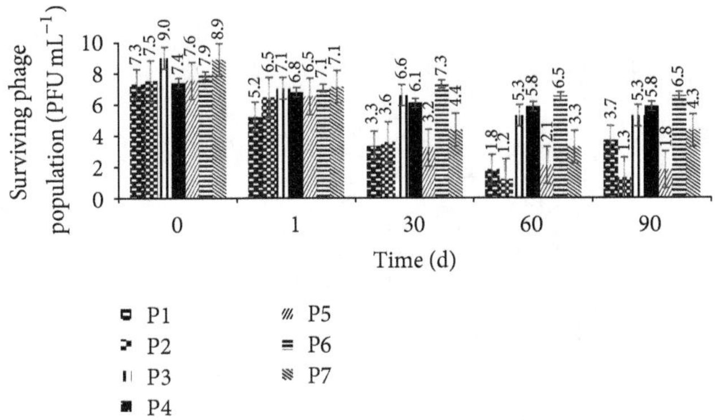

(b) Phage survival at −20°C over 90 d. Data was expressed as mean ± SD for three independent experiments

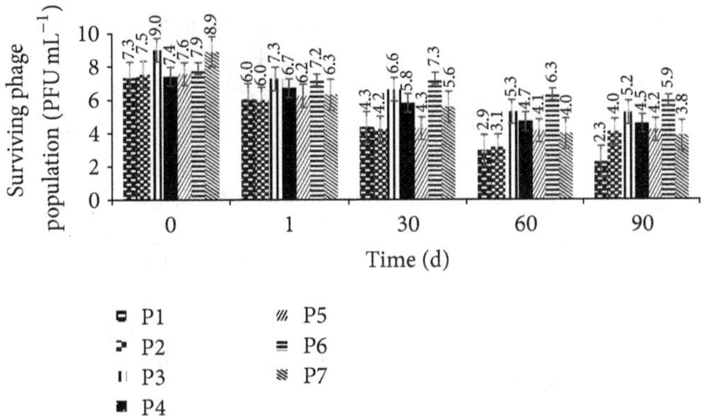

(c) Phage survival at −80°C over 90 d. Data was expressed as mean ± SD for three independent experiments

FIGURE 5

logs PFU mL^{-1}, respectively, after 10 min incubation. Phages P-3, P-4, and P-6 showed a population reduction between 6.4 and 7.3 logs PFU mL^{-1}, whereas phages P-1 and P-2 showed no viability after 10 min. All the other phages lost viability after 20 min of incubation at 70°C (Figure 4(c)). At 90°C, no viable phages were detected after 10 min of incubation (Figure 4(d)).

Phages were also tested for cold storage stability and found to be more stable at refrigeration temperature (4°C) than at frozen temperatures (−20 and −80°C) over a period of 90 d (Figures 5(a), 5(b), and 5(c)). At 4°C, minimal reduction in phage populations (0.1–2.3 logs PFU mL^{-1}) was observed over a period of 90 d. Phages P-6 and P-7 were the most stable, with minor (0.6 logs PFU mL^{-1}) loss in titers during the 90-day storage (Figure 5(a)). At −20°C, all the phages were stable at day 1 with a slight decrease (0.6–2.1 logs PFU mL^{-1}) in population (Figure 5(b)). Phages P-3, P-4, and P-6 maintained their titers over the 90-day period. The other phages showed a decrease in population (4.0–6.3 logs PFU mL^{-1}) but maintained their viability over a period of 90 d (Figure 5(b)). Phages P-1, P-2, and P-5 showed better survival at −80°C than at −20°C. All the other phages showed a decrease in their population (between 2.0 and 5.1 logs PFU mL^{-1}) after 90-day storage (Figure 5(c)). Compared to the other phages, P-6 was the most stable at all cold storage temperatures with a slight decrease (0.6–2.0 logs PFU mL^{-1}) in population after 90-day storage.

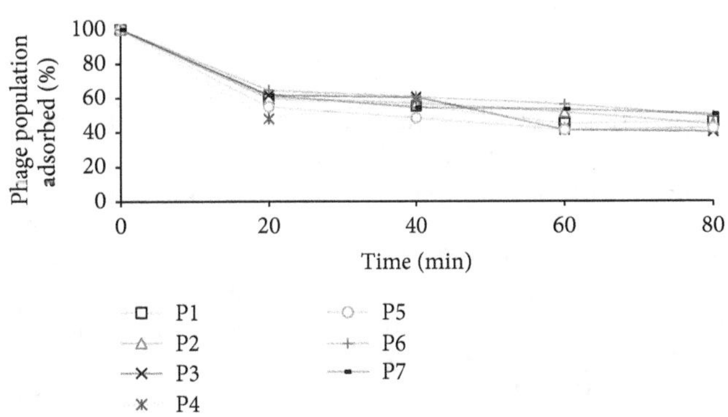

FIGURE 6: Rate of phage adsorption to host bacterial surface.

3.4. Phage Adsorption and One-Step Growth Kinetics. Phage growth properties were determined by one-step growth kinetics and phage adsorption rate. Results from the adsorption study revealed that, for all the phages, it took 20 min for approximately 41% of the initial population, and 80 min for approximately 55% of the initial population, to be adsorbed to the host cell surface (Figure 6). Of the seven phages, P-4 had the fastest adsorption rate taking only 20 min for 52% of the population to be adsorbed to the host cell surface. On the other hand, P-3 showed the maximum adsorption rate (60% population adsorbed) after 80 min of incubation.

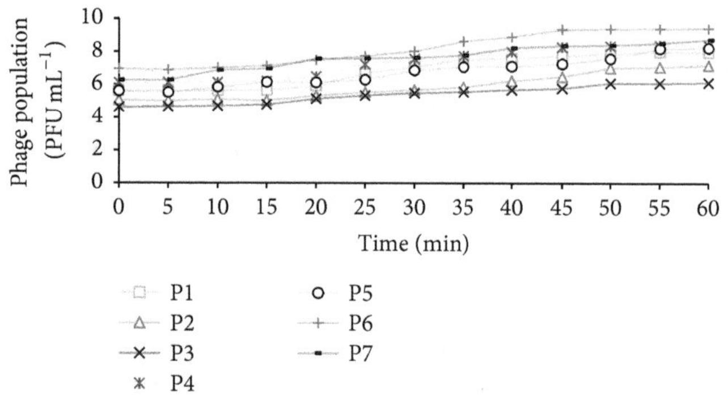

FIGURE 7: Experimental one-step growth curves of phages over a period of 60 min.

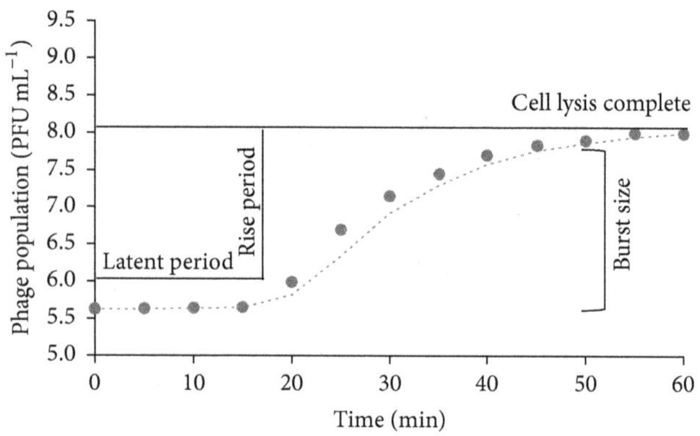

FIGURE 8: Predicted growth curve of phage P-1 using the 4-parameter sigmoidal function to obtain latent period (21 min), rise period (27 min), and burst size (224 virions per infected cell).

One-step growth kinetics of phages, observed experimentally over a period of 60 min, is shown in Figure 7. Phage growth parameters were calculated using the equation described previously. A representative, predicted one-step growth curve of phage P-1 is shown in Figure 8. Results revealed that phages belonging to the Siphoviridae family (P-3, P-4, P-6, and P-7) had a very short latent period of 12–23 min. On the other hand, the Myoviridae phages (P-1, P-2, and P-5) had a slightly longer latent period of 20–30 min. The rise period was between 19 and 40 min and the burst size was between 89 and 631 virions per infected cell for all the phages.

4. Discussion

Several studies have demonstrated the efficacy of bacteriophages against foodborne pathogens; however, limited information is available on their general characteristics and physiology. At the same time, the stability of bacteriophages under various stress conditions, such as acidic and alkaline pH, high temperatures, and cold storage, has not been very well studied. Phage application in the food industry or food-animal industry requires them to be stable in acidic, alkaline, and high or low temperature conditions. For example, food-animal production practices, such as high temperature processing of animal feed, acidic or alkaline environment of the animal GI tract, or antimicrobials used at the farm level, could affect the viability of phages during application. Similar challenges exist at the food-processing level such as high temperature processing, fermentation, and high pressure processing. Additionally, use of multiple-hurdle control strategies is gaining popularity, where phages with high lytic activity may be used in combination with commercial sanitizers, which would require them to be stable and active in concert with the sanitizers. It is therefore important to determine the stability of bacteriophages in adverse conditions to select highly effective phages for application in the food industry and food-animal industry. In the present study, physiomorphological characteristics of bacteriophages, isolated from beef cattle environment, and their stability at a wide pH and temperature range along with their growth and lytic properties were determined to identify and select effective phages against *E. coli* O157:H7.

All the isolated phages showed high lytic activity against 50 isolates of *E. coli* O157:H7, including the WT isolates (Table 1), suggesting that they have a broad host range. Studies have shown that phages specific to one *E. coli* O157:H7 strain can also infect other O157:H7 strains [39, 57, 58]. In a study by Raya et al. [39], phage AR1 lysed all O157:H7 tested strains, while in another study [59], phage CEV1 infected 17 of the tested 19 strains of *E. coli* O157:H7. Results from the current study are similar to these studies, revealing that isolated phages are virulent against a wide range of *E. coli* O157:H7 isolates with high target specificity.

Phages in the current study were also stable at a wide pH range (1–11) compared to the phages in other studies [49, 60]. Niu et al. [49] revealed that at pH 3 the titer of *E. coli* O157:H7 phage, AKFV33, dropped by 1.9 logs PFU mL^{-1} after 15 min and was undetectable after 2 h. In the current study, 3 of the 7 phages survived at pH 1, while all phages survived at pH 2 and 5 after 24 h, with minimal loss in viability. It is possible for phages to acquire nonreversible mutations at low pH [61–63], which could explain the survival of phages at low pH in the present study. Strack et al. [61] showed a linear relationship between phage mutation rate and incubation at low pH, suggesting that phages can acquire mutation to survive acidic environment. Phages in the current study were also stable at alkaline pH (7–11), showing no loss at pH 7 and 9 and minimal loss in titer at pH 11. In earlier studies by the authors, isolated phages were found to be stable in water (pH 7) for a period of 30 d (unpublished data). Reductions in phage titers at pH 11 could be due to the dissociation of the capsid protein, caused by high concentrations of hydrogen and hydroxyl ion in the solution [64]. The ability of isolated phages to survive extreme pH conditions could be utilized in a number of applications in the food and animal industry. Survival at low pH could be used to control foodborne pathogen colonization in the animal GI tract (pH 1–5), through oral administration [57, 65, 66]. It can also be used as biocontrol in acidic foods such as fruit juices, fermented products, and pickles. Furthermore, their ability to survive alkaline pH could be utilized during the multiple-hurdle approach in processing facilities or on food-contact surfaces [67, 68].

Studies have shown that high temperature inactivates phages due to nucleic acid and protein denaturation [50, 69]. Yamaki et al. [50] observed that Myoviridae phages drastically lost phage activity (3.5 logs PFU mL^{-1}) after 60 min of incubation at 60°C. However, in the current study, no significant decrease in phage activity was observed at 60°C for 60 min. Additionally, five of the seven phages (P-3, P-4, P-5, P-6, and P-7) also survived at 70°C for 10 min. Different phage strains behave differently to heat treatments [70], which could explain the difference in results in the two studies. Furthermore, Yamaki et al. [50] used the phage infecting *Morganella morganii*, which is sensitive to high temperature [71] and may therefore be sensitive to high temperatures. Additionally, some phages can develop heat resistance due to mutations or strong protein interactions [72], explaining the survival of phages at high temperatures in the current study.

Bacteriophage viability at refrigerated and frozen temperatures was determined to ensure their survival during storage or shipment. All the phage isolates maintained their viability over a storage period of 90 d at 4, −20, and −80°C. Previous studies have revealed that tailed phages are resistant to refrigeration temperatures and can retain viability for more than 10–12 years at 4°C [73, 74]. In the current study, phages survived very well at 4°C, with minimal activity loss during the 90-day storage. However, compared to 4°C, phages were less stable at −20 and −80°C and showed a decrease in population after 90-day storage. Compared to −20°C, phages P-2 and P-5 showed better survival at −80°C, while other phages showed similar survival trend at both storage temperatures. Loss in phage titers at −20 and −80°C during 90-day storage could be due to the formation of ice crystals at frozen temperatures [75, 76]. For long-term storage, it is recommended that phages be maintained at −80°C [74], which could be true for phages P-2 and P-5 in the current study.

The efficacy of phages to infect bacterial cells was measured using phage growth kinetics and adsorption rate. Isolated phages had a long rise period (19–40 min) and a short latent period (12–30 min) with a large burst size (89–631 virions per infecting cell), suggesting that they had a high progeny production rate. The average burst size of tested phages was about 329 phages per infected cell, which is more than the typical range of 50–100 PFU $cell^{-1}$ for many Myoviridae and Siphoviridae phages [77, 78]. Furthermore, the Siphoviridae phage, P-7, had the largest burst size of 631 virions per infected cell, which is much larger than other Siphoviridae phages [77, 79]. The latent period and rise period of the characterized phages were similar to those of the other Myoviridae and Siphoviridae phages [40, 78–81]. Bacteriophages with a short latent period and large burst size may have a selective advantage over other phages due to high lytic activity [78]. These characteristics of isolated phages could therefore be applied to develop control strategies for the transmission and survival of *E. coli* O157:H7 in the food chain. However, one concern with this approach is the emergence of phage-resistant bacteria, particularly with application of phages in animals or on the farm. Use of phage cocktails that are regularly updated with new or different phages could potentially address the issue, by maintaining selective pressure on bacterial host [43, 82].

5. Conclusions

This study provides a better understanding of the infection kinetics of seven phages, targeting *E. coli* O157, and their survivability under various stress conditions. This information will afford the ability to determine the application of phage-based interventions. Bacteriophages isolated in this study showed a wide range of host specificity towards *E. coli* O157:H7 isolates, along with high lytic activity, and pH and thermal stability and could therefore be used as biocontrol agents in the food industry.

Conflicts of Interest

The authors declare that they have no conflicts of interest.

Acknowledgments

This research was supported by the Technology and Business Development Program at Oklahoma State University. The authors would like to thank Joyjit Saha, Lisa Whitworth, and Brent Johnson for technical support.

References

[1] M. A. Rasmussen, W. C. Cray, T. A. Casey, and S. C. Whipp, "Rumen contents as a reservoir of enterohemorrhagic *Escherichia coli*," *FEMS Microbiology Letters*, vol. 114, no. 1, pp. 79–84, 1993.

[2] C. Su and L. J. Brandt, "*Escherichia coli* O157: H7 infection in humans," *Annals of Internal Medicine*, vol. 123, no. 9, pp. 698–707, 1995.

[3] I. T. Kudva, P. G. Hatfield, and C. J. Hovde, "Characterization of *Escherichia coli* O157:H7 and other shiga toxin-producing *E. coli* serotypes isolated from sheep," *Journal of Clinical Microbiology*, vol. 35, no. 4, pp. 892–899, 1997.

[4] D. Hancock, T. Besser, and D. Rice, "The ecology of *Escherichia coli* O157: H7 in cattle and the impact of management practices," in *Proceedings of the In:VTEC'97. Conference proceedings of the 3rd International Symposium and Workshop on Shiga Toxin (Verocytotoxin) Producing Escherichia coli Infections*, Baltimore, Maryland, 1998.

[5] L. J. Grauke, I. T. Kudva, J. W. Yoon, C. W. Hunt, C. J. Williams, and C. J. Hovde, "Gastrointestinal tract location of *Escherichia coli* O157:H7 in ruminants," *Applied and Environmental Microbiology*, vol. 68, no. 5, pp. 2269–2277, 2002.

[6] D. Rice, D. Hancock, and T. Besser, "Verotoxigenic *E coli* O157:H7 colonisation of wild deer and range cattle," *The Veterinary record*, vol. 137, no. 20, p. 524, 1995.

[7] J. S. Wallace, T. Cheasty, and K. Jones, "Isolation of Vero cytotoxin-producing *Escherichia coli* O157 from wild birds," *Journal of Applied Microbiology*, vol. 82, no. 3, pp. 399–404, 1997.

[8] W. E. Keene, E. Sazie, J. Kok et al., "An outbreak of *Escherichia coli* O157:H7 infections traced to jerky made from deer meat," *Journal of the American Medical Association*, vol. 277, no. 15, pp. 1229–1231, 1997a.

[9] E. Scallan, R. M. Hoekstra, F. J. Angulo et al., "Foodborne illness acquired in the United States—major pathogens," *Emerging Infectious Diseases*, vol. 17, no. 1, pp. 7–15, 2011.

[10] S. Hoffmann, B. Maculloch, and M. Batz, *Economic burden of major foodborne illnesses acquired in the*, United States , Economic Research Services, US Department of Agriculture, Washington, DC, EIB140, 2015.

[11] P. I. Tarr, "*Escherichia coli* O157:H7: Clinical, diagnostic, and epidemiological aspects of human infection," *Clinical Infectious Diseases*, vol. 20, no. 1, pp. 1–8, 1995.

[12] G. G. Greer, "Bacteriophage control of foodborne bacteria," *Journal of Food Protection*, vol. 68, no. 5, pp. 1102–1111, 2005.

[13] L. W. Riley, R. S. Remis, and S. D. Helgerson, "Hemmorhagic colitis associated with a rare *Escherichia coli* serotype," *New England Journal of Medicine*, vol. 308, no. 12, pp. 681–685, 1983.

[14] P. M. Griffin and R. V. Tauxe, "The epidemiology of infections caused by *Escherichia coli* O157: H7, other enterohemorrhagic *E. coli*, and the associated hemolytic uremic syndrome," *Epidemiologic Reviews*, vol. 13, no. 1, pp. 60–98, 1991.

[15] B. P. Bell, M. Goldoft, P. M. Griffin et al., "A Multistate Outbreak of *Escherichia coli* O157:H7—Associated Bloody Diarrhea and Hemolytic Uremic Syndrome From Hamburgers: The Washington Experience," *JAMA: The Journal of the American Medical Association*, vol. 272, no. 17, pp. 1349–1353, 1994.

[16] K. Como-Sabetti, S. Reagan, S. Allaire et al., "Outbreaks of *Escherichia coli* O157:H7 Infection Associated With Eating Alfalfa Sprouts—Michigan and Virginia, June-July 1997," *Morbidity Mortality Weekly Report*, vol. 46, pp. 741–744, 1997.

[17] S. H. Cody, M. K. Glynn, J. A. Farrar et al., "An outbreak of *Escherichia coli* O157:H7 infection from unpasteurized commercial apple juice," *Annals of Internal Medicine*, vol. 130, no. 3, pp. 202–209, 1999.

[18] K. C. Jinneman, K. J. Yoshitomi, and S. D. Weagant, "Multiplex Real-Time PCR Method To Identify Shiga Toxin Genes stx1 and stx2 and *Escherichia coli* O157:H7/H-Serotype," *Applied and Environmental Microbiology*, vol. 69, no. 10, pp. 6327–6333, 2003.

[19] M. A. Karmali, "Infection by shiga toxin-producing *Escherichia coli*," *Molecular Biotechnology*, vol. 26, no. 2, pp. 117–122, 2004.

[20] Center of Disease Control and Prevention (CDC), "Ongoing multistate outbreak of *Escherichia coli* serotype O157:H7 infections associated with consumption of fresh spinach United States," September 2006, https://www.cdc.gov/mmwr/preview/mmwrhtmL/mm55d926a1.htm.

[21] CDC, "Multistate outbreak of *E. coli* O157:H7 infections associated with cheese (final update)," 2010, https://www.cdc.gov/ecoli/2010/bravo-farms-cheese-11-24-10.htmL.

[22] CDC, "Multistate outbreak of *E. coli* O157:H7 Infections associated with lebanon bologna (final update)," 2011, https://www.cdc.gov/ecoli/2011/lebanon-bologna-3-23-11.htmL.

[23] CDC, "Multistate outbreak of *E. coli* O157:H7 infections linked to romaine lettuce (final update)," 2012a, https://www.cdc.gov/ecoli/2011/romaine-lettace-3-23-12.htmL.

[24] CDC, "Multistate outbreak of shiga toxin-producing *Escherichia coli* infections linked to organic spinach and spring mix blend (final update)," 2012b, https://www.cdc.gov/ecoli/2012/o157h7-11-12/.

[25] CDC, "Multistate outbreak of shiga toxin-producing *Escherichia coli* O157:H7 infections linked to ready-to-eat salads," 2013, http://www.cdc.gov/ecoli/2013/O157H7-11-13/index.htmL.

[26] CDC, "Multistate outbreak of shiga toxin-producing *Escherichia coli* O157:H7 infections linked to ground beef," 2014, https://www.cdc.gov/ecoli/2014/o157h7-05-14/index.htmL.

[27] CDC, "Multistate outbreak of shiga toxin-producing *Escherichia coli* O157:H7 infections linked to beef products produced by adams farm (final update)," 2016, https://www.cdc.gov/ecoli/2016/o157h7-09-16/index.htmL.

[28] J. T. McCollum, N. J. Williams, S. W. Beam et al., "Multistate outbreak of *Escherichia coli* O157:H7 infections associated with in-store sampling of an aged raw-milk Gouda cheese, 2010," *Journal of Food Protection*, vol. 75, no. 10, pp. 1759–1765, 2012.

[29] S. E. Murinda, R. F. Roberts, and R. A. Wilson, "Evaluation of colicins for inhibitory activity against diarrheagenic *Escherichia coli* strains, including serotype O157: H7," *Applied and Environmental Microbiology*, vol. 62, no. 9, pp. 3196–3202, 1996.

[30] W. C. Cray and H. W. Moon, "Experimental infection of calves and adult cattle with *Escherichia coli* O157:H7," *Applied and Environmental Microbiology*, vol. 61, pp. 1586–1590, 1995.

[31] R. E. Besser, S. M. Lett, J. T. Weber et al., "An outbreak of diarrhea and hemolytic uremic syndrome from *Escherichia coli* O157:H7 in fresh-pressed apple cider," *Journal of the American Medical Association*, vol. 269, no. 17, pp. 2217–2220, 1993.

[32] M. Bielaszewska, J. Janda, K. Bláhová et al., "Human *Escherichia coli* O157:H7 infection associated with the consumption of unpasteurized goat's milk," *Epidemiology and Infection*, vol. 119, no. 3, pp. 299–305, 1997.

[33] W. E. Keene, K. Hedberg, D. E. Herriott et al., "A prolonged outbreak of *Escherichia coli* O157:H7 infections caused by commercially distributed raw milk," *Journal of Infectious Diseases*, vol. 176, no. 3, pp. 815–818, 1997b.

[34] R. V. Tauxe, "Emerging Foodborne Diseases: An Evolving Public Health Challenge," *Emerging Infectious Diseases*, vol. 3, no. 4, pp. 425–434, 1997.

[35] K. E. Ashelford, M. J. Day, and J. C. Fry, "Elevated abundance of bacteriophage infecting bacteria in soil," *Applied and Environmental Microbiology*, vol. 69, no. 1, pp. 285–289, 2003.

[36] A. J. Synnott, Y. Kuang, M. Kurimoto, K. Yamamichi, H. Iwano, and Y. Tanji, "Isolation from sewage influent and characterization of novel *Staphylococcus aureus* bacteriophages with wide host ranges and potent lytic capabilities," *Applied and Environmental Microbiology*, vol. 75, no. 13, pp. 4483–4490, 2009.

[37] F. d'Herelle, "Sur le role du microbe bacteriophage dans la typhose aviare," *Comptesrendus Acadamy of Science Paris*, vol. 169, pp. 932–934, 1919.

[38] Y. Tanji, T. Shimada, H. Fukudomi, K. Miyanaga, Y. Nakai, and H. Unno, "Therapeutic use of phage cocktail for controlling *Escherichia coli* O157:H7 in gastrointestinal tract of mice," *Journal of Bioscience and Bioengineering*, vol. 100, no. 3, pp. 280–287, 2005.

[39] R. R. Raya, P. Varey, R. A. Oot et al., "Isolation and characterization of a new T-even bacteriophage, CEV1, and determination of its potential to reduce *Escherichia coli* O157:H7 levels in sheep," *Applied and Environmental Microbiology*, vol. 72, no. 9, pp. 6405–6410, 2006.

[40] T. Abuladze, M. Li, M. Y. Menetrez, T. Dean, A. Senecal, and A. Sulakvelidze, "Bacteriophages reduce experimental contamination of hard surfaces, tomato, spinach, broccoli, and ground beef by *Escherichia coli* O157:H7," *Applied and Environmental Microbiology*, vol. 74, no. 20, pp. 6230–6238, 2008.

[41] M. Sharma, J. R. Patel, W. S. Conway, S. Ferguson, and A. Sulakvelidze, "Effectiveness of bacteriophages in reducing *Escherichia coli* O157:H7 on fresh-cut cantaloupes and lettuce," *Journal of Food Protection*, vol. 72, no. 7, pp. 1481–1485, 2009.

[42] S. Viazis, M. Akhtar, J. Feirtag, and F. Diez-Gonzalez, "Reduction of *Escherichia coli* O157:H7 viability on leafy green vegetables by treatment with a bacteriophage mixture and trans-cinnamaldehyde," *Food Microbiology*, vol. 28, no. 1, pp. 149–157, 2011.

[43] G. O'Flynn, R. P. Ross, G. F. Fitzgerald, and A. Coffey, "Evaluation of a cocktail of three bacteriophages for biocontrol of *Escherichia coli* O157:H7," *Applied and Environmental Microbiology*, vol. 70, no. 6, pp. 3417–3424, 2004.

[44] M. Sharma, J. H. Ryu, and L. R. Beuchat, "Inactivation of *Escherichia coli* O157:H7 in biofilm on stainless steel by treatment with an alkaline cleaner and a bacteriophage," *Journal of Applied Microbiology*, vol. 99, no. 3, pp. 449–459, 2005.

[45] B. P. K. Jayasundera, *Identification of Contamination Sources and Occurrence of Shigatoxigenic E. coli on Small-Scale Cow/Calf Operations in Oklahoma and Louisiana*, Oklahoma State University, 2015.

[46] J. Sambrook, E. F. Fritsch, and T. Maniatis, *Molecular cloning*, Cold spring harbor laboratory press New York, 1989.

[47] M. Chandra, S. Thakur, D. Narang, and H. M. Saxena, "Isolation of a bacteriophage against *Salmonella* Dublin and determination of its physical resistance under varied in vitro conditions," *African Journal of Microbiology Research*, vol. 5, no. 15, pp. 2044–2047, 2011.

[48] M. H. Adams, *Bacteriophages*, Interscience, New York, USA, 1959.

[49] Y. D. Niu, K. Stanford, A. M. Kropinski et al., "Genomic, proteomic and physiological characterization of a T5-like bacteriophage for control of shiga toxin-producing *Escherichia coli* O157:H7," *PLoS ONE*, vol. 7, no. 4, Article ID e34585, 2012.

[50] S. Yamaki, T. Omachi, Y. Kawai, and K. Yamazaki, "Characterization of a novel *Morganella morganii* bacteriophage FSP1 isolated from river water," *FEMS Microbiology Letters*, vol. 359, no. 2, pp. 166–172, 2014.

[51] G. Di Lallo, M. Evangelisti, F. Mancuso et al., "Isolation and partial characterization of bacteriophages infecting *Pseudomonas syringae* pv. *actinidiae*, causal agent of kiwifruit bacterial canker," *Journal of Basic Microbiology*, vol. 54, no. 11, pp. 1210–1221, 2014.

[52] Z. Lu and F. Breidt, "*Escherichia coli* O157:H7 bacteriophage Φ241 isolated from an industrial cucumber fermentation at high acidity and salinity," *Frontiers in Microbiology*, vol. 6, 2015.

[53] P. Mudgal, F. Breidt Jr., S. R. Lubkin, and K. P. Sandeep, "Quantifying the significance of phage attack on starter cultures: A mechanistic model for population dynamics of phage and their hosts isolated from fermenting sauerkraut," *Applied and Environmental Microbiology*, vol. 72, no. 6, pp. 3908–3915, 2006.

[54] M. Delbrück, "Bacterial Viruses (Bacteriophages)," in *Advances in Enzymology and Related Areas of Molecular Biology*, vol. 2, pp. 1–32, 1942.

[55] Z. Lu, F. Breidt Jr., H. P. Fleming, E. Altermann, and T. R. Klaenhammer, "Isolation and characterization of a *Lactobacillus plantarum* bacteriophage, ΦJL-1, from a cucumber fermentation," *International Journal of Food Microbiology*, vol. 84, no. 2, pp. 225–235, 2003.

[56] E. L. Ellis and M. Delbrück, "The growth of bacteriophage," *The Journal of General Physiology*, vol. 22, no. 3, pp. 365–384, 1939.

[57] S. J. Bach, T. A. McAllister, D. M. Veira, V. P. J. Gannon, and R. A. Holley, "Effect of bacteriophage DC22 on *Escherichia coli* O157:H7 in an artificial rumen system (Rusitec) and inoculated sheep," *Animal Research*, vol. 52, no. 2, pp. 89–101, 2003.

[58] H. Sheng, H. J. Knecht, I. T. Kudva, and C. J. Hovde, "Application of bacteriophages to control intestinal *Escherichia coli* O157:H7 levels in ruminants," *Applied and Environmental Microbiology*, vol. 72, no. 8, pp. 5359–5366, 2006.

[59] L. Goodridge, A. Gallaccio, and M. W. Griffiths, "Morphological, host range, and genetic characterization of two coliphages," *Applied and Environmental Microbiology*, vol. 69, no. 9, pp. 5364–5371, 2003.

[60] J. H. Northrop, "The effect of proteolytic enzymes on *E. coli* phages and on native proteins," *The Journal of General Physiology*, vol. 48, no. 1, pp. 73–78, 1964.

[61] H. B. Strack, E. B. Freese, and E. Freese, "Comparison of mutation and inactivation rates induces by bacteriophage and transforming DNA by various mutagenes," *Mutation Research/Fundamental and Molecular Mechanisms of Mutagenesis*, vol. 1, no. 1, pp. 10–21, 1964.

[62] E. Jończyk, M. Kłak, R. Międzybrodzki, and A. Górski, "The influence of external factors on bacteriophages—review," *Folia Microbiologica*, vol. 56, no. 3, pp. 191–200, 2011.

[63] F. L. Nobrega, A. R. Costa, J. F. Santos et al., "Genetically manipulated phages with improved pH resistance for oral administration in veterinary medicine," *Scientific Reports*, vol. 6, Article ID 39235, 2016.

[64] Y. Y. Feng, S. L. Ong, J. Y. Hu, X. L. Tan, and W. J. Ng, "Effects of pH and temperature on the survival of coliphages MS2 and Qβ," *Journal of Industrial Microbiology and Biotechnology*, vol. 30, no. 9, pp. 549–552, 2003.

[65] T. Callaway, T. Edrington, R. Anderson et al., "Isolation of naturally occurring bacteriophage from sheep that reduce populations of *E. coli* O157: H7 in vitro and in vivo," in *Proceedings of the In:VTEC'03. Conference proceedings of the 5th International Symposium and Workshop on Shiga Toxin (Verocytotoxin) Producing Escherichia coli Infections*, Edinburgh, UK, 2003.

[66] R. L. Andreatti Filho, J. P. Higgins, S. E. Higgins et al., "Ability of bacteriophages isolated from different sources to reduce *Salmonella enterica* serovar Enteritidis in vitro and in vivo," *Poultry Science*, vol. 86, no. 9, pp. 1904–1909, 2007.

[67] K. Tait, L. C. Skillman, and I. W. Sutherland, "The efficacy of bacteriophage as a method of biofilm eradication," *Biofouling*, vol. 18, no. 4, pp. 305–311, 2002.

[68] S. Srey, I. K. Jahid, and S.-D. Ha, "Biofilm formation in food industries: A food safety concern," *Food Control*, vol. 31, no. 2, pp. 572–585, 2013.

[69] A. Wang, N. Cheng, Y. Liou, and K. Lin, "Inactivation of bacteriophage by microwave irradiation," *Journal of Experimental Microbiology and Immunology*, vol. 1, pp. 9–18, 2001.

[70] H. Wilkowske, F. Nelson, and C. Parmelee, "Heat inactivation of bacteriophage strains active against *Lactic Streptococci*," *Journal of Applied Microbiology*, vol. 2, no. 5, pp. 250–253, 1954.

[71] M. Zarei, S. Maktabi, M. Khezrzadeh, and A. Jamnejad, "Susceptibility of *Morganella morganii* to Various Environmental Stresses after Cold and Heat Shock Treatments," *Journal of Food Safety*, vol. 33, no. 2, pp. 107–113, 2013.

[72] K.-I. Kadowaki, T. Shibata, K. Takeuchi, M. Himeno, H. Sakai, and T. Komano, "Identification of a Temperature-resistant Bacteriophage φX174 Mutant," *Journal of General Virology*, vol. 68, no. 9, pp. 2443–2447, 1987.

[73] H. W. Ackermann, "Bacteriophage observations and evolution," *Research in Microbiology*, vol. 154, no. 4, pp. 245–251, 2003.

[74] M. R. Olson, R. P. Axler, and R. E. Hicks, "Effects of freezing and storage temperature on MS2 viability," *Journal of Virological Methods*, vol. 122, no. 2, pp. 147–152, 2004.

[75] J. C. Warren and M. T. Hatch, "Survival of T3 coliphage in varied extracellular environments," *Journal of Applied Microbiology*, vol. 17, pp. 256–261, 1969.

[76] E. A. Gould, "Methods for Long-Term Virus Preservation," *Molecular Biotechnology*, vol. 13, no. 1, pp. 57–66, 1999.

[77] H. Bao, H. Zhang, and R. Wang, "Isolation and characterization of bacteriophages of *Salmonella enterica* serovar Pullorum," *Poultry Science*, vol. 90, no. 10, pp. 2370–2377, 2011.

[78] M. Park, J.-H. Lee, H. Shin et al., "Characterization and comparative genomic analysis of a novel bacteriophage, SFP10, simultaneously inhibiting both *Salmonella enterica* and *Escherichia coli* O157:H7," *Applied and Environmental Microbiology*, vol. 78, no. 1, pp. 58–69, 2012.

[79] G. V. Carey-Smith, C. Billington, A. J. Cornelius, J. A. Hudson, and J. A. Heinemann, "Isolation and characterization of bacteriophages infecting *Salmonella* spp," *FEMS Microbiology Letters*, vol. 258, no. 2, pp. 182–186, 2006.

[80] J. Uchiyama, M. Rashel, Y. Maeda et al., "Isolation and characterization of a novel *Enterococcus faecalis* bacteriophage φEF24C as a therapeutic candidate," *FEMS Microbiology Letters*, vol. 278, no. 2, pp. 200–206, 2008.

[81] S.-E. Hsieh, H.-H. Lo, S.-T. Chen, M.-C. Lee, and Y.-H. Tseng, "Wide host range and strong lytic activity of *Staphylococcus aureus* lytic phage Stau2," *Applied and Environmental Microbiology*, vol. 77, no. 3, pp. 756–761, 2011.

[82] Y. Tanji, T. Shimada, M. Yoichi, K. Miyanaga, K. Hori, and H. Unno, "Toward rational control of *Escherichia coli* O157:H7 by a phage cocktail," *Applied Microbiology and Biotechnology*, vol. 64, no. 2, pp. 270–274, 2004.

15

Automated Flow Cytometry: An Alternative to Urine Culture in a Routine Clinical Microbiology Laboratory

Patricia Mejuto,[1] Mariam Luengo,[2] and Julio Díaz-Gigante[1]

[1] *Department of Microbiology, Hospital del Oriente de Asturias Francisco Grande Covián, Arriondas, Asturias, Spain*
[2] *Department of Biochemistry, Hospital del Oriente de Asturias Francisco Grande Covián, Arriondas, Asturias, Spain*

Correspondence should be addressed to Patricia Mejuto; patriciamejutolv@gmail.com

Academic Editor: Maurizio Sanguinetti

The urine culture is the "gold standard" for the diagnosis of urinary tract infections (UTI) but constitutes a significant workload in the routine clinical laboratory. Due to the high percentage of negative results, there is a need for an efficient screening method, with a high negative predictive value (NPV) that could reduce the number of unnecessary culture tests. With the purpose of improving the efficiency of laboratory work, several methods for screening out the culture-negative samples have been developed, but none of them has shown adequate sensitivity (SE) and high NPV. Many authors show data about the efficacy of flow cytometry in the routine clinical laboratory. The aim of this article is to review and discuss the current literature on the feasibility of urine flow cytometry (UFC) and its utility as an alternative analytical technique in urinalysis.

1. Introduction

Urinary tract infections (UTIs) are among the most common infections in both hospitalized patients and outpatients [1–4], and the health cost they cause is considerable. About 50% of women state that they have experienced one infection in their lifetime and 27–48% of them have recurrent infections [5], mainly in senior women living in the community [6]. Although for most patients the disease of the infection is minimal, in certain population groups like children, pregnant women, the elderly, and immunosuppressed patients, it is associated with severe complications [7]. In the USA, it is responsible for more than 7 million medical visits annually, and the most recent medical literature indicates a global prevalence of 0.7% of community-acquired UTI (19.6% in Europe, 12.9% in the USA, and 24% in developing countries) [8]. According to medical literature, *E. coli* more usually causes lower community-acquired UTI [9, 10]. Up to 15% of antibiotics prescribed in the community are due to these infections, with an annual cost of approximately 1.6 billion US dollars [11]. Taking these data into account, there is an enhanced risk of antibiotic resistance [12].

The diagnosis of UTI is based on clinical signs and symptoms of the patient and urinalysis results. Consequently, urine samples are the most numerous specimens received by clinical microbiology laboratories. Culturing of the urine samples is the "gold standard" method to exclude these infections, although it is a significant workload and time-consuming in the microbiology laboratory [13]. Nevertheless, a high percentage of these specimens will yield no growth, reaching even up to 60% of negative results [14, 15]. Also, other different urinalysis techniques are widely used in routine laboratories. The dipstick chemical test is based on strips that had reagent pads for semiquantitative assessment of nitrite (a product of common urinary pathogens), leukocyte esterase, protein, and blood (as a sign of inflammation). This method has a poor NPV, which makes it unsuitable to exclude the presence of urine infection. It is only useful in populations when the results of both nitrite and leukocyte esterase are positive [16]. However, with a negative result, UTI cannot be excluded mainly in children, who normally have low bacterial counts [17]. Another technique is the microscopic examination of urine sediment where cells, particles, and microorganism are counted and reported according to the number observed. Diverse factors can produce interpretation errors, and there are different results among laboratories, due to interindividual variability [18]. Although these screening methods are primarily used in general practice and microbiology laboratories, they are subjective and time-consuming or have a poor sensitivity and/or specificity. For ten years ago, the use of flow cytometry analyzers (FCA) based on detection and quantification both leukocytes and bacteria have been under evaluation. These techniques would reduce the number of samples cultured with a sharp decrease in workload, time, and costs especially in the bigger clinical laboratories [19]. In this way, negative results could be informed considerably earlier, which would imply a reduction of unnecessary empirical antibiotic prescriptions [20].

2. Review Methods

Data from 2009 to 2017 were assessed. The search process was conducted using Cochrane, the electronic databases Pubmed/NCBI, and Google Scholar. The terms "urine flow cytometer," "flow cytometry," "urinary tract infection," "screening urinary tract infection" and "flow cytometry analyzers" were used. We also searched reference lists of articles to identify supplementary information. Citations were limited to studies conducted in humans and using the UF1000i (Sysmex) analyzer for UTI screening and the urine culture as the reference standard. Conference abstracts were excluded due to lack of information provided. Our initial literature search yielded 605 unique citations. After a critical evaluation, based on study design, a total of 17 articles were selected, for review (Table 1).

3. Diagnosis of Urinary Tract Infection

The detection of the uropathogen by urine culture is the "gold standard" for the diagnosis of UTI. In this way, the level of the bacteriuria can be estimated and the antimicrobial susceptibility determined. However, the minimum number of bacteria demonstrating an infection of the urinary tract has not been specified in the scientific literature or fixed by microbiological laboratories [37, 38]. In 1960, Kass developed the concept of significant bacteriuria as more than 10^5 CFU/mL when pyelonephritis happens during pregnancy [39]. This assessment introduced quantitative microbiology into the diagnosis of infectious diseases and is therefore still of general importance. But, there is no standardized bacterial count that indicates significant bacteriuria, applicable in all types of UTIs.

The European urinalysis guidelines recommend the following limits of uropathogens counts, for a diagnosis of UTI [36, 40]: a threshold of $>10^5$ Colony Forming Unit (CFU) per ml in MSU (midstream urine) in women or $>10^4$ CFU/ml in MSU in men or straight urine catheter in complicated UTI in women; more than 10^4 CFU/ml in a MSU in uncomplicated pyelonephritis in women or $>10^3$ CFU/ml in MSU in uncomplicated cystitis in women. In a sample of urine obtained by bladder puncture, any count is significant. According to the Infectious Diseases Society of America (IDSA) guideline, for uncomplicated UTI, counts of more than 10^4 CFU/ml can be considered as the cut-off for positive culture [41, 42].

4. Urinary Flow Cytometer Principle

Flow cytometer analyzers are systems for the rapid, optical analysis of individual cells that allow simultaneous analysis of chemical and/or physical characteristics of single urinary particles [43]. Measurement is made by an array of detectors as the cells flow in a fluid stream through a laser beam. The urinary flow cytometer (UFC) can identify red blood cells (RBC), white blood cells (WBC), squamous epithelial cells, small round cells (renal tubular cells and transitional epithelial cells), hyaline casts, bacteria, yeast-like cells, spermatozoa, and crystals. The urine sample is automatically mixed, aspired, and stained using a specific fluorescent polymethine dye in two different analytical chambers: one for bacteria and another one for other urine forming elements. Once stained, the fluorescent particles are inserted into the cytometer pool and exposed to the semiconductor laser beam (λ 635 nm). For each particle the scattered light is detected at two positions: forward and side scatters and fluorescence intensities are produced [44, 45]. These signals are converted to optoelectronic signals so that components can be identified, counted, classified, and analyzed. Bacteria are counted via an individual bacterial examination channel, so any interference with RBCs is prevented. All the results are presented in histograms and scattergrams by the software. Each particle type is color coded for the ease of reading. The bacteria and WBC counts are recorded in a database for subsequent analysis to determine the optimal screening method.

In recent years, the analytical quality of these analyzers has been improved because bacteria are counted in an independent examination channel, and interference with red blood cells is prevented. Also, the latest models have a more powerful laser that provides a high sensitivity analytic as well as a flag which informs on the gram characteristics of bacteria detected.

5. Studies

Several studies reported different sensitivities and specificities about the application of flow cytometry in the diagnosis of suspected UTI. The results obtained depend on the definition used for gold standard positive and negative urines cultures, and these definitions vary among laboratories. For example, most often $>10^5$ CFU/ml [22, 25, 27, 29, 46] has been considered a threshold for positive urine cultures in the clinical, microbiological practice. Some authors used pathogenicity cut-offs of 10^3 CFU/ml or less [28, 47, 48] while others consider high counts of 10^5 CFU/ml that were still measured negative [49]. Taking into account these microbial counts limits, authors establish different cut-off values of bacteria and leukocytes per μl, below which the reported result of the urinalysis will be negative. We have observed that when increasing the definition of a negative urine culture, the cut-points that the authors use for screening are higher. In the same way, these combinations of cutoff values might improve the combination of sensitivity and specificity of detecting UTI by the analyzer. For example, Broeren et al. [14] reported a SE of 96%, a SP 78%, and a NPV of 99.7%, at a cutoff value of 230 bacteria/μl, when a definition of negative urine culture

Table 1: Summary of eligible studies.

Reference (country)	Patients and samples	Reference Counts	Cutoff Values	SE/SP/NPV/ER%
Monsen and Rydén 2015 (Turkey) [21]	4458 urine samples from adults	>10^4 CFU/mL	50 BAC/μL 30 WBC/μL	1/49.7/1/44.8
García-Coca et al., 2017 (Spain) [22]	17483 urine samples from 85% women and 14.7% (median age 47 years). 97.4% outpatients and 2.6 hospitalized subjects	>10^5 CFU/mL	>200 BAC/μL >40 WBC/μL	AUC (BAC) count: 0.79 AUC (WBC) count: 0.60 ER: 48
Kulkarni and Nigrin 2013 (Netherlands) [23]	7322 urinary samples from 60% inpatients and 40% outpatients (average age 66 years)	>10^5 CFU/mL	>200 BAC/μL	0.93/0.63/0.96/49
Íñigo et al. (2016) [24] (Spain)	1934 urine samples: MSU (89.14%), catheterized urine (9.98%), pediatric urine bags (0.72%), suprapubic aspiration (0.16%), belonging to 56% ♀ and 44% ♂ (51.19% hospitalized patients and 48.81% outpatients)	>10^2 CFU/mL	460 BAC/μL 40 WBC/μL	0.98/0.76/0.99/57
Gessoni et al., 2015 (Italy) [25]	2335 MSU from adult patients (41% inpatients and 59% outpatients), between 15 to 91 years old (♀: 34%, ♂: 66%)	≥10^5 CFU/mL	175 BAC/μL	0.95/0.80/—/50
Martín-Gutiérrez et al., 2015 (Spain) [26]	346 MSU collected from elderly outpatients with a median age of 76 years (♀: 43%, ♂: 56%)	≥10^5 CFU/mL for women ≥10^4 CFU/mL for men	200 BAC/μL	0.99/0.91/0.99/60
Manoni et al., 2009 (Italy) [27]	372 MSU specimens from 167 ♂ and 205 ♀ (57.37% outpatients and 42.63% inpatients)	>10^5 CFU/mL	14.2 BAC/μL 6.7 WBC/μL 14.2 BAC/μL	0.8/0.58/0.86/—/— 0.95/0.44/0.95/30.6
Marschal et al., 2012 (Germany) [28]	572 urine samples (31.8% catheterized, 65.5% MSU and 2.7% not specified), from 513 patients (27.9% outpatients, 72.1% inpatients) ♀: 48.9% and ♂: 51.1% (mean age of 54.2 years)	>10^2 CFU/mL	>30 BAC/μL >20 WBC/μL	0.80/0.78/—/—
Gutiérrez-Fernández et al., 2012 (Spain) [29]	1225 urine samples (urine catheter, MSU or pediatric bags) from 158 inpatients (27 catheterized adults, 33 children, 98 inmunocompromised adults) and 1067 outpatients (464 adults and 603 pregnant women).	>10^5 CFU/mL	690 BAC/μL 38 WBC/μL	0.92/0.65/0.97/—
Broeren et al. (2011) (Netherlands) [14]	1577 urine samples collected from 681 outpatients (196 ♂) and 896 (403 ♂) from hospitalized patients	<10^5 CFU/mL <10^4 CFU/mL	230 BAC/μL 39 BAC/μL	0.96/0.78/0.99/52 0.82/0.83/0.95/28

Table 1: Continued.

Reference (country)	Patients and samples	Reference Counts	Cutoff Values	SE/SP/NPV/ER%
Kadkhoda et al. 2011 (Canada) [30]	2496 (75%) MSU and 25% catheterized urine (patients origin not informed)	$\geq 10^4$ CFU/mL of up two uropathogens $\geq 10^5$ CFU/mL Non uropathogens	20 BAC/μL	0.93/0.69/—/35
Pieretti et al. 2010 (Italy) [31]	703 MSU from 18.2% hospitalized patients and 81.8% outpatients (70.6% females and 29.4% males)	$>10^4$ CFU/mL	65 BAC/μL 100 WCB/μL	0.98/0.62/0.98/43
van der Zwet et al. 2010 (Dutch) [32]	358 urine samples (patients origin not informed)	$>10^4$ CFU/mL	50 BAC/μL 20 WBC/μL	1/0.8/0.59/42
De Rosa et al. 2010 (Italy) [33]	1349 urine samples (patients origin not informed)	$>10^4$ CFU/mL for adult $>10^6$ CFU/mL for children.	170 BAC/μL 150 WBC/μL	0.98/0.76/0.99/57
Shang et al. 2013 (China) [34]	313 urine samples, belonging inpatients and outpatients	$>10^5$ CFU/mL for gram (−) $>10^4$ CFU/mL for gram (+)	100 BAC/μL 56 WBC/μL	0.86/0.95/0.94/50
Hu et al. 2010 [35] (China)	308 urine samples from outpatients	$>10^5$ CFU/mL for gram (−) $>10^4$ CFU/mL for gram (+)	160 BAC/μL	0.81/0.83/0.89/55
Kouri 2000 (Italy) [36]	1463 MSU samples from adults	$>10^5$ CFU/mL	>125 BAC/μL >40 WBC/μL	0.99/0.77/0.98/—

MSU: midstream urine sample; WBC: white blood cells; BAC: bacteria, NPV: negative predictive value; CFU: colony forming units; ER: elimination rate; AUC: area under the ROC curve.

of <10^5 CFU/ml was used. When these authors choose a less stringent gold standard definition (<10^4 CFU/ml), the cutoff decreases until 39 bacteria/μl, with a consequent decrease in SE (82%) and an increase of SP (83%). Other studies, with the same definition of negative urine culture (<10^5 CFU/ml), show different data. Manoni et al. [27] choose cut-off values for bacteria and leukocytes of 125/μl and 40/μl, respectively, to obtain a sensibility of 97% with a specificity of 94%. Gutiérrez-Fernández et al. [29] fix different counts limits (690 bacteria/μl and 38 leukocytes/μl), mainly due to differences in the population analyzed.

On the other hand, in two studies, we observed strikingly low cut-off points. Kadkhoda et al. [30] considered >20 bacteria/μl, using a definition of culture positive of >10^5 CFU/ml for uropathogens. Krongvoraluk et al. [50] showed the lowest cutoffs of the revised literature (bacteria >14.2/μl and leukocyte >6.7/μl), perhaps for the use of the filter paper method instead of the calibrated loop urine quantification technique.

Diverse papers have shown that the use of FCA reduces the number of processed negative urinary specimens. Authors demonstrate variably reduction in bacterial culture ranging from 28% to 60% [14, 22, 24–26, 29–33, 50]

We observed a wide variation in cut-off points applied, as well as in the SE and SP of the results obtained in the literature. Several authors did not recommend to use a standard, unique cutoff value of both bacteria and leukocytes counts. These will be chosen depending on the characteristics of the analyzed population, which is usually heterogeneous. Furthermore, published data recommended adjusting the cutoff values according to gender and age of the patients to increase detection reliability [15, 23, 31, 51].

If confirmed in studies with larges series, the UFC screening method would be more efficient when both bacterial and leukocyte counts were detected. Various authors demonstrate that this implies an increase in the sensitivity of the technique [24, 29, 33, 35, 49]. In contrast to these findings, other studies do show no benefit when combining the use of both counts for UTI screening [14, 26, 30].

Also, some authors suggest that, with the general definition of <10^5 CFU/ml for a negative urine culture, a significant amount of microorganisms would not be considered. But, down CFU counts (≤10^3 CFU/ml) in some specific groups of patients (urological patients, children, pregnant women, and catheterized patients or when "fastidious" microorganisms are involved) are relevant. For this reason, some authors suggest that these samples have to be evaluated individually and also cultivated [14, 27, 30].

On the other hand, diverse studies with similar patient population analyzed, and the same gold standard definition for negative culture, show differences in the number of false negative results [14, 27, 33]. Diverse aspects can explain these discrepancies. Defects in the cytometer detection of Gram-positive bacteria, due to phenomena of aggregation, have been described by several authors [47, 49, 52]. There are other factors than can lead to high count false negatives: antibiotic treatment before urine analysis, for nonviable bacteria detection [14] as well as different culture procedures, transportation times or the use of boric acid as preservative [19].

We checked that the majority of subjects analyzed were outpatients [14, 22, 26, 29, 31, 35], whereas other authors evaluated mainly samples of hospitalized patients [28, 53]. Besides, most studies did not establish the discrepancies between hospitalized patients and outpatients. But, it is important to distinguish them because there are substantial differences, in connection with the presence of catheters or the use of antimicrobials [19].

Concerning to microorganisms causing UTI, Gram-negative bacteria were reported more frequently, being *E. coli* the most prevalent microorganism isolated in samples with significant bacteriuria. In different series analyzed, it represents more than 50% of isolates [27, 30, 33, 52]. Curiously, Monsen and Rydén [21] confirmed that different microorganisms were associated with distinct parameters counts. *Proteus mirabilis*, *P. aeruginosa*, *S. aureus*, and group C streptococci showed high leukocyte counts. *Klebsiella* sp. had smaller leukocyte counts whereas the highest bacterial, leukocyte, and erythrocyte counts were found for *E. coli*. One study conducted in Spain [22] demonstrated that cultures with *E. coli* and *Klebsiella* spp. had less epithelial cells counts that contaminated urine samples. And specimens with *Enterococcus* spp. were associated with lower WBC counts than negative urines.

From an economical point of view, some studies have examined the possible savings that would be obtained with the use of this automated system in routine clinical microbiology laboratories. Marschal et al. [28] demonstrated money savings of 5965 euros every 12000 urine samples per year. And in one study conducted in Turkey, Yasuma et al. [54] showed savings of 239–306 US dollars per 100 samples. Ilki et al. confirmed that the method is cost efficient since it represents savings of at least 6798 euros [55].

Finally, in most of the eligible studies it was unknown if patient's clinical manifestations were representatives of UTI. On the other hand, criteria of inclusion and exclusion were not determined as well as information about preanalytical phase. Also, only a few studies reported the percentage of contaminated samples and its influence in UTI prevalence.

6. Conclusions

It is not possible to provide effective patient management if the results of urinary investigation are not interpreted considering patient's clinical characteristics. It is important to remember that the number of bacteria deemed relevant for the diagnosis of UTI depends on various factors. The type of specimen (mid-stream urine, bag urine, catheter urine, urostomy urine, and suprapubic aspiration urine), clinical situation of the patient (abnormalities of the genitourinary tract, presence of clinical manifestations, gender, age, pregnancy, and antibiotics previously taken), or variety of bacteria (category and isolated species) should be considered [19]. With a unique gold-standard definition for a negative culture of <10^5 UFC/ml, the presence of high bacteria amount is considered not significant. That is questionable, especially for particular patients as well as children, pregnant women, and older adults, for whom low counts are relevant.

Classical microbiology techniques are quite slow in comparison to new automated methods, based on advanced technologies for UTI screening, as they require the isolation of the organism before its identification and subsequently antibiotic susceptibility determination. With the implementation of FCA in clinical laboratories, the results are practically available in real time, within a few minutes of the sample admission and can be promptly informed in case of negative results.

It is vital to bear in mind that sensitivity is the most important requirement of a good screening process. In a screening test all positive specimens are cultured and, consequently, false-positive results will not be informed to the clinicians [27]. For FC to be a useful rapid test to rule out negative urine samples, the attention should be paid to obtaining the optimal results in comparison with the urine culture. This would require the highest sensibility and therefore a high negative predictive value, to minimize the number of false negative results.

Different limits for both leukocytes and bacteria detection were established in available reports. Each microbiology laboratory should adjust its bacteria and leukocytes cutoffs, depending on their UTI prevalence and the pathology of the patients analyzed as well as their origins, gender, and age.

In patients with UTI, the TAT (turnaround time) for a test result is approximately 48 h. For a patient with overt clinical manifestations, it is too long, because the introduction of antibiotic therapy should be early. Unfortunately, the most active antibiotics against Gram-positive bacteria are not useful in Gram-negative germs, which are the most frequent bacteria involved in urinary tract infections. Differentiating the type of bacteria involved in these infections would help to improve the efficacy of empirical therapy [49, 56, 57]. If information about the Gram properties of the suspected bacteria was available quickly, clinicians could select the most effective antibiotic for UTI infection.

Most revised articles have agreed that FC is a useful method for screening urinary samples if the optimal cut-off value is established for each group of patients. The adoption of correct criteria is crucial to improving the performance of the screening process. This will lead to significant reduction in TAT of negative cultures in bacterial culture and unnecessary empirical antibiotic prescription. However, in subjects at risk of UTI complications, its use is controversial, being the culture necessary.

In the future, perhaps the possibility of combining these systems with a rapid identification technique like matrix-assisted laser desorption ionization time-of-flight (MALDI-TOF) mass spectrometry (MS) can permit recognizing bacterial species directly in urine positive samples [58–60] and choosing the best antibiotics depending on the results provided by the analyzer. Further studies could be of interest to investigate the impact of these automatic techniques in both prescription of antibiotics and its relationship with patient care.

At present, there is a need to improve the quality and optimize the procedures performed in laboratories. Before deciding to automate a manual technique, a rigorous study should be conducted. In this way, microbiologists could offer the best diagnostic option to patients and physicians.

Disclosure

This research received no specific grant from any funding agency in the public, commercial, or not-for-profit sector

Conflicts of Interest

All authors declare that they have no conflicts of interest.

Authors' Contributions

Patricia Mejuto, Julio Díaz-Gigante, and Mariam Luengo participated in the planning of the study, wrote the first draft of the manuscript, revised it critically, and approved the final version.

Acknowledgments

The authors greatly appreciate the help received from Rosa F. Buisson for English language revision.

References

[1] K. G. Naber, "Urogenital infections: the pivotal role of the urologist," *European Urology*, vol. 50, no. 4, pp. 657–659, 2006.

[2] J. Santra, P. Eeva-Liisa, K. Pauliina, and et al., "Screening of urine samples by flow cytometry reduces the need for culture," *Journal of Clinical Microbiology*, vol. 48, no. 9, pp. 3117–3121, 2010.

[3] J. C. dos Santos, L. P. Weber, and L. R. R. Perez, "Evaluation of urinalysis parameters to predict urinary-tract infection," *Brazilian Journal of Infectious Diseases*, vol. 11, no. 5, pp. 479–481, 2007.

[4] B. Foxman, "Urinary tract infection syndromes. Occurrence, recurrence, bacteriology, risk factors, and disease burden," *Infectious Disease Clinics of North America*, vol. 28, no. 1, pp. 1–13, 2014.

[5] T. M. Hooton, "Pathogenesis of urinary tract infections: an update," *Journal of Antimicrobial Chemotherapy*, vol. 46, S1, pp. 1–7, 2000.

[6] S. J. Matthews and J. W. Lancaster, "Urinary tract infections in the elderly population," *American Journal Geriatric Pharmacotherapy*, vol. 9, no. 5, pp. 286–309, 2011.

[7] B. Foxman, "Epidemiology of urinary tract infections: incidence, morbidity, and economic costs," *The American Journal of Medicine*, vol. 113, no. 1, pp. 5–13, 2002.

[8] Z. Tandogdu and F. M. Wagenlehner, "Global epidemiology of urinary tract infections," *Current Opinion in Infectious Diseases*, vol. 29, no. 1, pp. 73–79, 2016.

[9] J. P. Horcajada, E. Shaw, B. Padilla et al., "Healthcare-associated, community-acquired and hospital-acquired bacteraemic urinary tract infections in hospitalized patients: A prospective multicentre cohort study in the era of antimicrobial resistance," *Clinical Microbiology and Infection*, vol. 19, no. 10, pp. 962–968, 2013.

[10] ECDC 2012, *Annual epidemiological report: reporting on 2010 surveillance data and 2011 epidemic intelligence data*, European Centre for Disease Prevention and Control, Stockholm, Sweden, 2012, http://www.ecdc.europa.eu/en/publications/publications/annual-epidemiological-report-2012.pdf.

[11] K. G. Naber et al., *EAU/International Consultation on Urological Infections*, The Netherlands European Association of urology, 2010.

[12] A. Sorlozano, A. Jimenez-Pacheco, J. de Dios Luna del Castillo et al., "Evolution of the resistance to antibiotics of bacteria involved in urinary tract infections: a 7-year surveillance study," *American Journal of Infection Control*, vol. 42, no. 10, pp. 1033–1038, 2014.

[13] EM. Davies and DA. Lewis, "Bacteriology of urine," in *Medical bacteriology Oxford*, P. Hawkey and DA. Lewis, Eds., pp. 1–25, Oxford University Press, 2004.

[14] M. A. C. Broeren, S. Bahçeci, H. L. Vader, and N. L. A. Arents, "Screening for urinary tract infection with the sysmex UF-1000i urine flow cytometer," *Journal of Clinical Microbiology*, vol. 49, no. 3, pp. 1025–1029, 2011.

[15] S. Jolkkonen, E.-L. Paattiniemi, P. Kärpänoja, and H. Sarkkinen, "Screening of urine samples by flow cytometry reduces the need for culture," *Journal of Clinical Microbiology*, vol. 48, no. 9, pp. 3117–3121, 2010.

[16] R. E. Berger, "The urine dipstick test useful to rule out infections. A meta-analysis of the accuracy," *The Journal of Urology*, vol. 174, no. 3, pp. 941-942, 2005.

[17] Y. Waisman, E. Zerem, L. Amir, and M. Mimouni, "The validity of the uriscreen test for early detection of urinary tract infection in children.," *Pediatrics*, vol. 104, no. 4, p. e41, 1999.

[18] V. Wiwanitkit, N. Udomsantisuk, and C. Boonchalermvichian, "Diagnostic value and cost utility analysis for urine Gram stain and urine microscopic examination as screening tests for urinary tract infection," *Urological Research*, vol. 33, no. 3, pp. 220–222, 2005.

[19] S. Grosso, G. Bruschetta, and R. De Rosa, "Improving the efficiency and efficacy of pre-analytical and analytical workflow of urine cultures with urinary flow citometry," *New Microbiol*, vol. 31, pp. 501–505, 2008.

[20] W. J. Isaac, D. E. Low, A. Biringer, N. Pimlott, M. Evans, and R. Glazier, "The impact of empirical management of acute cystitis on unnecessary antibiotic use," *Archives of Internal Medicine*, vol. 162, no. 5, pp. 600–605, 2002.

[21] T. Monsen and P. Rydén, "Flow cytometry analysis using sysmex UF-1000i classifies uropathogens based on bacterial, leukocyte, and erythrocyte counts in urine specimens among patients with urinary tract infections," *Journal of Clinical Microbiology*, vol. 53, no. 2, pp. 539–545, 2015.

[22] M. García-Coca, I. Gadea, and J. Esteban, "Relationship between conventional culture and flow cytometry for the diagnosis of urinary tract infection," *Journal of Microbiological Methods*, vol. 137, pp. 14–18, 2017.

[23] S. Kulkarni and J. Nigrin, "Letter to the Editor re: "Performance of flow cytometry to screen urine for bacteria and white blood cells prior to urine culture"," *Clinical Biochemistry*, vol. 46, no. 18, pp. 1914-1915, 2013.

[24] M. Íñigo, A. Coello, G. Fernández-Rivas et al., "Evaluation of the SediMax automated microscopy sediment analyzer and the Sysmex UF-1000i flow cytometer as screening tools to rule out negative urinary tract infections," *Clinica Chimica Acta*, vol. 456, pp. 31–35, 2016.

[25] G. Gessoni, G. Saccani, S. Valverde, F. Manoni, and M. Caputo, "Does flow cytometry have a role in preliminary differentiation between urinary tract infections sustained by gram positive and gram negative bacteria? An Italian polycentric study," *Clinica Chimica Acta*, vol. 440, pp. 152–156, 2015.

[26] G. Martín-Gutiérrez, A. Porras-González, C. Martín-Pérez, J. A. Lepe, and J. Aznar, "Evaluation and optimization of the Sysmex UF1000i system for the screening of urinary tract infection in primary health care elderly patients," *Enfermedades Infecciosas y Microbiologia Clinica*, vol. 33, no. 5, pp. 320–323, 2015.

[27] F. Manoni, L. Fornasiero, M. Ercolin et al., "Cutoff values for bacteria and leukocytes for urine flow cytometer Sysmex UF-1000i in urinary tract infections," *Diagnostic Microbiology and Infectious Disease*, vol. 65, no. 2, pp. 103–107, 2009.

[28] M. Marschal, M. Wienke, S. Hoering, I. B. Autenrieth, and J.-S. Frick, "Evaluation of 3 different rapid automated systems for diagnosis of urinary tract infections," *Diagnostic Microbiology and Infectious Disease*, vol. 72, no. 2, pp. 125–130, 2012.

[29] J. Gutiérrez-Fernández, A. Lara, M. Bautista et al., "Performance of the Sysmex UF1000i system in screening for significant bacteriuria before quantitative culture of aerobic/facultative fast-growth bacteria in a reference hospital," *Journal of Applied Microbiology*, vol. 113, no. 3, pp. 609–614, 2012.

[30] K. Kadkhoda, K. Manickam, P. DeGagne et al., "UF-1000i™ flow cytometry is an effective screening method for urine specimens," *Diagnostic Microbiology and Infectious Disease*, vol. 69, no. 2, pp. 130–136, 2011.

[31] B. Pieretti, P. Brunati, B. Pini et al., "Diagnosis of bacteriuria and leukocyturia by automated flow cytometry compared with urine culture," *Journal of Clinical Microbiology*, vol. 48, no. 11, pp. 3990–3996, 2010.

[32] W. C. van der Zwet, J. Hessels, F. Canbolat, and M. M. L. Deckers, "Evaluation of the Sysmex UF-1000i® urine flow cytometer in the diagnostic work-up of suspected urinary tract infection in a Dutch general hospital," *Clinical Chemistry and Laboratory Medicine*, vol. 48, no. 12, pp. 1765–1771, 2010.

[33] R. De Rosa, S. Grosso, G. Bruschetta et al., "Evaluation of the Sysmex UF1000i flow cytometer for ruling out bacterial urinary tract infection," *Clinica Chimica Acta*, vol. 411, no. 1-16, pp. 1137–1142, 2010.

[34] Y. Shang, Q. Wang, J. Zhang et al., "Systematic review and meta-analysis of flow cytometry in urinary tract infection screening," *Clinica Chimica Acta*, vol. 424, pp. 90–95, 2013.

[35] X. Hu, J. Zhang, and X. Zhang, "Evaluation of the sysmex UF-1000i urine analyzer as a screening test to reduce the need for urine cultures for urinary tract infection," *Laboratory Medicine*, vol. 41, no. 6, pp. 349–352, 2010.

[36] T. Kouri, G. Fogazzi, V. Gant et al., "European urinalysis guidelines," *Scandinavian Journal of Clinical and Laboratory Investigation*, vol. 60, supplement 231, pp. 1–96, 2000.

[37] G. Schmiemann, E. Kniehl, K. Gebhardt, M. M. Matejczyk, and E. Hummers-Pradier, "The diagnosis of urinary tract infection. A systematic review," *Deutsches Ärzteblatt International*, vol. 107, no. 21, pp. 361–367, 2010.

[38] L. S. Garcia, *Clinical Microbiology Procedures Handbook*, American Society for Microbiology Press, 3rd edition, 2010.

[39] E. H. Kass, "Bacteriuria and pyelonephritis of pregnancy," *A.M.A Archives of Internal Medicine*, vol. 105, no. 2, pp. 194–198, 1960.

[40] Guidelines on Urological Infections, "European Association of Urology," http://www.uroweb.org/nc/professional-resources/guidelines/online/.

[41] L. E. Nicolle, S. Bradley, R. Colgan, J. C. Rice, A. Schaeffer, and T. M. Hooton, "Infectious diseases society of America guidelines for the diagnosis and treatment of asymptomatic bacteriuria in adults," *Clinical Infectious Diseases*, vol. 40, no. 5, pp. 643–654, 2005.

[42] T. M. File Jr., "Highlights from international clinical practice guidelines for the treatment of acute uncomplicated cystitis and pyelonephritis in women: A 2010 update by the infectious diseases society of America and the european society for microbiology and infectious diseases," *Infectious Diseases in Clinical Practice*, vol. 19, no. 4, pp. 282-283, 2011.

[43] J. R. Delanghe, T. T. Kouri, A. R. Huber et al., "The role of automated urine particle flow cytometry in clinical practice," *Clinica Chimica Acta*, vol. 301, no. 1-2, pp. 1–18, 2000.

[44] H. M. Davey, A. S. Kaprelyants, D. H. Weichart et al., *Estimation of Microbial Viability Using Flow Cytometry. Current Protocols in Cytometry*, vol. 11, New York Wiley, New York, NY, USA, 1999.

[45] B. Moshaver, F. de Boer, H. van Egmond-Kreileman, E. Kramer, C. Stegeman, and P. Groeneveld, "Fast and accurate prediction of positive and negative urine cultures by flow cytometry," *BMC Infectious Diseases*, vol. 16, no. 1, article no. 211, 2016.

[46] E. Gieteling et al., "Accurate and fast diagnostic algorithm for febrile urinary tract infections in humans," *The Netherlands Journal of Medicine*, vol. 72, no. 7, pp. 356–362, 2014.

[47] Z. Zaman, S. Roggeman, and J. Verhaegen, "Unsatisfactory performance of flow cytometer UF-100 and urine strips in predicting outcome of urine cultures," *Journal of Clinical Microbiology*, vol. 39, no. 11, pp. 4169–4171, 2001.

[48] Y. K. Shine, J. K. Young, M. L. Sun et al., "Evaluation of the Sysmex UF-100 urine cell analyzer as a screening test to reduce the need for urine cultures for community-acquired urinary tract infection," *American Journal of Clinical Pathology*, vol. 128, no. 6, pp. 922–925, 2007.

[49] F. Manoni, S. Valverde, F. Antico, M. M. Salvadego, A. Giacomini, and G. Gessoni, "Field evaluation of a second-generation cytometer UF-100 in diagnosis of acute urinary tract infections in adult patients," *Clinical Microbiology and Infection*, vol. 8, no. 10, pp. 662–668, 2002.

[50] J. Krongvoraluk et al., "A flow cytometric urine analyzer for bacteria and white blood cell counts plus urine dispstick test for rapid screening of bacterial urinary tract infection," *Asian Biomedicine*, vol. 6, no. 4, pp. 601–608, 2012.

[51] N. Geerts, K. Boonen, A. Boer, and V. Scharnhorst, "Cut-off values to rule out urinary tract infection should be gender-specific," *Clinica Chimica Acta*, vol. 452, pp. 173–176, 2016.

[52] J. Wang, Y. Zhang, D. Xu, W. Shao, and Y. Lu, "Evaluation of the sysmex UF-1000i for the diagnosis of urinary tract infection," *American Journal of Clinical Pathology*, vol. 133, no. 4, pp. 577–582, 2010.

[53] H. Okada, T. Shirakawa, A. Gotoh et al., "Enumeration of bacterial cell numbers and detection of significant bacteriuria by use of a new flow cytometry-based device," *Journal of Clinical Microbiology*, vol. 44, no. 10, pp. 3596–3599, 2006.

[54] K. Yasuma, M. Nagao, Y. Matsumura et al., "[Evaluation of a UF-1000i screening method to identify the bacteriuria for cultures and susceptibility testing].," *Rinsho byori. The Japanese journal of clinical pathology*, vol. 60, no. 11, pp. 1070–1074, 2012.

[55] A. Ilki, R. Ayas, S. Ozsoy, and G. Soyletir, "Cost-effectiveness of a new system in ruling out negative urine cultures on the day of administration," *European Journal of Clinical Microbiology and Infectious Diseases*, pp. 1–5, 2017.

[56] G. Schmiemann, I. Gágyor, E. Hummers-Pradier, and J. Bleidorn, "Resistance profiles of urinary tract infections in general practice - An observational study," *BMC Urology*, vol. 12, article no. 33, 2012.

[57] J. M. Haslund, M. Rosborg Dinesen, A. B. Sternhagen Nielsen, C. Llor, and L. Bjerrum, "Different recommendations for empiric first-choice antibiotic treatment of uncomplicated urinary tract infections in Europe," *Scandinavian Journal of Primary Health Care*, vol. 31, no. 4, pp. 235–240, 2013.

[58] X.-H. Wang, G. Zhang, Y.-Y. Fan, X. Yang, W.-J. Sui, and X.-X. Lu, "Direct identification of bacteria causing urinary tract infections by combining matrix-assisted laser desorption ionization-time of flight mass spectrometry with UF-1000i urine flow cytometry," *Journal of Microbiological Methods*, vol. 92, no. 3, pp. 231–235, 2013.

[59] L. Ferreira, F. Sánchez-Juanes, J. L. Muñoz-Bellido, and J. M. González-Buitrago, "Rapid method for direct identification of bacteria in urine and blood culture samples by matrix-assisted laser desorption ionization time-of-flight mass spectrometry: Intact cell vs. extraction method," *Clinical Microbiology and Infection*, vol. 17, no. 7, pp. 1007–1012, 2011.

[60] M. Fritzenwanker, C. Imirzalioglu, T. Chakraborty, and F. M. Wagenlehner, "Modern diagnostic methods for urinary tract infections," *Expert Review of Anti-Infective Therapy*, vol. 14, no. 11, pp. 1047–1063, 2016.

16

Evaluation of Antimicrobial Activity of Four Organic Acids Used in Chicks Feed to Control *Salmonella typhimurium*: Suggestion of Amendment in the Search Standard

Aicha El Baaboua[1,2], **Mohamed El Maadoudi**[2], **Abdelhakim Bouyahya**[3], **Omar Belmehdi**[1], **Ayoub Kounnoun**[2,4], **Rajae Zahli**[1], **and Jamal Abrini**[1]

[1]*Biology and Health Laboratory, Biotechnology and Applied Microbiology Team, Department of Biology, Faculty of Science, Abdelmalek Essaadi University, Tetouan, Morocco*
[2]*Regional Laboratory for Analysis and Research, National Office for Food Safety, Tangier, Morocco*
[3]*Laboratory of Human Pathology Biology, Faculty of Sciences, and Genomic Center of Human Pathology, Mohammed V University, Rabat, Morocco*
[4]*Laboratory of Applied Biology and Pathology, Department of Biology, Faculty of Science, Abdelmalek Essaadi University, Tetouan, Morocco*

Correspondence should be addressed to Aicha El Baaboua; elbaaboua.aisha14@gmail.com

Academic Editor: Giuseppe Comi

Today, the general public has become increasingly aware of salmonellosis problems. Organic acids are known by their antimicrobial potential and commonly used for improving the quality of poultry feed. In this context, the present work evaluated the inhibitory effect of four organic acids, namely, acetic acid, citric acid, lactic acid, and tartaric acid, at different levels of contamination by *Salmonella typhimurium*. The neutralization of these organic acids *in vitro* and in the presence of one-day-old chick's organs was also investigated during the search for *Salmonella* serovars in birds as described in the Moroccan standard "NM 08.0.550." The effect of four organic acids on *Salmonella typhimurium* was tested *in vitro* and in the presence of chick's organs at different concentrations set of strain and organic acids tested. The MIC results demonstrated that tartaric acid, citric acid, and acetic acid inhibited *Salmonella typhimurium* at concentrations of 0.312%, 0.625%, and 0.512% for the three levels of strain: 10, 100, and 10^3 CFU/ml, respectively, while lactic acid and depending on the amount of the strain introduced acts differently: 0.078% for 10 CFU/ml and 0.156% for 100 and 10^3 CFU/ml. The concentration of 0.04M of Na_2HPO_4 solution has proved, *in vitro*, in caecums and organs of chicks (in presence of organic acids) that strain introduced, even at low concentrations, can be recovered. The use of additives has beneficial effects in *Salmonella* control program. However, the present results recommend the amendment of *Salmonella* research standard, taking into account the probable presence of organic acids in digestive content of one-day-old chicks.

1. Introduction

Salmonellosis is a zoonotic disease that causes a serious issue in both human health and animal health. It is the major problem of morbidity and mortality that are regularly detected by official inspection authorities. The most reported serovars in *Salmonella* taxonomy are *Salmonella* Pullorum, *Salmonella* Gallinarum, and *Salmonella* Arizonae [1]. For instance, *Salmonella* is the second commonly reported bacteria causing gastrointestinal infection in the EU after *Campylobacter* [2, 3]. Each year, over 90,000 foodborne illness cases caused by *Salmonella* have been reported in the EU. This pathogen can have a large socioeconomic impact due to illness, medical costs, lost of productivity, disability, and deaths. In addition, the European Food Safety Authority (EFSA) has estimated that the overall economic burden of human salmonellosis could be as high as 3 billion dollars per year [4]. That is why *Salmonella* is an important target for the food industry, owing to their wide spread through international trade in animal feed, live animals, and food.

It is well documented that human *Salmonella* illnesses are caused by consumption of contaminated raw or undercooked food from animal origin, and it is known that the contamination process can happen during transportation, during packinghouse operations of meat products, or by cross contamination throughout processing steps such as evisceration or by slaughtering tools [5, 6]. In fact, chicken skin was reported to be an essential part of contamination and the most difficult site to control *Salmonella* [7, 8]. Many authors agree that animal feed is the first step to prevent contamination from the entrance, particularly in poultry [9–11].

Organic acids are designated and approved by the Federal Drug Administration (FDA) as safe substances (GRAS). Citric, lactic, tartaric, and acetic acids are the most known organic acids used, and their mechanism of action is previously mentioned [12–15]. They have been proved as effective sanitizers in reducing the bacterial charge (antimicrobial capacity and preservation of product quality), cost-effectiveness, and their simplicity. Indeed, the uses of these solvents differ from country to another depending on differences in legislation. Therefore, the use of organic acids is one of the most old and popular in decontamination procedures worldwide.

In Morocco, population growth rate has been trending up in last decade [16] and chicken meat is still the highest meat consumed in poultry sector because the prices are much cheaper than those of red meats or seafood [17, 18].

For this reasons, Morocco imports, from Spain and France, a considerable amount of birds to satisfy the population needs. The objective of the present work was to study the inhibitory effect of four organic acids on *Salmonella typhimurium*, to evaluate the efficacy of an alternative neutralizer tested *in vitro* and on the digestive contents of chicks imported by Moroccan companies, and to suggest amendment in the Moroccan standard of *Salmonella* research in one-day-old chicks.

2. Materials and Methods

2.1. Bacterial Strain and Condition. A pellet of *Salmonella typhimurium* ATCC® 14028™ was grown overnight in brain heart infusion broth (Biokar, Beauvais, France) at 37°C and isolated on trypticase soy agar (TSA) (Becton Dickson Microbiology Systems, Cockeysville, Md.) plates. The pure colonies were multiplied on TSA and stored in physiological solution with 25% glycerol at −80°C until further use. The bacterial strain was enumerated using standard serial 10-fold dilutions in physiological saline solution.

2.2. Minimal Inhibitory Concentrations (MIC) and Minimal Bactericides Concentrations (MBC) of Organic Acids. The choice of organic acids is based on the list of food additives authorized for use in animal feed. The four acids used have a respective purity of 99.8% acetic acid, 99.5% tartaric acid, 99% citric acid, and 90% lactic acid. The MIC is studied at three different concentrations of the test strain: 10, 10^2, and 10^3 CFU/ml. To achieve the dilution of the organic acids, we need three sets of test tubes. Each set consists of 8 test tubes containing 9 ml of buffered peptone water (BPW). All organic acids tested in this study were diluted at a concentration of 5%. The measurements used for each acid were as follows: 0.5 ml of acetic acid in 10 ml of BPW, 0.56 ml of lactic acid (90%) in 10 ml of BPW, 0.5 g of tartaric acid in 10 ml of BPW, and 0.5 g of citric acid in 10 ml of BPW.

Nine milliliters of the organic acids stock solution was taken, and the dilutions were carried out at a half dilution in each case, so as to obtain the corresponding concentrations ranging from 2.5% to 0.0198%. In each series, the same dilution protocol of stock solution was carried out. Only strain concentration inoculated in each series has been changed. The tubes were incubated at 37°C for 18 to 24 h.

The study of the MBC was thus made, in our case, from the tubes used for the determination of the MIC and incubated for 18 to 24 h at 37°C. For each of the three levels of inoculums tested, and each of the organic acids tested, 1 ml of each tube was taken and placed in a Petri dish, followed by about 15 ml of TSA. The mixture was homogenized by orbital shaking by hand. The cultures were incubated at 37°C for 18 to 24 hours and colonies were counted the next day.

2.3. Neutralization of Inhibitory Effect of Acids by Using Universal Neutralizer. The choice of the neutralization solution was based on a universal neutral recommended in the standards by studying the bactericidal activity of disinfectants and antiseptics and also by the norm concerning the search for *Salmonella* in animal production environment "NM 08.0.549."

The protocol consists in measuring the MIC of the studied acids in the presence of the neutralizing solution. The test was carried out in three successive steps:

Dilution of Acid. A half dilution of the acid is made by taking 8 ml of the 5% preparation which was added to the first tube set containing 8 ml of BPW for obtaining 2.5% acid. The serial acid dilution was prepared until the last test concentration.

Addition of Neutralizer. 1 ml of neutralizer stock solution was added to the series of organic acid dilution tubes.

Addition of Strain. After 5 to 10 min, the tubes of each organic acid dilution set tested were supplemented with 1 ml of one of the three bacterial suspensions prepared: 10^2 CFU/ml, 10^3 CFU/ml, and 10^4 CFU/ml to each series, in order to obtain a final titer of 10 CFU/ml, 10^2 CFU/ml, and 10^3 CFU/ml, respectively.

2.4. Use of Buffer Solution Neutralizer. The choice of an alternative neutralizing solution has been taken into account, the fact that organic acids' action are mainly known by their capacity to acidify the medium matters that inhibit the bacterial growth. Starting from principle, neutralization of the acidity will make it possible to neutralize the inhibitory effect of the test dose under consideration. It was also the choice of a bearing on the components of the universal neutral, namely, the disodium hydrogen phosphate dodecahydrate (Na_2HPO_4). This solution was used at the same concentration in the universal buffer (0.01 M) = (1X concentrated solution) and at the concentrations 0.015 M, 0.02 M, and 0.04 M. This amounts to weighing, respectively, 17.14 g, 25.71 g, 34.28 g, and 68.56 g of Na_2HPO_4 in 200 ml of distilled water. The dissolution of the crystals was completed with magnetic stirring on a hot plate at 50°C.

2.4.1. pH Measurement of Organic Acids before and after Neutralization. Four sets of polypropylene 15 ml conical bottom tube containing 8 ml of BWP were prepared for each organic acid handled. Then, the same protocol of acid dilution was carried out to have the test doses: 2.5%, 1.25%, 0.625%, 0.312%, and 0.0098%. Using a pH meter, the pH of each dilution acid was measured. Subsequently, 1 ml of phosphate buffer concentration tested was added and the pH was measured after stirring.

2.4.2. Determination of MIC of Organic Acids Tested after Neutralization. The same protocol was repeated for the study of the MIC of organic acids for *Salmonella typhimurium* in the presence of the greatest phosphate buffer concentration as neutralizing solution (0.04 M). The final test doses of the organic acids were as follows: 2.5%, 1.25%, 0.625%, and 0.312%. The three inoculum levels of the tested strain were again studied: 10, 10^2, and 10^3 CFU/ml. The aim was to demonstrate that the greatest neutralizing solution has no toxic effect on the *Salmonella* strain, and also to determine the largest dose of acid that can be neutralized under the test conditions.

2.5. Neutralization of Organic Acids in the Presence of Chick's Organ Extracts: Organ Extract Preparation. After the euthanasia of eight one-day-old chicks, the chicks were undressed by removing the skin previously soaked by alcohol 70°. Then, openings were made with the scissors in the thorax to remove the lungs, liver, cecum, and yolk sacs as well. All organs were pooled in a 50 ml sterile flask and mixed until a homogenate was obtained. The choice of the doses tested of each organic acid was based on the results of their MIC in the presence of the neutralizing solution.

Three sets were prepared for each acid corresponding to the strain concentrations of 10, 10^2, and 10^3 CFU/ml. Each tube contains 8 ml of organic acid-BPW, 1 g of organ homogenate, and then 1 ml of the neutralizing solution and was allowed to react for 5 to 10 minutes. At the end, 1 ml of the tested bacterial suspension was added. After exposure, incubation was carried out at 37°C for 18 h to 24 h. After that, the search for *Salmonella* was done according to the protocol described in Moroccan standard "NM 08.0.550."

In order to confirm colonies obtained on selective isolation media of *Salmonella*, hektoen and xylose lysine deoxycholate agar (XLD), two boxes containing characteristic colonies were chosen for each organic acid, on which 5 colonies were taken from each box to undergo differentiation test between *Salmonella* and *Proteus*: test of tryptophan deaminase.

2.6. Application of the Methodological Parameters Studied on Laboratory Samples. The *Salmonella* research was carried out on the samples of 10 batches of 1-day-old chicks, each batch was composed of 5 chicks, and each of them was separated from the ceca in 50 ml sterile vials. The neutralizing solution was added after BWP dilution and incubation as described in the standard. The same protocol described in the standard of *Salmonella* research in foods has been used for the detection of *Salmonella* in organ extracts, adding 1 ml the neutralization solution (0.04 M), before preenrichment incubation, in order to neutralize any organic acid compound.

3. Results

3.1. MIC and MBC of Organic Acids. Table 1 shows the results obtained for the inoculum suspension count in each case of the four organic acids and the results of the MIC and the MBC studied at the different inocula. For acetic acid, the MIC was the same for the three levels of inoculums and it was equal to 0.312%, while the MBC determination shows that the first two inocula of 10 and 10^2 CFU/ml have an equal value to MIC (0.312%). Contrary to the result obtained for the inoculum of 10^3 CFU/ml, an increase in MBC was observed (0.625%).

In the case of tartaric acid, from these results, it appears that the strain tested has a MIC of 0.312% for the three levels of inoculums tested. The study of MBC of tartaric acid indicates that it has a similar value to MIC (0.312%) for the inoculum of 10 CFU/ml. On the other hand, the inoculums of 10^2 and 10^3 CFU/ml demonstrate the bactericidal action at a concentration of 0.625%.

For citric acid, as the table shows, the MIC was 0.625% for inoculums of 10, 10^2, and 10^3 CFU/ml. In addition, the bactericidal action was obtained only at a concentration of 1.25% for the three levels of inoculum.

Lactic acid, compared to the other three acids studied, has a very strong inhibitory action. At 10 CFU/ml, the inhibition of *Salmonella* occurred at a concentration of 0.078%. However, for the two inoculums of 10^2 and 10^3 CFU/ml, a MIC was obtained at 0.156%. Furthermore, the study of the MBC of the three series of inoculum showed that the bactericidal action was achieved at the same concentration (0.156%).

3.2. Neutralization of Organic Acids Inhibitory Effect by Using Universal Neutralizer. By using universal neutralizer, it was possible to neutralize, partially, the organic acids selected for this study. Table 2 summarizes the results obtained from enumeration and the MIC study of the organic acids studied by using the universal neutralizer at the three levels of inocula studied.

Table 1: MIC and MBC enumeration of the four organic acids tested on *Salmonella typhimurium* ATCC® 14028™.

	Strain titer (CFU/ml)	SC (CFU/ml)	Test doses of organic acids (%)							
			2.5	1.25	0.625	0.312	0.156	0.078	0.039	0.0195
Acetic acid	10	8	--- (0)	--- (0)	--- (0)	--- (0)	+++ (>300)	+++ (>300)	+++ (>300)	+++ (>300)
	10^2	27	--- (0)	--- (0)	--- (0)	--- (22)	+++ (>300)	+++ (>300)	+++ (>300)	+++ (>300)
	10^3	243	--- (0)	--- (0)	--- (104)	--- (>300)	+++ (>300)	+++ (>300)	+++ (>300)	+++ (>300)
Citric acid	10	9	--- (0)	--- (0)	--- (31)	+++ (>300)	+++ (>300)	+++ (>300)	+++ (>300)	+++ (>300)
	10^2	133	--- (0)	--- (0)	--- (>300)	+++ (>300)	+++ (>300)	+++ (>300)	+++ (>300)	+++ (>300)
	10^3	>300	--- (0)	--- (0)	--- (>300)	+++ (>300)	+++ (>300)	+++ (>300)	+++ (>300)	+++ (>300)
Tartaric acid	10	28	--- (0)	--- (0)	--- (0)	--- (6)	+++ (>300)	+++ (>300)	+++ (>300)	+++ (>300)
	10^2	114	--- (0)	--- (0)	--- (4)	--- (>300)	+++ (>300)	+++ (>300)	+++ (>300)	+++ (>300)
	10^3	>300	--- (0)	--- (0)	--- (1)	--- (>300)	+++ (>300)	+++ (>300)	+++ (>300)	+++ (>300)
Lactic acid	10	17	--- (0)	--- (0)	--- (0)	--- (0)	--- (0)	--- (>300)	+++ (>300)	+++ (>300)
	10^2	185	--- (0)	--- (0)	--- (0)	--- (0)	--- (0)	+++ (>300)	+++ (>300)	+++ (>300)
	10^3	>300	--- (0)	--- (0)	--- (0)	--- (0)	--- (0)	+++ (>300)	+++ (>300)	+++ (>300)

---: absence of strain, +++: presence of strain; SC: strain control.

Table 2: MIC of organic acids in the presence of neutralizer for inocula of 10, 10^2, and 10^3 of *S. typhimurium*.

	Strain titer (CFU/ml)	SC (CFU/ml)	Test doses of organic acids (%)							
			2.5	1.25	0.625	0.312	0.156	0.078	0.039	0.0195
Acetic acid	10 10^2 10^3	19 220 >300	---	---	---	---	+++	+++	+++	+++
Citric acid	10 10^2 10^3	25 286 >300	---	---	--- +++	+++	+++	+++	+++	+++
Tartaric acid	10 10^2 10^3	17 263 >300	---	---	--- +++	+++	+++	+++	+++	+++
Lactic acid	10 10^2 10^3	34 286 >300	---	---	---	+++	+++	+++	+++	+++

---: absence of *Salmonella typhimurium*, +++: presence of *Salmonella typhimurium*, SC: strain control.

It emerges that the neutralization of acetic acid was not effective, and the MIC was found to remain the same as before the addition of the neutralizer (0.312%).

The inhibitory action of tartaric acid was conditioned by the titer of the studied strain. It could be seen that for the inoculums of 10 and 10^2 CFU/ml using the neutralizer, the MIC was 0.625%. It was also noted that, for 10^3 CFU/ml, the MIC increases from 0.312% before neutralization to 1.25% after the addition of neutralizer.

In citric acid, by resemblance to tartaric acid, the titer of inoculum introduced plays a decisive factor in the MIC. At 10 CFU/ml, the MIC for citric acid was 0.625, while at 10^2 and 10^3 CFU/ml, the MIC was the same as that present in tartaric acid at the same inoculum (1.25%). However, the MIC of lactic acid in the presence of neutralizer is 0.625% for the three inoculums introduced.

The results of the pH measurements of the four organic acids before and after the neutralization were collected and are presented in Figure 1. These results show that, before neutralization, the pH was clearly acidic at the first test doses used and the values obtained increased exponentially with dilution of acids to reach neutrality at the last two test doses studied. Generally, the pH values obtained were overall less than or equal to 5 for the test

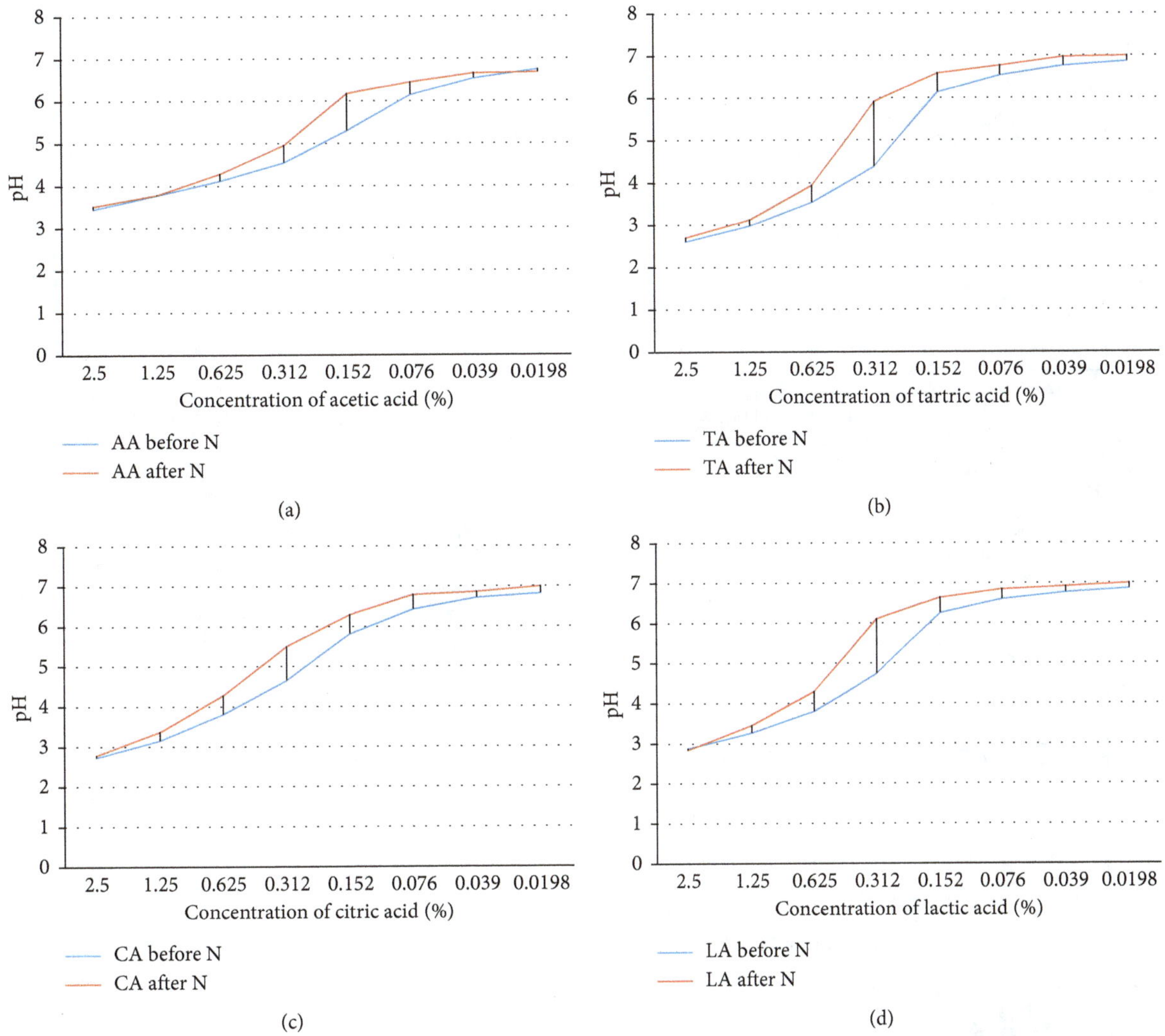

FIGURE 1: pH measurement curves of the four organic acids' concentrations using universal neutralizer.

doses of 2.5 to 0.312% and greater than or equal to 5 for the other doses.

After neutralization, the values recorded were less than or equal to 5 for only the first three test doses (2.5% to 0.625%), in the case of citric, lactic, and tartaric acids, whereas these values were greater than or equal to 6 for the other doses (0.312% to 0.0198%). Acetic acid showed resistance to neutralization with $pH \leq 5$ for doses of 2.5 to 0.312% and $pH \geq 6$ for doses of 0.152 to 0.0198%.

3.3. *Use of Buffer Solution Neutralizer*

3.3.1. pH Measurement. The results of pH measurements of the four organic acids after neutralization, using disodium phosphate buffer solution at four different concentration levels: 0.01 M, 0.015 M, 0.02 M, and 0.04 M are summarized in Figure 2. It emerges that the results for the concentration of the 0.01 M buffer solution, the concentration of the universal neutralizer, are the same as when using this latter.

In the case of acetic acid, the neutralization ($pH \geq 6$) is obtained with the 0.02 M and 0.04 M of solution from the concentrations of 0.312% and 0.625%, respectively.

For tartaric and citric acids, neutralization by 0.02 M and 0.04 M solutions makes it possible to have pH values ≥ 5 from 0.625%. A $pH \approx 5$ is also obtained for 1.25% for acetic, tartaric, and citric acids after neutralization by 0.04 M of buffer solution.

In the case of lactic acid, the neutralization appears to be easier, pH values ≥ 6, by using the buffer solutions from 0.625%. The values of $pH \geq 5$ are also obtained at a dose of 1.25% by 0.015 M, 0.02 M, and 0.04 M of buffer solutions.

3.3.2. Organic Acids Neutralization Test. From the results obtained (Table 3), it is clear that neutralizing solution at 0.04 M did not interfere with *Salmonella* culture and the three organic acids, tartaric, citric, and lactic acids, could be neutralized up to a concentration of 1.25% for the three

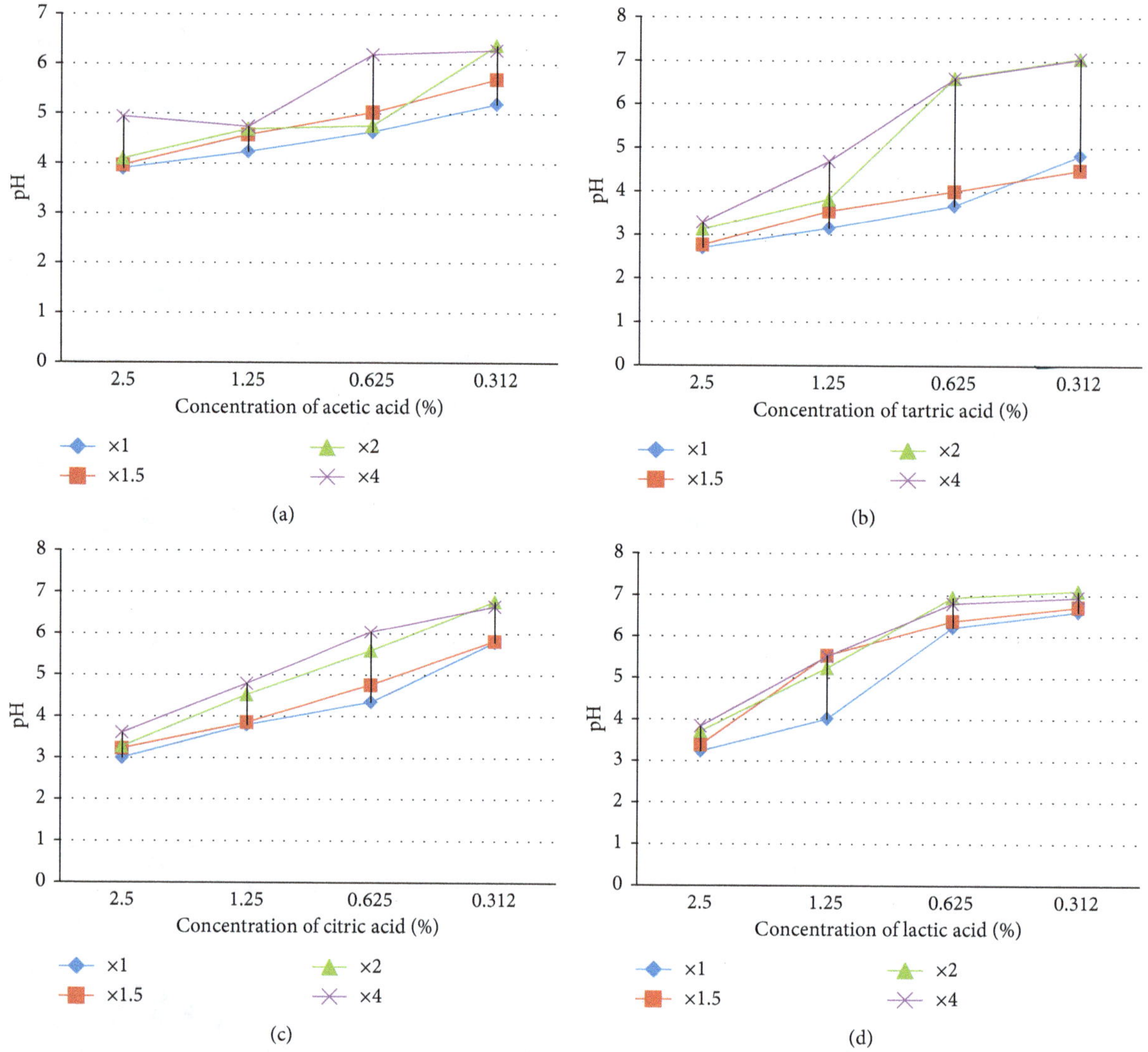

FIGURE 2: pH measurement curves of organic acids studied at different concentrations of the disodium hydrogen phosphate dodecahydrate (Na_2HPO_4).

TABLE 3: MIC of the organic acids using 0.04 M of Na_2HPO_4 solution, for inoculums of 10, 10^2, and 10^3 of tested strain.

	Inoculum (CFU/ml)	Test doses of organic acids 2.5%	1.25%	0.625%	0.312%
Tartaric acid	10 10^2 10^3	---	+++	+++	+++
Citric acid	10 10^2 10^3	---	+++	+++	+++
Lactic acid	10 10^2 10^3	---	+++	+++	+++
Acetic acid	10 10^2 10^3	---	---	--- +++	+++

inoculums tested, whereas for acetic acid, this neutralization did not occur. It is obtained only for maximum concentrations of 0.32% for the three inocula tested and 0.625% for inoculums of 10^2 and 10^3 CFU/ml.

3.4. *Study of Inhibitory Effect and Neutralization Test of Organic Acids in the Presence of Organ Extracts.*

For tartaric acid, the tested strain was reisolated after exposure to the three doses tested: 2.5, 1.25, and 0.625% of acid, using the two enrichment (Selenite Cystine broth and MRSV) and isolation media used (XLD and Hektoen).

In the enrichment step on Selenite Cystine broth, the strain was recovered in citric acid case at a dose of 2.5% only for an inoculum of 103 CFU/ml and also at a dose of 1.25% for inocula of 102 and 103 CFU/ml. While it was recovered for the three inocula tested at the dose of 0.625%. In addition, it was very interesting to note that the MRVS medium makes it possible to scour the strain, at all doses and for the three inocula.

For lactic acid, the strain was reisolated at a dose of only 2.5% for the inoculum of 10^3 CFU/ml after enrichment and isolation. However, it was recovered at test doses of 0.625% and 1.25% for the three inocula in all isolation media.

In the case of acetic acid, the reisolation of *Salmonella typhimurium* was obtained just at the test doses of 0.312% and 0.625% for the three levels in all media used.

It should be noted that the confirmation of the reisolation of *Salmonella typhimurium* was done during the neutralization tests in the presence of organ extracts, by carrying out the phenylalanine deaminase test on the characteristic colonies of *Salmonella* in order to differentiate them from *Proteus* sp. 10 colonies of isolates made for each acid were 100% negative for the test performed. Therefore, all colonies tested were *Salmonella* strains.

3.5. Application of Methodological Parameters Studied in Salmonella Detection on Laboratory Samples. After preenrichment incubation, the samples enriched in Selenite Cystine broth and modified semisolid Rapport-Vassiliadis (MRVS) and isolated on Hektoen and XLD medium showed that all tested samples were negative (absence of *Salmonella* characteristic colonies) even after further incubation (up to 48 h) of isolation media.

4. Discussion

Organic acids offer several advantages as antimicrobials molecule, because of their low cost and the easy manipulation [11]. The early use of organic acids in the production chain helps to prevent the spread of *Salmonella* contamination in all production devices as well as during feedstuff storage [19].

Significant differences in the effects of acids on reduction of *Salmonella* in poultry feed have been observed [13, 20, 21]. The results of the MIC and MBC study were compared to the other three acids studied; lactic acid has the strongest inhibitory action on the *Salmonella typhimurium* strain. The inhibitory dose for the three levels of inoculum was 0.156%, while citric acid had a lower action, with an increased MIC at different levels of inoculums (0.625%). Tartaric acid, as well as acetic acid, had a mean action with a MIC of 0.312% for the three levels of inocula tested. These results were in agreement with the results obtained in a research work entitled the use of acetic acid and citric acid to control of *Salmonella typhimurium* in tahini [22]. This can be explained by the pH-acid inhibitor of *Salmonella*, produced by these organic acids as demonstrated in pH results obtained from the acids solutions and their neutralization (Figure 1). Moreover, the efficacy differences of organic acids against *Salmonella* have been determined by various factors, such as the nature and physical form of the preparation and composition of the treated feed [11, 12].

For the neutralization of the inhibitory effects of acids, it should be noted that the universal neutralizer has a rather limited efficiency, since its neutralizing effect is obtained for lactic, tartaric, and citric acids only at a dose of 0.312% for the three levels of inoculums tested, whereas it is completely ineffective towards the citric acid (MIC), thing that makes use of this neutralizer for the recovery of *Salmonella typhimurium* exposed to this acid was critical, especially for acetic acid. When verifying the pH of test doses of solutions before and after the addition of the universal neutralizer, we found that, before neutralization, the pH values obtained were less than or equal to 5 for the test doses of 2.5 to 0.312% and greater than or equal to 5 for other doses, with an absence of bacterial growth in the first case and its presence in the second. As a result, the inhibition of bacterial culture coincides with pH values ≤ 5. This is all the more true as the neutralization of the inhibitory effect is accompanied by an increase in pH values, which testifies the close relationship between the decrease in pH and the inhibition of bacterial growth described in the literature [11, 12].

These results could be confirmed when using the alternative neutralizing solutions prepared with Na_2HPO_4, one of the main components of the universal neutralizer, with different molarities. For 0.01 M buffer solution concentration, the concentration of the universal neutralizer, the results were the same as when using the latter. This goes in the direction of the affirmation that the inhibitory action of these organic acids is mainly through acidification of the medium, which leads to a stop of multiplication or even a bacterial mortality. For 0.04 M concentration of the neutralizing solution, the three tartaric, citric, and lactic acids could be neutralized to a concentration of 1.25% for the three inocula tested. For acetic acid, this neutralization is obtained for the maximum concentrations of 0.32% for the three inocula tested and 0.625% for inoculums of 10^2 and 10^3 CFU/ml. Neutralizing solution at 0.04 M has no toxic effect on the *Salmonella typhimurium*. These results are complemented by those of the study of the inhibitory effect and the neutralization test of organic acids in the presence of organ extracts that show the capability of recuperating the strain at test doses of 0.625% and 1.25% for the three inoculums and in different media for lactic acid.

The strain was recovered from the organ extracts, in the case of enrichment on Selenite Cystine broth, for the three inocula tested at a dose of 0.625%. In addition, it was very interesting to note the high sensitivity of Selenite broth to the detection of small numbers of *Salmonella* [23–25]. The test strain was reisolated after exposure to the three doses tested: 2.5, 1.25, and 0.625% of acid on the enrichment media (Selenite broth and MRVS) and on XLD and Hektoen. The reisolation of the strain of organ extracts was obtained at a dose of 0.625% for the three levels of inoculums. Despite the fact that, during neutralization in the absence of the organ extracts, the 0.04 M neutralizing solution was not able to recover the strain at an inoculum of 10 CFU/ml. This could be explained by the technique used in this study, after neutralization, cannot recover the bacterial cells injured by organic acids. This makes it possible to amplify the neutralization of organic acids (in the case of acetic acid at a dose of 0.625% for the inoculum of 10 CFU/ml).

To our knowledge, the present work is the first study about the neutralization of organic acid in chick's organs.

The results obtained allow us to propose an amendment to *Salmonella* search method "NM 08.0.550" in the chick used by the Moroccan surveillance laboratory (taking into account the possible presence of authorized organic acids). This amendment consists of proceeding in parallel to the standardized method and in the same way by preenrichment of the samples analyzed, supplemented by 1 ml of 0.04 M disodium phosphate buffer neutralizer. Thus, about 10 samples of imported chicks were tested for *Salmonella* by this method, in parallel to the official control carried out by the laboratory, and were found negative.

5. Conclusion

Salmonella is the leading pathogen responsible for food poisoning worldwide and causes an economic burden, through both healthcare use and lost productivity. The Moroccan standard of *Salmonella* research based on three steps (preenrichment, enrichment, and isolation) remains the method of choice used in national programs for surveillance and control of salmonellosis. Nowadays, an impressive range of organic acids is used in poultry feed to reduce *Salmonella* number and other bacteria in digestive tract of animals. The intensive use of organic acids can interfere with the *Salmonella* detection method by inhibiting their multiplication during the preenrichment and enrichment steps without eliminating the bacteria that could be an attack for *Salmonella* control program.

Our work has clearly established that the four organic acids tested interfered effectively with the *Salmonella* search standard from 0.078 to 0.312% of acid, but also depending on acid nature. Based on the results, it was concluded that the use of a universal neutralizer as stipulated in the *Salmonella* standard is not effective in neutralization of the organic acids tested. Moreover, this interference can be controlled by making an amendment to the standardized method, by introducing a neutralization step, at the time of preenrichment in peptone water (the addition of 1 ml of neutralizer based on 0.04 M disodium phosphate buffer). Consequently, the *Salmonella* methods will be optimized to detect the minimum level of contamination in chick's samples.

Conflicts of Interest

The authors declare that they have no conflicts of interest.

References

[1] EFSA Panel on Animal Health and Welfare (AHAW), S. More, A. Bøtner et al., "Assessment of listing and categorisation of animal diseases within the framework of the Animal Health Law (Regulation (EU) No 2016/429): Salmonella infection in poultry with serotypes of animal health relevance (S. Pullorum, S. Gallinarum and S. Arizonae)," *EFSA Journal*, vol. 15, no. 8, 2017.

[2] A. E. Baaboua, M. E. Maadoudi, A. Bouyahya, and J. Abrini, "Intestinal infections of campylobacter: a review," *Microbiology Research Journal International*, vol. 18, no. 6, pp. 1–8, 2017.

[3] S. Shea, K. A. Kubota, H. Maguire et al., "Clinical microbiology laboratories' adoption of culture-independent diagnostic tests is a threat to foodborne-disease surveillance in the United States," *Journal of Clinical Microbiology*, vol. 55, no. 1, pp. 10–19, 2017.

[4] EFSA explains zoonotic diseases: Salmonella | European Food Safety Authority, 2017, https://www.efsa.europa.eu/en/corporate/pub/factsheetsalmonella.

[5] H. K. Crabb, J. L. Allen, J. M. Devlin, S. M. Firestone, M. A. Stevenson, and J. R. Gilkerson, "The use of social network analysis to examine the transmission of Salmonella spp. within a vertically integrated broiler enterprise," *Food Microbiology*, vol. 71, pp. 73–81, 2017.

[6] B. Karraouan, N. Ziyate, A. Ed-dra et al., "Salmonella Kentucky: antimicrobial resistance and molecular analysis of clinical, animal and environment isolates, Morocco," *Journal of Infection in Developing Countries*, vol. 11, no. 4, pp. 368–370, 2017.

[7] Q. Li, K. Yin, X. Xie et al., "Detection and CRISPR subtyping of Salmonella spp. isolated from whole raw chickens in Yangzhou from China," *Food Control*, vol. 82, pp. 291–297, 2017.

[8] K. Tozzo, A. F. G. Neto, K. M. Spercoski, M. Ronnau, V. M. Soares, and L. S. Bersot, "Migration of Salmonella serotypes Heidelberg and Enteritidis in previously frozen chicken breast meat," *Food Microbiology*, vol. 69, pp. 204–211, 2018.

[9] V. Economou and P. Gousia, "Agriculture and food animals as a source of antimicrobial-resistant bacteria," *Infection and Drug Resistance*, vol. 8, pp. 49–61, 2015.

[10] F. T. Jones, "A review of practical Salmonella control measures in animal feed," *Journal of Applied Poultry Research*, vol. 20, no. 1, pp. 102–113, 2011.

[11] H. Lynch, F. C. Leonard, K. Walia et al., "Investigation of in-feed organic acids as a low cost strategy to combat Salmonella in grower pigs," *Preventive Veterinary Medicine*, vol. 139, pp. 50–57, 2017.

[12] B. Amrutha, K. Sundar, and P. H. Shetty, "Effect of organic acids on biofilm formation and quorum signaling of pathogens from fresh fruits and vegetables," *Microbial Pathogenesis*, vol. 111, pp. 156–162, 2017.

[13] E. Mani-López, H. S. García, and A. López-Malo, "Organic acids as antimicrobials to control Salmonella in meat and poultry products," *Food Research International*, vol. 45, no. 2, pp. 713–721, 2012.

[14] H. M. H. Mohamed and H. H. S. Abdel-Naeem, "Enhancing the bactericidal efficacy of lactic acid against *Salmonella typhimurium* attached to chicken skin by sodium dodecyl sulphate addition," *LWT—Food Science and Technology*, vol. 87, pp. 464–469, 2018.

[15] H. M. B. A. Zaki, H. M. H. Mohamed, and A. M. A. El-Sherif, "Improving the antimicrobial efficacy of organic acids against Salmonella enterica attached to chicken skin using SDS with acceptable sensory quality," *LWT—Food Science and Technology*, vol. 64, no. 2, pp. 558–564, 2015.

[16] H. Alaoui, "Recensement. La population marocaine a augmenté de 3,9 millions en dix ans," 2017, http://fr.le360.ma/societe/recensement-la-population-marocaine-a-augmente-de-39-millions-en-dix-ans-35177.

[17] J. Bounouar, "Morocco targets expansion in poultry sector,"

2017, https://www.globalmeatnews.com/Article/2016/12/13/Morocco-targets-expansion-in-poultry-sector.

[18] B. Khalid, "Poultry production in Morocco," *World's Poultry Science Journal*, vol. 39, no. 1, pp. 52–60, 1983.

[19] G. Kleter, S. McFarland, A. Bach et al., "Surveying selected European feed and livestock production chains for features enabling the case-specific post-market monitoring of livestock for intake and potential health impacts of animal feeds derived from genetically modified crops," *Food and Chemical Toxicology*, vol. 117, pp. 66–78, 2017.

[20] A. M. Burns, P. G. Lawlor, G. E. Gardiner et al., "Salmonella occurrence and Enterobacteriaceae counts in pig feed ingredients and compound feed from feed mills in Ireland," *Preventive Veterinary Medicine*, vol. 121, no. 3-4, pp. 231–239, 2015.

[21] S. Koyuncu, M. G. Andersson, C. Löfström et al., "Organic acids for control of Salmonella in different feed materials," *BMC Veterinary Research*, vol. 9, no. 1, p. 81, 2013.

[22] A. A. Al-Nabulsi, A. N. Olaimat, T. M. Osaili et al., "Use of acetic and citric acids to control *Salmonella typhimurium* in tahini (sesame paste)," *Food Microbiology*, vol. 42, pp. 102–108, 2014.

[23] N. Daquigan, C. J. Grim, J. R. White, D. E. Hanes, and K. G. Jarvis, "Early recovery of Salmonella from food using a 6-hour non-selective pre-enrichment and reformulation of tetrathionate broth," *Frontiers in Microbiology*, vol. 7, 2016.

[24] R. Kumar, P. K. Surendran, and N. Thampuran, "Evaluation of culture media for selective enrichment and isolation of Salmonella in seafood," *Journal of AOAC International*, vol. 93, no. 5, pp. 1468–1471, 2010.

[25] Q. Zheng, M. Mikš-Krajnik, C. D'Souza et al., "Growth of healthy and sanitizer-injured Salmonella cells on mung bean sprouts in different commercial enrichment broths," *Food Microbiology*, vol. 52, pp. 159–168, 2015.

17

Characterization of Carbapenem-Resistant *Acinetobacter baumannii* Strains Isolated from Hospitalized Patients in Palestine

Regeen Handal,[1,2] Lulu Qunibi,[3] Ibrahim Sahouri,[4] Maha Juhari,[5] Rula Dawodi,[1] Hiyam Marzouqa,[1] and Musa Hindiyeh[1,2,6]

[1]*Caritas Baby Hospital, Bethlehem, State of Palestine*
[2]*Bethlehem University, Bethlehem, State of Palestine*
[3]*Alia Hospital, Hebron, State of Palestine*
[4]*Hussein Hospital, Bethlehem, State of Palestine*
[5]*Rafidia Hospital, Nablus, State of Palestine*
[6]*Palestinian Forum for Medical Research, Ramallah, State of Palestine*

Correspondence should be addressed to Musa Hindiyeh; mousa.hindiyeh@crb-mail.org

Academic Editor: Dulal Borthakur

The American Centers for Disease Control and Prevention (CDC) recognizes *Acinetobacter baumannii* as a source of global outbreaks and epidemics especially due to its increasing resistance to commercially available antibiotics. In this study, 69 single patient multidrug resistant isolates collected from all over Palestine, except Gaza, were studied. All the isolates were resistant to all the β–lactam antibiotics including the carbapenems. Of the 69 isolates, 82.6% were positive for $bla_{\text{OXA-23}}$, 14.5% were positive for $bla_{\text{OXA-24}}$, and 3% were positive for $bla_{\text{OXA-58}}$. None were positive for $bla_{\text{OXA-143}}$ and $bla_{\text{OXA-235}}$. In addition, 5.8% and 0% were positive for bla_{NDM} and bla_{KPC}, respectively. Of the 69 isolates, none were positive for the aminoglycoside *aphA6* gene while 93% were positive for the *aphA1* gene. The acetyltransferases *aacC1* and *aacA4* genes tested positive in 22% and 13% of the isolates, respectively. The *ompA* biofilm-producing virulence gene was detected in all isolates. Finally, Multilocus Sequence Typing (MLST) of 13 isolates revealed that more than one strain of *A. baumannii* was circulating in Palestinian hospitals as results revealed that 7 isolates were of ST208, 2 isolates ST218, 1 isolate ST231, 1 isolate ST348, and 2 new Sequence Types. The detection of these drug resistant pathogens is a reminder of the importance of active surveillance for resistant bacteria in order to prevent their spread in hospital settings.

1. Introduction

Acinetobacter baumannii is a gram-negative, nonfermentative, oxidase-negative coccobacillus. The Centers for Disease Control and Prevention (CDC) recognizes multidrug resistant (MDR) *A. baumannii* as a source of global outbreaks and epidemics especially due to its effectiveness in colonizing hospital environments and due to its increasing resistance to commercially available antibiotics, including β-lactams, fluoroquinolones, tetracyclines, and aminoglycosides [1–3]. *A. baumannii* is associated with hospital-acquired infections which include ventilator-associated pneumonia, bloodstream infections, meningitis, and urinary tract infections (UTI's) [3].

Mechanisms of resistance in *Acinetobacter* strains include efflux pumps, β-lactamases, and modifications in porin proteins. *A. baumannii* expresses aminoglycoside-modifying enzymes (AMEs) making them resistant to aminoglycoside antibiotics. Moreover, mutations in the *gyrA* and *parC* genes make them resistant to quinolones [1]. Worthy of mention are the findings of Fournier et al. (2006) who identified AbaRl resistance islands (86 kb region) encompassing a cluster of resistance genes, namely, ones coding for tetracycline efflux pumps, several AMEs, AmpC, and OXA-10 β-lactamases.

Genetic analysis of this region also indicated the presence of transposons and genes formerly identified in *Salmonella* spp. and *E. coli* [4].

Enzymatic degradation by β-lactamases is the most prevalent mechanism of β-lactam resistance in *A. baumannii*. Serine oxacillinases (Ambler class D OXA-type) and metallo-β-lactamases (MBLs) (Ambler class B) are β-lactamases with carbapenemase activity. OXA-type carbapenemase genes in *Acinetobacter* can be classified into different phylogenetic subgroups: $bla_{\text{OXA-23}}$, $bla_{\text{OXA-24}}$, $bla_{\text{OXA-51}}$, $bla_{\text{OXA-58}}$, $bla_{\text{OXA-143}}$, and $bla_{\text{OXA-235}}$ [5, 6]. The emergence of carbapenem-resistant strains associated with OXA-type β-lactamases is increasing as well as their contribution to outbreaks and patient mortality [7]. Another group of recently discovered β-lactamases in *Acinetobacter* is the class B β-lactamase bla_{NDM} and the class A serine carbapenemase bla_{KPC} both have a great opportunity to spread to other bacteria due to their location on mobile genetic elements [8].

In addition, *A. baumannii* has other virulence factors that aid in its protection and survival. It has the ability to form biofilms, which are essentially aggregates in which the cells adhere to each other and to a surface in a self-produced matrix of extracellular DNA, polysaccharides and proteins [1]. In *A. baumannii*, outer membrane protein A (ompA) is involved in biofilm formation on abiotic surfaces [9]. Biofilm formation not only helps protect the bacteria against disinfection for instance, but also helps in trading resistance genes between the participating cells.

In this study, we performed in-depth characterization of extremely resistant nonduplicated *A. baumannii* strains isolated from samples of hospitalized patients from all over Palestine except Gaza.

2. Materials and Methods

2.1. Ethical Considerations. This study was approved by the Medical Research Committee at Caritas Baby Hospital (MRC-014) in Bethlehem, Palestine.

2.2. Study Population. Extremely drug resistant *A. baumannii* isolates, all resistant to carbapenems, were collected from 69 patients from Palestinian hospitals all over Palestine with the exception of the Gaza strip. Isolates were collected from five Palestinian districts as follows: Jerusalem (N = 2), Bethlehem (N = 22), Hebron (N = 34), Nablus (N = 8), and Ramallah (N = 3). The isolates were collected between January 2006 and February 2014 from blood, urine, sputum, and rectal samples.

2.3. A. baumannii Species Identification. In order to identify *A. baumannii* from the collected *Acinetobacter* spp. isolates, all isolates were identified biochemically and screened for the naturally occurring carbapenemase gene ($bla_{\text{OXA-51}}$) intrinsic to *A. baumannii* using polymerase chain reaction (PCR) as previously reported [10, 11].

2.4. Presence of ISAba1 Element and the Expression of $bla_{\text{OXA-51}}$ Gene. The isolates were screened for the presence of the IS element, ISAba1, and analyzed for the potential expression of the $bla_{\text{OXA-51}}$ gene by investigating the presence of the ISAba1 element upstream of the gene [12].

2.5. Antibiotic Susceptibility Testing. Antibiotic susceptibility testing (AST) by disk diffusion test was performed for the following antibiotics (ceftazidime 30 μg, cefotaxime 30 μg, ceftriaxone 30 μg, piperacillin-tazobactam 100/10 μg, cefepime 30 μg, sulfamethoxazole 25 μg, amikacin 30 μg, gentamycin 10 μg, ciprofloxacin 5 μg, meropenem 10 μg, and imipenem 10 μg). In addition, the minimal inhibitory concentration of the two antibiotics meropenem and colistin were determined by Etest (BioMérieux, Marcy-l'Étoile, France). AST was performed and interpreted according to the Clinical and Laboratory Standards Institute (CLSI) guidelines [13].

2.6. Detection of Antibiotics Resistant Genes and the Biofilm-Producing ompA Gene. Carbapenem- and aminoglycoside-resistant determinants as well as biofilm-producing virulence factors were screened for using standard PCR assay using a PTC-100 Peltier Thermal Cycler (Bio-Rad, USA). The PCR products were then subjected to gel electrophoresis on 1.5% agarose gel for 45 minutes at 85 V and viewed under UV light using the Transilluminator imaging device (Dinco & Rhenium Industries Ltd., Israel). Target genes and their corresponding primers for PCR amplification are mentioned in Table 1. All PCR reactions were done in singleplex with the exception of the OXA-type class D carbapenemases ($bla_{\text{OXA-23}}$, $bla_{\text{OXA-24}}$, and $bla_{\text{OXA-58}}$) which were done in a multiplex PCR reaction as reported in Mostachio et al. (2009). The assays were held in 25 μl volume reactions containing 2x ReddyMix PCR Master Mix with 1.5 mM $MgCl_2$ (Thermo Fisher Scientific, USA) with a primer concentration of 20 pmol/μl at standard PCR conditions. A PTC-100 Peltier Thermal Cycler (Bio-Rad, USA) was used for the PCR amplification reactions.

2.7. Multilocus Sequence Typing (MLST). *A. baumannii* bacterial DNA was extracted from 13 representative *A. baumannii* isolates using the High Pure Nucleic Acid Extraction Kit (Roche, Basel, Switzerland). The isolates covered the time period of the study (between 2006 and 2014) such that one isolate was from the year 2006, two were from the year 2007, one was from the year 2008, one was from the year 2010, two were from the year 2011, two were from the year 2012, and four isolates were from the year 2013. PCR reactions were performed for the seven housekeeping genes: *gltA*, *gyrB*, *gdhB*, *recA*, *cpn60*, *gpi*, and *rpoD* as previously reported by Bartual et al. (2005). PCR reactions were performed in 50 μl volume reactions containing 2x ReddyMix PCR Master Mix with 1.5 mM $MgCl_2$ (Thermo Fisher Scientific, USA) with a primer concentration of 20 pmol/μl. The PCR conditions were those described by Bartual et al. with an annealing temperature of 55°C for all genes. A PTC-100 Peltier Thermal Cycler (Bio-Rad, USA) was used for the PCR amplification reactions. PCR products were analyzed on a 1.5% agarose gel and visualized using UV light using the Transilluminator imaging device (Dinco & Rhenium Industries Ltd., Israel). Amplified PCR products with the appropriate size were purified using the High Pure PCR Product Purification

Table 1: Summary of the target genes screened for using polymerase chain reaction (PCR).

Target gene	Primer name	Sequence (5′ to 3′)	Annealing temperature* (°C)	Amplicon size (bp)
Class D carbapenemases	OXA-23	F: GATCGGATTGGAGAACCAGA R: ATTTCTGACCGCATTTCCAT	58	501
	OXA-24	F: GGTTAGTTGGCCCCCTTAAA R: AGTTGAGCGAAAAGGGGATT	58	246
	OXA-58	F: AAGTATTGGGGCTTGTGCTG R: CCCCTCTGCGCTCTACATAC	58	599
	OXA-51	F: TAATGCTTTGATCGGCCTTG R: TGGATTGCACTTCATCTTGG	58	353
	OXA-143	F: TGGCACTTTCAGCAGTTCCT R: TAATCTTGAGGGGGCCAACC	58	180
	OXA-235	F: TTGTTGCCTTTACTTAGTTGC R: CAAAATTTAAGACGGATCG	58	700
Class B carbapenemases	NDM	F: CCAATATTATGCACCCGGTCG R: ATGCGGGCCGTATGAGTGATTG	58	812
Class A carbapenemases	KPC	F: ATGTCACTGTATCGCCGTCT R: TTTTCAGAGCCTTACTGCCC	58	538
AME's	aphA6	F: ATGGAATTGCCCAATATTATTC R: TCAATTCAATTCATCAAGTTTTA	55	797
	aphA1	F: CAACGGGAAACGTCTTGCTC R: ATTCGTGATTGCGCCTGAG	58	455
	aacA4	F: ATGACTGAGCATGACCTTGCG R: TTAGGCATCACTGCGTGTTCG	65	518
	aadB	F: ATGGACACAACGCAGGTCGC R: TTAGGCCGCATATCGCGACC	65	524
	aadA1	F: ATGAGGGAAGCGGTGATCG R: TTATTTGCCGACTACCTTGGTG	65	254
	aacC1	F: ATGGGCATCATTCGCACATGTAGG R: TTAGGTGGCGGTACTTGGGTC	65	456
Biofilm-producing virulence factors	ompA	F: CGCTTCTGCTGGTGCTGAAT R: CGTGCAGTAGCGTTAGGGTA	58	531
	epsA	F: AGCAAGTGGTTATCCAATCG R: ACCAGACTCACCCATTACAT	58	451

*Annealing temperature (°C) used in this study.

Kit (Roche) and sequenced using the BigDye® Terminator v3.1 Cycle Sequencing Kit (Life Technology, Carlsbad, USA). The sequences were analyzed and cleaned using Sequencher® software (Gene Codes, MI, USA) and the PubMLST database. Isolates were assigned sequence types (STs) as per the allelic profiles in the *A. baumannii* MLST database (https://pubmlst.org/abaumannii/). The eBURST V3 algorithm (http://eburst.mlst.net/) was used to show phylogenetic relationships among related STs.

3. Results

3.1. Acinetobacter Species Identification. $bla_{OXA\text{-}51}$ PCR amplification confirmed that all 69 isolates were *A. baumannii*.

3.2. Presence of ISAba1 Element and the Expression of $bla_{OXA\text{-}51}$ Gene. Of the 69 isolates, 68 (99%) carried the ISAba1 element. Of these, only 30 (43.5%) had the IS element upstream of the $bla_{OXA\text{-}51}$ gene.

3.3. A. baumannii Susceptibility Testing. All isolates had a multidrug resistant profile, and they were all resistant to all β-lactam antibiotics including the carbapenems (Figure 1). The carbapenem resistance profile of all isolates was confirmed by determining the MIC value of meropenem by Etest method (BioMérieux, Marcy-l'Étoile, France). The only class of antibiotics to which the isolates were susceptible was colistin sulfate (100%) (Figure 1). In addition, the majority of *A. baumannii* isolates (>90%) were resistant to the aminoglycosides, macrolides, and cotrimoxazole.

3.4. Molecular Characterization

3.4.1. Detection of Carbapenem-Resistant Genes. Of the 69 *A. baumannii* isolates tested, 57 (82.6%) were positive for $bla_{OXA\text{-}23}$, 10 (14.5%) tested positive for $bla_{OXA\text{-}24}$, and 2 (3%) tested positive for $bla_{OXA\text{-}58}$ (Figure 2). None of the isolates were positive for the class D β-lactamases $bla_{OXA\text{-}143}$ and $bla_{OXA\text{-}235}$. In addition, 3 (5.8%) tested positive for the class B carbapenemase bla_{NDM}. Sequencing analysis of the bla_{NDM}

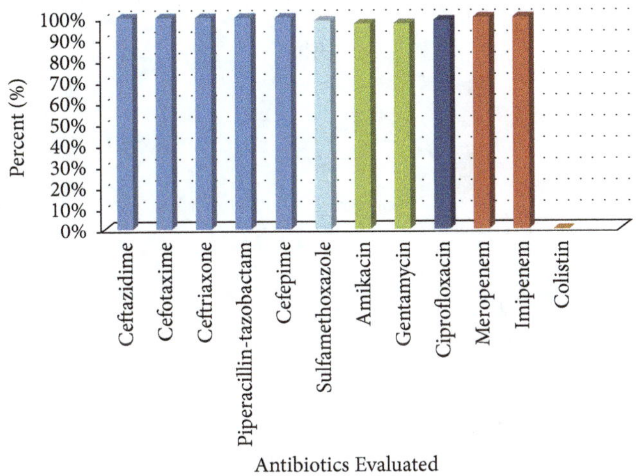

FIGURE 1: Histogram showing the percent resistance of *Acinetobacter* isolates to each antibiotic tested according to the CLSI guidelines.

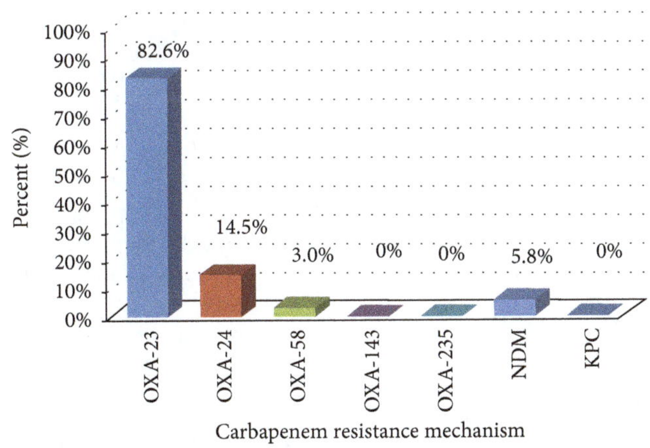

FIGURE 2: Histogram showing the distribution of carbapenemases in the *Acinetobacter* isolates tested.

gene revealed that the three isolates carried the bla_{NDM-1} gene. None were positive for the class A carbapenemase bla_{KPC}. It is worthy to note that one MDR *A. baumannii* isolate was a combination OXA-23/OXA-24.

3.4.2. Detection of Aminoglycoside-Resistant Genes. Of the 69 isolates, 64 (93%) were positive for the *aphA1* gene while none were positive for the *aphA6* gene (Figure 3). On the other hand, the isolates were positive for the acetyltransferases *aacC1* and *aacA4* genes in 15 (22%) and 9 (13%) of the isolates, respectively. None were positive for the nucleotidyltransferases *aadB* or *aadA1* genes (Figure 3).

3.4.3. Detection of Biofilm-Producing Virulence Factors. Of the 69 *A. baumannii* isolates tested, all (100%) tested positive for *ompA* gene.

3.4.4. Multilocus Sequence Typing (MLST). Of the 13 *A. baumannii* strains evaluated by MLST, the following Sequence Types (STs) were obtained: ST208, ST218, ST231, ST348, and

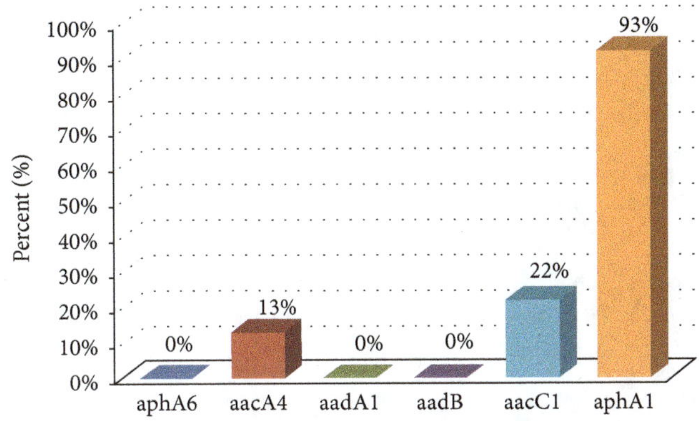

FIGURE 3: Histogram showing the distribution of AME's genes in the *Acinetobacter* isolates tested.

two new Sequence Types suggesting that more than one clone is circulating Palestinian hospitals. Two isolates did not have an exact ST match in the database. The nearest ST match yielded three and four possible STs for the *A. baumannii* isolates CBH|4248|2007 and CBH\2D2\2013, respectively. After assigning sequence types for each isolate, phylogenetic relationships among related STs were established using the eBURST V3 algorithm (http://eburst.mlst.net/), and a dendrogram was generated (Figure 4). The central primary founder is ST208 [CBH|152|2012] (bootstrap = 1000). There were 13 linked single locus variants (SLVs) of the founder (ST75, ST544, ST138, ST21, ST3, ST584, ST467, ST218, ST92, ST238, ST69, CBH|2D2|2013, and ST533). Six of the SLVs of the founder (ST533, ST218, ST238, ST584, ST21, and ST3) have diversified to produce double locus variants (DLVs). There were a total of five subgroup founders (ST533, ST1, ST187, ST109, and ST231 [CBH|3750|2007]) that diversified and produced their own SLVs. One of the new STs [CBH|2D2|2013] has only one out of seven of the MLST loci altered when compared to the primary founder genotype, ST208. Its allelic profile has a single locus variant (SLV) from that of the founder's, ST208. On the other hand, the other new ST [CBH|4248|2007] has two out of seven MLST loci altered in comparison to the founder ST208. Hence, its allelic profile has a double locus variant (DLV) from that of the founder's.

4. Discussion

A. baumannii has acquired a huge genetic repertoire via horizontal gene transfer that makes it virulent and resistant to any environmental pressures [1, 4]. Antibiotic susceptibility testing in this study showed that all *A. baumannii* isolates were resistant to the commercially available antibiotics with the exception of colistin. All isolates had an extremely drug resistant profile and they were resistant to carbapenems. MDR *A. baumannii* outbreaks were reported in hospitals worldwide including the Middle East [14, 15] and this is consistent with the increase in the incidence of health care associated *A. baumannii* infections reported in Palestinian and Israeli hospitals [2, 14, 15]. *A. baumannii* isolates from other geographical regions had a similar antimicrobial profile [16]. MIC tests also showed a similar result since most isolates were

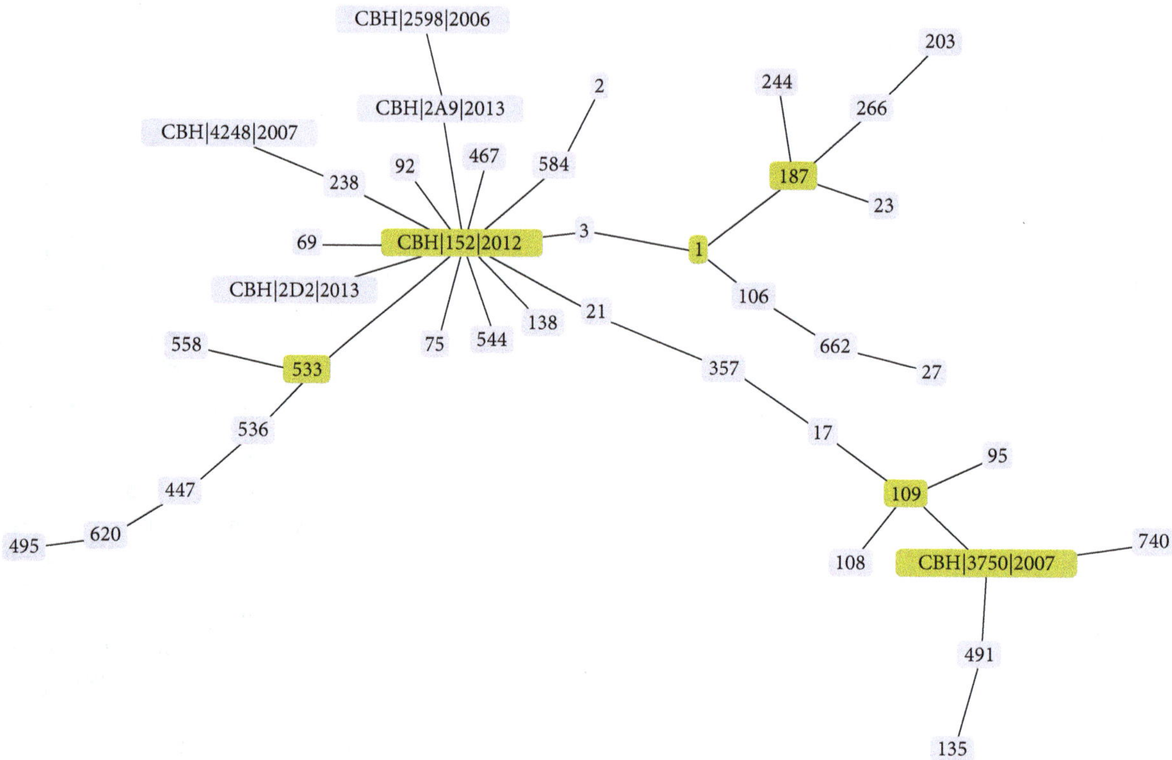

FIGURE 4: Population snapshot of the 13 *A. baumannii* isolates and existing reference isolates obtained from the PubMLST database, represented by the eBURST algorithm (http://eburst.mlst.net/). CBH|152|2012 (ST208) represents the primary founder with a bootstrap value = 1000, where CBH|2598|2006: ST348, CBH|4248|2007: New proposed ST, CBH|3750|2007: ST231, CBH|2A9|2013: ST218, CBH|152|2012: ST208, CBH|2D2|2013: New proposed ST.

resistant to meropenem and sensitive to colistin. There are, however, *A. baumannii* isolates that are becoming resistant to colistin [3].

Studying the carbapenem-resistant mechanisms of this pathogen was necessary in order to better control the spread of *A. baumannii* and devise a better antibiotics treatment plan. Carbapenem resistance was mainly attributed to the presence of bla_{OXA} genes that produce carbapenem-hydrolyzing enzymes [5]. Since the $bla_{\mathrm{OXA\text{-}51}}$ gene is an oxacillinase naturally occurring mainly in *A. baumannii*, it was present in all *A. baumannii* isolates tested, and the potential expression of this gene in 43.5% of the isolates is facilitated by the IS*Aba1* element present upstream of the gene. The production of $bla_{\mathrm{OXA\text{-}23}}$ by an *A. baumannii* strain is enough to confer resistance to the carbapenems, and 82.6% of the isolates in this study carried this gene. This gene is plasmid-borne suggesting the mobility of this genetic segment facilitating horizontal gene transfer [5]. $bla_{\mathrm{OXA\text{-}23}}$ was also reported in neighboring countries including Iraq, Greece, Italy, and Turkey [16]. Molecular class D β-lactamases confer resistance to the carbapenems and narrow-spectrum cephalosporins. 14.5% of the isolates also carried the $bla_{\mathrm{OXA\text{-}24}}$ gene, and 3% of them carried the recently identified $bla_{\mathrm{OXA\text{-}58}}$ gene. This is also the case in neighboring countries as those class D β-lactamases have been reported in Iraq, Kuwait, Lebanon, Turkey, Italy, Greece, and the UK [16]. Enzymatic degradation by β-lactamases is the most prevalent mechanism of β-lactam resistance in *A. baumannii* [7], and this is clearly shown in our sampled *A. baumannii* isolates in the West Bank.

About 6% of our *A. baumannii* isolates tested positive for the class B carbapenemase $bla_{\mathrm{NDM\text{-}1}}$. This is consistent with regional and global results as there are reports in Israel, India, China, United Kingdom, Canada, France, Sweden, Morocco, South Africa, United Arab Emirates, and Iran among many others [8].

The multidrug resistant profile of the isolates was also heightened by the presence of aminoglycoside-resistant determinants, namely, the phosphotransferase *aphA1* and the acetyltransferases *aacC1* and *aacA4*. Almost all our isolates carried the *aphA1* gene, and 22% and 13% were positive for *aacC1* and *aacA4* genes, respectively. In comparison to other studies, Aliakbarzade et al. (2014) reported the following aminoglycoside-resistant genes in Iran: *aacC1, aadA1, aphA6,* and *aadB*. Only *aacC1* was detected in our Palestinian isolates, and none of our isolates carried the phosphotransferase *aphA6*. The aminoglycoside-resistant gene *aphA6* was reported in Egypt [17]. On a more international scale, the following aminoglycoside-resistant genes were reported in Poland in 2013: *aphA1, aphA6,* and *aacC1* [18]. Our isolates are equipped with not only enzymatic resistance mechanisms, but also the ompA biofilm-producing virulence factor. It is unfortunate that our *A. baumannii* isolates in the West Bank produce biofilms: 100% tested positive for *ompA*.

There is an apparent global outbreak of carbapenem-resistant *A. baumannii* especially in the last decade. MLST is a discriminative typing method whose data are in concordance with other typing results produced by PFGE and AFLP analysis [5]. We sequence typed 13 isolates that are representative samples from different areas and different time periods throughout the West Bank. Sequence analysis of Palestinian genotypes showed a similarity with the global genotypes. Our MLST results reveal that more than one strain of *A. baumannii* is circulating in our Palestinian hospitals; the resulting STs were ST208, ST218, ST231, ST348, and two new STs. We correlated the STs with the timeline of their isolation and preliminary results show that there is a successful spread of clone. To illustrate, ST208 is evident throughout the studied time period, starting in 2008 up till 2014, indicating a successful spread of that particular strain. Internationally, ST208 was reported in the US in 2009, then in Egypt the following year, and then in China in 2012 [19]. Interestingly, the seven ST208 isolates came from the same Palestinian geographical district, Hebron. There are reports of ST218 in the PubMLST database in the year 2000. But, locally in Palestine, our MLST results show that the two isolates, isolated in 2013 from urine and sputum samples, were of ST218. ST231 was first reported in Brazil in 2000 and later again in 2007 in the same country [19]. Our results report ST231 that belongs to a 2007 isolate. Internationally, there are reports of ST348 in the US in 2008 according to the PubMLST database. But this particular ST was detected in Palestine earlier in 2006 as per our MLST results. Worthy of mention is the fact that the isolates carrying the $bla_{NDM\text{-}1}$ gene did not belong to the same clone.

Moreover, new strains are emerging. According to our MLST results, there is a new strain that emerged in 2007, and another new one in 2013. Over time, the frequent use of antibiotics will result in the emergence of new resistant strains of *A. baumannii* bacteria.

The eBURST V3 algorithm [20] shows the relationship between the closely related isolates in which the founding genotype starts to diversify producing a cluster of closely related genotypes as it increases in frequency in the population. In summary, there is more than one clone of *A. baumannii* circulating worldwide as shown by the sequence analysis of our Palestinian genotypes and the global genotypes.

Resistance to antibiotics, resistance to disinfectants, and resistance to desiccation accompanied by the presence of drug resistant and biofilm-producing genes all contribute to the persistence of *A. baumannii* as a hospital pathogen. Resistant *A. baumannii* strains have a selective advantage in environments like the ICU's where antimicrobials and surface disinfectants are extensively used. Selection for resistant forms can take place during antimicrobial treatment or after treatment. There is a sense of altered microbial ecology with regard to resistant and susceptible bacteria as well as the types of microorganisms surviving after treatment.

Epidemiologically, *A. baumannii* is of concern due to its dissemination mostly in a clonal fashion whether it is within health care institutions and cities or globally between countries. Since *A. baumannii* is easily transmitted and can, therefore, cause outbreaks, certain control measures have to be taken into account. To start with, health care facilities and personnel should follow the CDC's infection control guidelines and protocols, which attempt to control and prevent the transmission of MDR organisms, such as active surveillance, hand hygiene, and contact precautions [19]. Molecular epidemiological studies should be implemented to determine the presence or absence of a clonal outbreak strain such as *A. baumannii* ST208 which appears to be a very successful clone. Proper environmental cleaning of potential reservoir sources present in hospital environments such as ventilators, catheters, containers, and moist articles is highly recommended to prevent the spread of *Acinetobacter*. Moreover, antibiotic stewardship programs must be initiated in order to properly manage treatment and spread of this deadly pathogen.

Conflicts of Interest

The authors declare that there are no conflicts of interest regarding the publication of this paper.

References

[1] L. S. Munoz-Price and R. A. Weinstein, "Acinetobacter infection," *The New England Journal of Medicine*, vol. 358, no. 12, pp. 1214–1281, 2008.

[2] F. Perez, A. M. Hujer, K. M. Hujer, B. K. Decker, P. N. Rather, and R. A. Bonomo, "Global challenge of multidrug-resistant *Acinetobacter baumanni*," *Antimicrobial Agents and Chemotherapy*, vol. 51, no. 10, pp. 3471–3484, 2007.

[3] Y. Cai, D. Chai, R. Wang, B. Liang, and N. Bai, "Colistin resistance of Acinetobacter baumannii: clinical reports, mechanisms and antimicrobial strategies," *Journal of Antimicrobial Chemotherapy*, vol. 67, no. 7, pp. 1607–1615, 2012.

[4] P.-E. Fournier, D. Vallenet, V. Barbe et al., "Comparative genomics of multidrug resistance in *Acinetobacter baumannii*," *PLoS Genetics*, vol. 2, no. 1, 2006.

[5] A. Y. Peleg, H. Seifert, and D. L. Paterson, "*Acinetobacter baumannii*: emergence of a successful pathogen," *Clinical Microbiology Reviews*, vol. 21, no. 3, pp. 538–582, 2008.

[6] P. G. Higgins, F. J. Pérez-Llarena, E. Zander, A. Fernández, G. Bou, and H. Seifert, "OXA-235, a novel class D β-lactamase involved in resistance to carbapenems in *Acinetobacter baumannii*," *Antimicrobial Agents and Chemotherapy*, vol. 57, no. 5, pp. 2121–2126, 2013.

[7] S. Brown and S. Amyes, "OXA β-lactamases in Acinetobacter: the story so far," *Journal of Antimicrobial Chemotherapy*, vol. 57, no. 1, pp. 1–3, 2006.

[8] M. Berrazeg, S. Diene, L. Medjahed et al., "New Delhi metallo-beta-lactamase around the world: an ereview using Google Maps," *Eurosurveillance*, vol. 19, no. 20, article 20809, 2014.

[9] A. H. Delcour, "Outer membrane permeability and antibiotic resistance," *Biochim Biophys Acta*, vol. 1794, no. 5, pp. 808–816, 2009.

[10] A. K. Mostachio, I. Van Der Heidjen, F. Rossi, A. S. Levin, and S. F. Costa, "Multiplex PCR for rapid detection of genes encoding oxacillinases and metallo-β-lactamases in carbapenem-resistant *Acinetobacter* spp," *Journal of Medical Microbiology*, vol. 58, no. 11, pp. 1522–1524, 2009.

[11] B. A. Evans and S. G. B. Amyes, "OXA β-lactamases," *Clinical Microbiology Reviews*, vol. 27, no. 2, pp. 241–263, 2014.

[12] J. F. Turton, M. E. Ward, N. Woodford et al., "The role of IS*Aba1* in expression of OXA carbapenemase genes in *Acinetobacter baumannii*," *FEMS Microbiology Letters*, vol. 258, no. 1, pp. 72–77, 2006.

[13] Clinical and Laboratory Standards Institute (CLSI), "Performance Standards for Antimicrobial Susceptibility," January 2014, http://ncipd.org/control/images/NCIPD_docs/CLSI_M100-S24.pdf.

[14] A. M. Al Jarousha, A. H. El Jadba, A. S. Al Afifi, and I. A. El Qouga, "*Acinetobacter baumannii* infection in the neonatal intensive care unit," *Iranian Journal of Public Health*, vol. 37, no. 3, pp. 107–112, 2008.

[15] A. Abbo, S. Navon-Venezia, O. Hammer-Muntz, T. Krichali, Y. Siegman-Igra, and Y. Carmeli, "Multidrug-resistant *Acinetobacter baumannii*," *Emerging Infectious Diseases*, vol. 11, no. 1, pp. 22–29, 2005.

[16] R. Zarrilli, M. Crispino, M. Bagattini et al., "Molecular epidemiology of sequential outbreaks of *Acinetobacter baumannii* in an intensive care unit shows the emergence of carbapenem resistance," *Journal of Clinical Microbiology*, vol. 42, no. 3, pp. 946–953, 2004.

[17] M. El-Sayed-Ahmed, M. A. Amin, W. M. Tawakol, L. Loucif, S. Bakour, and J.-M. Rolain, "High prevalence of bla_{NDM-1} carbapenemase-encoding gene and 16S rRNA *armA* methyltransferase gene among *Acinetobacter baumannii* clinical isolates in Egypt," *Antimicrobial Agents and Chemotherapy*, vol. 59, no. 6, pp. 3602–3605, 2015.

[18] P. Nowak, P. M. Paluchowska, and A. Budak, "Co-occurrence of carbapenem and aminoglycoside resistance genes among multidrug-resistant clinical isolates of Acinetobacter baumannii from Cracow, Poland," *Medical science monitor basic research*, vol. 20, pp. 9–14, 2014.

[19] C. Camp and O. L. Tatum, "A review of *Acinetobacter baumannii* as a highly successful pathogen in times of war," *Laboratory Medicine*, vol. 41, no. 11, pp. 649–657, 2010.

[20] eBURST V3, August 2014, Available from http://eburst.mlst.net.

18

Prevalence and Antimicrobial Susceptibility of Bacterial Uropathogens Isolated from Pediatric Patients at Yekatit 12 Hospital Medical College, Addis Ababa, Ethiopia

Yamirot Merga Duffa,[1,2] Kumera Terfa Kitila,[3] Dereje Mamuye Gebretsadik,[3] and Adane Bitew[2]

[1]*Yekatit 12 Hospital Medical College, Departments of Microbiology, Addis Ababa, Ethiopia*
[2]*Addis Ababa University, College of Health Sciences, Department of Medical Laboratory Sciences, Addis Ababa, Ethiopia*
[3]*Ethiopia Public Health Institute (EPHI), Addis Ababa, Ethiopia*

Correspondence should be addressed to Yamirot Merga Duffa; ymerga@gmail.com

Academic Editor: Joseph Falkinham

Background. Urinary tract infection (UTI) is considered as the most common bacterial infection seen among the pediatric patients. *Objective.* This study was carried out in order to determine the prevalence of urinary tract infection in pediatric patients, identify bacterial uropathogens responsible for the infection, and study the antibiotic sensitivity patterns of bacterial isolates. *Materials and Methods.* A cross-sectional study designed and conducted from January to April 2014. Clean-voided midstream urine specimens were obtained from 384 pediatric patients less than or equal to 15 years in sterile universal bottles. Urine collected from each patient was inoculated onto CLED and blood agar plates using a calibrated inoculating loop with a capacity of 0.001 ml. Inoculated plates were incubated for 24–48 hours at 37°C at inverted position aerobically. Bacterial isolates were indentified and characterized by Gram stain and by using an array of standard routine biochemical tests. The antimicrobial susceptibility test was carried out by using the Kirby–Bauer disc diffusion method. Frequency distribution tables were used to describe the findings. Logistical regression was also used to estimate crude odds ratio (COR) with 95% confidence interval (CI) of positive responses to the different variables, and P values less than 0.05 were taken as statistically significant. *Result.* In this study, a total of 384 patients (199 males and 185 females) aged less than or equal to 15 years from whom urine samples were collected were enrolled. Of these patients, 61 (15.9%) had significant bacteriuria. Of the 185 females, 36 (19.5%) came up with positive cultures, while 25 (12.6%) of the 199 males had significant bacteriuria, and the largest number of study subjects were below the age of 3 years, and the largest positive culture was obtained from this age group, accounting for 35 (57.4%.) out of 61 positive cultures. Bacterial species belonging to six genera were isolated and identified from 61 positive cultures, and the genera were *Escherichia*, *Klebsiella*, *Staphylococcus*, *Proteus*, *Acinetobacter*, and *Enterococcus*. *E. coli* was isolated in 28 cases (49.5 %), followed by *Klebsiella* spp. in 17 cases (27.9%), *Staphylococcus* spp. in 5 patients (8.2%.) (*S. aureus* in one and coagulase-negative staphylococci in 4 cases), *Enterococcus* in 7 cases (11.5%), *Proteus* spp. in 3 cases (4.9%), and *Acinetobacter* in one case (1.6%). Of the bacterial isolates, *E. coli* was found out to be the most common pathogen followed by *Klebsiella* spp. Furthermore, *E. coli* and *Klebsiella* spp. were the most common pathogens in female patients accounting for 71.4% and 64.7%, respectively. Regarding susceptibility tests, *E. coli* and *Klebsiella* spp. were not 100% susceptible to any of the 11 antibiotics tested. *Acinetobacter* spp. had 100% resistance to three antibiotics: gentamicin (GN), trimethoprim-sulfamethoxazole (SXM), and augmentin (AMP). But they were 100% susceptible to ciprofloxacin (CIP), cefuroxime (CXM), norfloxacin (NOR), and ceftazidime (CAZ). On the contrary, *Proteus* spp. was 100% sensitive to all drugs except to nitrofurantoin. Species of *Enterococcus* had resistance of 71.4% to chloramphenicol (C) and 85.7% to both SXM and erythromycin. *S. aureus* was 100% susceptible to almost all drugs, while coagulase-negative staphylococci were not as susceptible as *S. aureus*. Multidrug resistance to two or more drugs was observed in 73.7% of the bacterial isolates. *Conclusion.* This study determined the prevalence of urinary tract infection in pediatric patients and highlighted the major bacterial uropathogens involved in UTI for the first time in the country. Furthermore, bacterial pathogen species and their frequency was consistent with the usually reported pattern, with *E. coli* being the most common organism isolated in cases of urinary tract infections followed by *Klebsiella* spp. Most of the bacterial isolates were multidrug resistant, and it is therefore suggested that appropriate antimicrobials should be administered to reduce the risk of multidrug resistant organisms developing and avert ineffectiveness of antibiotics. This condition indicates that antibiotic selection should be based on knowledge of the local prevalence of bacterial organisms and antibiotic sensitivities rather than empirical treatment. The present study indicated that ciprofloxacin (CIP), ceftazidime (CAZ), cefotaxime (CTX), cefuroxime (CXM), clindamycin (DA), and ceftriaxone (CRO) were the best antibiotics for the treatment of Gram-negative and Gram-positive bacterial uropathogens, respectively, in the study area relatively.

1. Background

Urinary tract infections (UTIs) are a common and important clinical problem in childhood and may lead to systemic illness and renal injury in the short term, and repeated infections, renal scarring, hypertension, and end-stage renal dysfunction may develop [1]. It is a serious health problem, and it has been estimated that about six million patients visit outpatient departments and about 300,000 are treated in the wards every year for UTI worldwide [2]. UTI is also a common and important clinical problem in pediatrics, as recurrent UTIs may lead to renal scarring, hypertension, and end-stage renal dysfunction later in life [3]. UTI occurs in 3% to 5% of girls and 1% of boys during childhood, with the first attack occurring in girls by 5 years, peaking during infancy and toilet training, while it is common in boys during the first year of life, especially among those who are uncircumcised [4]. UTI is caused mainly by bacteria, although viruses, fungi, and parasites can also cause urinary tract infections. Among bacteria, Gram-negative organisms are most commonly isolated from urine samples of children with *Escherichia coli* (*E. coli*) accounting for 70% to 90% of infections [5]. *Klebsiella* species, *Proteus mirabilis*, *Pseudomonas aeruginosa*, *Acinetobacter*, and *Serratia* are other Gram-negative bacteria isolated from pediatric patients. However, only 10% of the cases are caused by Gram-positive bacteria and include *Enterococcus*, *Staphylococcus*, and *Streptococcus agalactiae* [6].

In recent years, widespread use of antibiotics has resulted in an increasing incidence of antibiotic resistance among the urinary tract pathogens all over the world. Worldwide, emergence of antibiotic resistance is increasing among the urinary pathogens [7]. More than 80% of bacterial strains causing urinary tract infections in developing countries are now resistant to trimethoprim or trimethoprim-sulfamethoxazole [8].

Urinary tract infection (UTI) is one of the most common bacterial infections encountered by clinicians in developing countries and the cause of significant morbidity and mortality [9]. Several studies from the African continent have investigated the profile of common uropathogens and the pattern of their susceptibility to commonly used antimicrobial agents in order to guide the choice of empiric therapy. These studies reported the emergence of antibiotic-resistant Gram-negative bacilli with special emphasis on ESBL-producing isolates [10–12].

However, to the best of our knowledge, no such studies have been published describing the spectrum of uropathogens in pediatric patients and their antimicrobial susceptibility patterns in Ethiopia.

Therefore, this research was conducted with the aim of investigating the prevalence and patterns of antibiotic resistance of uropathogens among pediatric patients less than or equal to fifteen years at Yekatit 12 Teaching Medical Hospital, Addis Ababa, Ethiopia.

2. Materials and Methods

2.1. Study Area. The study was conducted at Yekatit 12 Hospital Medical College, which is located in the center of Addis Ababa. Yekatit 12 Teaching Hospital is one of the hospitals under Addis Ababa City Administration Health Bureau that has been giving routine health services for the city community and other referral cases from different regional states of Ethiopia. It is a tertiary level referral and teaching hospital in Addis Ababa that serves for a large number of people from the surrounding zones and nearby regions both for inpatient and outpatient treatment. This teaching hospital consists of an operating room, intensive care unit (ICU), and pediatric department with 12 beds, 13 wards with 327 beds, and outpatient departments.

2.2. Study Design and Period. It was a cross-sectional study designed and conducted with the objective of isolating, characterizing uropathogens in pediatric patients, and determining the antimicrobial susceptibility pattern of the isolates. The study was conducted prospectively from January to April 2014.

2.3. Source of Population. All pediatric patients not greater than 15 years of age attending Yekatit 12 Hospital Medical College were the source of population.

2.4. Study Population. All pediatric patients less than or equal to 15 years of age who were sent to the laboratory for urinalysis were included in the study.

2.5. Inclusion and Exclusion Criteria. Pediatric patients less than or equal to 15 years of age with urinalysis were included in the study, whereas pediatric patients who were taking antibiotics and mothers/guardians of pediatrics who did not volunteer for written consents were excluded.

2.6. Sample Size Determination and Sampling Technique. Since data were not available on pediatric bacterial UTI in Ethiopia, 50% of population proportion was used to determine the sample size based on single population proportion, and the level of precision (d) was 0.05. Urine specimens were collected from each pediatric patient. The study participants' parents or guardians were given appropriate instructions of clean catch midstream urine sample collection before providing urine samples, and immediately after collection, the samples were brought to microbiology laboratory for further process. Sterile plastic bags were used for admitted pediatric patients, and the bags were checked every 15 minutes for urine samples. Accordingly, about 10 ml urine specimen was collected from each patient in a sterile screw-capped, wide-mouthed container labelled with unique sample numbers, date, and time of collection. Immediately, they were delivered to the bacteriology laboratory and processed.

The antimicrobial susceptibility test was carried out by the Kirby–Bauer disc diffusion method as per Clinical Laboratory Standards Institute (CLSI—formerly NCCLS) guidelines on Mueller–Hinton agar (Oxoid, Basingstoke, Hampshire, England [13]). Briefly, four to five bacterial colonies were inoculated into 5 ml of Soyabean Casein Digest Broth (Oxoid, Basingstoke, Hampshire, England) and incubated at 37°C for 12 hours. After an overnight incubation, the growth suspension was prepared in 0.5 ml of the same broth medium, and the turbidity was adjusted to match that of 0.5 McFarland standards to obtain approximately the organism number of 1×10^6 colony-forming units (CFU) per ml. Then, a sterile swab was dipped into this suspension, and the excess of inocula was removed by pressing it against the sides of the tube. Then, the swab was applied to the center of Mueller–Hinton agar plate (Oxoid, Basingstoke, Hampshire, England) and evenly spread on the medium. Antibiotic discs were placed after 15 minutes of inoculation to Mueller–Hinton agar seeded with each isolate and were incubated for 24 hours at 37°C. The diameter of the zone of inhibition around the disc was measured using a sliding metal caliper. Interpretation criteria were those of Clinical Laboratory Standards Institute (CLSI) guidelines. The following drugs and concentrations were used to determine the antibiogram of the strains: penicillin (10 U), augmentin (30 μg), trimethoprim-sulfamethoxazole (1.25/23.75 μg), clindamycin (30 μg), gentamicin (30 μg), ciprofloxacin (5 μg), erythromycin (15 μg), chloramphenicol (30 μg), cefuroxime (30 μg), cephalothin (10 μg), ceftazidime (30 μg), cefotaxim (30 μg), ceftriaxone (10 μg), oxacillin (1 μg), nitrofurantoin (30 μg), nalidixic acid (30 μg), and norfloxacin (30 μg). All antibiotic disks were purchased from Oxoid Limited Company, United Kingdom.

2.7. Quality Control Assurance. The reliability of the study findings was guaranteed by implementing quality control measures throughout the whole process of the laboratory work. Staining reagents, culture media, and antibiotic discs were checked for their normal shelf life before use. All culture plates and antibiotic discs were stored at recommended refrigeration temperature (2–8°C) after preparation. Reference strains of *S. aureus* ATCC 25923, *E. coli* ATCC 25922, and *P. aeruginosa* ATCC 27853 were used as controls. In general, all laboratory procedures were done based on recommended standard laboratory procedures by strictly following preanalytical, analytical, and postanalytical stages of quality assurance that are incorporated in standard operational procedures (SOPs) of microbiology laboratory of Yekatit 12 Hospital Medical College.

2.8. Data Analysis. Data were cleaned, entered, and analyzed using SPSS, version 17. Descriptive statistics was computed for most of the study variables, and frequency distribution tables were used to describe the findings. Logistical regression was also used to estimate crude odds ratio (COR) with 95% confidence interval (CI) of positive responses to the different variables, and P values less than 0.05 were taken as statistically significant when looking for associations between dependent and independent variables.

3. Results

3.1. Sociodemographic Characteristics of Patients. There were 384 (199 males and 185 females) pediatric patients less than or equal to 15 years, from whom urine samples were collected. Of these patients, 61 (15.9%) had significant bacteriuria. Of the 185 females, 36 (19.5%) came up with positive cultures, while 25 (12.6%) of the 199 males had significant bacteriuria, indicating that female patients were more affected than male patients. As can be seen in Table 1, the largest numbers of study subjects were below the age of 3 years, and the largest positive culture was obtained from this age group accounting for 35 (57.4%.) positive cultures out of 61 positive cultures (Table 1).

Bacterial species belonging to six genera were isolated and identified from 61 positive cultures, and the genera were *Escherichia*, *Klebsiella*, *Staphylococcus*, *Proteus*, *Acinetobacter*, and *Enterococcus*; *E. coli* was isolated from 28 cases (49.5 %), followed by *Klebsiella* spp. from 17 cases (27.9%), *Staphylococcus* spp. from 5 patients (8.2 %.), *S. aureus* from one case and coagulase-negative staphylococci from 4 cases, *Enterococcus* from 7 cases (11.5%), *Proteus* spp. from 3 cases (4.9%), and *Acinetobacter* from one case (1.6%). Of the bacterial isolates, *E. coli* was found to be the most common pathogen followed by *Klebsiella*. Furthermore, *E. coli* and *Klebsiella* were the most common pathogens in female patients accounting for 71.4% and 64.7%, respectively (Table 2).

Isolation rate of UTI with regard to circumcision revealed that out of 178 circumcised male study subjects, 10.1% (18/178) were culture-positive, but from a total of 21 uncircumcised study subjects, 33.3% (7/21) were culture-positive. Circumcision was significantly associated (COR, 95% CI: 0.225 (0.080–0.630), $P = 0.005$) with urinary tract infection (Table 3).

3.2. Antibiotic Sensitivity Testing. The drug susceptibility profile of bacterial isolates is given in Tables 4 and 5. *E. coli* and *Klebsiella* spp. were not 100% susceptible to any of the 11 antibiotics tested. On the one hand, *Acinetobacter* spp. had 100% resistance to three antibiotics: gentamicin, trimethoprim-sulfamethoxazole, and augmentin, but they were 100%

TABLE 1: Distribution of bacterial isolates in relation to age group among pediatric patients at Yekatit 12 Hospital Medical College, Addis Ababa, Ethiopia, 2014.

Age in years	Males with positive culture	Females with positive culture	Total	%
<3	16 (*n* = 110)	19 (*n* = 83)	35	57.4
3–6	1 (*n* = 13)	3 (*n* = 20)	4	6.6
6–9	1 (*n* = 26)	8 (*n* = 30)	9	14.8
9–12	2 (*n* = 24)	4 (*n* = 29)	6	9.8
12–15	5 (*n* = 22)	2 (*n* = 33)	7	11.5
Total	25 (40.9%)	36 (59.01%)	*N* = 61	100

n: number of patients in each age group; *N*: total number of positive male and female patients in each age group.

TABLE 2: Distribution of bacterial isolates in relation to sex among pediatric patients at Yekatit 12 Hospital Medical College, Addis Ababa, Ethiopia, 2014.

Bacterial isolates	Female	Male	Total
E. coli	20/28 (71.4%)	8/28 (28.6)	28/61 (45.9%)
Klebsiella spp.	11/17 (64.7%)	6/17 (35.3)	17/61 (27.9%)
Proteus spp.	—	3/3 (100%)	3/61 (4.9%)
Acinetobacter spp.	—	1/1 (100%)	1/61 (1.6%)
Enterococcus spp.	4/7 (57.1)	3/7 (42.9%)	7/61 (11.5)
S. aureus	1/1 (100%)	—	1/61 (1.6%)
CONS	—	4/4 (100%)	4/61 (6.6%)
Total	36 (59.0)	25 (40.9%)	61 (100%)

CONS: coagulase-negative staphylococci.

TABLE 3: Association of UTI in relation to circumcision among pediatric patients at Yekatit 12 Hospital Medical College, Addis Ababa, Ethiopia, 2014 (*n* = 199).

Circumcision	UTI		COR 95% CI	*P* value
	No	Yes		
No	14 (66.6%)	7 (33.3%)	1	—
Yes	160 (89.9%)	18 (10.1%)	0.225 (0.080–0.630)*	0.005

*Significant at *P* value <0.05; COR: crude odds ratio.

susceptible to ciprofloxacin, cefuroxime, norfloxacin, and ceftazidime. On the other hand, *Proteus* spp. were 100% sensitive to all drugs except nitrofurantoin (Table 4).

Also, a high degree of resistance was observed in *Enterococcus* spp. Species of *Enterococcus* had resistance of 71.4 to chloramphenicol and 85.7% to both trimethoprim-sulfamethoxazole and erythromycin (Table 5). As shown in Table 5, *S. aureus* was 100% susceptible to almost all drugs while coagulase-negative staphylococci were not as susceptible as *S. aureus*.

Multidrug resistance (MDR) to two or more drugs was observed in 11/12 (91.7%) and 34/49 (69.4%) of Gram-positive and Gram-negative bacteria, respectively, (Tables 6 and 7). The overall prevalence of MDR in both groups was 45/61 (73.7%).

4. Discussion

The present study was conducted to determine the prevalence of UTIs, to evaluate the bacterial agents involved in UTI, and to determine the drug susceptibility profile of bacterial uropathogens. A total of 384 study participants were enrolled in the present study, of whom, 199 (51.8%) were males and 185 (48.2%) were females. The ages of study subjects ranged from 0 to 15 years.

Out of the 384 study subjects (185 female and 199 males between the ages of 0 and 15 years) who participated in this study, only 61 (15.8%) had urine samples with significant bacteriuria. The findings of this study were similar to those reported in Nigeria [14], Iran [15], and Egypt [16], where prevalence rates of UTI documented were 11.96%, 16.2%, and 15.05%, respectively. However, the result obtained in this study (15.8%) appeared to be higher when compared with those reported in Philadelphia [17], USA [6], India [18], and Nigeria [19]. The prevalence rates reported were 3.3%, 7.0%, 4%, and 3.0%, respectively. On the contrary, the prevalence rate of UTI obtained in this study was much less than that reported by Olowu (1996), which was 28.1%.

The present study revealed a higher incidence of urinary tract infection in females than in males. Of the 185 females, 36 (36/185; 19.5%) came up with positive cultures, while 25 (25/199; 12.6%) of the 199 males had significant bacteriuria indicating that female patients were more affected than male patients. Furthermore, the largest numbers of study subjects were patients below the age of 3 years. Among the culture-positive study subjects, 57.4% of them were below 3 years of age. Our result in this regard was comparable to that reported in Nigeria [14]. The higher incidence of urinary tract infections in females might be a result of shorter urethra and the proximity of their reproductive organ to the anus. UTI occurs in 3% to 5% of girls and 1% of boys during childhood, with the first attack occurring in girls by 5 years, peaking during infancy and toilet training, while it is common in boys during the first year of life, especially among those who are uncircumcised because during this time of age, they get exposed to pathogens easily [15, 18].

Sixty-one bacterial isolates were recovered from this study. They were *E. coli*, *Klebsiella* spp., *Proteus* spp., *Enterococcus* spp., *Acinetobacter* spp., *S. aureus*, and coagulase-negative staphylococci. Of these, *E. coli* was the most common organism isolated from patients with significant bacteriuria and was isolated from 28 (45.9%) followed by *Klebsiella* spp. from 17 cases (27.9%), *Enterococcus* spp from 7 cases (11.5%), coagulase-negative staphylococci from 4 cases (6.6%), *Proteus* spp. from 3 cases (4.9%), and *S. aureus* from 1 case (1.6%). The pattern and the frequency of bacterial isolates obtained in this study were comparable with different study findings done in different countries. For

Table 4: The number and percent of resistant strains of Gram-negative uropathogens among pediatric patients at Yekatit 12 Hospital Medical College, Addis Ababa, Ethiopia, 2014.

Bacteria isolated	Total no.	Antimicrobial agents tested										
		CIP No. (%)	CXM No. (%)	GN No. (%)	NA No. (%)	SXT No. (%)	F No. (%)	AMP No. (%)	C No. (%)	NOR No. (%)	CTX No. (%)	CAZ No. (%)
E. coli	28	6 (21.4)	9 (32.1)	8 (28.6)	9 (32.1)	19 (67.9)	5 (17.9)	11 (39.3)	8 (28.6)	7 (25.0)	9 (32.1)	9 (32.1)
Klebsiella spp.	17	5 (29.4)	15 (88.2)	14 (82.4)	9 (52.9)	15 (88.2)	5 (29.4)	12 (70.6)	10 (58.8)	6 (35.3)	15 (88.2)	15 (88.2)
Acinetobacter spp.	1	0	0	1 (100)	1 (100)	1 (100)	0	1 (100)	0	0	0	0
Proteus spp.	3	0	0	0	0	0	3 (100)	0	0	0	0	0
Total	**49**	**11 (22.5)**	**24 (48.9)**	**23 (46.9)**	**19 (38.7)**	**35 (71.4)**	**13 (26.5)**	**24 (48.9)**	**18 (36.7)**	**13 (26.5)**	**24 (48.9)**	**24 (48.9)**

AMP: augmentin (30 μg); NOR: norfloxacin (30 μg); GN: gentamicin (30 μg); CIP: ciprofloxacin (5 μg); NA: nalidixic acid (30 μg); F: nitrofurantoin (30 μg); CAZ: ceftazidime (30 μg); CTX: cefotaxim (30 μg); C: chloramphenicol (30 μg); SXT: trimethoprim-sulfamethoxazole (1.25/23.75 μg); CXM: cefuroxime (30 μg).

Table 5: The number and percent of resistant strains of Gram-positive uropathogens among pediatric patients at Yekatit 12 Hospital Medical College, Addis Ababa, Ethiopia, 2014.

Bacteria isolated	Total no.	Antimicrobial agents tested												
		P No. (%)	CIP No. (%)	C No. (%)	E No. (%)	OX No. (%)	CXM No. (%)	GN No. (%)	KF No. (%)	SXT No. (%)	F No. (%)	DA No. (%)	AMP No. (%)	CRO No. (%)
Enterococcus spp.	7	0	1 (14.3)	5 (71.4)	6 (85.7)	—	3 (42.9)	0	3 (42.9)	6 (85.7)	1 (14.3)	0	3 (42.9)	3 (42.9)
CONS	4	0	1 (25.0)	1 (25.0)	4 (100)	—	4 (100)	0	0	4 (100)	0	0	1 (25.0)	1 (25.0)
S. aureus	1	1 (100)	0	0	1 (100)	0	0	0	1 (100)	0	0	0	1 (100)	0
Total	**12**	**1 (8.3)**	**2 (16.6)**	**6 (50)**	**11 (91.6)**	**0**	**7 (58.3)**	**0**	**4 (33.3)**	**10 (83.3)**	**1 (8.3)**	**0**	**5 (41.6)**	**4 (33.3)**

CONS: coagulase-negative staphylococci; P: penicillin (10 U); AMP: augmentin (30 μg); KF: cephalothin (10 μg); DA: clindamycin (30 μg); CRO: ceftriaxone (10 μg); GN: gentamicin (30 μg); E: erythromycin (15 μg); CIP: ciprofloxacin (5 μg); OX: oxacillin (1 μg); F: nitrofurantoin (30 μg); C: chloramphenicol (30 μg); SXT: trimethoprim-sulfamethoxazole (1.25/23.75 μg); CXM: cefuroxime (30 μg).

Table 6: Multidrug resistance pattern of Gram-negative bacteria isolated from pediatric patients at Yekatit 12 Hospital Medical College, Addis Ababa, Ethiopia, 2014.

Combination of antibacterial agent	*E. coli* no. (%)	*Klebsiella* spp. no. (%)	*Acinetobacter* spp. no. (%)	Total no. (%)
AMP, SXT	3 (16.6)	—	—	3 (8.8)
C, SXT	3 (16.6)	—	—	3 (8.8)
CIP, NOR, SXT	1 (5.6)	—	—	1 (2.94)
AMP, SXT, NA, GN	1 (5.6)	—	1 (100)	2 (5.9)
AMP, CIP, SXT, NOR, GN, CAZ, CTX, CXM	4 (22.2)	—	—	4 (11.8)
AMP, SXT, NA, GN, CAZ, CTX, CXM, C	4 (22.2)	4 (26.7)	—	8 (23.5)
CIP, SXT, NA, GN, CAZ, CTX, CXM, NOR, F	1 (5.6)	5 (33.3)	—	6 (17.6)
AMP, SXT, NA, GN, CAZ, CTX, CXM, NOR, F	1 (5.6)	6 (40)	—	7 (20.6)
Total	**18 (100)**	**15 (100)**	**1 (100)**	**34 (100)**

AMP: augmentin (30 μg); NOR: norfloxacin (30 μg); GN: gentamicin (30 μg); CIP: ciprofloxacin (5 μg); NA: nalidixic acid (30 μg); F: nitrofurantoin (30 μg); CAZ: ceftazidime (30 μg); CTX: cefotaxim (30 μg); C: chloramphenicol (30 μg); SXT: trimethoprim-sulfamethoxazole (1.25/23.75 μg); CXM: cefuroxime (30 μg).

example, a study conducted in Iran [15] showed that, among bacterial isolates, 40% accounted for *E. coli*, and this was followed by *Klebsiella* spp. (17.9%), coagulase-negative staphylococci (12.3%), *Enterococcus* spp. (8.7%), *Pseudomonas* spp. (6.7%), and *S. aureus* (3.6%). A study conducted in Nigeria [14] reported a total of 36 bacterial isolates, of which, *E. coli* was the predominant organism and was responsible for 52.77% of the cases of UTI, and this was followed by *Klebsiella* spp. (25%), *Proteus mirabilis* (13.89%), *Streptococcus faecalis* (5.56%), and *Pseudomonas aeruginosa* (2.78%). Similarly, a study conducted in Egypt [16] indicated that, among all the bacteria isolated in their study, *E. coli* and *Klebsiella* spp. were the most commonly isolated bacteria accounting for 58.1% and 41.9%, respectively This study was

Table 7: Multidrug resistance pattern of Gram-positive bacteria isolated from pediatric patients at Yekatit 12 Hospital Medical College, Addis Ababa, Ethiopia, 2014.

Combination of antibacterial agent	*Enterococcus* spp. no. (%)	*S. aureus* no. (%)	CONS no. (%)	Total no. (%)
P, AMP, KF	—	1 (100)	—	1 (9.09)
P, SXT, E, CXM	—	—	3 (75)	3 (27.2)
P, SXT, E, C	2 (33.3)	—	—	2 (18.2)
P, SXT, E, C, AMP	1 (16.7)	—	1 (25)	2 (18.2)
CXM, E, F, AMP, SXT	1 (16.7)	—	—	1 (9.09)
CXM, E, F, KF, CRO, SXT, C	1 (16.7)	—	—	1 (9.09)
C, E, F, KF, CRO, SXT	1 (16.7)	—	—	1 (9.09)
Total	6 (100)	1 (100)	4 (100)	11 (100)

CONS: coagulase-negative staphylococci; P: penicillin (10 U); AMP: augmentin (30 μg); KF: cephalothin (10 μg); DA: clindamycin (30 μg); CRO: ceftriaxone (10 μg); GN: gentamicin (30 μg); E: erythromycin (15 μg); CIP: ciprofloxacin (5 μg); OX: oxacillin (1 μg); F: nitrofurantoin (30 μg); C: chloramphenicol (30 μg); SXT: trimethoprim-sulfamethoxazole (1.25/23.75 μg); CXM: cefuroxime (30 μg).

comparable with the study conducted in different areas [20, 24, 26].

A study conducted in Philadelphia [17] indicated uncircumcision as one of the risk factors for UTI; a study in USA [6] also showed that the prevalence of UTI was less in circumcised males (2.4%) than in uncircumcised males (20.1%). This study has also showed that circumcised (10.1%) male patients were less susceptible to bacterial pathogens than were uncircumcised (33.3%) patients.

The drug susceptibility profile of Gram-negative and Gram-positive bacteria tested in the present study was variable. For instance, *E. coli* and *Klebsiella* spp. were not 100% susceptible to any of the 11 antibiotics tested. However, percentage resistance of *Klebsiella* was much higher when compared with that of *E. coli*. Eighty-eight percent of *Klebsiella* spp. was resistant to cefotaxim, ceftazidime, trimethoprim-sulfamethoxazole, and cefuroxime. The drug susceptibility profile of *E. coli* and *Klebsiella* spp. obtained in the present study was similar with the findings in Ilorin [20]. However, drug susceptibility profile of *E. coli* in this study strongly contradicts with the findings reported in Tanzania [21]. On the one hand, *Acinetobacter* spp. had 100% resistance to four antibiotics: gentamicin, trimethoprim-sulfamethoxazole, augmentin, and nalidixic acid and 100% susceptible to ciprofloxacin, cefuroxime, norfloxacin, cefotaxim, chloramphenicol, and ceftazidime. The present finding with regard to these isolated pathogens strongly contradicted with the findings in Iran [15]. On the other hand, *Proteus* spp. was 100% sensitive to all drugs except nitrofurantoin. Also, a high degree of resistance was observed in *Enterococcus* spp. *Enterococcus* spp. were 100% sensitive to penicillin but were resistant to chloramphenicol by 71.4% and by 85.7% to both trimethoprim-sulfamethoxazole and erythromycin. Similarly, *S. aureus* was 100% susceptible to almost all drugs, while coagulase-negative staphylococci were not as susceptible as *S. aureus*. This study also showed that both Gram-positive and Gram-negative bacteria had multidrug resistance (MDR) to two or more drugs with an overall prevalence of 45/61 (73.7%) in both groups. This finding was also supported by different study findings conducted in different areas [27–30]. Finally, as a limitation, the lack of such similar studies in Ethiopia for the purposes of comparison and absence of some biochemical reagents and antibiotic discs might influence the appreciation of complete profile of isolates. Serotyping and genotyping of bacterial isolates especially multidrug-resistant organisms were not possible due to limited laboratory facility.

5. Conclusion and Recommendation

This study determined the prevalence of urinary tract infection in pediatric patients and highlighted the major bacterial uropathogens involved in UTI for the first time in the country. Furthermore, bacterial pathogen species and their frequency was consistent with the usually reported pattern, with *E. coli* being the most common organism isolated in cases of urinary tract infections followed by *Klebsiella* spp. Most of the bacterial isolates were multidrug resistant, and it is therefore suggested that appropriate antimicrobials should be administered to reduce the risk of multidrug-resistant organisms developing and avert ineffectiveness of antibiotics. A widespread screening program for UTI in pediatric should be implemented to know the exact prevalence and to measure resistance rate of uropathogens to commonly used antibiotics.

Finally, the authors recommended that treatments of bacterial uropathogens indicate that antibiotic selection should be based on knowledge of the local prevalence of bacterial organisms and antibiotic sensitivities rather than empirical treatment. The present study indicated that ciprofloxacin (CIP), ceftazidime (CAZ), cefotaxime (CTX), cefuroxime (CXM), clindamycin (DA), and ceftriaxone (CRO) were the best antibiotic treatment against Gram-negative and Gram-positive bacterial uropathogens, respectively, in the study area relatively.

Abbreviations

AST: Antimicrobial susceptibility testing
BA: Blood agar
CLED: Cystine-lactose-electrolyte deficient
CI: Confidence interval
CONS: Coagulase-negative staphylococci
CLSI: Clinical Laboratory Standards Institute
DMLT: Department of Medical Laboratory Technology
ESBL: Extended-spectrum β-lactamase
KIA: Kligler iron agar

IRB: Institutional Review Board
MSU: Midstream urine samples
NCCLS: National Committee for Clinical Laboratory Standards
OPD: Outpatient Department
QC: Quality control
SOPs: Standard operating procedures
Spp: Species
SPSS: Statistical Package for Social Sciences
UTI: Urinary tract infection
USA: United States of America
WHO: World Health Organization.

Consent

Informed written consents were obtained from mothers/guardians of pediatric patients before data collection. The respondent was given the right to refuse to take part in the study as well as to withdraw at any time during the study period. All the information obtained from the study subjects was coded to maintain confidentially. When the participants are found to be positive for urine culture, they were informed by the hospital clinician and received proper treatment.

Conflicts of Interest

The authors declare that they have no conflicts of interest.

Authors' Contributions

YMD conceived and designed the study protocol, collected data, performed analysis, and interpreted data. All of the authors critically reviewed and approved the manuscript.

Acknowledgments

The authors would like to acknowledge Addis Ababa University, School of Medical Laboratory Sciences, for creating and facilitating this research work opportunity. The authors express their heartfelt thanks to the study participants for their willingness and being involved in this study. Lastly, the authors also acknowledge staff members of the hospitals that helped them in data collection, preparation, and laboratory examination activities.

References

[1] B. Mathur, H. S. Agarwal, and A. Maria, "Acute renal failure in neonatal sepsis," *Indian Journal of Pediatrics*, vol. 73, no. 6, pp. 499–502, 2006.

[2] E. Akortha and K. Ibadin, "Incidence and antibiotic susceptibility pattern of *Staphylococcus aureus* among patients with urinary tract infection in UBTH Benin City, Nigeria," *African Journal of Biotechnology*, vol. 7, no. 11, pp. 1637–1640, 2008.

[3] J. Joseph, A. Darcie, and N. Kiddoo, "Diagnosis and management of pediatric urinary tract infections," *Clinical Microbiology Reviews*, vol. 18, no. 2, pp. 417–422, 2005.

[4] J. A. Owa, J. C. Azubuike, and K. E. Nkangineme, "Urinary tract infections in children," in *Paediatrics and Child Health in a Tropical Region*, pp. 480-481, African Educational Services, Owerri, Nigeria, 1999.

[5] R. E. Behrman, R. M. Kliegman, and H. B. Jenson, "Urinary tract infection," in *Nelson Textbook of Pediatrics*, vol. 17, pp. 1621–1625, Elsevier, Philadelphia, PA, USA, 2004.

[6] N. Shaikh, N. E. Morone, J. E. Bost, and M. H. Farrell, "Prevalence of urinary tract infection in childhood: a meta-analysis," *Pediatric Infectious Disease Journal*, vol. 27, no. 4, pp. 302–308, 2008.

[7] R. Runehagen and G. Kahlmeter, "A 10- year study of the consumption of quinolones trimethoprim and mecillinam in relation to the development of antimicrobial resistance in a large number of species," in *Proceeding of European Congress of Clinical Microbiology and Infectious Diseases, Poster 417*, Amsterdam, Netherlands, July 2002.

[8] G. Kahlmeter, "An international survey of the antimicrobial susceptibility of pathogens from uncomplicated urinary tract infections: the ECO-SENS project," *Journal of Antimicrobial Chemotherapy*, vol. 51, no. 1, pp. 69–76, 2003.

[9] B. Tessema, A. Kassu, A. Mulu, and G. Yismaw, "Predominant isolates of urinary tract pathogens and their antimicrobial susceptibility in Gondar University Teaching Hospital, northwest Ethiopia," *Ethiopian Medical Journal*, vol. 45, no. 1, pp. 61–67, 2007.

[10] O. A. Aboderin, A. R. Abdu, B. W. Odetoyin, and A. Lamikanra, "Antimicrobial resistance in *Escherichia coli* strains from urinary tract infections," *Journal of the National Medical Association*, vol. 101, no. 12, pp. 1268–1273, 2009.

[11] R. Bercion, D. Mossoro-Kpinde, A. Manirakiza, and A. L. Faou, "Increasing prevalence of antimicrobial resistance among *Enterobacteriaceae* uropathogens in Bangui, Central African Republic," *Journal of Infection in Developing Countries*, vol. 3, no. 3, pp. 187–190, 2009.

[12] T. M. Habte, S. Dube, N. Ismail, and A. A. Hoosen, "Hospital and community isolates of uropathogens at a tertiary hospital in South Africa," *South African Medical Journal*, vol. 99, no. 8, pp. 584–588, 2009.

[13] Clinical Laboratory Standards Institute, *2009 Guidelines by CLSI/NCCLS Information Supplement*, Approved Standard M 100-S15, Wayne, PA, USA, 2000..

[14] O. A. Aiyegoro, O. O. Igbinosa, I. N. Ogunmwonyi et al., "Incidence of urinary tract infections (UTI) among children and adolescents in Ile-Ife, Nigeria," *African Journal of Microbiology Research*, pp. 13–19, 2007.

[15] R. Ranjbar, M. T. Haghi-Ashtiani, N. Jonaidi Jafari, and M. Abedini, "The prevalence and antimicrobial susceptibility of bacterial uropathogens isolated from pediatric patients," *Iranian Journal of Public Health*, vol. 38, no. 2, pp. 134–213, 2009.

[16] D. M. Yousssef, H. A. Elfateh, R. Sedeck, and S. Seleem, "Epidemiology of urinary tract infection in neonatal intensive care unit: a single center study in Egypt," *Journal of Academy of Medical Sciences*, vol. 2, no. 1, pp. 25–29, 2012.

[17] K. N. Shaw, M. Gorelick, K. L. McGowan, N. M. Yakscoe, and

J. S. Schwartz, "Prevalence of urinary tract infection in febrile young children in the emergency department," *Pediatrics*, vol. 102, no. 2, p. e16, 1998.

[18] C. Ashok, G. V. Kumar, and H. M. Viswanathakumar, "Study of the prevalence and clinical profile of urinary tract infection in febrile children aged 3–6 years who attended pediatric outpatient department in a tertiary care hospital," *International Journal of Health Sciences and Research*, vol. 3, no. 2, pp. 1–5, 2013.

[19] V. U. Muoneke, M. U. Ibekwe, and R. C. Ibekwe, "Childhood urinary tract infection in Abakaliki: etiological organisms and antibiotic sensitivity pattern," *Annals of Medical and Health Sciences Research*, vol. 2, no. 1, pp. 29–32, 2012.

[20] O. T. Adedoyin, B. O. Oyeymi, and O. V. Aiyedehin, "Screening of febrile children on hospital admission for urinary tract infections (UTI)," *African Journal of Clinical and Experimental Microbiology*, vol. 4, no. 1, pp. 56–62, 2003.

[21] B. P. Msaki, E. Stephen, S. E. Mshana, A. Hokororo, H. D. Mazigo, and M. D. Domenica, "Prevalence and predictors of urinary tract infection and severe malaria among febrile children attending Makongoro health centre in Mwanza city, North-Western Tanzania," *Archives of Public Health*, vol. 70, no. 1, pp. 1–8, 2012.

[22] S. K. Abdulhadi, A. H. Yashua, and A. Uba, "Organisms causing urinary tract infection in paediatric patients at Murtala Muhammad Specialist Hospital, Kano, Nigeria," *Journal of Internal of Biomedical and Health Science*, vol. 4, no. 4, pp. 165–167, 2008.

[23] B. Fantahun and A. Bayeh, "Antimicrobial resistance of bacterial isolates from urinary tract infections at Felge Hiwot Referral Hospital, Ethiopia," *Ethiopian Journal of Health Development*, vol. 23, no. 3, pp. 236–238, 2009.

[24] G. Beyene and T. W. Wondewosen, "Bacterial uropathogens in urinary tract infection and antibiotic susceptibility pattern in Jimma University Specialized Hospital, Southwest Ethiopia," *Ethiopian Journal of Health Sciences*, vol. 21, no. 2, pp. 141–146, 2011.

[25] A. Hoberman, H. P. Chao, D. M. Keller, R. Hickey, H. W. Davis, and D. Ellis, "Prevalence of urinary tract infection in febrile infants," *Journal of Pediatrics*, vol. 123, no. 1, pp. 17–23, 1993.

[26] C. W. Lin, Y. H. Chiou, Y. Y. Chen et al., "Urinary tract infection in neonates," *Journal of Clinical Neuroscience*, vol. 6, no. 2, pp. 1–4, 1999.

[27] E. J. Schoen, C. J. Colby, and G. T. Ray, "Newborn circumcision decreases incidence and costs of urinary tract infections during the first year of life," *Pediatrics*, vol. 105, no. 4, pp. 789–793, 2000.

[28] M. Barton, Y. Bell, M. Thame, A. Nicholson, and H. Trotman, "Urinary tract infection in neonates with serious bacterial infections admitted to the university hospital of the West Indies," *West Indian Medical Journal*, vol. 57, no. 2, pp. 101–105, 2008.

[29] B. T. Van der Meeren, K. D. Chhaganlal, A. Pfeiffer et al., "Extremely high prevalence of multi-resistance among uropathogens from hospitalised children in Beira, Mozambique," *South African Medical Journal*, vol. 103, no. 6, pp. 382–386, 2013.

[30] W. A. Olowu, "Symptomatic bacteriuria in children, Nigeria," *African Journal of Microbiology Research*, vol. 27, pp. 76–80, 1996.

19

Prevalence of Bacterial Vaginosis and Associated Risk Factors among Women Complaining of Genital Tract Infection

Adane Bitew,[1] Yeshiwork Abebaw,[2] Delayehu Bekele,[3] and Amete Mihret[4]

[1] *Department of Medical Laboratory Sciences, College of Health Sciences, Addis Ababa University, P.O. Box 1176, Addis Ababa, Ethiopia*
[2] *Clinical Laboratory, Fitche Hospital, P.O. Box 46, Oromia Administrative Region, Ethiopia*
[3] *Department of Obstetrics and Gynecology, St Paul's Hospital Millennium Medical College, P.O. Box 1271, Addis Ababa, Ethiopia*
[4] *Department of Clinical Bacteriology and Mycology, Ethiopian Public Health Institute, P.O. Box 1242, Addis Ababa, Ethiopia*

Correspondence should be addressed to Adane Bitew; bitewadane@gmail.com

Academic Editor: Karl Drlica

Background. Bacterial vaginosis is a global concern due to the increased risk of acquisition of sexually transmitted infections. *Objectives.* To determine the prevalence of bacterial vaginosis and bacteria causing aerobic vaginitis. *Methods.* A cross-sectional study was conducted among 210 patients between September 2015 and July 2016 at St. Paul's Hospital. Gram-stained vaginal swabs were examined microscopically and graded as per Nugent's procedure. Bacteria causing aerobic vaginitis were characterized, and their antimicrobial susceptibility pattern was determined. *Results.* The overall prevalence of bacterial vaginosis was 48.6%. Bacterial vaginosis was significantly associated with number of pants used per day ($p = 0.001$) and frequency of vaginal bathing ($p = 0.045$). Of 151 bacterial isolates, 69.5% were Gram-negative and 30.5% were Gram-positive bacteria. The overall drug resistance level of Gram-positive bacteria was high against penicillin, tetracycline, and erythromycin. Cefoxitin and tobramycin were the most active drugs against Gram-positive bacteria. The overall drug resistance level of Gram-negative bacteria was high against tetracycline, ampicillin, and amoxicillin. Amikacin and tobramycin were the most active drugs against Gram-negative bacteria. *Conclusions.* The prevalence of bacterial vaginosis was high and was affected by individual hygiene. Routine culture of vaginal samples should be performed on patients with vaginitis and the drug susceptibility pattern of each isolate should be determined.

1. Introduction

Vaginitis is an inflammation of the vagina in which bacterial vaginosis, vulvovaginal candidiasis, and trichomoniasis are the most vaginitides [1]. Nearly 5–10 million females seek gynecologic advice for vaginitis every year worldwide [2].

Bacterial vaginosis is described as a shift in the balance of the vaginal microflora characterized by an increase in the vaginal pH, a reduction in lactobacilli, predominantly hydrogen peroxide producing species, and an increase in facultative and anaerobic bacteria in number and/or type [3]. Although the prevalence of bacterial vaginosis differs widely from country to country within the same region and even within similar population groups, it has been estimated to be in the range of 8% to 75% [4]. Bacterial vaginosis can occur in any age group, but globally it is more prevalent in females of reproductive age [1].

For many years, bacterial vaginosis has received little attention, since it is considered to be a trivial disease. However, it is a morbid disease in terms of loss of working days and treatment cost [2]. Furthermore, it increases the risk of acquiring (i) human immunodeficiency virus [5] and other sexually transmitted infections (STI), such as gonorrhea, trichomoniasis, and herpes simplex virus type 2 (HSV-2) [6, 7], and (ii) miscarriage, preterm labor, preterm delivery, and postpartum complications such as endometritis and wound infections in pregnant women [8–11]. It also increases HIV viral shedding [12, 13].

The diagnosis of bacterial vaginosis can be made using clinical criteria [14] or in the laboratory by scoring bacterial morphotypes from a Gram stain of vaginal fluid, where a score of 0–3 represented normal vaginal flora, a score of 4–6 represented intermediate vaginal flora, and a score of 7–10 was considered as diagnostic for bacterial vaginosis [15]. A few studies [16, 17] have also isolated and characterized aerobic microorganisms identified as major causes of aerobic vaginitis from cultures of vaginal swabs. *E. coli, Pseudomonas* spp., *S. aureus, Mycoplasma hominis, and Ureaplasma urealyticum* have been reported as the most frequently isolated microorganisms from patients with aerobic vaginitis.

Although bacterial vaginosis is associated with numerous health problems and is a major global concern, it has been the focus of neither intensive study nor active control programs in Ethiopia. Therefore, the purpose of the present study was to determine the prevalence of bacterial vaginosis and associated risk factors among women attending gynecology and antenatal clinics.

2. Materials and Methods

2.1. Study Design, Period, and Area. A hospital-based cross-sectional study was conducted from September 2015 to July 2016 at Department of Obstetrics and Gynecology, St Paul's Hospital Millennium Medical College, Addis Ababa, Ethiopia. The hospital has 1486 professional and supporting staff members. It provides health services for about 700 patients daily. It has 13 departments and 340 beds offering various specialized services. The Gynecology Department has 5 OPD clinics; and many patients are referred from all over the country to this hospital.

2.2. Inclusion and Exclusion Criteria. All women willing to participate in the study with presumptive diagnosis of vaginal infections and no history of antibacterial therapy within two weeks prior to their attendance constituted the inclusion criteria. Women with genital malignancy were excluded from the study. The requisition form filled out by physicians was used as standard proforma to document clinical information and previous treatment history.

2.3. Collection of Sociodemographic Characteristics, Sexual Behavior, and Reproductive Health Information. Sociodemographic characteristics, sexual behavior, and reproductive health characteristics such as age, educational and marital status, number of lifetime male sex partners, history of abortion, previous history of genital tract infection, frequency of vaginal bathing, and pant change were collected by face-to-face interviews using a structured questionnaire.

2.4. Specimen Collection and Transportation. Upon admission to the study, physicians performed clinical examination of each participant and recorded signs of vaginal abnormalities. During examinations, vaginal specimens were collected aseptically from the study participants using sterile rayon-tipped applicator stick swabs by experienced nurses. All vaginal swabs were then transferred without delay to the microbiology laboratory of the Ethiopian Public Health Institute.

2.5. Microscopic Examination. For diagnosis of bacterial vaginosis, slide smears were prepared from vaginal swabs, and the slides were heat-fixed, Gram-stained, and examined under oil immersion objective. Each slide was then graded as per the standardized quantitative morphological classification method developed by Nugent et al. [15] which assigns a score between 0 and 10 based on the following various bacterial morphotypes: large Gram-positive rods (*Lactobacillus* morphotypes), small Gram-variable rods (*G. vaginalis* morphotypes), small Gram-negative rods (*Bacteroides* spp. morphotypes), curved Gram-variable rods (*Mobiluncus* spp. morphotypes), and Gram-positive cocci. Each morphotype was quantitated from 1 to 4+ with regard to the number of morphotypes per oil immersion field (0, no morphotypes; 1+, less than 1 morphotype; 2+, 1 to 4 morphotypes; 3+, 5 to 30 morphotypes; and 4+, 30 or more morphotypes). Scores between 0 and 3 represented "normal vaginal flora," scores between 4 and 6 represented "intermediate vaginal flora," and scores between 7 and 10 were considered diagnostic for bacterial vaginosis.

2.6. Inoculation and Incubation. Each vaginal swab was inoculated onto Blood Agar base (Oxoid, Basingstoke, Hampshire, UK) to which 10% sheep blood was incorporated, MacConkey agar (Oxoid, Basingstoke, Hampshire, UK), and chocolate agar (Oxoid, Basingstoke, Hampshire, UK) before slide preparation for the isolation and characterization of aerobic bacteria. Blood agar and chocolate agar plates were incubated at 35–37°C up to 48 hrs in a 5% CO_2 incubator. MacConkey agar was incubated at 35–37°C up to 48 hrs aerobically. Preparation and performance evaluation of culture media were done as per the instruction of the manufacturer.

2.7. Bacterial Identification. Pure isolates of bacterial pathogen were preliminarily characterized by colony morphology, Gram stain, and hemolytic reactions on blood agar plates. Identification of bacteria to genus and/or species level was done by employing an array of routine biochemical tests such as DNase, catalase, optochin, bacitracin, CAMP, and bile-esculin tests for Gram-positive bacteria and indole production, H_2S production, gas production, motility, urease, citrate utilization tests, and fermentation of various carbohydrates for Gram-negative bacteria.

2.8. Antimicrobial Susceptibility Testing. The in vitro antibacterial susceptibility testing of bacterial isolates was performed by the Kirby-Bauer disc diffusion method. The following antimicrobial agents were employed: penicillin (10 μg), cefoxitin (30 μg), trimethoprim/sulfamethoxazole (25/23.75 μg), ceftriaxone (30 μg), clindamycin (2 μg), erythromycin (15 μg), gentamycin (10 μg), ciprofloxacin (5 μg), tobramycin (10 μg), vancomycin (10 μg), tetracycline (30 μg), amoxicillin (10 μg), amoxicillin/clavulanate (20/10 μg), and amikacin (30 μg). Sensitivity test results were interpreted according to the Clinical and Laboratory Standards Institute. Reference strains, *E. coli* ATCC 25922 and *S. aureus* ATCC 25923, were used for quality control for antimicrobial susceptibility testing. All drugs were generously provided by the Ethiopian Public Health Institute.

2.9. Statistical Analysis. All data from the investigation were coded, double-entered, and analyzed using SPSS version 20. Descriptive statistics and logistical regressions were used to estimate crude and adjusted crude ratio with 95% confidence interval to the different variables. p value < 0.05 was considered significant.

2.10. Ethical Clearance. All ethical considerations and obligations were duly addressed, and the study was conducted after the approval of the Department Research and Ethical Review Committee (DRERC) of the Department of Medical Laboratory Sciences, College of Health Sciences, Addis Ababa University, and Ethical Review Board of the hospital. Written informed consent was obtained from the participants before data collection. Each respondent was given the right to refuse to take part in the study and to withdraw at any time during the study period. All information obtained from the study subjects was coded to maintain confidentially. When the participants were found to be positive for a bacterial pathogen, they were informed by the hospital clinician and received proper treatment. An assent form was completed and signed by a family member and/or adult guardian for participants under the age of 16 years.

3. Results

3.1. Prevalence of Bacterial Vaginosis with Sociodemographic Characteristics, Sexual Behavior, Reproductive Health Information, and Personal Hygiene. A total of 210 women were included in the study. The overall prevalence of bacterial vaginosis was 48.6%. Subgroup prevalence of bacterial vaginosis is presented in Table 1. Younger women, 15 to 24 years old, had somewhat lower prevalence (41.5%) of bacterial vaginosis, while in the 25 years and older group, the prevalence was between 47.8% and 60.0%. As shown in Table 1, the adjusted odds ratio depicted that bacterial vaginosis was not significantly associated with age.

The prevalence of bacterial vaginosis varied with education and marital status. Women with a college-level education were less likely to be positive for bacterial vaginosis than those with a high school education or less (35.9% versus 44.7–55.3%). Similarly, the prevalence of bacterial vaginosis was high among unmarried study subjects (53.8%) compared to those who were married (44.8%) or divorced (50.0%). As depicted in Table 1, neither marital status nor education was statistically associated with bacterial vaginosis.

The prevalence of bacterial vaginosis varied with selected reproductive health history. For example, it was less prevalent in patients who had previous bacterial vaginosis (46.3%) than in patients with no previous bacterial vaginosis (50.4%). The prevalence rate of bacterial vaginosis was higher in women with a history of abortion (53.8%) than in women with no history of abortion (46.8%). These two variables were not significantly associated with bacterial vaginosis (Table 1). The prevalence of bacterial vaginosis also varied with number of lifetime male sex partners. Women who reported 1–3 lifetime male sex partners had prevalence rate of 43.4%, while those who reported ≥4 lifetime male sex partners had prevalence rate of 58%. The prevalence of bacterial vaginosis was lower among patients who changed pants more frequently (two per day; 36.9%) than among those who changed their pants less frequently (one pant for 2–4 days; 57.6%). Statistical analysis showed a significant correlation between bacterial vaginosis and number of pants used per day ($p = 0.001$). Similarly, patients who bathed their vaginal region more frequently were less affected than those who did not bath their vaginal area as much (prevalence rate of 40.2% versus 53.9%). The association of frequency of vaginal bathing and bacterial vaginosis was statistically significant ($p = 0.045$).

3.2. Spectrum of Bacteria Causing Aerobic Vaginosis. A total of 151 bacterial isolates were recovered from vaginal swabs, of which 105 (69.5%) were Gram-negative and 46 (30.5%) were Gram-positive bacteria. Of the Gram-negative bacteria, *E. coli* and *Klebsiella* spp. were dominant. *S. aureus* and *S. agalactiae* were the dominant Gram-positive bacteria (Table 2).

3.3. Drug Susceptibility Pattern of Bacterial Isolates. Table 3 summarizes the overall drug susceptibility pattern of the Gram-positive bacteria against the eleven antibacterial drugs tested. Among the agents tested, the highest overall resistance rate of Gram-positive bacteria was observed against penicillin (67.4%), followed by tetracycline (58.7%) and erythromycin (45.6%). Cefoxitin and tobramycin were the most active of the drugs tested against Gram-positive bacteria. *S. aureus*, the most frequently isolated Gram-positive bacterium, was 97.2%, 88.8%, and 86.1% sensitive to cefoxitin, tobramycin, and clindamycin, respectively.

The overall drug susceptibility pattern of Gram-negative bacteria against the nine antibacterial agents tested is summarized in Table 4. Tetracycline exhibited the highest overall drug resistance rate (77.3%) against Gram-negative bacteria, followed by ampicillin (77.1) and amoxicillin (70.6%). Amikacin with an overall sensitivity rate of 85.7% and tobramycin with an overall sensitivity rate of 82.8% were better active against Gram-negative bacteria. As far as species-specific antimicrobial resistance rates are concerned, *E. coli*, the most frequently isolated bacterium, showed 76.7% resistance to both ampicillin and tetracycline. The lowest resistance rate was observed with amikacin and tobramycin. Amikacin, tobramycin, and gentamycin were the most active drugs against *K. pneumoniae*, the second most commonly isolated Gram-negative bacterium.

4. Discussion

The overall prevalence rate of bacterial vaginosis in the present study as determined by Gram-stain Nugent scoring criteria was 48.6%. Although the prevalence rate of bacterial vaginosis in the present study was well within the reported range, that is, 8%–75% [4], it was higher than the prevalence rates of bacterial vaginosis reported by similar local studies [16, 18]. Local studies reported prevalence rates of bacterial vaginosis in the range of 15.4% [16] to 19.4% [18]. Lower prevalence rates of bacterial vaginosis than those in the present study were also reported from other sub-Saharan countries, such as Kenya (37%), Botswana (38%),

Table 1: Prevalence of bacterial vaginosis (Nugent's Gram Stain Score: 7–10) by selected characteristics in relation to sociodemographic characteristics, sexual behavior, reproductive health information, and personal hygiene (n = 210).

Characteristics	Number	Bacterial vaginosis (n, %)		p value	COR	95% CL	p value	AOR	95% CL
		Yes	No						
Age in years									
15–24	53 (25.2)	22 (41.5)	31 (58.4)		1			1	
25–44	117 (55.7)	56 (47.8)	61 (52.1)	0.416	.763	0.178–0.955	0.379	0.735	0.371–1.459
45–64	40 (19.04)	24 (60)	16 (40)	0.039	.413	0.397–1.466	0.078	0.457	0.191–1.093
Total	210	102 (48.6)	108 (51.4)						
Marital status									
Unmarried	65 (30.9)	35 (53.8)	30 (46.1)		1				
Married	107 (50.9)	48 (44.8)	59 (55.1)	0.254	1.434	0.772–2.663			
Divorced	38 (18.1)	19 (50)	19 (50)	0.706	1.167	0.524–2.600			
Total	210 (100)	102 (48.6)	108 (51.4)						
Education									
Illiterate	47 (22.4)	21 (44.7)	26 (55.3)	0.410	0.693	0.290–1.657			
Primary school	59 (28.1)	31 (52.5)	28 (47.5)	0.108	0.506	0.221–1.160			
Secondary school	65 (31.0)	36 (55.3)	29 (44.6)	0.056	0.451	0.199–1.021			
College	39 (18.6)	14 (35.9)	25 (64.1)		1				
Total	210 (100)	102 (48.6)	108 (51.4)						
Number of lifetime male sex partners									
1–3	136 (64.8)	59 (43.4)	77 (56.6)		1			1	
≥4	74	43 (58)	31 (41.8)	0.042	1.810	1.021–3.210	0.103	1.645	.904–2.995
Total	210	102 (48.6)	108 (51.4)						
History of abortion									
Yes	52 (24.7)	28 (53.8)	24 (46.1)	0.381	0.755	0.403–1.416			
No	158 (75.2)	74 (46.8)	84 (53.1)						
Previous BV/GTI					1				
Yes	95 (45.2)	44 (46.3)	51 (53.6)	0.552	1.179	0.684–2.033			
No	115 (54.7)	58 (50.4)	57 (49.5)		1				
Total	210	102 (48.6)	108 (51.4)						
Vaginal bathing/day									
1–3	128 (60.9)	69 (53.9)	59 (46)		1			1	
≥4	82 (39)	33 (40.2)	49 (59.7)	0.054	1.737	0.990–3.045	0.045	1.847	1.013–3.370
Total	120	102	108						
Number of pants used/day									
1-2 pants/a day	92 (43.8)	34 (36.9)	58 (63)		1				
1 pant for 2–4 days	118 (56.1)	68 (57.6)	50 (42.3)	0.003	0.431	0.236–0.754	0.001	0.367	0.201–0.672
Total	*102*	*108*							

TABLE 2: Distribution of bacterial isolates (n = 151).

Species	Number of isolates	% of the total isolates
S. aureus	36	23.8
S. agalactiae	7	4.6
S. pyogenes	3	2.0
E. coli	43	28.5
Klebsiella pneumoniae	28	18.5
Klebsiella ozaenae	4	2.6
Enterobacter aerogenes	17	11.3
Citrobacter freundii	6	4.0
Citrobacter diversus	2	1.3
Proteus mirabilis	1	0.7
Providencia rettgeri	4	2.6
Total	151	100%

and Zimbabwe (32.5%) [19–21]. Sociodemographic characteristics, sexual activity, reproductive health information, and behavioral and genital hygiene have been identified as causes of variation in the prevalence rates of bacterial vaginosis [22–26].

The present study revealed that the proportion of bacterial vaginosis was the highest in age groups above 45 years. Our finding was comparable to the studies of Fang et al. [27], Ocviyanti et al. [28], and Yusuf et al. [29]. Similarly, a study conducted in a population of individuals seeking STD treatment showed that 23% of women aged 14–24 years exhibited bacterial vaginosis compared to 33% of women aged 25 years and older [30]. An elevation of pH in women above age of 45 years has been identified as a cause of a decline in the level of estrogen, which in turn creates an optimal condition for the growth of bacteria other than lactobacilli.

Lack of education has been found to be significantly associated with bacterial vaginosis [26]. However, our finding like other studies [27] contradicted this conclusion. In the present study, bacterial vaginosis was higher among subjects having an education level of primary and secondary school compared to illiterate patients.

The role of sexual activity in the acquisition of bacterial vaginosis is not clear. Bacterial vaginosis' prevalence rates of 18.8%, 18%, and 12% were reported among women who reported that they have never had sex by Koumans et al. [25], Yen et al. [24], and Bump and Buesching [23], respectively. Contrary to this, sexual behavior-related characteristics, including number of lifetime male sex partners, multiple male sex partners, and a recent history of new sex partners, have been consistently associated with bacterial vaginosis. In support of a role of sexual transmission, Bump and Buesching [23], Yen et al. [24], and Allsworth and Peipert [31] found that multiple or new sex partners increased the risk of acquiring bacterial vaginosis by a factor of 1.6–2.5. These studies also suggested that condom use may be protective. In the present study, the adjusted odds ratio analysis revealed that bacterial vaginosis and the number of lifetime male sex partners were not statistically associated (p = 0.103). Thus, our result did not support previous studies that reported that multiple or new sex partners increased the risk of acquiring bacterial vaginosis [2, 23, 24, 31]. Verstraelen et al. [32] argue that bacterial vaginosis may be considered as a sexually enhanced disease rather than sexual transmitted infection, with the frequency of sexual intercourse being a critical factor.

In contrast to other findings, no significant correlation was observed between bacterial vaginosis and number of abortions [24]. As far as personal hygiene is concerned, statistical analysis showed a significant correlation between frequency of vaginal bathing (p = 0.047) and number of pants used per day (p = 0.001). Our result was in good agreement with the findings of Bahram et al. [26] who reported that bacterial vaginosis is significantly associated with individual hygiene.

Bacterial vaginosis is a situation that occurs when lactobacilli are replaced by the overgrowth of anaerobic bacteria, primarily *G. vaginalis* and *Mobiluncus* spp. Other bacteria (*Escherichia coli, Klebsiella* spp., *Acinetobacter* spp., *Staphylococcus* spp., enterococci, and *Streptococcus agalactiae* (group B streptococci), however, have been termed "intermediate flora" in some studies or have been included with bacterial vaginosis in others [8, 33]. Still others consider them as distinct bacterial floras that cause aerobic vaginitis which has been thought to be a better candidate than bacterial vaginosis as a cause of pregnancy complications such as preterm rupture of the membranes and preterm delivery [2, 34, 35]. With attention to the above findings, in the present study, vaginal swabs were cultured and 151 bacterial isolates of Gram-positive and Gram-negative bacteria were recovered. The type of bacterial isolates and their frequency recorded in the present study were more diverse than in the study of Lakshmi et al. [1, 16]. Therefore, further studies to differentiate the effects of bacterial vaginosis and aerobic vaginitis on the outcome of pregnancy should be conducted.

The overall drug resistance rates of Gram-negative bacterial isolates ranged from 14.3% for amikacin to 77.3% for tetracycline. *E. coli,* the most frequently isolated Gram-negative bacterium, showed a high level of resistance to tetracycline and ampicillin. Contrary to this, 86% of *E. coli* were susceptible to amikacin and tobramycin. The drug resistance of level of *K. pneumoniae,* the second frequent isolate, was high against ampicillin, amoxicillin, and tetracycline.

Similarly, the overall drug resistance rates of Gram-positive bacterial isolates ranged from 2.8% for cefoxitin to 67.4% for penicillin. *S. aureus,* the most frequently isolated Gram-positive bacterium, revealed a high level of resistance to the commonly prescribed drugs, penicillin, tetracycline, and erythromycin. Our result was consistent with studies conducted in Ethiopia [16, 36, 37] and Pakistan [38, 39]. Availability of antimicrobials without prescription and inappropriate dosing schedules may explain the isolation of high level of drug resistance in the present study.

5. Conclusions

The prevalence of bacterial vaginosis was relatively high and was affected by individual hygiene. Therefore, comprehensive healthcare education aimed at reducing bacterial vaginosis

TABLE 3: Percentage of in vitro antibacterial susceptibility pattern of Gram-positive bacteria isolates (n = 46).

Species	P	Antibacterial drugs										
		Pen	FOX	E	DA	SXT	TE	CN	VA	CRO	CIP	TOB
	S	13.9	97.2	41.7	86.1	58.3	11.1	80.6	—	—	77.7	88.8
S. aureus (36)	I	—	0	8.3	2.8	22.2	22.2	8.3	—	—	0	0
	R	86.1	2.8	50%	11.1	19.4	63.8	11.1	—	—	22.2	12
	S	100	100	85.7	100	57.1	42.8	ND	100	100	—	—
S. agalactiae (7)	I	—	0	0	0	0	28.0	—	0	0	—	—
	R	0	0	14.2	0	42.9	28.0	—	0	0	—	—
	S	100	100	66.67	66.67	33.3	33.3	—	100	100	—	—
S. pyogenes (3)	I	—	0	0	0	0	0	—	0	0	—	—
	R	0	0	33.3	33.3	66.67	66.67	—	0	0	—	—
	S	32.6	97.2	50.0	86.9	56.5	17.3	80.6	100	100	77.7	88.8
Total isolates (46)	I	—	0	6.0	2.0	17.3	21.7	8.3	0	0	0	0
	R	67.4	2.8	45.6	10.8	26	58.7	11.1	0	0	22.2	12.0

Pen: penicillin; Fox: cefoxitin; SXT: trimethoprim/sulfamethoxazole; CRO: ceftriaxone; CIP: ciprofloxacin; Da: clindamycin; E: erythromycin; CN: gentamicin; TE: tetracycline; TOB: tobramycin; VA: vancomycin; S: sensitive; R: resistance; P: pattern; —: not tested.

TABLE 4: Percentage of in vitro antibacterial susceptibility pattern of Gram-negative bacteria isolates (n = 105).

Species	Pattern	Antibacterial drugs								
		AM	AMC	SXT	TE	CN	CRO	CIP	TOB	AK
	S	23.3	32.6	60.5	18.6	83.7	76.7	79.1	86.04	86.04
E. coli (43)	I	0	7	16.3	4.7	0	0	7.0	0	0
	R	76.7	60.5	23.3	76.7	16.3	23.3	14.0	13.96	13.96
	S	14.3	14.3	14.3	17.8	78.5	60.7	57.1	78.5	82.1
K. pneumoniae (28)	I	7.1	7.1	0	0	0	0	0	0	0
	R	78.6	78.6	85.7	82.1	21.1	39.0	42.8	21.1	17.8
	S	0	0	0	0	50.0	50.0	50.0	75.0	100
K. ozaenae (4)	I	0	0	0	0	50.0	0	0	0	0
	R	100	100	100	100	0	50.0	50.0	25.0	0
	S	0	17.6	28.1	5.9	70.6	70.6	58.8	76.7	82.35
E. aerogenes (17)	I	17.6	11.8	18.8	17.6	5.9	11.8	0	0	0
	R	82.4	70.6	53.1	76.5	23.5	17.6	41.2	23.5	17.64
	S	0	0	50.0	33.33	83.3	66.6	83.3	100	100
C. freundii (6)	I	0	0	0	0	0	0	0	0	0
	R	100	100	50.0	66.6	16.67	33.33	16.67	0	0
	S	0	0	0	0	50.	50.0	50.0	50.0	50.0
C. diversus (2)	I	0	0	0	0	0	0	0	0	0
	R	100	100	100	100	50.0	50.0	50.0	50.0	50.0
	S	100	100	100	100	100	100	100	100	100
P. mirabilis (1)	I	0	0	0	0	0	0	0	0	0
	R	0	0	0	0	0	0	0	0	0
	S	100	50.0	50.0	50.0	100	75.0	100	100	100
P. rettgeri (4)	I	0	0	0	0	0	0	0	0	0
	R	0	50.0	50.0	50.0	0	25.0	0	0	0
	S	18	22.8	39	18	79	68.5	69.5	82.8	85.7
Total isolates	I	4.7	6.6	9.5	4.7	2.8	1.9	2.8	0	0
	R	77.1	70.6	51.5	77.3	18.2	29.6	27.7	17.2	14.3

AMP: ampicillin; AMC: amoxicillin/clavulanate; SXT: trimethoprim/sulfamethoxazole; CRO: ceftriaxone; CIP: ciprofloxacin; CN: gentamicin; TE: tetracycline; TOB: tobramycin; AK: amikacin; S: sensitive; R: resistance; P: pattern.

is needed. Isolation and characterization of aerobic bacteria implicated in causing aerobic vaginitis initiate inclusion of vaginal culture and sensitivity testing along with microscopic and clinical diagnosis of bacterial vaginitis.

Conflicts of Interest

The authors declare that there are no conflicts of interest regarding the publication of this paper.

Acknowledgments

The authors would like to acknowledge the College of Health Sciences, Addis Ababa University, and the Ethiopian Public Health Institute for allowing them to use laboratory facilities and consumable materials for free. The authors would like to extend their thanks to all patients who participated in this study and to Mintwab Hussien for editing the manuscript.

References

[1] K. Lakshmi, S. Aishwarya Chitralekha Jr., and G. A. Menezes, "Review on infectious vaginitis," *Research Journal of Pharmaceutical, Biological and Chemical Sciences*, vol. 4, article 679, 2013.

[2] G. G. G. Donders, A. Vereecken, E. Bosmans, A. Dekeersmaecker, G. Salembier, and B. Spitz, "Definition of a type of abnormal vaginal flora that is distinct from bacterial vaginosis: Aerobic vaginitis," *BJOG: An International Journal of Obstetrics and Gynaecology*, vol. 109, no. 1, pp. 34–43, 2002.

[3] D. A. Eschenbach, S. S. Thwin, D. L. Patton et al., "Influence of the normal menstrual cycle on vaginal tissue, discharge, and microflora," *Clinical Infectious Diseases*, vol. 30, no. 6, pp. 901–907, 2000.

[4] E. F. C. Murta, A. O. Silva, E. A. C. Silva, and S. J. Adad, "Frequency of infectious agents for vaginitis in non- and hysterectomized women.," *Archives of gynecology and obstetrics.*, vol. 273, no. 3, pp. 152–156, 2005.

[5] L. Myer, L. Denny, R. Telerant, M. De Souza, T. C. Wright Jr., and L. Kuhn, "Bacterial vaginosis and susceptibility to HIV infection in South African women: A nested case-control study," *Journal of Infectious Diseases*, vol. 192, no. 8, pp. 1372–1380, 2005.

[6] H. L. Martin Jr., B. A. Richardson, P. M. Nyange et al., "Vaginal lactobacilli, microbial flora, and risk of human immunodeficiency virus type 1 and sexually transmitted disease acquisition," *Journal of Infectious Diseases*, vol. 180, no. 6, pp. 1863–1868, 1999.

[7] T. L. Cherpes, L. A. Meyn, M. A. Krohn, J. G. Lurie, and S. L. Hillier, "Association between acquisition of herpes simplex virus type 2 in women and bacterial vaginosis," *Clinical Infectious Diseases*, vol. 37, no. 3, pp. 319–325, 2003.

[8] P. E. Hay, R. F. Lamont, D. Taylor-Robinson, D. J. Morgan, C. Ison, and J. Pearson, "Abnormal bacterial colonisation of the genital tract and subsequent preterm delivery and late miscarriage," *BMJ*, vol. 308, no. 6924, p. 295, 1994.

[9] Centers for Disease Control and Prevention, "Sexually transmitted diseases treatment guidelines 2006," *MMWR*, vol. 55, no. RR–11, pp. 50–54, 2006.

[10] H. Minkoff, A. N. Grunebaum, R. H. Schwarz et al., "Risk factors for prematurity and premature rupture of membranes: A prospective study of the vaginal flora in pregnancy," *American Journal of Obstetrics and Gynecology*, vol. 150, no. 8, pp. 965–972, 1984.

[11] M. G. Gravett, H. P. Nelson, T. Derouen, C. Critchlow, D. A. Eschenbach, and K. K. Holmes, "Independent Associations of Bacterial Vaginosis and Chlamydia trachomatis Infection With Adverse Pregnancy Outcome," *JAMA: The Journal of the American Medical Association*, vol. 256, no. 14, pp. 1899–1903, 1986.

[12] S. Cu-Uvin, J. W. Hogan, A. M. Caliendo, J. Harwell, K. H. Mayer, and C. C. J. Carpenter, "Association between bacterial vaginosis and expression of human immunodeficiency virus type 1 RNA in the female genital tract," *Clinical Infectious Diseases*, vol. 33, no. 6, pp. 894–896, 2001.

[13] J. A. Cohn, F. B. Hashemi, M. Camarca et al., "HIV-inducing factor in cervicovaginal secretions is associated with bacterial vaginosis in HIV-1-infected women," *Journal of Acquired Immune Deficiency Syndromes*, vol. 39, no. 3, pp. 340–346, 2005.

[14] R. Amsel, P. A. Totten, C. A. Spiegel, K. C. Chen, D. Eschenbach, and K. K. Holmes, "Nonspecific vaginitis," *The American Journal of Medicine*, vol. 74, no. 1, pp. 14–22, 1983.

[15] R. P. Nugent, M. A. Krohn, and S. L. Hillier, "Reliability of diagnosing bacterial vaginosis is improved by a standardized method of gram stain interpretation," *Journal of Clinical Microbiology*, vol. 29, no. 2, pp. 297–301, 1991.

[16] M. Wondemagegn, M. Yimer, Y. Zenebe, and B. Abera, "Common causes of vaginal infections and antibiotic susceptibility of aerobic bacterial isolates in women of reproductive age attending at Felegehiwot referral Hospital, Ethiopia: a cross sectional study," *BMC Women's Health*, vol. 15, no. 42, pp. 1–9, 2015.

[17] Z. A. Abdelaziz, M. E. Ibrahim, N. E. Bilal, and M. E. Hamid, "Vaginal infections among pregnant women at Omdurman Maternity Hospital in Khartoum, Sudan," *Journal of Infection in Developing Countries*, vol. 8, no. 4, pp. 490–497, 2014.

[18] Z. Mengistie, Y. Woldeamanuel, D. Asrat, and A. Adera, "Prevalence of bacterial vaginosis among pregnant women attending antenatal care in Tikur Anbessa University Hospital, Addis Ababa, Ethiopia," *BMC Research Notes*, vol. 7, article 822, 2014.

[19] G. Marx, G. John-Stewart, R. Bosire, D. Wamalwa, P. Otieno, and C. Farquhar, "Diagnosis of sexually transmitted infections and bacterial vaginosis among HIV-1-infected pregnant women in Nairobi," *International Journal of STD & AIDS*, vol. 21, no. 8, pp. 549–552, 2010.

[20] M. Romoren, M. Velauthapillai, M. Rahman, J. Sundby, E. Klouman, and P. Hjortdahl, "Trichomoniasis and bacterial vaginosis in pregnancy: inadequately managed with the syndromic approach," *Bulletin of the World Health Organization*, vol. 85, no. 4, pp. 297–304, 2007.

[21] N. E. Kurewa, M. P. Mapingure, M. W. Munjoma, M. Z. Chirenje, S. Rusakaniko, and B. Stray-Pedersen, "The burden and risk factors of Sexually Transmitted Infections and Reproductive Tract Infections among pregnant women in Zimbabwe," *BMC Infectious Diseases*, vol. 10, article 127, 2010.

[22] J. R. Schwebke, C. M. Richey, and H. L. Weiss, "Correlation of behaviors with microbiological changes in vaginal flora," *Journal of Infectious Diseases*, vol. 180, no. 5, pp. 1632–1636, 1999.

[23] R. C. Bump and W. J. Buesching III, "Bacterial vaginosis in virginal and sexually active adolescent females: Evidence against exclusive sexual transmission," *American Journal of Obstetrics and Gynecology*, vol. 158, no. 4, pp. 935–939, 1988.

[24] S. Yen, M.-A. Shafer, J. Moncada, C. J. Campbell, S. D. Flinn, and C. B. Boyer, "Bacterial vaginosis in sexually experienced and

non-sexually experienced young women entering the military," *Obstetrics and Gynecology*, vol. 102, no. 5, pp. 927–933, 2003.

[25] E. H. Koumans, M. Sternberg, C. Bruce et al., "The prevalence of bacterial vaginosis in the United States, 2001-2004; associations with symptoms, sexual behaviors, and reproductive health," *Sexually Transmitted Diseases*, vol. 34, no. 11, pp. 864–869, 2007.

[26] A. Bahram, B. Hamid, and T. Zohre, "Prevalence of bacterial vaginosis and impact of genital hygiene practices in non-pregnant women in Zanjan, Iran," *Oman Medical Journal*, vol. 24, no. 4, pp. 288–293, 2009.

[27] X. Fang, Y. Zhou, Y. Yang, Y. Diao, and H. Li, "Prevalence and risk factors of trichomoniasis, bacterial vaginosis, and candidiasis for married women of child-bearing age in rural Shandong," *Japanese Journal of Infectious Diseases*, vol. 60, no. 5, pp. 257–261, 2007.

[28] D. Ocviyanti, Y. Rosana, S. Olivia, and F. Darmawan, "Risk factors for bacterial vaginosis among Indonesian women," *Medical Journal of Indonesia*, vol. 19, pp. 130–135, 2010.

[29] M. A. Yusuf, M. Chowdhury, K. S. Islam et al., "Common microbial aetiology of abnormal vaginal discharge among sexually active women in Dhaka, Bangladesh," *South East Asia Journal of Public Health*, vol. 1, no. 1, pp. 35–39, 2013.

[30] H. Moi, "Prevalence of bacterial vaginosis and its association with genital infections, inflammation, and contraceptive methods in women attending sexually transmitted disease and primary health clinics," *International Journal of STD and AIDS*, vol. 1, no. 2, pp. 86–94, 1990.

[31] J. E. Allsworth and J. F. Peipert, "Prevalence of bacterial vaginosis: 2001–2004 National Health and Nutrition Examination Survey data," *Obstetrics and Gynecology*, vol. 109, no. 1, pp. 114–120, 2007.

[32] H. Verstraelen, R. Verhelst, M. Vaneechoutte, and M. Temmerman, "The epidemiology of bacterial vaginosis in relation to sexual behaviour," *BMC Infectious Diseases*, vol. 10, article 81, 2010.

[33] H. M. Mc Donald, J. A. O'loughlin, P. T. Jolley, R. Vigneswaran, and P. J. Mc Donald, "Changes in vaginal flora during pregnancy and association with preterm birth," *Journal of Infectious Diseases*, vol. 170, no. 3, pp. 724–728, 1994.

[34] G. Donders, H. G. De Wet, P. Hooft, and J. Desmyter, "Lactobacilli in papanicolaou smears, genital infections, and pregnancy," *American Journal of Perinatology*, vol. 10, no. 5, pp. 358–361, 1993.

[35] G. G. G. Donders, A. Odds, A. Vereecken et al., "Abnormal vaginal flora in the first trimester, but not full-blown bacterial vaginosis, is associated with preterm birth," *Prenatal and Neonatal Medicine*, vol. 3, no. 6, pp. 588–593, 1998.

[36] B. Abera and M. Kibret, "Bacteriology and antimicrobial susceptibility of otitis media at dessie regional health research laboratory, Ethiopia," *Ethiopian Journal of Health Development*, vol. 25, no. 2, pp. 161–167, 2011.

[37] W. Mulu, G. Kibru, G. Beyene, and M. Damtie, "Postoperative nosocomial infections and antimicrobial resistance pattern of bacteria isolates among patients admitted at Felege hiwot referral Hospital, Bahir Dar, Ethiopia," *Ethiopian Journal of Health Sciences*, vol. 22, pp. 7–18, 2012.

[38] N. Tariq, T. Jaffery, R. Ayub, AY. Alam, MH. Javid, and S. Shafique, "Frequency and antimicrobial susceptibility of aerobic vaginal isolates," *Journal of the College of Physicians and Surgeons—Pakistan*, pp. 16–196, 2006.

[39] S. Mumtaz, M. Ahmad, I. Aftab, N. Akhtar, M. ul Hassan, and A. Hamid, "Aerobic vaginal pathogens and their sensitivity pattern.," *Journal of Ayub Medical College, Abbottabad : JAMC*, vol. 20, no. 1, pp. 113–117, 2008.

20

Syphilis among Female Sex Workers: Results of Point-of-Care Screening during a Cross-Sectional Behavioral Survey in Burkina Faso, West Africa

Henri Gautier Ouedraogo,[1,2,3] Ivlabehire Bertrand Meda,[1,3] Issaka Zongo,[1] Odette Ky-Zerbo,[4] Ashley Grosso,[5] Benoit Cesaire Samadoulougou,[4] Grissoum Tarnagda,[1] Kadari Cisse,[1] Apoline Sondo,[2] Nongoba Sawadogo,[6] Yves Traoré,[2] Nicolas Barro,[2] Stefan Baral,[5] and Seni Kouanda[1,3]

[1]*Institut de Recherche en Sciences de la Santé (IRSS/CNRST), Ouagadougou, Burkina Faso*
[2]*Université Ouaga I Professeur Joseph KI-ZERBO, Ouagadougou, Burkina Faso*
[3]*Institut Africain de Santé Publique (IASP), Ouagadougou, Burkina Faso*
[4]*Programme d'Appui au Monde Associatif et Communautaire (PAMAC), Ouagadougou, Burkina Faso*
[5]*Department of Epidemiology, Johns Hopkins Bloomberg School of Public Health, Baltimore, MD, USA*
[6]*Centre Hospitalier Régional de Kaya, Kaya, Burkina Faso*

Correspondence should be addressed to Henri Gautier Ouedraogo; gouedraogo@irss.bf

Academic Editor: Maurizio Sanguinetti

Background. Syphilis among female sex workers (FSW) remains a public health concern due to its potential impact on their health and the possibility of transmission to their clients, partners, and children. Recent data on the prevalence of syphilis in the population in West Africa are scarce. The objective of this study was to measure the seroprevalence of syphilis serological markers among female sex workers in Burkina Faso. *Methods.* We conducted a cross-sectional survey among FSW between February 2013 and May 2014. Participants were recruited using respondent-driven sampling (RDS) methods in five cities of Burkina Faso (Ouagadougou, Bobo-Dioulasso, Koudougou, Ouahigouya, and Tenkodogo). FSW were enrolled and screened for syphilis using a syphilis serological rapid diagnostic test. Data from all cities were analyzed with Stata version 14.0. *Results.* A total of 1045 FSW were screened for syphilis. Participants' mean age was 27.2 ± 0.2 years. The syphilis serological markers were detected in 5.6% (95% CI: 4.4–7.2) of the participants whereas active syphilis was seen in 1.4% (95% CI: 0.9–2.4). RDS weighted prevalence of syphilis serological markers and active syphilis by city were, respectively, estimated to be 0.0% to 11.0% (95% CI: 8.1–14.7) and 0.0% to 2.2% (95% CI: 1.1–4.4). No syphilis markers were found among Ouahigouya FSW. Low education level and high number of clients were factors associated with syphilis markers among the FSW. *Conclusion.* The prevalence of syphilis markers was high during this study among FSW. This highlights the need to reinforce the comprehensive preventive measures and treatment of syphilis in this population.

1. Background

Syphilis is a sexually transmitted disease (STD) due to *Treponema pallidum*, a spirochete bacteria belonging to the *Spirochaetaceae* family [1, 2]. The infection is often asymptomatic but can cause genital ulceration associated with enhanced risk of transmission or acquisition of HIV infection [3–7]. Syphilis remains a worldwide public health problem. Despite the availability of effective preventive measures and relatively low treatment cost, the World Health Organization (WHO) estimates that more than 10 million persons are newly infected each year [8].

A number of studies around the world have reported high prevalence of syphilis among risk groups including female sex workers (FSW) [9–12]. In Africa, the prevalence of syphilis among FSW was 2.2% in Togo [13] and between 1.5 and 89% in Sudan [14]. Studies conducted in the 1990s have reported a prevalence of 42.1% in South Africa [15] and 15% in Burkina Faso [16]. Despite this high prevalence in Africa, FSW have limited access to STIs/STDs like syphilis screening and care [17, 18]. Stigma and criminalization of sex work in some countries have complicated their access to care and have prevented the planning of efficient health systems [19–22]. FSW are identified as a population at high risk for HIV acquisition [23, 24]. Biologic risk factors for HIV include physiological factors [25] such as the presence of STDs, including syphilis [26, 27]. Recent data on syphilis prevalence in FSW are scarce in West Africa. The main objective of this study was to determine the serological prevalence and correlates of syphilis with rapid diagnostic testing among FSW in Burkina Faso.

2. Methods

2.1. Study Design. We conducted a cross-sectional survey between February 2013 and May 2014 among FSW in five cities in Burkina Faso. Participants were recruited using respondent-driven sampling (RDS) to be part of HIV seroprevalence and behavioral studies. RDS is an approach to recruit hidden and hard-to-reach populations [28].

2.2. Setting. The study was conducted in five cities of Burkina Faso: Ouagadougou (the capital city in the centre), Bobo-Dioulasso (in the Hauts-Bassins region), Koudougou (in the middle west), Ouahigouya (in the north), and Tenkodogo (in the middle east). These areas were selected based on the urbanization level, the prevalence of HIV, and the geographical location.

2.3. Study Participants. Inclusion criteria were aged ≥ 18 years, assigned female sex at birth, earned at least 50% of her income in the last 12 months from sex work, lived in the selected city during the last three months, and provided informed consent. Recruitment methods were fully described elsewhere [29]. The enrolment of FSW in each city started with an eligible seed who was enrolled and then trained to recruit other participants.

2.4. Sample Size. The sample size was based on expected HIV prevalence and calculated to recruit 345 FSW in the two major cities of Ouagadougou and Bobo-Dioulasso and 126 FSW in the other medium cities of Koudougou, Ouahigouya, and Tenkodogo. To calculate the sample size, we hypothesized that populations who use condoms would have 75% lower HIV prevalence than that of those not using condoms, assuming condom efficacy as 80% (73% to be cautious). A design effect of 1.5 was used associated with the RDS methods; we set 0.05 for type I errors and 80% power.

2.5. Data Collection. Data were collected from February to August 2013 in Ouagadougou and Bobo-Dioulasso and from February to May 2014 in Koudougou, Ouahigouya, and Tenkodogo. Upon informed consent and completion of a behavioral questionnaire, participants received counseling for HIV and syphilis screening. Participant information was anonymized and identified with a unique code number to protect confidentiality. Data collected in the questionnaire included demographic and socioeconomic information, the number of sexual partners, sexual behavior (including condom use), HIV-related knowledge, and STD screening and symptoms. In each city, one site was used for data collection. The team was trained on the study procedures and on health research ethics prior to the onset of the survey. The staff included one site manager, one coupon manager, two data collectors for questionnaire administration, a nurse for HIV and syphilis pre- and posttest counseling, and a lab technician for blood collection and sample processing for HIV and syphilis screening. Participants with positive syphilis test results were treated in situ while those with positive HIV test results were referred to dedicated staff for appropriate care.

2.6. Syphilis Testing. Figure 1 shows the flowchart of the syphilis screening among study participants. Blood samples were centrifuged, and serum obtained was used in situ for syphilis screening. *Alere Determine™ Syphilis TP* ("Alere Ltd.," USA) rapid diagnostic test was used to screen the syphilis serological markers as per manufacturer recommendations. Positive samples were consecutively tested with a Rapid Plasma Reagin (RPR-CHARBON®) test kit (*BIOLABO SAS, French*), a nontreponemic test [30]. *Alere Determine™ Syphilis TP* is a qualitative in vitro immunechromatographic test detecting antibodies against the *Treponema pallidum* bacteria in patient's serum, plasma, and whole blood [31]. Positivity to the *Alere Determine™ Syphilis TP* test reveals the presence of antibodies against the *Treponema pallidum* bacteria indicating the presence of either previous, acute, or treated syphilis. The nontreponemic rapid plasma reagin (RPR) test was used to determine the syphilis stage (active or not).

2.7. Data Entry and Statistical Analysis. Data were double entered into EpiData 3.1 (the EpiData Association, Odense, Denmark) software and analysed with Stata 14 (StataCorp, College Station, TX). RDS-weighted analysis was only used to estimate active syphilis and syphilis serological markers seroprevalence for each city. To have sufficient number of positive cases of syphilis serological markers for statistical analysis, data from all study cities were pooled. The pooled data were not analyzed using RDS weighting because there was no overlap in recruitment between study cities. Baseline survey data and biological test results of participants were summarised with descriptive statistics. Bivariate and multivariate logistic regression analyses were conducted to identify factors associated with syphilis serological markers. Associations were expressed as odds ratios (ORs) and adjusted odds ratios (aORs) with 95% confidence intervals (CIs). Results significant at the $p < 0.2$ level in the bivariate analyses were included in the multivariate analysis. Results were considered statistically significant at the $p < 0.05$ level.

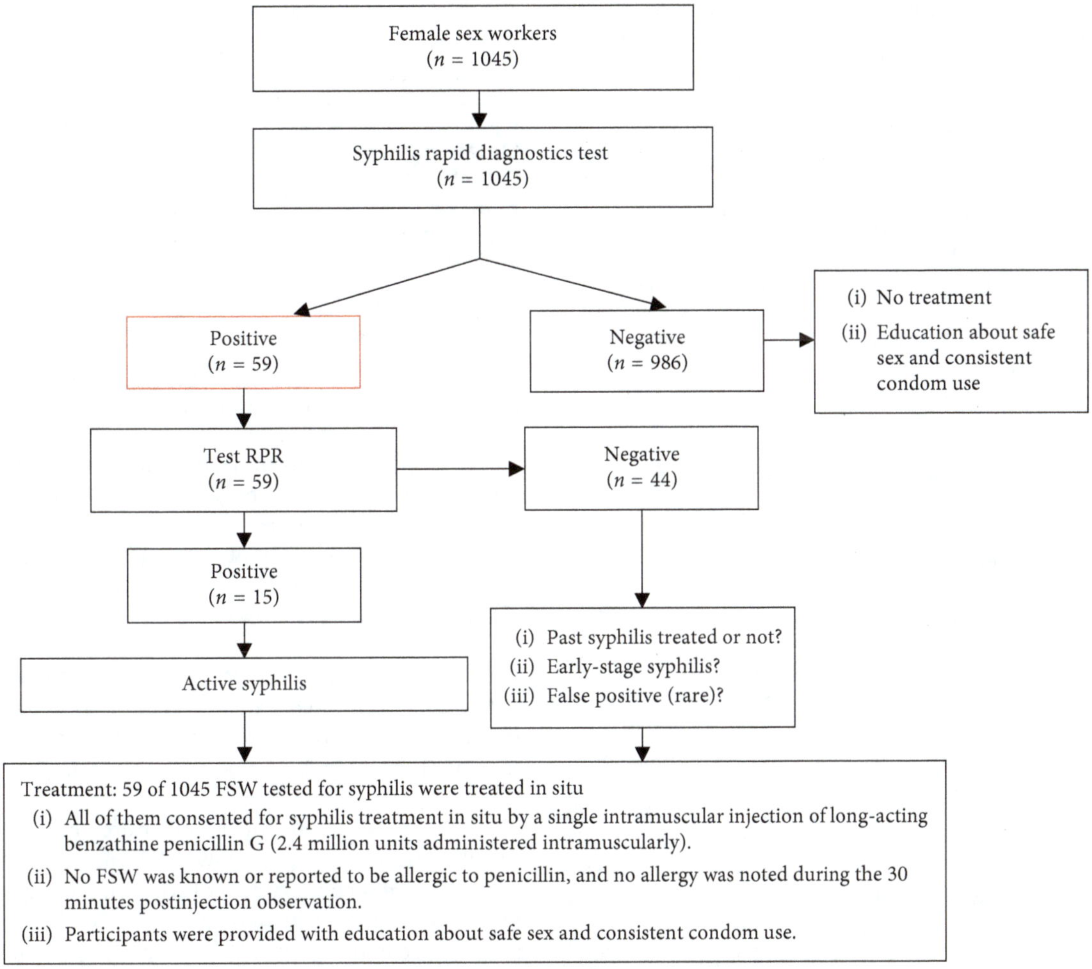

FIGURE 1: Flowchart of syphilis screening among study participants.

2.8. Ethical Issues. The National Ethics Committee for Health Research in Burkina Faso reviewed and approved the study protocol prior to participant recruitment. Collected data were anonymized with a unique identifier code so that no participant could be identified by their name. All FSW detected with positive antibodies were treated in situ with one intramuscular dose of benzathine benzylpenicillin G 2.4 MUI and kept under observation over 30 minutes to capture any immediate adverse event. Positive cases presenting allergy received doxycycline 100 mg tablet over 2 weeks. FSW were sensitized on safe sex (protection measures against STDs), were encouraged to undergo a serological test for syphilis every 3 to 6 months, and were provided with condoms and lubricants free of charge.

3. Results

A total of 1045 FSW were screened for syphilis in the five study cities: 349 FSW in Ouagadougou, 350 in Bobo-Dioulasso, 115 in Koudougou, 102 in Ouahigouya, and 129 in Tenkodogo. Approximately 44.7% (463/1036) of those interviewed self-reported a history of STD-related symptoms during the last twelve months prior to the survey. Only 1.3% (14/1042) underwent syphilis serological testing during the last twelve months.

3.1. FSW Characteristics. The mean age of the study population was 27.2 ± 0.2 years; 59.8% (95% CI: 56.8–62.7) were single and 67.8% (95% CI: 64.9–70.6) came from Burkina Faso. Approximately 31.1% (95% CI: 28.4–34.0) of FSW had no formal education while more than 36% (95% CI: 33.3–39.2) had a primary school education. Unemployment outside of sex work was common among FSW, 58% (95% CI: 54.9–60.9). Only 4.7% (95% CI: 3.6–6.2) were students. The proportion of FSW who were divorced, widowed, or separated was 30.4% (95% CI: 27.7–33.3); more than 70% of the participants were mothers (Table 1).

3.2. Prevalence of Syphilis Serological Markers. The prevalence of syphilis serological markers was 5.6% (95% CI: 4.4–7.2) among FSW overall, and active syphilis was found among 1.4% (95% CI: 0.9–2.4) of the participants. RDS weighted prevalence of syphilis serological markers and active syphilis by city were, respectively, estimated between 0.0% and 11.0% (95% CI: 8.1–14.7) and 0.0% and 2.2% (95% CI: 1.1–4.4) (Table 2). No syphilis markers were found among FSW in Ouahigouya. As shown in Table 3, the prevalence of syphilis markers increased with age ($p = 0.025$) and decreased with education level ($p < 0.001$).

TABLE 1: Characteristics of female sex workers in Burkina Faso.

Characteristics	Number (n)	Proportion (%)	95% CI
Age (years)			
18-19	118	11.3	9.5–13.4
20–24	364	34.8	32.0–37.8
25–29	245	23.4	21.0–26.1
≥30	318	30.4	27.7–33.3
Education level			
None	324	31.1	28.4–34.0
Primary	377	36.2	33.3–39.2
Secondary	340	32.7	29.9–35.6
Occupation other than sex work			
Student	49	4.7	3.6–6.2
Worker	390	37.4	34.5–40.4
Unemployed	604	57.9	54.9–60.9
Marital status			
Single	625	59.8	56.8–62.7
Married or cohabitating	102	9.8	8.1–11.7
Divorced/ separated/widowed	318	30.4	27.7–33.3
Number of biological children			
0	277	26.5	23.9–29.3
1	397	38.0	35.1–41.0
2+	371	35.5	32.7–38.4
Duration in sex work			
<1 year	237	23.0	20.5–25.7
1–5 years	506	49.2	46.1–52.2
6+ years	286	27.8	25.1–30.6
Number of clients per week			
1–14	451	45.9	42.8–49.1
15–21	241	24.5	21.9–27.3
22–28	71	7.2	05.7–09.0
29–80	219	22.3	19.8–25.0
Has been pregnant?			
No	188	18.0	15.8–20.5
Yes	856	82.0	79.5–84.2
Past 12 months consistent condom use			
No	403	39.5	36.5–42.5
Yes	618	605.	57.5–63.5
Past 12 months STD symptoms			
No	573	55.3	52.1–58.1
Yes	463	44.7	41.5–47.6
Past 12 months STD screening			
No	829	79.6	77.0–81.9
Yes	213	20.4	18.1–23.0
HIV status			
Positive	168	16.1	13.9–18.4
Negative	877	83.9	81.7–86.0
Syphilis serological marker status			
Positive	59	5.6	4.4–7.2
Negative	986	94.4	92.8–95.6
Active syphilis status			
Positive	15	1.4	0.9–2.4
Negative	1030	98.6	97.6–99.1

TABLE 1: Continued.

Characteristics	Number (n)	Proportion (%)	95% CI
Study site			
Ouagadougou	349	33.4	30.6–36.3
Bobo-Dioulasso	350	33.5	30.9–36.4
Koudougou	115	11.0	9.2–13.0
Ouahigouya	102	9.8	8.1–11.7
Tenkodogo	129	12.3	10.5–14.5
Country of origin			
Burkina Faso	709	67.8	64.9–70.6
Other countries	336	32.2	29.4–35.0
Overall	1045	100.0	

STD = sexually transmitted disease; CI = confidence interval.

Active syphilis was more prevalent among FSW with no education (2.8%) compared to those with a primary school education (0.8%) or secondary school education (0.6%) ($p = 0.037$). The country of origin (Burkina Faso and other countries) was not significantly associated with the presence of serological markers among FSW.

3.3. Factors Associated with Syphilis Serological Markers. The factors associated with the syphilis serological markers among FSW in bivariate and multivariate analyses are presented in Table 4.

3.4. Bivariate Analysis. In the bivariate analysis, the risk of syphilis serological markers was 80% lower in participants with at least a secondary school education compared to FSW with no education (OR = 0.20; $p < 0.001$). Participants with at least 22 clients per week were significantly more likely to have syphilis serological markers. FSW who reported 22 to 28 clients or at least 29 clients per week were, respectively, 2.97 (OR = 2.97; $p = 0.021$) and 3.34 (OR = 3.34; $p < 0.001$) times more at risk of serological marker carriage than those who reported less than 15 clients per week. FSW with a history of pregnancy were 4.31 times more likely to carry the markers compared to those who did not (OR = 4.31; $p = 0.014$). Participants who sold sex for at least 6 years were more likely to carry syphilis markers than those who sold sex for less than one year (OR = 3.85; $p = 0.004$). FSW in Bobo-Dioulasso had higher risk of syphilis serological marker carriage (OR = 2.87; $p = 0.001$).

3.5. Multivariate Analysis. In multivariate analysis, after adjustment for age, education level, marital status, number of clients per week, history of pregnancy, use of condoms, and the study city, the number of clients and education level were independently associated with syphilis serological marker carriage. No education was associated with an 77% increase in the risk of syphilis serological marker carriage as compared to a secondary school level of education (aOR = 0.23; $p = 0.002$). FSW having more than 29 clients per week were 2.62 times more likely to be syphilis serological marker positive compared to those with less than 15 clients per week (aOR = 2.62; $p = 0.008$).

Table 2: Active syphilis and syphilis serological markers prevalence among female sex workers by city.

Variables	All cities (n = 1045)	Ouagadougou (n = 349)		Bobo-Dioulasso (n = 350)		Koudougou (n = 115)		Ouahigouya (n = 102)		Tenkodogo (n = 129)	
	%	Unadjusted %	RDS-adjusted % (95% CI)	Unadjusted %	RDS-adjusted % (95% CI)	Unadjusted %	RDS-adjusted % (95% CI)	Unadjusted %	RDS-adjusted % (95% CI)	Unadjusted %	RDS-adjusted % (95% CI)
Syphilis serological markers	**5.6**	4.3	4.4 (2.7–7.2)	**11.4**	11.0 (8.1–14.7)	**1.7**	1.7 (0.4–6.6)	**0.0**	—	**1.6**	1.5 (4.4–7.2)
Active syphilis	**1.4**	1.4	1.4 (0.6–3.3)	**2.3**	2.2 (1.1–4.4)	**1.7**	1.7 (0.4–6.6)	**0.0**	—	**0.0**	—

TABLE 3: Prevalence of syphilis serological markers and of active syphilis among female sex workers in Burkina Faso.

FSW characteristics	n	Syphilis serological markers (%)	p	Active syphilis (%)	p
Age (years)					
18-19	118	3.4		0.0	
20–24	364	3.8		1.6	
25–29	245	5.3	0.025*	1.2	0.558
≥30	318	8.8		1.9	
All ages	**1045**	**5.6**		**1.4**	
Education level					
None	324	9.3		2.8	
Primary	377	5.4	<0.001***	0.8	0.037*
Secondary	340	2.1		0.6	
Occupation other than sex work					
Student	49	8.2		2.0	
Worker	390	4.2	0.216	1.3	0.723
Unemployed	604	5.7		1.5	
Marital status					
Single	625	4.5		0.8	
Married or cohabitating	102	6.9	0.133	1.0	0.038*
Divorced/separated/widowed	318	7.5		2.8	
Number of biological children					
0	277	26.5		0.36	
1	397	38.0	0.005	1.52	0.162
2+	371	35.5		2.16	
Duration in sex work					
<1 year	237	2.5		0.4	
1–5 years	506	5.3	0.005**	1.4	0.169
6+ years	286	9.1		2.5	
Number of clients per week					
1–14	451	3.5		0.9	
15–21	241	4.6	0.001***	0.4	0.014*
22–28	71	9.9		2.8	
29–80	219	11.0		3.7	
Has been pregnant					
No	188	1.6	0.008**	0.5	0.494
Yes	856	6.5		1.6	
Past 12 months consistent condom use					
No	403	6.9	0.124	1.2	0.418
Yes	618	5.0		1.6	
Past 12 months STD symptoms					
No	573	5.2	0.876	1.6	0.806
Yes	463	6.3		1.3	
Past 12 months STD screening					
No	829	6.1	0.177	1.8	0.051
Yes	213	3.8		0.0	
HIV status					
Positive	168	7.1	0.359	0.6	0.488
Negative	877	5.4		1.6	
Country of origin					
Burkina Faso	709	5.9	0.572	1.4	1.000
Other countries	336	5.1		1.5	

STD = sexually transmitted disease; $^{*}p < 0.05$, $^{**}p < 0.01$, and $^{***}p < 0.001$.

4. Discussion

This study found that the prevalence of syphilis serological markers and active syphilis were, respectively, 5.6% and 1.4% among FSW in Burkina Faso. The prevalence varied between study cities. To our knowledge, there are no recent data on syphilis among FSW in Burkina Faso. Overall, it appears that the prevalence of active syphilis (acute) was lower than results found in previous studies in Sub-Saharan Africa. A systematic review of congenital syphilis reported a decrease of 3.01% in syphilis prevalence between 1990 and 1999 and 1.48% between 2000 and 2009 [32]. Other studies have reported a similar reduction from 8.9% in 1993 to 1.5% in 1999 in Benin [33] and from 12.5% in 1991 to 0.8% from 1998 to 2000 in Burkina Faso [34]. This decrease of syphilis prevalence could be explained by several factors including

Table 4: Factors associated with syphilis serological markers among female sex workers in Burkina Faso.

Characteristics	FSW (n)	Syphilis serological markers (%)	Bivariate analysis			Multivariate analysis		
			OR	95% CI	p	aOR	95% CI	p
Age (years)								
18-19	118	3.4	1	—	—	—	—	—
20–24	364	3.8	1.14	0.36–3.53	0.820	—	—	—
25–29	245	5.3	1.59	0.50–5.00	0.422	—	—	—
≥30	318	8.8	2.75	0.94–8.02	0.064	1.18	0.30–4.63	0.805
All ages	1045	5.6						
Education level								
None	324	9.3	1	—	—	—	—	—
Primary	377	5.6	0.57	0.32–1.03	0.063	0.62	0.33–1.13	0.085
Secondary	340	2.1	0.20	0.09–0.47	<0.001	0.23	0.09–0. 60	0.002
Occupation other than sex work								
Student	49	8.2	1	—	—	—	—	—
Worker	390	4.1	0.48	0.15–1.50	0.208	—	—	—
Unemployed	604	5.7	0.77	0.26–2.27	0.644	—	—	—
Marital status								
Single	625	4.5	1	—	—	—	—	—
Married or cohabitating	102	6.9	1.57	0.66–3.69	0.301	—	—	—
Divorced/separated/widowed	318	7.5	1.74	0.99–3.05	0.054	0.83	0.40–1.72	0.631
Number of biological children								
0	277	2.17	1					
1	397	5.79	2.78	1.11–6.91	0.028	1.69	0.56–5.14	0.351
2+	371	8.09	3.97	1.63–9.68	0.002	1.26	0.38–4.15	0.706
Duration in sex work								
<1 year	237	2.5	1	—	—	—	—	—
1-5 years	506	5.3	2.17	0.88–5.32	0.091	1.04	0.37–2.92	0.933
6+ years	286	9.1	3.85	1.56–9.51	0.004	1.12	0.38–3.31	0.833
Number of clients per week								
1–14	451	3.5	1	—	—	—	—	—
15–21	241	4.6	1.30	0.59–2.84	0.512			
22–28	71	9.9	2.97	1.18–7.50	0.021	1.97	0.73–5.26	0.177
29–80	219	11.0	3.34	1.73–6.44	<0.001	2.62	1.28–5.35	0.008
Has been pregnant								
No	188	1.6	1	—	—	—	—	—
Yes	856	6.5	4.31	1.33–3.94	0.014	1.68	0.38–7.38	0.493
Past 12 months consistent condom use								
No	403	6.9	1	—	—	—	—	—
Yes	618	5.0	0.71	0.41–1.19	0.190	0.65	0.35–1.19	0.163
Past 12 months STD symptoms								
No	573	5.2	1					
Yes	463	6.3	1.21	0.71–2.04	0.710	—	—	—
Past 12 months STD screening								
No	829	6.1	1	—	—	—	—	—
Yes	213	3.8	1.20	0.71–2.04	0.478	—	—	—
HIV status								
Negative	877	5.4	1					
Positive	168	7.1	1.35	0.70–2.61	0.360	—	—	—
Study site								
Ouagadougou	349	4.3	1					
Bobo-Dioulasso	350	11.4	2.87	1.56–5.30	0.001	1.74	0.85–3.56	0.130
Koudougou	115	1.7	0.39	0.09–1.74	0.221	0.40	0.09–1.88	0.248
Ouahigouya	102	0.0	—			—		
Tenkodogo	129	1.5	0.35	0.08–1.55	0.168	0.19	0.02–1.53	0.120
Country of origin								
Burkina Faso	709	5.9	1	—	—	—	—	—
Other countries	336	5.1	0.82	0.47–1.46	0.515	—	—	—

STD = sexually transmitted disease; OR = odds ratio; aOR = adjusted odds ratio; CI = confidence interval.

the impact of sustained campaigns on the prevention of HIV and STDs over the last two decades in most countries [35]. This can also be explained by the introduction of syndromic management of STDs [36–38], and the widespread use of antibiotics in Africa may have contributed to the decrease. In addition to penicillin, the *Treponema pallidum* (bacterial cause of infectious syphilis) is also treated with other antibiotics such as cephalosporins (for example, cephaloridin), tetracyclins, macrolids (erythromycin and spiramycin), and chloramphenicol [2]. These antibiotics probably used for other indications have significantly contributed to reducing the incidence and evolution of infectious syphilis, including symptomatic and severe chronic forms [1].

Paradoxically, the prevalence of syphilis in our sample was increased, compared to the 2000 data [34]. This could be related to increase in sexual risk-taking, especially unprotected sex following high-active antiretroviral therapy availability and its effectiveness in HIV prevention that could lead to increase acquisition of other sexual-transmitted diseases such as syphilis [39, 40].

In addition, the prevalence of active syphilis in our study remains high in comparison with the estimate of 0.24% found in pregnant women in 2000 [41] and 1.2% in 2011 [42]. FSW are a leading group at high risk of syphilis infection due to several factors such as the multiple sexual partners and barriers to condom use [43]. Similar to prior research, in our study, FSW reporting more than 29 clients per week were three times more likely to be infected compared to those having less than 14 clients per week. Lower education was also a risk factor for syphilis infection in our study and more generally, for STDs in prior research [44, 45]. The control of STDs and particularly syphilis infection is a public health challenge [8]. Indeed, syphilis infection is very often asymptomatic and therefore untreated; thus, the potential for transmission is enhanced, and the risk of complication (including cervical cancer, higher genital tract infections, sterility, chronic pain, ectopic pregnancy, congenital abnormalities, spontaneous abortion, and associated neonatal and maternal mortality) is increased [46]. Syphilis-related genital ulcerations among FSW are important risk factors for HIV transmission [3], highlighting the relevance of early screening and treatment. In this study, all FSW who screened positive marker test were treated with a single intramuscular dose of benzathine benzylpenicillin G slow release 2.4 MUI [47] as per WHO recommendations for syphilis screening and treatment in high-risk groups, including FSW [30].

4.1. Limitations. In our study, the number of cases of active syphilis was low, which limited the ability to further identify factors associated with the acute infection through multivariate analysis. We instead used syphilis serological markers for analysis purposes. Furthermore, self-reporting sexual behavior in a context of social stigmatization of and discrimination toward FSW may have led to social desirability bias; however, this did not have impact on the prevalence of syphilis, as this was evaluated through syphilis serological markers. Our study provides updated data on syphilis serological markers in Burkina Faso.

5. Conclusion

This study provided updated epidemiological data on syphilis based on rapid diagnostic testing in situ among sex workers in Burkina Faso. We found that syphilis prevalence among FSW was relatively high and requires targeted preventive measures. Screening with rapid diagnostic tests followed by a single-dose administration of penicillin could be an effective approach to control syphilis among FSW and thus contribute to the prevention of HIV transmission. Safer sex and HIV prevention activities for FSW offer an opportunity to integrate screening and early treatment of syphilis.

Conflicts of Interest

The authors declare that there are no conflicts of interest regarding the publication of this article.

Authors' Contributions

S. Baral, S. Kouanda, H. G. Ouedraogo, O. Ky-Zerbo, and A. Grosso conceptualized the study and contributed to study design, implementation, and manuscript writing; I.B. Meda and I. Zongo assisted in manuscript writing; B. C. Samadoulougou and G. Tarnagda helped in study implementation, data cleaning, and manuscript review; K. Cissé, A. Sondo, N. Sawadogo, Y. Traore, and Nicolas Barro conducted the manuscript review.

Acknowledgments

The authors acknowledged the USAID-Project SEARCH, Task Order No. 2, the Global Fund to Fight AIDS, Tuberculosis, and Malaria, the Permanent Secretary of the National Council against AIDS (SP/CNLS), Burkina Faso, the Yerelon Association in Bobo-Dioulasso and all other associations that were involved in this study. They are grateful to all the study participants. The USAID and Project SEARCH, Task Order no. 2, are funded by the US Agency for International Development under Contract no. GHH-I-00-0700,032-00, beginning from September 30, 2008, and supported by the President's Emergency Plan for AIDS Relief. The Research to Prevention Project is led by the Johns Hopkins Center for Global Health and managed by the Johns Hopkins Bloomberg School of Public Health Center for Communication Programs.

References

[1] E. W. Hook, "Syphilis," *The Lancet*, vol. 389, no. 10078, pp. 1550–1557, 2017.

[2] A. E. Singh and B. Romanowski, "Syphilis: review with emphasis on clinical, epidemiologic, and some biologic features,"

Clinical Microbiology Reviews, vol. 12, no. 2, pp. 187–209, 1999.

[3] R. M. Greenblatt, S. A. Lukehart, F. A. Plummer et al., "Genital ulceration as a risk factor for human immunodeficiency virus infection," *AIDS*, vol. 2, no. 1, pp. 47–50, 1988.

[4] Syphilis and HIV, March 2017, http://hivinsite.ucsf.edu/InSite?page=kb-00&doc=kb-05-01-04.

[5] K. Salado-Rasmussen, "Syphilis and HIV co-infection. Epidemiology, treatment and molecular typing of *Treponema pallidum*," *Danish Medical Journal*, vol. 62, no. 12, p. B5176, 2015.

[6] J. Stevenson and M. Heath, "Syphilis and HIV infection: an update," *Dermatologic Clinics*, vol. 24, no. 4, pp. 497–507, 2006.

[7] N. M. Zetola and J. D. Klausner, "Syphilis and HIV infection: an update," *Clinical Infectious Diseases*, vol. 44, no. 9, pp. 1222–1228, 2007.

[8] World Health Organization, *Prevention and Treatment of HIV and other Sexually Transmitted Infections for Sex Workers in Low- and Middle-Income Countries: Recommendations for a Public Health Approach*, World Health Organization, Geneva, Switzerland, 2012.

[9] D. Kang, M. Liao, Z. Jiang et al., "Commercial sex venues, syphilis and methamphetamine use among female sex workers," *AIDS Care*, vol. 23, no. 1, pp. 26–36, 2011.

[10] C. Zhou, K. Rou, W. M. Dong et al., "High prevalence of HIV and syphilis and associated factors among low-fee female sex workers in mainland China: a cross-sectional study," *BMC Infectious Diseases*, vol. 14, no. 1, p. 225, 2014.

[11] X. H. Tao, T. Jiang, D. Shao et al., "High prevalence of syphilis among street-based female sex workers in Nanchang, China," *Indian Dermatology Online Journal*, vol. 5, no. 4, pp. 449–455, 2014.

[12] S. Kakchapati, D. R. Singh, B. B. Rawal, and A. Lim, "Sexual risk behaviors, HIV, and syphilis among female sex workers in Nepal," *HIV/AIDS-Research and Palliative Care*, vol. 9, pp. 9–18, 2017.

[13] W. A. Halatoko, D. E. Landoh, B. Saka et al., "Prevalence of syphilis among female sex workers and their clients in Togo in 2011," *BMC Public Health*, vol. 17, no. 1, 2017.

[14] M. Elhadi, A. Elbadawi, S. Abdelrahman et al., "Integrated bio-behavioural HIV surveillance surveys among female sex workers in Sudan, 2011-2012," *Sexually Transmitted Infections*, vol. 89, no. 3, pp. iii17–iii22, 2013.

[15] G. Ramjee, S. S. Karim, and A. W. Sturm, "Sexually transmitted infections among sex workers in KwaZulu-Natal, South Africa," *Sexually Transmitted Diseases*, vol. 25, no. 7, pp. 346–349, 1998.

[16] S. L. Nicolas, M. L. Sangare, I. P. Compaore et al., "Prevalence and risk of HIV infection among female sex workers in Burkina Faso," *International Journal of STD & AIDS*, vol. 9, no. 3, pp. 146–150, 1998.

[17] F. Scorgie, D. Nakato, E. Harper et al., "'We are despised in the hospitals': sex workers' experiences of accessing health care in four African countries," *Culture, Health & Sexuality*, vol. 15, no. 4, pp. 450–465, 2013.

[18] R. K. Wanyenze, G. Musinguzi, J. Kiguli et al., ""When they know that you are a sex worker, you will be the last person to be treated": perceptions and experiences of female sex workers in accessing HIV services in Uganda," *BMC International Health and Human Rights*, vol. 17, no. 1, p. 11, 2017.

[19] H.-Y. Kim, A. Grosso, O. Ky-Zerbo et al., "Stigma as a barrier to health care utilization among female sex workers and men who have sex with men in Burkina Faso," *Annals of Epidemiology*, vol. 28, no. 1, pp. 13–19, 2018.

[20] A. Davis, B. E. Meyerson, B. Aghaulor et al., "Barriers to health service access among female migrant Ugandan sex workers in Guangzhou, China," *International Journal for Equity in Health*, vol. 15, no. 1, 2016.

[21] Y. Lafort, F. Lessitala, B. Candrinho et al., "Barriers to HIV and sexual and reproductive health care for female sex workers in Tete, Mozambique: results from a cross-sectional survey and focus group discussions," *BMC Public Health*, vol. 16, no. 1, p. 608, 2016.

[22] T. Wahed, A. Alam, S. Sultana et al., "Barriers to sexual and reproductive healthcare services as experienced by female sex workers and service providers in Dhaka city, Bangladesh," *PLoS One*, vol. 12, no. 7, Article ID e0182249, 2017.

[23] S. Baral, C. Beyrer, K. Muessig et al., "Burden of HIV among female sex workers in low-income and middle-income countries: a systematic review and meta-analysis," *The Lancet Infectious Diseases*, vol. 12, no. 7, pp. 538–549, 2012.

[24] S. Baral, S. Ketende, J. L. Green et al., "Reconceptualizing the HIV epidemiology and prevention needs of female sex workers (FSW) in Swaziland," *PLoS One*, vol. 9, no. 12, Article ID e115465, 2014.

[25] C. Kaushic, K. L. Roth, V. Anipindi, and F. Xiu, "Increased prevalence of sexually transmitted viral infections in women: the role of female sex hormones in regulating susceptibility and immune responses," *Journal of Reproductive Immunology*, vol. 88, no. 2, pp. 204–209, 2011.

[26] HIV Prevention through Early Detection and Treatment of Other Sexually Transmitted Diseases—United States Recommendations of the Advisory Committee for HIV and STD Prevention, March 2017, https://www.cdc.gov/mmwr/preview/mmwrhtml/00054174.htm.

[27] Syphilis and HIV: the intersection of two epidemics, NEJM Journal Watch, http://www.jwatch.org/ac201009030000001/2010/09/03/syphilis-and-hiv-intersection-two-epidemics.

[28] D. Heckathorn Douglas, "Respondent-driven sampling: a new approach to the study of hidden populations," *Social Problems*, vol. 44, no. 2, pp. 174–199, 1997.

[29] H. G. Ouedraogo, O. Ky-Zerbo, A. Baguiya et al., "HIV among female sex workers in five cities in Burkina Faso: a cross-sectional baseline survey to inform HIV/AIDS Programs," *AIDS Research and Treatment*, vol. 2017, Article ID 9580548, 11 pages, 2017.

[30] World Health Organisation/TDR, *The Use of Rapid Syphilis Test*, World Health Organisation/TDR, Geneva, Switzerland, 2006.

[31] F. Bocoum, H. Ouedraogo, G. Tarnagda et al., "Evaluation of the diagnostic performance and operational characteristics of four rapid immunochromatographic syphilis tests in Burkina Faso," *African Health Sciences*, vol. 15, no. 2, p. 360, 2015.

[32] C. R. Kenyon, K. Osbak, and A. Tsoumanis, "The global epidemiology of syphilis in the past century—a systematic review based on antenatal syphilis prevalence," *PLoS Neglected Tropical Diseases*, vol. 10, no. 5, article e0004711, 2016.

[33] M. Alary, L. Mukenge-Tshibaka, F. Bernier et al., "Decline in the prevalence of HIV and sexually transmitted diseases among female sex workers in Cotonou, Benin, 1993-1999," *AIDS*, vol. 16, no. 3, pp. 463–470, 2002.

[34] N. Nagot, N. Meda, A. Ouangre et al., "Review of STI and HIV epidemiological data from 1990 to 2001 in urban Burkina Faso: implications for STI and HIV control," *Sexually Transmitted Infections*, vol. 80, no. 2, pp. 124–129, 2004.

[35] P. D. Ghys, M. O. Diallo, V. Ettiègne-Traoré et al., "Increase in condom use and decline in HIV and sexually transmitted

diseases among female sex workers in Abidjan, Côte d'Ivoire, 1991–1998," *AIDS*, vol. 16, no. 2, pp. 251–258, 2002.

[36] B. Vuylsteke, "Current status of syndromic management of sexually transmitted infections in developing countries," *Sexually Transmitted Infections*, vol. 80, no. 5, pp. 333-334, 2004.

[37] L. F. Johnson, R. E. Dorrington, D. Bradshaw, and D. J. Coetzee, "The effect of syndromic management interventions on the prevalence of sexually transmitted infections in South Africa," *Sexual & Reproductive Healthcare*, vol. 2, no. 1, pp. 13–20, 2011.

[38] C. R. Kenyon, K. Osbak, and R. M. Chico, "What underpins the decline in syphilis in Southern and Eastern Africa? An exploratory ecological analysis," *International Journal of Infectious Diseases*, vol. 29, pp. 54–61, 2014.

[39] S. Scheer, P. L. Chu, J. D. Klausner, M. H. Katz, and S. K. Schwarcz, "Effect of highly active antiretroviral therapy on diagnoses of sexually transmitted diseases in people with AIDS," *The Lancet*, vol. 357, no. 9254, pp. 432–435, 2001.

[40] M. L. Rekart, W. Ndifon, R. C. Brunham et al., "A double-edged sword: does highly active antiretroviral therapy contribute to syphilis incidence by impairing immunity to *Treponema pallidum*?," *Sexually Transmitted Infections*, vol. 93, no. 5, pp. 374–378, 2017.

[41] I. Sombie, N. Meda, M. Cartoux et al., "Seroprevalence of syphilis among women attending urban antenatal clinics in Burkina Faso, 1995–8," *Sexually Transmitted Infections*, vol. 76, no. 4, pp. 314–316, 2000.

[42] F. Kirakoya-Samadoulougou, M.-C. Defer, S. Yaro et al., "Low seroprevalence of syphilis in Burkina Faso," *Sexually Transmitted Infections*, vol. 1, no. 87, pp. 35–37, 2010.

[43] R. Cai, J. G. Tan, L. Chen, J. H. Richardus, and S. J. de Vlas, "Prevalence and risk factors of syphilis infection among female sex workers in Shenzhen, China: an observational study (2009–2012)," *Tropical Medicine & International Health*, vol. 18, no. 12, pp. 1531–1538, 2013.

[44] M. M. Solomon, M. J. Smith, and C. Del Rio, "Low educational level: a risk factor for sexually transmitted infections among commercial sex workers in Quito, Ecuador," *International Journal of STD & AIDS*, vol. 19, no. 4, pp. 264–267, 2008.

[45] D. D. Celentano, K. E. Nelson, S. Suprasert et al., "Epidemiologic risk factors for incident sexually transmitted diseases in young Thai men," *Sexually Transmitted Diseases*, vol. 23, no. 3, pp. 198–205, 1996.

[46] L. Newman, M. Kamb, S. Hawkes et al., "Global estimates of syphilis in pregnancy and associated adverse outcomes: analysis of multinational antenatal surveillance data," *PLoS Medicine*, vol. 10, no. 2, article e1001396, 2013.

[47] M. E. Clement, N. L. Okeke, and C. B. Hicks, "Treatment of syphilis: a systematic review," *JAMA*, vol. 312, no. 18, pp. 1905–1917, 2014.

21

Vancomycin-Resistant Enterococci and Its Associated Risk Factors among HIV-Positive and -Negative Clients Attending Dessie Referral Hospital, Northeast Ethiopia

Seid Ali,[1] Martha Alemayehu,[2] Mulat Dagnew,[2] and Teklay Gebrecherkos[2]

[1]*Microbiology Laboratory Unit, Dessie Referral Hospital, Amhara Region, Dessie, Ethiopia*
[2]*Department of Medical Microbiology, School of Biomedical and Laboratory Sciences, College of Medicine and Health Sciences, University of Gondar, Gondar, Ethiopia*

Correspondence should be addressed to Teklay Gebrecherkos; estiftg17@gmail.com

Academic Editor: Todd R. Callaway

Background. Enterococci are becoming the most important public health concern and emerging as multidrug-resistant organisms around the world including Africa particularly in Ethiopia where there is a lack of availability of effective antimicrobial drugs. However, there is a paucity of data on the prevalence and associated risk factors of vancomycin-resistant enterococci in Ethiopia. *Objective.* This study was aimed to assess the prevalence of vancomycin-resistant enterococci and its associated risk factors among HIV-positive and -negative clients. *Methods.* A comparative cross-sectional study was conducted from February to May, 2017, on 300 participants at Dessie Referral Hospital. Data were gathered using a pretested structured questionnaire, stool samples were collected and inoculated on to bile esculin agar, and presumptive colonies were inoculated in brain-heart infusion broth containing 6.5% NaCl for selective identification of enterococci. Antibiotic susceptibility tests were done using the Kirby–Bauer disk diffusion method. Data were analyzed using SPSS version 22 software package. *Results.* A total of 300 study participants were enrolled in this study, of which 57.7% were females with a mean age of 34.4, a range of 19–73 years. The overall prevalence of enterococci was 37.3%. The prevalence of VRE was 6.3%. From all isolates, the prevalence of VRE among HIV-positive and -negative clients was 5.9% and 7.4%, respectively. Resistance gentamicin, ampicillin, penicillin, and erythromycin was 37.5%, 34.8%, 34.8%, and 22.3%, respectively. Prevalence of multidrug resistance was (29.5%). Being low in hemoglobin content was significantly associated with VRE. *Conclusion.* The high prevalence of VRE and multidrug-resistant enterococci in this study signals the emergence of VRE. Detection of VRE in this study indicates decreased antibiotic treatment options of multidrug-resistant enterococci. Therefore, there should be a need to perform continuous surveillance, rational use of antibiotics, and more detailed study using phenotypic and genotypic methods.

1. Introduction

Enterococci are Gram-positive bacteria arranged in pairs and short chains, and they are part of facultative anaerobic normal flora of both humans and animals. They are normally found in the oropharynx, vagina, urethra, and skin of healthy persons [1]. Previously, it is considered as bacteria of minimal clinical impact, particularly *Enterococcus faecium* and *Enterococcus faecalis*, and has now emerged as one of the major causes of human clinical infections [2]. Enterococci are productive colonizers, with genome flexibility and a tendency for persistency in hospital environments. It causes urinary tract infections mainly followed by intra-abdominal and pelvic infections. They also cause surgical wound infections, bacteremia, endocarditis, neonatal sepsis, and rarely meningitis [3].

The growing evidences show that enterococci are among the third and fourth leading cause of nosocomial infections worldwide [4]. Studies indicate that, in hospitals, it is the second most common cause of catheter-associated bloodstream infections [5]. For instance in the United States, 18% of all bloodstream infections were due to enterococci, ranking as second from overall infections [6]. In Africa, enterococci prevalence was documented in different clinical samples ranging from 5.7% to 88.9% [7–9].

Moreover, enterococci acquire resistance to currently available drugs either by mutation or receipt of foreign genetic materials through the transfer of plasmids and transposons, or they have a selective pressure over other microorganisms of the intestinal flora, allowing for overgrowth of intrinsic vancomycin-resistant enterococci (VRE). Despite the fact that the antibiotics are no longer effective due to the emergence of VRE is becoming a grave public health concern especially in developing countries where there is a lack of availability of effective antibiotics [10]. Furthermore, VRE have been emerging and posing a therapeutic challenge to physicians due to the ease of acquiring and transferring antimicrobial-resistant genes and presence of different selection pressures for VRE proliferation and rapid expansion of resistant populations.

Evidences showed that previous administration of antibiotic treatment, concurrent infections, surgery, catheterizations, duration of hospital stay, presence of pervious hospitalization, and underlying diseases like cancer, HIV, and diabetics are among the risk factors associated with the spread of enterococci infections [4, 7, 8].

There were published studies concerning the burden of enterococci, drug-resistant enterococci, and VRE worldwide and in Africa. However, little is known about the prevalence of VRE in the susceptible group of individuals like HIV-infected patients in Ethiopia. Thus, the aim of this study was to determine the prevalence of enterococci, VRE, and other antimicrobial-resistant enterococci among HIV-positive and -negative clients at Dessie Referral Hospital.

2. Materials and Methods

2.1. Study Area, Design, and Period. A hospital-based comparative cross-sectional study was conducted from February to May, 2017, at Dessie Referral Hospital, South Wollo zone of Amhara regional state, Northeast Ethiopia. This hospital is found in Dessie town with a distance of 400 km from the capital (Addis Ababa) and 471 km far from Bahir Dar, which is the capital city of the Amhara regional state. According to 2007 population and housing census of Ethiopia, the town had a total population of 151,094; among these, 72,891 were males, and 78,203 were females.

In Dessie town, there are one referral hospital, three private general hospitals, five health centers, eight private higher clinics, and one regional health and research laboratory. Dessie Referral Hospital provides emergency, ART services, chronic care, surgical, dental, medical, pediatric, gynecologic, obstetric, and other services for more than 4 million clients annually.

2.2. Study Population, Sample Size, and Sampling Procedure. HIV-positive patients from ART clinic and healthy blood donors attending Dessie Referral Hospital during the study period were enrolled in this study. The sample size was determined by using EPI INFO version 7 in double population proportion. Since there was no previous study done in this area, 50% prevalence was given to HIV-negative controls, while the proportion given for HIV-positive cases was 70%. This was depending on an assumption that HIV-positive clients have 20% more VRE carriage rate than HIV-negative clients [7]. EpiInfo soft was used to determine the sample size by considering power 80% and confidence interval at 95% for the two sample sizes with a 1 : 1 ratio; the initial sample size was 206, by considering 10% nonresponse rate with a total of 300 study participants (150 from HIV positives and 150 from HIV negatives). A systematic random sampling technique was used to select study participants.

2.3. Definition of Terms

(i) Body mass index (BMI): BMI is a simple index of weight-for-height that is commonly used and is defined as the weight in kilograms divided by the square of the height in meters (kg/m^2). It is used to assess the physical and nutritional status.

(ii) Hospitalization: Admission and waiting in the hospital for more than 48 hours or visiting health institutions repeatedly.

(iii) MDR: Bacteria that is nonsusceptible to ≥3 antimicrobial class.

3. Data Collection and Laboratory Methods

3.1. Data Collection. Data were collected using a short interview guided by a pretested structured questionnaire consisting of the client's information like demographic characteristics, dates of recent hospitalization, immune status, duration of hospitalization before the onset of the sample collection, and invasive procedures which was obtained for each study subject either from the clinical record or from the interview. Five milliliters of blood sample was collected by experienced laboratory technologists. The HIV status of the clients was determined using rapid test kits following the current Ethiopian algorithm (for healthy blood donors), and CD4 count, WHO staging of HIV, and ART linkage (for HIV positives) were collected from their charts/request papers. Data were collected by trained nurses with previous experience and speaking local language.

3.2. Stool Sample Collection. Each patient was instructed how to collect stool specimens, and about 5 mg of the fecal specimen was collected in a sterile wide-mouth screw-capped container from each patient and labeled with the unique sample number, date, and time of collection. Immediately, it was delivered to bacteriology laboratory (Dessie Regional Health and Research Laboratory) for culturing within 30 minutes. The collected samples were kept in the refrigerator for 1-2 hours if unprocessed immediately [11].

3.3. Sample Processing and Culture Identification. The media were prepared as per the manufacturers' instructions. The stool sample collected from each client was streaked on sterile bile esculin agar media (BEA) (OXOID, Hardy Diagnostics, Santa Maria, CA, USA) and incubated for 24–48 hours at 37°C. Plates were observed for appearance of characteristic colonies of growth and blackening. Typical characteristic colonies were selected randomly for characterization and presumptive identification of enterococci. Each isolate was also assessed using Gram staining, catalase test, and its growth in 6.5% NaCl broth. Presumptive pure colonies were picked and inoculated into brain-heart infusion (BHI) broth and incubated at 45°C for 24 hours, and growth in the medium was assessed by its turbidity. An isolate fulfilling the above criteria was assumed to be *Enterococcus* species [12].

3.4. Antimicrobial Susceptibility Testing. The antimicrobial susceptibility testing of the isolates was performed on Mueller–Hinton agar (OXOID, UK) by the Kirby–Bauer disk diffusion technique as modified by the Clinical and Laboratory Standard Institute in 2017 [13]. After a pure culture was obtained, a loop full of bacteria was taken from a colony and was transferred to a tube containing 5 ml of sterile normal saline (0.85% NaCl) and mixed gently until it formed a homogenous suspension. The turbidity of the suspension was determined in comparison with 0.5 McFarland standards. A sterile swab was dipped into the suspension, and excess suspension was removed by pressing the swab against the wall of the tube. The swab was used to distribute the bacteria suspension evenly over the entire surface of MHA.

The inoculated plates were left at room temperature for 3–5 minutes until it dried. The antimicrobial impregnated disks were placed with sterile forceps on the agar surface at least 24 mm away from each other to avoid the overlapping zone of inhibition. After placing the disk, the plate was allowed to stand for 30 minutes to dissolve the antibiotic in the media. Then, the plates were inverted and incubated at 37°C for 18–24 hrs, and the diameter of the zone of inhibition was determined using a ruler. Susceptibility pattern was categorized as sensitive, intermediate, and resistant by comparison of the zone of inhibition with a standard indicated in the manufacturer's guide.

The isolated enterococci were tested against penicillin (10 IU), ampicillin (10 μg), vancomycin (30 μg), gentamicin (30 μg), ciprofloxacin (5 μg), erythromycin (15 μg), and chloramphenicol (30 μg).

Then, the plates were incubated at 37°C for 24 hours, and the results were interpreted according to the most recent version of Clinical Laboratory Standard Institute (CLSI 2017) [13].

3.5. Data Quality Assurance. The standard reference bacterial strains such as the reference strain of *S. aureus* ATCC25923 and *E. faecalis* ATCC51299 were used as negative and positive controls, respectively, and tested weekly as controls on the biochemical tests and agar plates including Mueller–Hinton agar with antimicrobial discs to ensure testing performance of the potency of antimicrobial discs. Sterility of the culture media was tested by incubating 5% of the batch at 35–37°C overnight and was evaluated for possible contamination. In addition, the whole procedures and result interpretation were cross-checked by senior laboratory professionals of the Dessie regional health and research laboratory.

3.6. Data Processing and Analysis. Data were checked for completeness, cleaned manually, entered, and analyzed using the Statistical Package for Social Science (SPSS) version 22. Frequencies and cross tabulations were used to summarize descriptive statistics. The association of dependent and independent variables was measured by using bivariate and multivariate analysis (logistic regression). The strength of the association was measured by the odds ratio. At 95% of confidence interval, a p value < 0.05 was considered as statistically significant.

3.7. Ethical Consideration. The study was conducted after getting ethical approval and support letter from the ethical Committee of School of Biomedical and Laboratory Sciences, College of Medicine and Health Sciences, University of Gondar. Written informed consent was obtained from each of the study participants. For each confirmed resistant case, the responsible clinicians of the patient were informed about the results of the clients using letters that contain results of all clients.

4. Results

4.1. Sociodemographic Characteristics. A total of 300 study participants were enrolled in this study. The mean age and standard deviation was 34.4 ± 10.01, and age range was from 19 to 73 years. Of the 300 study participants, 150 were HIV positive (with mean age and standard deviation of 38.50 ± 10.95, age range 19–73) and 150 were HIV negative (with mean and standard deviation of 30.35 ± 6.91, age range 19–56). Majority of the study participants were females, 173/300 (57.3%), and their educational level was mainly primary education, 89/300 (29.7%). In addition, most of the study participants were married, and 166/300 (55.3%) and 255/300 (85%) participants were urban dwellers (Table 1).

4.2. Prevalence of Enterococci and VRE among HIV-Positive and HIV-Negative Clients. From the total of 300 study participants, colonization of enterococci species was seen on 112 (37.33%) stool specimens (95% CI = 31.2–42.8%). Of this, 85 (56.7%) (95% CI = 48–64%) of HIV-positive clients and 27 (18%) (95% CI = 11.7–24) of HIV-negative clients were colonized with isolates of enterococci species (Figure 1).

From all 112 isolates, 7 (6.3%) were vancomycin-resistant enterococci (VRE) (95% CI = 1.8–10.7). Of all enterococci isolates, VRE among HIV-positive and -negative clients were 5/85 (5.9%) (CI = 2.4–10.7) and 2/27 (7.4%) (95% CI = 0–18.5), respectively.

4.3. Antimicrobial-Resistant Patterns of Enterococci to Other Antibiotics. The resistance pattern of enterococci for each ampicillin and penicillin was 39 (34.8%); of these, 32 (37.6%) and 7 (25.9%) isolates were resistant to HIV-positive and HIV-negative clients, respectively. Forty-two (37.5%) isolates were resistant to gentamicin, of which 38 (44.7%) were from HIV-positive clients and 4 (14.8%) were from HIV-negative clients. Twenty-five (22.3%) isolates were resistant

Table 1: Sociodemographic characteristics of HIV-positive and -negative clients at Dessie Referral Hospital, Northeast Ethiopia, 2017.

Sociodemographic characteristics	HIV status		Total (n = 300), N (%)
	HIV positive (n = 150), N (%)	HIV negative (n = 150), N (%)	
Sex			
Male	52 (34.7)	75 (50.0)	127 (42.3)
Female	98 (65.3)	75 (50.0)	173 (57.7)
Age group			
18–30	44 (29.3)	95 (63.3)	139 (46.3)
31–42	52 (34.7)	47 (31.3)	99 (33.0)
43–54	41 (27.3)	7 (4.7)	48 (16)
>54	13 (4.3)	1 (0.7)	14 (4.7)
Marital status			
Single	25 (8.3)	66 (44)	91 (30.3)
Married	83 (55.3)	83 (55.3)	166 (55.3)
Divorced	26 (17.3)	1 (0.7)	27 (9.0)
Widowed	16 (10.7)	0	16 (5.3)
Current			
Rural	28 (18.7)	17 (11.3)	45 (15)
Residence			
Urban	122 (81.3)	133 (88.7)	255 (85)
Education			
Cannot read and write	19 (12.7)	5 (3.3)	24 (8.0)
Primary	55 (36.7)	34 (22.7)	89 (29.7)
Secondary	52 (34.7)	40 (26.7)	92 (30.7)
College+	24 (16.0)	71 (47.3)	95 (31.6)

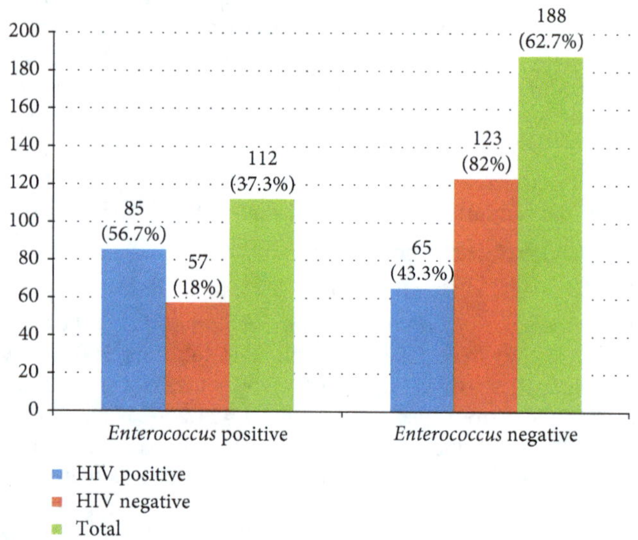

Figure 1: Prevalence of enterococci among HIV-positive and -negative clients at Dessie Referral Hospital, Northeast Ethiopia, 2017.

to erythromycin; all of the isolates (29.4%) were from HIV-positive clients only.

Statistically significant difference of resistant patterns was seen on gentamicin and erythromycin between isolates among HIV-positive and HIV-negative clients ($p < 0.05$). However, no statistically significant difference was observed in resistant patterns for ampicillin, penicillin, ciprofloxacin, chloramphenicol, and vancomycin among HIV-positive and HIV-negative clients ($p > 0.05$) (Table 2).

From the total enterococci isolates, 60 (53.6%) were resistant to one or more antibiotics; of those isolates, 21 (18.8%), 14 (12.5%), 20 (17.9%), and 5 (4.5%) were resistant to one, two, three, and four antibiotics, respectively. In this study, the overall prevalence of multiple drug resistance (MDR) enterococci was found to be 25/112 (22.3%) (Table 3).

In this study, the most sensitive antibiotic for enterococci isolates was vancomycin (93.7%), while ciprofloxacin and erythromycin were 100% susceptible in HIV-negative individuals. General susceptibility pattern for all antibiotics was shown in (Table 4).

4.4. Factors Associated with the Emergence of Vancomycin-Resistant Enterococci. Multivariate analysis was done on those variables that were in significant association with bivariate analysis and $n = 27$ less than 0.2. Accordingly, being low in hemoglobin content was significantly associated with vancomycin-resistant enterococci ((AOR = 19.18, 95% CI = 1.08–89.36), P value = 0.04). Clients with low-hemoglobin level were 19.2 times more affected by VRE than normal ones; however, other sociodemographic characteristics and risk factors were not statistically significant with VRE in multivariate analysis (Table 5).

5. Discussion

Enterococci are productive colonizers with a tendency for persistency in hospital environments. Likewise, VRE have been emerging and posing a therapeutic challenge to physicians due to the ease of acquiring and transferring antimicrobial-resistant genes [14].

Table 2: Antimicrobial-resistant patterns among HIV-positive and -negative clients at Dessie Referral Hospital, Northeast Ethiopia, 2017.

Resistant pattern		HIV status		Total (n = 112), N (%)	X^2	p value
		HIV positive (n = 85), N (%)	HIV negative (n = 27), N (%)			
Penicillin	R	32 (37.6)	7 (25.9)	39 (34.8)	1.30	0.52
Ampicillin	R	32 (37.6)	7 (25.9)	39 (34.8)	1.30	0.52
Vancomycin	R	5 (5.9)	2 (7.4)	7 (6.3)	0.08	0.78
Gentamicin	R	38 (44.7)	4 (14.8)	42 (37.5)	8.4	0.02*
Ciprofloxacin	R	8 (9.4)	0 (0%)	8 (7.1)	2.7	0.09
Chloramphenicol	R	7 (8.2)	3 (11.1)	10 (8.9)	0.99	0.61
Erythromycin	R	25 (29.4)	0 (0%)	25 (22.3)	11.05	0.00*

R = resistance. * = significantly associated.

Table 3: Profile of multidrug-resistant enterococci isolates at Dessie Referral Hospital, Northeast, Ethiopia, 2017.

Resistant pattern	Combination of antibiotics	Number (%) of isolates tested
R3	PEN, AMP, VAN, GENTA	4 (16)
	PEN, AMP, GENTA, ERY	8 (32)
	PEN, AMP, GENTA, CIP	5 (20)
	PEN, AMP, GENTA, CAF	1 (4)
	GENTA, CAF, ERY	2 (8)
R4	PEN, AMP, VAN, GENTA, ERY	3 (12)
	PEN, AMP, GENTA CAF, ERY,	1 (4)
	PEN, AMP, GENTA, CIPR, ERY	1 (4)
Total MDR	25 (100)	—

MDR = multidrug resistance; PEN = penicillin; AMP = ampicillin; GENTA = gentamicin; CIPR = ciprofloxacin; CAF = chloramphenicol; VAN = vancomycin; ERY = erythromycin; R2–R4 = number of antibiotics category resistance from 3 to 4, respectively.

Table 4: Antimicrobial susceptibility patterns among HIV-positive and -negative clients at Dessie Referral Hospital Northeast, Ethiopia, 2017.

Antibiotics pattern		Enterococci		Total (n = 112), N (%)
		Positive (n = 85), N (%)	Negative (n = 27), N (%)	
Penicillin	S	51 (60)	19 (70.4)	70 (62.5)
	I	2 (2.4)	1 (3.7)	3 (2.7)
	R	32 (37.6)	7 (25.9)	39 (34.8)
Ampicillin	S	51 (60)	19 (70.4)	70 (62.5)
	I	2 (2.4)	1 (3.7)	3 (2.7)
	R	32 (37.6)	7 (25.9)	39 (34.8)
Vancomycin	S	80 (94.1)	25 (92.6)	105 (93.7)
	I	0	0	0
	R	5 (5.9)	2 (7.4)	7 (6.3)
Gentamicin	S	46 (54.1)	23 (85.2)	69 (61.6)
	I	1 (1.2)	0	1 (0.9)
	R	38 (44.7)	4 (14.8)	42 (37.5)
Ciprofloxacin	S	77 (90.6)	27 (100)	104 (92.9)
	I	0	0	0
	R	8 (9.4)	0	8 (7.1)
Chloramphenicol	S	77 (90.6)	23 (85.2)	100 (89.3)
	I	1 (1.2)	1 (3.7)	2 (1.8)
	R	7 (8.2)	3 (11.1)	10 (8.9)
Erythromycin	S	58 (68.2)	25 (92.6)	83 (74.1)
	I	2 (2.4)	2 (7.4)	4 (3.4)
	R	25 (29.4)	0	25 (22.3)

S = susceptible; I = intermediate; R = resistance.

The overall prevalence of enterococci in this study was found to be 37.33%. This finding is comparable with the previous report of India, which was reported to be 39.3% (19), and Shiraz, Iran, 39.5% [15]. However, it is higher than the previous results reported from Nigeria, 5.9% [16], and Gondar, Ethiopia, 6.2% [8]. The difference might be due to the variation of the sample type, which were clinical samples including blood, urine, wound swabs, and sputum in the

Table 5: Multivariate analysis of sociodemographic characteristics and risk factors for VRE among HIV-positive and -negative clients at Dessie Referral Hospital, Northeast Ethiopia, 2017.

	Enterococci (112)		p value	AOR (95% CI)
	VRE (n = 7), N (%)	VSE (n = 105), N (%)		
Current residence				
Rural	4 (57.1)	15 (14.3)	0.17	5.97 (0.45–78.25)
Urban	3 (42.9)	90 (85.7)	NA	1
Duration of antitreatment				
<2 weeks	2 (28.6)	10 (0.9)	NA	1
≥2 weeks	4 (57.1)	19 (18.1)	0.29	0.24 (0.01–3.42)
Low hemoglobin level				
No	3 (42.9)	1 (0.9)	0.04	19.18 (1.08–89.36)
Yes	4 (57.1)	104 (99.0)	NA	1

VRE = vancomycin-resistant enterococci; VSE = vancomycin-susceptible enterococci; AOR = adjusted odds ratio.

previous study in Gondar, but the stool sample in the current study, in fact as enterococci are common normal flora of the gastrointestinal tract and undergone selective pressure over other normal flora in immune compromised individuals at the time of antibiotic treatment, their colonization has increased gradually.

Furthermore, the clients in the present study were HIV-positive cases, using repeated antibiotics and more hospitalized, while in the previous study, the clients were more of out-patients, could not use antibiotics regularly, and their hospitalization was not as much to HIV clients. It is evidenced that, in the health care facilities, enterococci may acquire resistance to all currently available antibiotics either by mutation or receipt of foreign genetic materials through the transfer of plasmids and transposons, or they have a selective pressure over other microorganisms of the intestinal flora, allowing for overgrowth of VRE, as it is intrinsically resistant to several antibiotics [10].

On the other hand, the present study showed that the prevalence of enterococci in HIV-positive patients was 56.7%; this is comparable with the result reported from Italy, 62% [17], and the Netherlands, 49% [18], on hospitalized patients. However, this result is lower than that of the previous report on hospitalized patients from Jimma, Ethiopia, 76% [9]. This discrepancy could be due to the difference in geographical location, and the method used in the previous study used the advanced molecular technique.

However, the prevalence of enterococci in HIV-negative controls was found to be 18%, and this is consistence with the result reported from USA, 18% [6]. However, it is lower than that of the previous finding of Egypt, 75% [16]. This might be due to difference in geographical location, and study subjects in the previous study were pediatrics/oncology patients.

The overall prevalence of VRE in this study was found to be 6.3%. This finding is in agreement with reports from Egypt, 6.8% [19], Nigeria, 5.9% [16], and Jimma, Ethiopia, 5% [9]. However, this result is lower than from the previous findings of Benin, 67.5% [20], Iran, 24.8% [21], and Brazil, 23.4% [22]. The lower prevalence of VRE in this study might be due to the study design and study participants (pediatric age and intensive care unit) in the previous study and the methods used for detection of vancomycin-resistant enterococci isolates.

This is true that VRE are circulating in the hospital settings and can be disseminated to the environment in the community. Vancomycin-resistant genes are easily transmitted through transposons which contain van genes and have ability to carry and transmit transferable genetic elements. In addition, in recent years, these genes have been transferring resistant genes to methicillin-resistant *Staphylococcus aureus* to form VRSA, thereby causing failure to treatment [23, 24]. On the contrary, the present study was higher than previous reports conducted from Saudi Arabia, 1.2% [25]. This difference could be because of small sample size, variation of study period, and absence of environmental contamination in previous studies, and clinical samples such as blood, pus, and urine were tested in Saudi Arabia.

The present finding showed that, of the 112 enterococci isolates, the prevalence of VRE among HIV-positive clients was 5.9%, which is in consistence with research done on HIV-positive clients from Gondar, Ethiopia, 7.8% [7]. However, the prevalence of VRE in this study is higher than that reported from California, 0% [26]. The higher prevalence of VRE among HIV positives might be due to repeated exposure to different antibiotics and immune suppression which are important risk factors for colonization or infection with VRE. This variation might also be due to individuals having high-animal contact in Ethiopia; this is evidenced from high prevalence of VRE (100%) in chickens and cattle, which was reported by Bekele et al. from Addis Ababa, Ethiopia [27].

In this study, enterococci tested against commonly used antibiotics showed that vancomycin was highly susceptible than other tested antibiotics. This result is in line with the study conducted in Gondar, Ethiopia, with a susceptibility of vancomycin greater than ampicillin, penicillin, ciprofloxacin, and chloramphenicol [8]. In this study, prevalence of penicillin and ampicillin resistant patterns was 34.8% each; this is in consistence with reports from Egypt, 44.4% for penicillin [19], and Jimma, Ethiopia, 36% for ampicillin [7]. This finding is lower than previous reports of the study from Iran, 72.7% and 69.7% for penicillin and ampicillin, respectively [28]. This variation might be due to the characteristics of the study participants; the study subjects in the previous studies were patients who underwent hemodialysis, who were more intensive, hospitalized, and immune compromised than the clients in the present study, who were HIV-positive clients with no patients from ICU.

On the contrary, the prevalence of gentamicin resistance was 37.5%, and this finding is more likely agreed with result reported from Turkey, 35.7% [29], Egypt, 35.5% [30], and

Jimma, Ethiopia, 34.2% [7]. However, it is lower than that reported from Iran, 86% [31]. This might be due to the method used different from the present study; in the previous study, the method used molecular system, which is more sensitive and specific to detect VRE in the gene level.

The prevalence of chloramphenicol in the present study was 8.9%. This finding is more likely comparable with the result reported from Brazil, 10.9% [22]. Although, it is lower than the report of Egypt, 31.6% [30], this difference might be due to study participants and sample type; the sample types in the previous study were various clinical samples including blood, urine, wound swabs, and sputum, and study participants were from pediatric ICU oncology/hematology patients.

From the total of 112 (37.4%) enterococci isolates identified, 25 (29.5%) were multiple drug resistance (MDR), this is comparable with the study done in Gondar, Ethiopia, 33.3% [7], and Addis Ababa, Ethiopia, 29% [32]. However, this finding is less than that reported from Jimma, Ethiopia, 89.5% [9]. This might be due to variation in the study subjects and the method of susceptibility testing. The method used in Jimma, Ethiopia, was minimum inhibitory concentrations (MIC) by using E-test strips for the determination of vancomycin.

Among sociodemographic characteristics in the present study, the prevalence of VRE was high among females (85.7%) than males (14.3%). This was evidenced from various epidemiological studies which showed that prevalence of VRE was more common in females than males of their counterparts, which is in agreement with two reports from Ethiopia [7, 9]. This is because of the fact enterococci carriage rate and VRE are more common on females than males.

The current finding showed that being HIV positive was not significantly associated ($p = 0.78$) with the occurrence of VRE; however, there are reports showing that enterococci and VRE have been more common in immune compromised clients in hospital settings, especially in HIV-positive patients [33]. The reason could be due to the fact that the number of isolates in our study was too small.

In this study, VRE was high among the age groups 43–54 (57%); this is in consistence with reports from Ethiopia, 45.5% [9]. In our study, low hemoglobin level is likely to develop VRE 19.2 times more compared to normal individuals, which is in consistence with the study conducted in Japan in which low hemoglobin level was significantly associated with the emergence of VRE [34]. However, previous antibiotic use, nutritional status, hospital admission, CD+ cell count level, ART linkage, and WHO staging were not significantly associated factors, and this is also consistent with the previous report from Gondar, Ethiopia [7].

Abbreviations

AOR: Adjusted odds ratio
ART: Antiretroviral therapy
ATCC: American type culture collection
BHI: Brain-heart infusion
CD4: Cluster of differentiation 4
CDC: Centers for disease control and prevention
HI: Human immune deficiency virus
IAIs: Intra-abdominal infections
ICU: Intensive care unit
MDR: Multiple drug resistance
VAN: Vancomycin-resistant gene
VRE: Vancomycin-resistant *Enterococcus*
VRSA: Vancomycin-resistant *Staphylococcus aureus.*

Additional Points

Species identification of enterococci was not investigated due to resource shortage and inaccessibility of testing materials. In conclusion, most of the isolates were resistant to one or more commonly used antibiotics. Prevalence of VRE and MDR enterococci in the study area was high, and this indicates that the rate of fecal colonization with VRE and MDR enterococci has increased through time, especially in HIV-positive clients than healthy individuals. The findings of this study are very crucial for indicating preferential spread of VRE and MDR in health care settings, where study subjects are readily exposed to VRE and MDR enterococci isolates; however, being HIV positive or negative was not statistically associated with the occurrence of vancomycin-resistant enterococci. Low hemoglobin level (being anemic) was significantly an associated factor for the emergence of vancomycin-resistant enterococci. Therefore, there should be a need to perform continuous surveillance, rational use of antibiotics, and more detailed study using phenotypic and genotypic methods.

Conflicts of Interest

The authors declare that that there are no conflicts of interest.

Authors' Contributions

Seid Ali designed the study and budget plan for all logistics, collected and analyzed the data, and participated in the draft and final write-up of the manuscript. Teklay Gebrecherkos participated in reviewing the initial and final drafts of the manuscript, data analysis, writing and interpretation of results, and final write-up of the manuscript. Martha Alemayehu and Mulat Dagnew assisted during the proposal write-up, interpreted the results, and reviewed the initial and final drafts of the manuscript. All authors reviewed and approved the final manuscript.

Acknowledgments

The authors would like to acknowledge Amhara Regional Health Bureau for sponsoring this study financially. The authors are very grateful to the staff of Dessie Referral Hospital, study participants, and data collectors. The authors are also grateful to all laboratory staff of Dessie Regional Health and Research Laboratory Center for their invaluable assistance. This project was funded by Amhara Regional Health Bureau.

References

[1] L. Teixeira and V. Merquior, "*Enterococcus*," in *Molecular Typing in Bacterial Infections*, Springer, New York, NY, USA, 2013.

[2] N. Silva, G. Igrejas, A. Gonçalves, and P. Poeta, "Commensal gut bacteria: distribution of *Enterococcus* species and prevalence of *Escherichia coli* phylogenetic groups in animals and humans in Portugal," *Annals of Microbiology*, vol. 62, no. 2, pp. 449–459, 2012.

[3] S. Sood, M. Malhotra, B. K. Das, and A. Kapil, "Enterococcal infections & antimicrobial resistance," *Indian Journal of Medical Research*, vol. 128, no. 2, p. 111, 2008.

[4] W. Bereket, K. Hemalatha, B. Getenet et al., "Update on bacterial nosocomial infections," *European Review for Medical and Pharmacological Sciences*, vol. 16, no. 8, pp. 1039–44, 2012.

[5] A. I. Hidron, J. R. Edwards, J. Patel et al., "Antimicrobial-resistant pathogens associated with healthcare-associated infections: annual summary of data reported to the National Healthcare Safety Network at the Centers for Disease Control and Prevention, 2006–2007," *Infection Control and Hospital Epidemiology*, vol. 29, no. 11, pp. 996–1011, 2008.

[6] S. B. Vakulenko, S. M. Donabedian, A. M. Voskresenskiy, M. J. Zervos, S. A. Lerner, and J. W. Chow, "Multiplex PCR for detection of aminoglycoside resistance genes in enterococci," *Antimicrobial Agents and Chemotherapy*, vol. 47, no. 4, pp. 1423–1426, 2003.

[7] W. Abebe, M. Endris, M. Tiruneh, and F. Moges, "Prevalence of vancomycin resistant enterococci and associated risk factors among clients with and without HIV in northwest Ethiopia: a cross-sectional study," *BMC Public Health*, vol. 14, no. 1, p. 185, 2014.

[8] A. Yilema, F. Moges, S. Tadele et al., "Isolation of enterococci, their antimicrobial susceptibility patterns and associated factors among patients attending at the University of Gondar Teaching Hospital," *BMC Infectious Diseases*, vol. 17, no. 1, p. 276, 2017.

[9] A. Abamecha, B. Wondafrash, and A. Abdissa, "Antimicrobial resistance profile of *Enterococcus* species isolated from intestinal tracts of hospitalized patients in Jimma, Ethiopia," *BMC Research Notes*, vol. 8, no. 1, p. 213, 2015.

[10] R. Leclercq, E. Derlot, J. Duval, and P. Courvalin, "Plasmid-mediated resistance to vancomycin and teicoplanin in *Enterococcus faecium*," *New England Journal of Medicine*, vol. 319, no. 3, pp. 157–161, 1988.

[11] M. Cheesbrough, *District Laboratory Practice in Tropical Countries*, Cambridge University Press, Cambridge, UK, 2006.

[12] R. Facklam, "Streptococcus and related catalase negative Gram-positive cocci," in *Manual of Clinical Microbiology*, pp. 238–257, ASMscience, Bel Air, MD, USA, 1992.

[13] Clinical and Laboratory Standards Institute, *Performance Standards for Antimicrobial Susceptibility Testing*, CLSI, Wayne, PA, USA, 27th edition, 2017.

[14] C. A. Arias and B. E. Murray, "Emergence and management of drug-resistant enterococcal infections," *Expert Review of Anti-Infective Therapy*, vol. 6, no. 5, pp. 637–655, 2008.

[15] M. Anvarinejad, G. Pouladfar, A. Japoni, S. Bolandparvaz, Z. Satiary, and J. Mardaneh, "Diabetic foot infections: antibiotic susceptibility patterns and determination of antibiotic cross-resistance in clinical isolates of *Enterococcus* species during 2012–2014 in Shiraz, Iran," *Archives of Pediatric Infectious Diseases*, vol. 5, no. 2, article e37680, 2017.

[16] K. O. Olawale, S. O. Fadiora, and S. S. Taiwo, "Prevalence of hospital acquired enterococci infections in two primary-care hospitals in Osogbo, Southwestern Nigeria," *African Journal of Infectious Diseases*, vol. 5, no. 2, 2011.

[17] C. G. Nobile, C. Ceravolo, L. Rizzo, and A. Bianco, "Prevalence of glycopeptide resistant enterococci in hospitalized patients and in the community in Calabria (Italy)," *Annali di Igiene: Medicina Preventivae Dicomunita*, vol. 15, no. 5, pp. 515–520, 2003.

[18] H. P. Endtz, N. van den Braak, A. van Belkum et al., "Fecal carriage of vancomycin-resistant enterococci in hospitalized patients and those living in the community in The Netherlands," *Journal of Clinical Microbiology*, vol. 35, no. 12, pp. 3026–3031, 1997.

[19] Y. A. Al-Tonbary, O. E. Soliman, M. M. Sarhan et al., "Nosocomial infections and fever of unknown origin in pediatric hematology/oncology unit: a retrospective annual study," *World Journal of Pediatrics*, vol. 7, no. 1, pp. 60–64, 2011.

[20] T. A. Ahoyo, H. S. Bankolé, F. M. Adéoti et al., "Prevalence of nosocomial infections and anti-infective therapy in Benin: results of the first nationwide survey in 2012," *Antimicrobial Resistance and Infection Control*, vol. 3, no. 1, p. 17, 2014.

[21] M. Nasaj, S. M. Mousavi, S. M. Hosseini, and M. R. Arabestani, "Prevalence of virulence factors and vancomycin-resistant genes among *Enterococcus faecalis* and *E. faecium* isolated from clinical specimens," *Iranian Journal of Public Health*, vol. 45, no. 6, p. 806, 2016.

[22] D. W. Batistão, P. P. Gontijo-Filho, N. Conceição, A. G Oliveira, and R. M. Ribas, "Risk factors for vancomycin-resistant enterococci colonisation in critically ill patients," *Memórias do Instituto Oswaldo Cruz*, vol. 107, no. 1, pp. 57–63, 2012.

[23] J. M. Oyce, L. A. Mermel, M. J. Zervos et al., "Controlling vancomycin-resistant enterococci," *Infection Control and Hospital Epidemiology*, vol. 16, no. 11, pp. 634–637, 1995.

[24] S. Chang, D. M. Sievert, J. C. Hageman et al., "Infection with vancomycin-resistant *Staphylococcus aureus* containing the vanA resistance gene," *New England Journal of Medicine*, vol. 348, no. 14, pp. 1342–1347, 2003.

[25] M. Salem-Bekhit, I. Moussa, M. A. Elsherbini, and S. AlRejaie, "Increasing prevalence of high-level gentamicin resistant enterococci: an emerging clinical problem," *African Journal of Microbiology Research*, vol. 5, no. 31, pp. 5713–5720, 2011.

[26] S. C. Pan, J. T. Wang, Y. C. Chen, Y. Y. Chang, M. L. Chen, and S. C. Chang, "Incidence of and risk factors for infection or colonization of vancomycin-resistant enterococci in patients in the intensive care unit," *PLoS One*, vol. 7, no. 10, Article ID e47297, 2012.

[27] B. Bekele and M. Ashenafi, "Distribution of drug resistance among enterococci and *Salmonella* from poultry and cattle in Ethiopia," *Tropical Animal Health and Production*, vol. 42, no. 5, pp. 857–864, 2010.

[28] S. Shaghaghian, B. Pourabbas, A. Alborzi, M. Askarian, and J. Mardaneh, "Vancomycin-resistant enterococci colonization in chronic hemodialysis patients and its risk factors in southern Iran (2005-2006)," *Iranian Red Crescent Medical Journal*, vol. 14, no. 10, p. 686, 2012.

[29] M. Butcu, S. S. Akcay, A. S. Inan, S. Aksaray, D. O. Engin, and G. Calisici, "In vitro susceptibility of enterococci strains isolated from urine samples to fosfomycin and other antibiotics," *Journal of Infection and Chemotherapy*, vol. 17, no. 4, pp. 575–578, 2011.

[30] I. O. H. El-Ghazawy and S. Mazloum, "A study of high level aminoglycoside resistant enterococci," *African Journal of Microbiology Research*, vol. 10, no. 16, pp. 572–577, 2016.

[31] H. S Kafil and M. Asgharzadeh, "Vancomycin-resistant *Enterococcus faecium* and *Enterococcus faecalis* isolated from education hospital of Iran," *Maedica*, vol. 9, no. 4, p. 323, 2014.

[32] A. M. D. Worku and A. Mache, *Isolation of Multiple Drug Resistant (MDR) Enterococci in Patients and Health Care Workers at Tikur Anbessa Hospital*, MSc thesis, 2005, In press.

[33] C. F. E. Achenbach, P. Ferrell, D. Pitrak, and S. Weber, "Prevalence of and risk factors for colonization with vancomycin-resistant *Enterococcus* among human immunodeficiency virus-positive outpatient," *Infection Control and Hospital Epidemiology*, vol. 27, no. 1, pp. 102–104, 2006.

[34] V. R. Deshpande, M. G. Karmarkar, and P. R. Mehta, "Prevalence of multidrug-resistant enterococci in a tertiary care hospital in Mumbai, India," *Journal of Infection in Developing Countries*, vol. 7, no. 2, pp. 155–158, 2013.

Permissions

All chapters in this book were first published in IJM, by Hindawi Publishing Corporation; hereby published with permission under the Creative Commons Attribution License or equivalent. Every chapter published in this book has been scrutinized by our experts. Their significance has been extensively debated. The topics covered herein carry significant findings which will fuel the growth of the discipline. They may even be implemented as practical applications or may be referred to as a beginning point for another development.

The contributors of this book come from diverse backgrounds, making this book a truly international effort. This book will bring forth new frontiers with its revolutionizing research information and detailed analysis of the nascent developments around the world.

We would like to thank all the contributing authors for lending their expertise to make the book truly unique. They have played a crucial role in the development of this book. Without their invaluable contributions this book wouldn't have been possible. They have made vital efforts to compile up to date information on the varied aspects of this subject to make this book a valuable addition to the collection of many professionals and students.

This book was conceptualized with the vision of imparting up-to-date information and advanced data in this field. To ensure the same, a matchless editorial board was set up. Every individual on the board went through rigorous rounds of assessment to prove their worth. After which they invested a large part of their time researching and compiling the most relevant data for our readers.

The editorial board has been involved in producing this book since its inception. They have spent rigorous hours researching and exploring the diverse topics which have resulted in the successful publishing of this book. They have passed on their knowledge of decades through this book. To expedite this challenging task, the publisher supported the team at every step. A small team of assistant editors was also appointed to further simplify the editing procedure and attain best results for the readers.

Apart from the editorial board, the designing team has also invested a significant amount of their time in understanding the subject and creating the most relevant covers. They scrutinized every image to scout for the most suitable representation of the subject and create an appropriate cover for the book.

The publishing team has been an ardent support to the editorial, designing and production team. Their endless efforts to recruit the best for this project, has resulted in the accomplishment of this book. They are a veteran in the field of academics and their pool of knowledge is as vast as their experience in printing. Their expertise and guidance has proved useful at every step. Their uncompromising quality standards have made this book an exceptional effort. Their encouragement from time to time has been an inspiration for everyone.

The publisher and the editorial board hope that this book will prove to be a valuable piece of knowledge for researchers, students, practitioners and scholars across the globe.

List of Contributors

Brice E. N. Wamba, Armelle T. Mbaveng, Paul Nayim, Joachim K. Dzotam, Ornella J. T. Ngalani and Victor Kuete
Department of Biochemistry, Faculty of Science, University of Dschang, Dschang, Cameroon

Manoela Brandão, Mariana Coutinho Hennemann, Denise Tonetta, Mônica Hessel Silveira and Mara Bedin
Departamento de Ecologia e Zoologia, Centro de Ciências Biológicas, Universidade Federal de Santa Catarina, Trindade, 88040-970 Florianópolis, SC, Brazil

Maria L. Schmitz Fontes
Departamento de Ecologia e Zoologia, Centro de Ciências Biológicas, Universidade Federal de Santa Catarina, Trindade, 88040-970 Florianópolis, SC, Brazil
PPGOCEANO, Departamento de Geociências, Centro de Filosofia e Ciências Humanas, Universidade Federal de Santa Catarina, Trindade, 88040-970 Florianópolis, SC, Brazil

Heloísa Fernandes
Departamento de Engenharia Sanitária, Centro Tecnológico, Universidade Federal de Santa Catarina, Trindade, 88040-970 Florianópolis, SC, Brazil

Raquel Aparecida Loss, Karina Cesca and Derce Recouvreux
Departamento de Engenharia Química e Engenharia de Alimentos, Centro Tecnológico, Universidade Federal de Santa Catarina, Trindade, 88040-970 Florianópolis, SC, Brazil

Valdelúcia Maria Alves de Souza Grinevicius
Departamento deBioquímica, Centro de Ciências Biológicas, Universidade Federal de Santa Catarina, Trindade, 88040-970 Florianópolis, SC, Brazil

Regina Vasconcellos Antônio
Campus Araranguá, Universidade Federal de Santa Catarina, Trindade, 88000-000 Araranguá, SC, Brazil

S. K. Kimaru and R. C. Cheruiyot
Department of Plant Sciences, Kenyatta University, Nairobi, Kenya

E. Monda
Department of Microbiology, Kenyatta University, Nairobi, Kenya

J. Mbaka
Kenya Agricultural and Livestock Research Organisation, Thika, Kenya

A. Alakonya
Jomo Kenyatta University of Agriculture and Technology, Nairobi, Kenya

Sally Ali Tawfik, Marwa Mohamed Azab and Ali Abdellah Abdelrahman Ahmed
Department of Microbiology & Immunology, Faculty of Pharmacy, Suez Canal University, Ismailia, Egypt

Dalia Mukhtar Fayyad
Department of Endodontics, Faculty of Dentistry, Suez Canal University, Ismailia, Egypt

Michael J. LaGier
Department of Biology, Grand View University, Des Moines, IA, USA

Thomas E. Bohl and Hideki Aihara
Department of Biochemistry, Molecular Biology, & Biophysics, University of Minnesota, 321 Church Street, 6-155 Jackson Hall, Minneapolis, MN 55455, USA

Elizaveta Anisimova and Dina Yarullina
Department of Microbiology, Kazan Federal University, Kremlevskaya Str. 18, Kazan 420008, Russia

Alejandro Llanes and Carlos Mario Restrepo
Centro de Biología Celular y Molecular de Enfermedades, Instituto de Investigaciones Científicas y Servicios de Alta Tecnología (INDICASAT AIP), Ciudad del Saber, Panama

Pablo Riesgo-Ferreiro
Guanábana Biodata Analytics Inc., Panama, Panama

Seid Ali
Microbiology Laboratory Unit, Dessie Referral Hospital, Amhara Region, Dessie, Ethiopia

Martha Alemayehu, Mulat Dagnew and Teklay Gebrecherkos
Department of Medical Microbiology, School of Biomedical and Laboratory Sciences, College of Medicine and Health Sciences, University of Gondar, Gondar, Ethiopia

Sreekumari Rajeev
School of Veterinary Medicine, Ross University, Basseterre, Saint Kitts and Nevis

Ricardo Silva Araujo and André Shigueyoshi Nakatani
Total Biotecnologia Ind´ustria e Com´ercio S/A, Rua Emilio Romani, No. 1190, CIC, Curitiba, PR, Brazil

Sonia Purin da Cruz
Universidade Federal de Santa Catarina, Campus Curitibanos, Rodovia Ulisses Gaboardi, Km 3, Curitibanos, SC, Brazil

Edson Luiz Souchie
Instituto Federal Goiano, Campus Rio Verde, Rodovia Sul Goiana, Km 01, Caixa Postal 66, Rio Verde, GO, Brazil

Thomas Newton Martin
Universidade Federal de Santa Maria, Avenida Roraima, No. 1000, Santa Maria, RS, Brazil

Marco Antonio Nogueira and Mariangela Hungria
Embrapa Soja, CP 231, 86001-970 Londrina, PR, Brazil

Viviane Santos de Sousa, Ana Paula de Souza da-Silva, Raphael Paiva Paschoal, Eloiza Helena Campana, Lyssa Oliveira Ferreira dos Santos, Natacha Martins, Ana Carolina Nunes Botelho, Renata Cristina Picão, Sérgio Eduardo Longo Fracalanzza and Beatriz Meurer Moreira
Laboratório de Investigação em Microbiologia Médica, Universidade Federal do Rio de Janeiro, Rio de Janeiro, RJ, Brazil

Leif Sorenson, Lee Woodland Riley and George Sensabaugh
School of Public Health, University of California, Berkeley, Berkeley, CA, USA

Renata Fernandes Rabello and Márcia Soares Pinheiro
Instituto Biomédico, Universidade Federal Fluminense, Niterói, RJ, Brazil

Abdulaziz Alqasim and Abdullah A. Alyousef
Microbiology Research Group, Department of Clinical Laboratory Sciences, College of Applied Medical Sciences, King Saud University, Riyadh 11433, Saudi Arabia

Ahmad Abu Jaffal
Department of Clinical Laboratory Sciences, King Saud Bin Abdulaziz University for Health Sciences, Riyadh, Saudi Arabia

Susan Michelz Beitel, Luciana Fontes Coelho and Daiane Cristina Sass
Department of Biochemistry and Microbiology, Institute of Bioscience, São Paulo State University (UNESP), São Paulo, SP, Brazil

Jonas Contiero
Department of Biochemistry and Microbiology, Institute of Bioscience, São Paulo State University (UNESP), São Paulo, SP, Brazil
Associate Laboratory IPBEN, UNESP, Av. 24A 1515 CEP, 13506-900, Rio Claro, SP, Brazil

Abebaw Bitew Kifilie
Department of Medical Laboratory Science, College of Medicine and Health Sciences, Debre Markos University, Debre Markos, Ethiopia

Mulat Dagnew and Birhanemeskel Tegenie
Department of Medical Microbiology, School of Biomedical and Laboratory Science, College of Medicine and Health Science, University of Gondar, Gondar, Ethiopia

Biruk Yeshitela and Rawleigh Howe
Armauer Hansen Research Institute, Addis Ababa, Ethiopia

Ebba Abate
Department of Immunology and Molecular Biology, School of Biomedical and Laboratory Science, College of Medicine and Health Science, University of Gondar, Gondar, Ethiopia

Pushpinder Kaur Litt and Divya Jaroni
Department of Animal Science and Food and Agricultural Products Center, Oklahoma State University, 148 FAPC, Monroe St., Stillwater, OK 74078, USA

Patricia Mejuto and Julio Díaz-Gigante
Department of Microbiology, Hospital del Oriente de Asturias Francisco Grande Covián, Arriondas, Asturias, Spain

Mariam Luengo
Department of Biochemistry, Hospital del Oriente de Asturias Francisco Grande Covián, Arriondas, Asturias, Spain

Rajae Zahli, Jamal Abrini and Omar Belmehdi
Biology and Health Laboratory, Biotechnology and Applied Microbiology Team, Department of Biology, Faculty of Science, Abdelmalek Essaadi University, Tetouan, Morocco

Aicha El Baaboua
Biology and Health Laboratory, Biotechnology and Applied Microbiology Team, Department of Biology, Faculty of Science, Abdelmalek Essaadi University, Tetouan, Morocco
Regional Laboratory for Analysis and Research, National Office for Food Safety, Tangier, Morocco

Ayoub Kounnoun
Regional Laboratory for Analysis and Research, National Office for Food Safety, Tangier, Morocco
Laboratory of Applied Biology and Pathology, Department of Biology, Faculty of Science, Abdelmalek Essaadi University, Tetouan, Morocco

Mohamed El Maadoudi
Regional Laboratory for Analysis and Research, National Office for Food Safety, Tangier, Morocco

Abdelhakim Bouyahya
Laboratory of Human Pathology Biology, Faculty of Sciences, and Genomic Center of Human Pathology, Mohammed V University, Rabat, Morocco

Rula Dawodi and Hiyam Marzouqa
Caritas Baby Hospital, Bethlehem, State of Palestine

Regeen Handal
Caritas Baby Hospital, Bethlehem, State of Palestine
Bethlehem University, Bethlehem, State of Palestine

Lulu Qunibi
Alia Hospital, Hebron, State of Palestine

Ibrahim Sahouri
Hussein Hospital, Bethlehem, State of Palestine

Maha Juhari
Rafidia Hospital, Nablus, State of Palestine

Musa Hindiyeh
Caritas Baby Hospital, Bethlehem, State of Palestine
Bethlehem University, Bethlehem, State of Palestine
Palestinian Forum for Medical Research, Ramallah, State of Palestine

Yamirot Merga Duffa
Yekatit 12 Hospital Medical College, Departments of Microbiology, Addis Ababa, Ethiopia
Addis Ababa University, College of Health Sciences, Department of Medical Laboratory Sciences, Addis Ababa, Ethiopia

Adane Bitew
Addis Ababa University, College of Health Sciences, Department of Medical Laboratory Sciences, Addis Ababa, Ethiopia

Kumera Terfa Kitila and Dereje Mamuye Gebretsadik
Ethiopia Public Health Institute (EPHI), Addis Ababa, Ethiopia

Adane Bitew
Department of Medical Laboratory Sciences, College of Health Sciences, Addis Ababa University, Addis Ababa, Ethiopia

Yeshiwork Abebaw
Clinical Laboratory, Fitche Hospital, Oromia Administrative Region, Ethiopia

Delayehu Bekele
Department of Obstetrics and Gynecology, St Paul's Hospital Millennium Medical College, Addis Ababa, Ethiopia

Amete Mihret
Department of Clinical Bacteriology and Mycology, Ethiopian Public Health Institute, Addis Ababa, Ethiopia

Issaka Zongo, Grissoum Tarnagda and Kadari Cisse
Institut de Recherche en Sciences de la Santé (IRSS/CNRST), Ouagadougou, Burkina Faso

Apoline Sondo, Yves Traoré and Nicolas Barro
Université Ouaga I Professeur Joseph KI-ZERBO, Ouagadougou, Burkina Faso

Henri Gautier Ouedraogo
Institut de Recherche en Sciences de la Santé (IRSS/CNRST), Ouagadougou, Burkina Faso
Université Ouaga I Professeur Joseph KI-ZERBO, Ouagadougou, Burkina Faso
Institut Africain de Santé Publique (IASP), Ouagadougou, Burkina Faso

Ivlabehire Bertrand Meda and Seni Kouanda
Institut de Recherche en Sciences de la Santé (IRSS/CNRST), Ouagadougou, Burkina Faso
Institut Africain de Santé Publique (IASP), Ouagadougou, Burkina Faso

Odette Ky-Zerbo and Benoit Cesaire Samadoulougou
Programme d'Appui au Monde Associatif et Communautaire (PAMAC), Ouagadougou, Burkina Faso

Ashley Grosso and Stefan Baral
Department of Epidemiology, Johns Hopkins Bloomberg School of Public Health, Baltimore, MD, USA

Nongoba Sawadogo
Centre Hospitalier Régional de Kaya, Kaya, Burkina Faso

Seid Ali
Microbiology Laboratory Unit, Dessie Referral Hospital, Amhara Region, Dessie, Ethiopia

Martha Alemayehu, Mulat Dagnew and Teklay Gebrecherkos
Department of Medical Microbiology, School of Biomedical and Laboratory Sciences, College of Medicine and Health Sciences, University of Gondar, Gondar, Ethiopia

Gour Gopal Satpati and Ruma Pal
Phycology Laboratory, Department of Botany, University of Calcutta, 35 Ballygunge Circular Road, Kolkata, West Bengal 700019, India

Kyle R. Baum and Vineet K. Singh
Department of Microbiology and Immunology, Kirksville College of Osteopathic Medicine, A.T. Still University of Health Sciences, Kirksville, MO 63501, USA

Zulfiqar Ahmad
Department of Biochemistry, Kirksville College of Osteopathic Medicine, A.T. Still University of Health Sciences, Kirksville, MO 63501, USA

Index

S

T

U

V

Z

www.ingramcontent.com/pod-product-compliance
Lightning Source LLC
LaVergne TN
LVHW081031210926
848485LV00004B/999